Geology of Gas and Oil under the Netherlands

Geology of Gas and Oil under the Netherlands

Selection of papers presented at the 1993 International Conference of the American Association of Petroleum Geologists, held in The Hague

Edited by

H.E. Rondeel
Institute of Earth Sciences, Vrije Universiteit, Amsterdam, the Netherlands

D.A.J. Batjes
The Hague, the Netherlands

W.H. Nieuwenhuijs
Delft University of Technology, Delft, the Netherlands

The Royal Geological and Mining Society of the Netherlands - KNGMG

Kluwer Academic Publishers
DORDRECHT / BOSTON / LONDON

Library of Congress Cataloging-in-Publication Data

Geology of gas and oil under the Netherlands / edited by H.E. Rondeel, D.A.J. Batjes and W.H. Nieuwenhuijs.
p. cm.
ISBN 0-7923-3749-2
1. Natural gas--Geology--Netherlands. 2. Petroleum--Geology--Netherlands. I. Rondeel, H. E. II. Batjes, D. A. J. III. Nieuwenhuijs, W. H.
TN897.N4G46 1996
553.2'85'09492--dc20 95-25226

ISBN 0-7923-3749-2

Published by Kluwer Academic Publishers,
P.O. Box 17, 3300 AA Dordrecht, The Netherlands

Kluwer Academic Publishers incorporates
the publishing programmes of
D. Reidel, Martinus Nijhoff, Dr W. Junk and MTP Press.

Sold and distributed in the U.S.A. and Canada
by Kluwer Academic Publishers,
101 Philip Drive, Norwell, MA 02061, U.S.A.

In all other countries, sold and distributed
by Kluwer Academic Publishers Group,
P.O. Box 322, 3300 AH Dordrecht, The Netherlands

Printed on acid-free paper

Printed in the Netherlands

Contents

Preface

In October 1993 an International Conference was sponsored in The Hague by the American Association of Petroleum Geologists, the European Association of Petroleum Scientists and Engineers and the Geological Survey of the Netherlands to elicit *'New Views on Old World Oil — Technology Leads the Way'*.

Together with the hosts, the Royal Geological and Mining Society of the Netherlands, we had planned from the outset with the support of Shell to select for publication a compendium of papers on the Geology of gas and oil in the Netherlands. The conference was a real milestone, and we have embodied the keynote addresses together with the wide-ranging views on production and opportunities from the Carboniferous to the Cretaceous herein. From the first producible oil found near the Royal Dutch/Shell offices followed the highlights of discovering the Schoonebeek Oil Field in 1943 and the Groningen Gas Field in 1959; today tenacious exploration harnesses ever smarter technology. Enhanced imaging with 3D, direct hydrocarbon indicators, basin modelling, high-resolution sequence prediction and advanced drilling techniques all play a part. Technology is moving quickly and here in the Netherlands it is being utilised to good effect. The stratigraphic sequences, the structural complexity, new niches for exploration, are all intricately hidden below a Quaternary cover.

This volume makes a noteworthy addition to the oil and gas story of the Netherlands. It was a privilege to be part of the organising community for the conference. Many thanks to all the geoscientists who made it possible, and congratulations to the editors for putting together a volume that every discerning bookshelf would like to embrace.

Dave Loftus
Technical Program Chairman

Introduction

The exploitation of natural gas resources in the Netherlands contributes nowadays a direct revenue to the State of seven to nine milliard guilders per year. The present Special Publication of the Royal Geological and Mining Society of the Netherlands (KNGMG) provides an insight into the subsurface geological conditions that allow to produce the gas. It also shows how these conditions are investigated and how the knowledge thus gained governs the production, not only of gas but also of oil.

The book finds its origin in the 1993 International Conference of the American Association of Petroleum Geologists, held in The Hague. On the occasion of this Conference, the Petroleum Geological Circle (PGK) of the KNGMG compiled the 'Synopsis: Petroleum geology of the Netherlands – 1993' of which 4000 copies were distributed. It is this 'Synopsis' that initiated the idea to assemble the contributions on the country's petroleum geology presented at the Conference. The Synopsis now constitutes the first of the articles collected. Where possible, these are arranged in stratigraphic order. One article on a related subject was added.

Invitations to authors were aimed from the outset at coverage of the petroleum geology of the Netherlands. No wonder that most contributions come from the Nederlandse Aardolie Maatschappij (NAM) and the Geological Survey of the Netherlands (RGD). Papers were further handed in by scientists of Amoco, Continental Netherlands Oil Co., Exxon Exploration Co., Shell Research and the Delft University of Technology. These contributions are placed in a wider economic and technological context by the keynote addresses of the Conference, which with the Synopsis now form the first part of the book. These addresses originated at the Ministry of Economic Affairs of the Netherlands, at Elf Aquitaine and at the Royal Dutch/Shell Group of Companies.

The book fills a gap in the geological literature on the Netherlands by bringing together up-to-date contributions to the petroleum geology of the country. It addresses a wide range of subjects. These include the characterization of reservoirs in single gas and oil fields by means of 3D seismic and borehole logging, the hydrocarbon habitat of the West Netherlands Basin and the regional Rotliegend facies distribution.

The stratigraphic nomenclature in the book reflects the transition during 1993 from the 'NAM & RGD' nomenclature published in 1980, to the revised nomenclature compiled by the Geological Survey.

The editors owe many thanks to the authors for their contributions and to the reviewers for their comments. They particularly mention the staff members of the NAM and RGD, who produced and reviewed the bulk of the present volume's content.

They gratefully acknowledge the financial support by Shell Internationale Petroleum Maatschappij, which allowed for most of the colour prints in this publication. They thank Exxon Exploration Company for financing the printing of the 'Rotliegend' maps.

Finally, they mention J. Halfon and P.J. Grantham, who did not live to see their contributions published.

Harm Rondeel
Dick Batjes
Willem Nieuwenhuijs

Synopsis:
Petroleum geology of the Netherlands – 1993

by
Petroleum Geological Circle

c/o Royal Geological and Mining Society of the Netherlands
P.O. Box 157, 2000 AD Haarlem, the Netherlands

PGK

This publication is financed by
Shell Internationale Petroleum Maatschappij B.V.

This preprint is issued on the occasion of the
International Conference and Exhibition of the
American Association of Petroleum Geologists,
The Hague, the Netherlands, 17–20 October 1993

Preface

It is a pleasure for the Royal Geological and Mining Society of the Netherlands (KNGMG) as the host society of the AAPG International Conference, to present to you the first-ever Synopsis on the petroleum geology of the Netherlands. The initiative for this publication came from the very active Petroleum Geological Circle (PGK) of our Society, notably from Dick Batjes and Willem Nieuwenhuijs.

Why did it take a conference of the *American* Association of Petroleum Geologists to make the Dutch write a synopsis of their own petroleum geology? I think there are several reasons for this. The first might be that in a small country like ours, almost all geologists know each other. Some, therefore, think they know everything their colleagues know. And others who know nothing always have a friend who knows something. So why write it down at all? The second reason is the opposite of the first: maybe you know a few things you do not want your colleagues to know. So why make them aware? Or maybe we were just too busy with our own wells to see the whole of the hydrocarbon province. The important thing about the AAPG Conference is that it has provided the incentive to take off for a bird's eye view of our own achievements. And as it happens so often, when you try to explain to others who you are, you discover yourself.

This booklet offers you a taste of the extensive knowledge of our subsurface geology, obtained by a fortunate mix of chance discoveries and tenacious exploration, and of course, aided by an enormous technological development. How far have we got since the years around 1910, when the first shows of oil were discovered in the Netherlands, although "there is some doubt whether these shows represent natural oil or are grease used for drilling" as W.A. Knaap and M.J. Coenen put it in our Society's 1987 anniversary volume *Seventy-five years of Geology and Mining in the Netherlands (1912–1987)*. The first discovery of oil in the western Netherlands was made by accident during demonstration drilling in your host town The Hague in 1938, and the first real oil field in this area is located only eight kilometres from the head office of Royal Dutch/Shell. At present, 114 gas fields and 19 oil fields are producing. How long will we be able to keep up the race between new discoveries and exhaustion? We certainly will need the *New Views on Old World Oil* and *Technology leading the way.*

The first step in developing new views is the compilation of an up-to-date overview. This booklet serves that purpose. We are grateful to Ab van Adrichem Boogaert, Dick Batjes, Manfred Epting, Jan de Jager, Willem Nieuwenhuijs and Ed van Riessen for compiling this fine piece of work. We thank the Geological Survey of The Netherlands and the Nederlandse Aardolie Maatschappij for preparing the figures, and Shell Internationale Petroleum Maatschappij for their generous financial support, which made printing possible.

Salomon B. Kroonenberg
Chairman Royal Geological and Mining Society of the Netherlands

Rondeel et al. (eds), Geology of gas and oil under the Netherlands, 1996 / *Geologie en Mijnbouw* Vol. 74, No. 4, 1996.

Synopsis: Petroleum geology of the Netherlands - 1993

Introduction

This Synopsis is intended to give a concise overview of the petroleum geology of the Netherlands. It is meant primarily to provide the participants of the AAPG Conference, to be held in The Hague in October 1993, with the basics of this geology. Those who wish to know more than these basics, either with the intention of extracting additional oil and gas, or out of a more academically oriented interest, may benefit from the bibliography.

These days, most new reserves are found in and around existing fields. If we wish to increase recoverable reserves we will first need to understand the reservoirs that contain our current assets. Only then can technology be applied and lead the way.

This Synopsis addresses first the mining legislation and the history of exploration and production, from the earliest oil shows to cumulative production and remaining reserves, and then the surface geology and the hydrocarbon plays. These subjects are presented in tables, figures and a brief accompanying text. All but one of the figures have been prepared by the Nederlandse Aardolie Maatschappij and the Geological Survey of The Netherlands. Much of the information given is based on the report "Oil and gas in the Netherlands, exploration and production 1992", issued by the Ministry of Economic Affairs.

At a later stage, this Synopsis will be published in 'Geologie en Mijnbouw', the journal of the Royal Geological and Mining Society of the Netherlands. It is intended to be the opening article of a special issue on the petroleum geology of the Netherlands, containing a selection of contributions presented at this AAPG Conference.

Mining legislation (adapted from Schierbeek (1987))

The exploration for and production of oil and gas are covered by two different regimes, one for the onshore ('Territory', including the territorial waters to the 3-nautical-mile boundary) and one for the offshore (Dutch part of the continental shelf).

The onshore regime is governed by:

- the Napoleonic Mining Law of 1810 (Mijnwet 1810), still in its original French text, no authorised translation being available (relic of the French occupation); amended in 1988,
- the Mining Law 1903 (Mijnwet 1903), on which the Mining Regulations 1964 (Mijnreglement 1964) are based; amended in 1988,
- the Minerals Exploration Act 1967 (Wet opsporing delfstoffen 1967).

The offshore regime is governed by the Mining Law Continental Shelf 1965 (Mijnwet continentaal plat) and its three most important general administrative orders (Koninklijke Besluiten):

- ex article 12, the terms and conditions under which non-exclusive reconnaissance permits as well as exclusive permits for exploration and exploitation of hydrocarbons may be granted; specified in general administrative orders of 1967 and 1976,
- ex article 26, the Mining Regulations 1967 (Mijnreglement 1967).

The sole political responsibility under both regimes rests since 1946 with the Minister of Economic Affairs.

The main features of the licence types which can be granted under the on- and offshore regimes are summarised in Table 1. They were described in detail by Roggenkamp (1991). Figure 6 shows the licence situation.

Exploration and production history

The history of exploration and of the beginning of the production of oil and gas is summarised in Table 2 and Figure 2. A detailed account was presented by Knaap & Coenen in 1987. Highlights were the discovery of the Schoonebeek oil field in 1943 and of the Groningen gas field in 1959. The Schoonebeek field, which has produced 39 million cubic metres of oil up to 1993, is the largest onshore oil accumulation in continental western Europe. The Groningen gas field is a giant with an expected ultimate recovery of some 2.74 trillion (10^{12}) cubic metres of gas. At the end of 1992, ca. 1.37 trillion had been produced.

The Ministry of Economic Affairs publishes an annual review of the licence situation, of seismic surveys and drilling activities, of volumes of oil and gas produced and of estimated reserves. Table 3 and Figures 3 to 6 have been prepared on the basis of the 1992 review. The survey and drilling statistics given in this review are summarised below. Figure 3 shows the 3D seismic coverage.

Seismic surveys 1992	Territory	Cont. shelf
2D line km	388	1799
3D sq. km	1307	4173
Drilling 1992 (wells/m)		
Exploration	11 } / 36900	19 } / 76331
Appraisal	1 }	1 }
Production	12 / 32892	15 / 61095
Total	24 / 69792	35 / 137426

Of the thirty exploration wells drilled, thirteen found gas and two oil and gas. The other fifteen are recorded as dry.

Table 3 lists the volumes of oil and gas produced in 1992 by eleven operating companies. Figure 4 shows the annual production volumes.

The country's cumulative production at the end of 1992 amounted to 97.9 million cubic metres of oil and 1.85 trillion cubic metres of gas (Fig. 5).

Remaining 'expected' reserves, according to the Ministry's review, are estimated at 61 million cubic metres of oil and 2.06 trillion cubic metres of gas (Fig. 5). Of these reserves, 21 million cubic metres of oil and 1.93 trillion cubic metres of gas are considered 'proven'.

Surface geology

The Netherlands is located in the southeastern part of the Cenozoic North Sea Basin. The edges of this basin are close to the country's eastern and southern borders (Fig. 1). The sediments at the surface are almost exclusively Quaternary (Fig. 7). The thickest Quaternary (600 m) occurs in the northwest. Tertiary and older sediments are only exposed in the extreme east and south of the country, where the edges of the basin were uplifted and eroded. The southeast of the Netherlands, moreover, is affected by a SE-NW striking fault system, which formed a number of horst and graben blocks during the Tertiary and Quaternary (Fig. 10). These faults are still active.

The landscape essentially consists of a Holocene coastal barrier and coastal plain, and inland of a low-lying, mostly flat area of Pleistocene deposits cut by a Holocene fluvial system.

The coastal barrier is interrupted in the south by the estuary of the Rhine, Meuse and Scheldt, and in the north by the tidal inlets of the Wadden Sea. The barrier bears dunes and is locally up to ten kilometres wide. In places it had to be reinforced by dikes.

The coastal plain covers about half of the country and consists mainly of clay and peat. Much of it would be flooded in the absence of dikes. Not only the distribution of land and water is strongly influenced by man, but also the present-day limited extent of peat, for instance, is artificial. In the past, peat was exploited as fuel, both in the coastal plain and further inland where moors covered parts of the Pleistocene.

At the surface, the Pleistocene is largely sandy and of glacial, fluvial and aeolian origin. Ice-pushed ridges locally reach heights of a hundred metres, but most of the Pleistocene occurs as flat-lying land. The Holocene alluvial valleys of the Rhine and Meuse systems, so clearly expressed in the Pleistocene area, merge downstream with the coastal plain. In many

Fig. 1. Provinces, towns and waterways. Translations: Noord = North, Zuid = South, Zee = Sea, Maas = Meuse, Rijn = Rhine, Schelde = Scheldt.

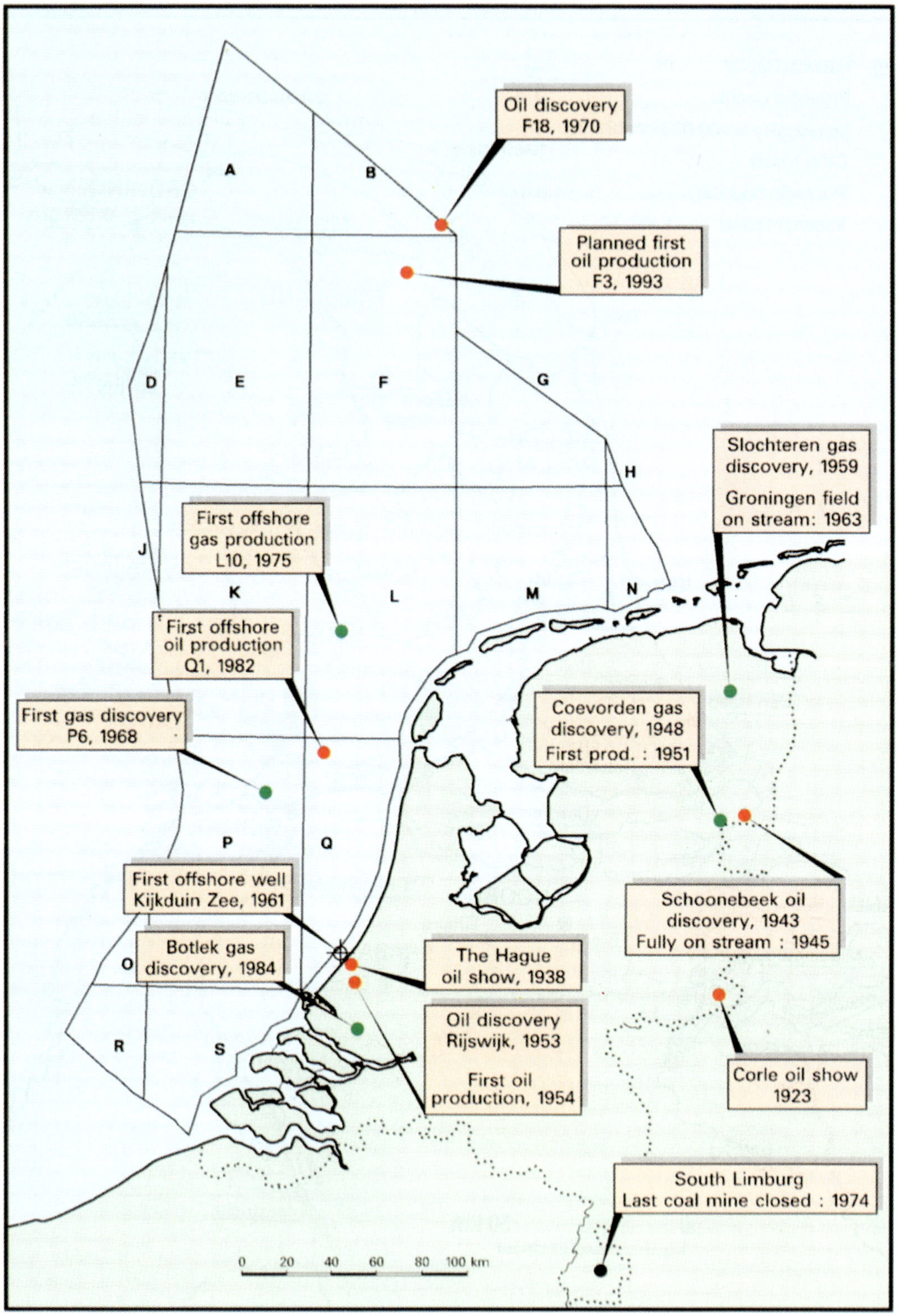

Fig. 2. Location map, history of oil and gas exploration and production.

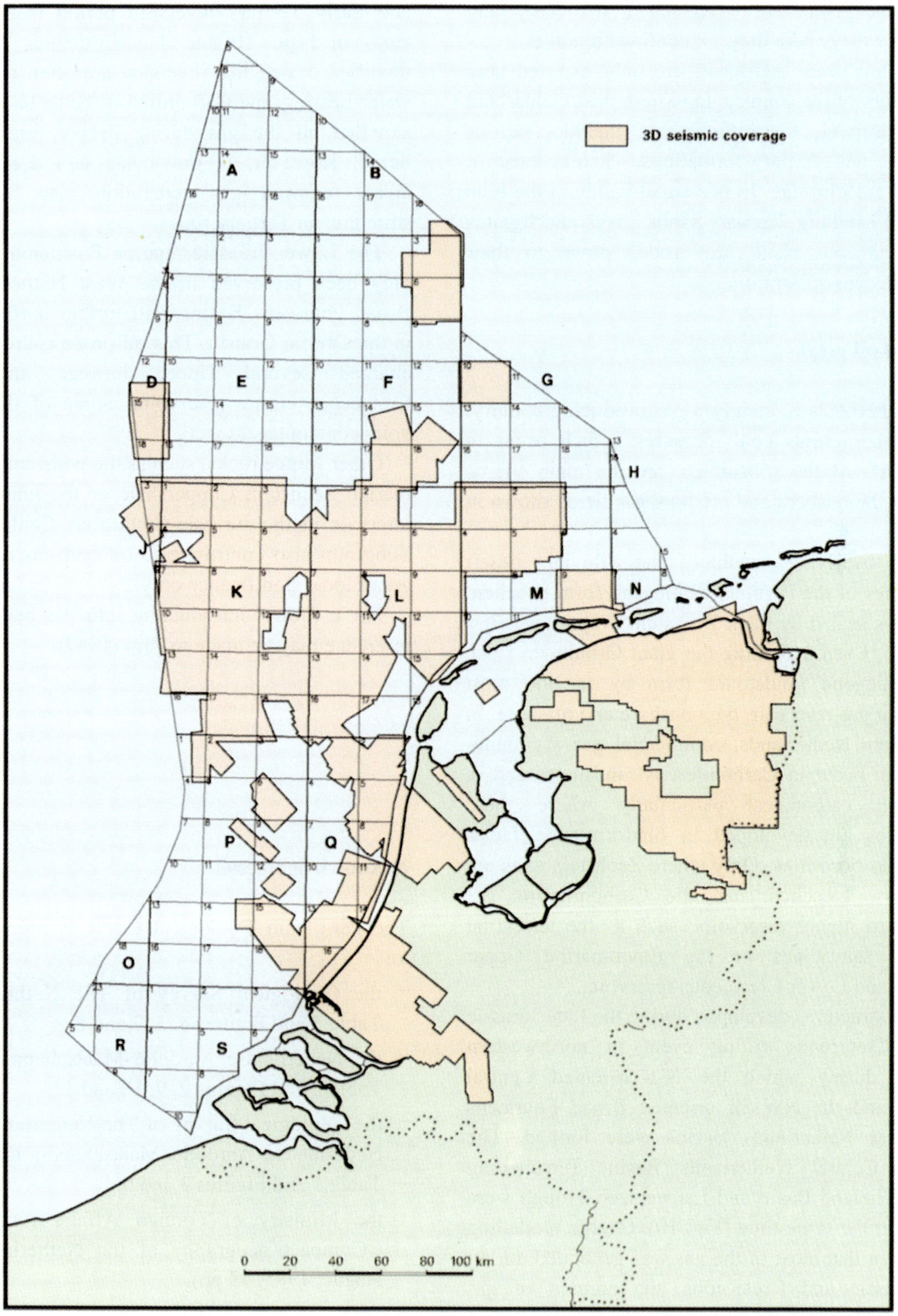

Fig. 3. 3D seismic coverage.

places the rivers are straightened artificially and virtually everywhere they are confined by dikes.

Pre-Pleistocene sediments are only exposed near the borders of the country. In easternmost Gelderland and Overijssel, these sediments include various Mesozoic and Tertiary formations, whereas those in Zeeland are Pliocene. In one small valley in the hills of South Limburg, Tertiary sands, clays and lignites and Cretaceous chalk are eroded down to their Carboniferous substratum.

Gas and oil plays

The Netherlands is mainly a gas-producing country. The Carboniferous coal measures, which occur in large parts of the subsurface, are the main source rock for the onshore and offshore gas fields shown in Figure 6.

The overlying aeolian and fluvial desert sandstones of the Permian Rotliegend form excellent reservoirs sealed by thick Zechstein evaporites (Figs. 8 and 9). Even excluding the giant Groningen field, the Rotliegend sandstones form by far the most important gas reservoir, both onshore and offshore. In the eastern Netherlands, commercial gas accumulations also occur in Carboniferous sandstones and in Zechstein carbonates, particularly where these carbonates are developed in platform-edge facies with good porosities. Only where Zechstein salts are absent or breached, can the Carboniferous gas migrate to higher reservoirs, such as the terrestrial Triassic sandstones or the fluviomarine Upper Jurassic and Lower Cretaceous reservoirs.

Most structures developed during the Late Jurassic -Early Cretaceous rifting event in northwestern Europe, during which the N-S oriented Central Graben and the NW-SE oriented Broad Fourteens and West Netherlands Basins were formed. The smaller Central Netherlands Basin, Terschelling Basin, Vlieland Basin and Lauwerszee Trough were formed at the same time (Fig. 10). Charge modelling has shown that most of the gas was generated during the Jurassic and Cretaceous, the amount of gas greatly exceeding the capacity of available traps.

Charge was interrupted during the Late Cretaceous and Early Tertiary inversion period (see inversion areas in Figs. 10 and 11). Structures formed or modified during this inversion are often found to be water-bearing; the Carboniferous coals had generally expelled all the gas during deeper, pre-inversion, burial. However, in previously less deeply buried areas, renewed gas generation may have filled structures in Tertiary times.

The Lower Jurassic, marine Posidonia Shale has only been preserved in the West Netherlands and Broad Fourteens Basins, and further north offshore, in the Central Graben. This oil-prone source rock has charged several Upper Jurassic and Lower Cretaceous clastic reservoirs. Some of these reservoirs contain gas as well.

Other source rocks, such as the wide-spread, basal, marine Zechstein Coppershale or the mainly humic intervals in the Jurassic-Cretaceous Central Graben Subgroup also contribute to oil and gas charge, but are of local significance only.

For a further introduction into gas and oil plays reference may be made to Zijp (1987).

Acknowledgements

The Petroleum Geological Circle is much indebted to:

- the Geological Survey of The Netherlands for Table 3 and Figures 4, 5, 7 and 8.
- the Nederlandse Aardolie Maatschappij B.V. for Table 2 and Figures 2, 9, 10 and 11.
- the Geological Survey of The Netherlands and the Nederlandse Aardolie Maatschappij together for Table 1 and Figures 3 and 6.
- the Ministry of Foreign Affairs for Figure 1 ('Compact geography of the Netherlands', The Hague, 1985, 43 pp).
- Shell Internationale Petroleum Maatschappij B.V. for financing the publication of this Synopsis.

Bibliography

Bless, M.J.M., J. Bouckaert & E. Paproth 1983 Recent exploration in Pre-Permian rocks around the Brabant massif in Belgium, the Netherlands and the Federal Republic of Germany – Geol. Mijnbouw 62: 51–62.

Bodenhausen, J.W.A. & W.F. Ott 1981 Habitat of the Rijswijk oil province, onshore, The Netherlands. In: Illing, L.V. & G.D. Hobson (eds.) Petroleum Geology of the Continental Shelf of NW Europe, Inst. of Petroleum, London: 301–309.

Burgers, W.F.J. & G.G. Mulder 1991 Aspects of the Late Jurassic and Cretaceous history of The Netherlands – Geol. Mijnbouw 70: 347–354, 8 encls.

De Jong, M.G.G. & N. Laker 1992 Reservoir modelling of the Vlieland Sandstone of the Kotter Field (Block K18b), offshore, The Netherlands – Geol. Mijnbouw 71: 173–188.

Doornhof, D. 1992 Surface subsidence in The Netherlands: the Groningen gas field – Geol. Mijnbouw 71: 119–130.

Dronkers, A.J. & F.J. Mrozek 1991 Inverted basins of The Netherlands – First Break 9: 409–425.

Gdula, J.E. 1983 Reservoir geology, structural framework and petrophysical aspects of the De Wijk gas field – Geol. Mijnbouw 62: 191–202.

Glennie, K.W. (ed.) 1990 Introduction to the petroleum geology of the North Sea – Blackwell Scientific Publications, Oxford, 3rd ed., 402 pp.

Harmsen, G.J. 1980 Steamflooding in a water drive reservoir in the Schoonebeek field in the Netherlands – Proc. Tenth World Petroleum Congr. 3: 275–282.

Herngreen, G.F.W., R. Smit & Th.E. Wong 1991 The stratigraphy and tectonics of the Vlieland Basin, The Netherlands. In: Spencer, A.M. (ed.) Generation, accumulation, and production of Europe's hydrocarbons – Spec. Publ. Europ. Assoc. of Petroleum Geoscientists Engineers, No. 1, Oxford Univ. Press: 175–192.

Herngreen, G.F.W. & Th.E. Wong 1989 Revision of the "Late Jurassic" stratigraphy of the Dutch Central North Sea Graben – Geol. Mijnbouw 68: 73–105.

Hoetz, H.L.J.G. & D.G. Waters 1992 Seismic horizon attribute mapping for the Annerveen Gasfield, The Netherlands – First Break 10: 41–51.

Kaasschieter, J.P.H. & T.J.A. Reijers (eds.) 1983 Petroleum Geology of the southeastern North Sea and the adjacent onshore areas (The Hague, 1982) – Geol. Mijnbouw 62: 1–239.

Knaap, W.A. & M.J. Coenen 1987 Exploration for oil and natural gas. In: Visser, W.A., J.I.S. Zonneveld & A.J. van Loon (eds.) Seventy-five years of geology and mining in The Netherlands (1912–1987) – Royal Geol. and Mining Soc. of The Netherlands (KNGMG), The Hague: 207–242.

Letsch, W.J. & W. Sissingh 1983 Tertiary stratigraphy of The Netherlands – Geol. Mijnbouw 62: 305–318.

Ministerie van Economische Zaken (Ministry of Economic Affairs) 1993 Olie en gas in Nederland; opsporing en winning 1992/Oil and gas in the Netherlands; exploration and production 1992 , 105 pp. 's-Gravenhage (The Hague; published yearly).

Nederlandse Aardolie Maatschappij & Rijks Geologische Dienst 1980 Stratigraphic nomenclature of The Netherlands – Verh. Kon. Ned. Geol. Mijnbouwk. Gen., 32: 77 pp, 36 encls.

Oele, J.A., A.C.P.J. Hol & J. Tiemens 1981 Some Rotliegend gasfields of the K and L blocks, Netherlands offshore (1968–1978) – a case history. In: Illing, L.V & G.D. Hobson (eds.) Petroleum Geology of the Continental Shelf of NW Europe, Inst. of Petroleum, London: 289–300.

Perrot, J. & A.B. van der Poel 1987 Zuidwal – a Neocomian gas field. In: Brooks, J. & K. Glennie (eds.) Petroleum Geology of North West Europe. Graham & Trotman: 325–335.

Ramaekers, J.J.F. 1992 The Netherlands. In: Hurtig, E., V. Cermák, R. Haenel & V. Zui (eds.) Geothermal Atlas of Europe. Geoforschungszentrum Potsdam, Publ. 1. Hermann Haack Verlagsgesellschaft, Gotha: 81–83.

Roelofsen, J.W. 1991 Geology of the Lower Cretaceous Q1 oil fields, Broad Fourteens Basin, The Netherlands. In: Spencer, A.M. (ed.) Generation, accumulation, and production of Europe's hydrocarbons. Spec. Publ. European Assoc. of Petroleum Geoscientists Engineers, No. 1, Oxford Univ. Press: 203–216.

Roggenkamp, M. 1991 Oil & Gas: Netherlands Law and Practice. Chancery Law Publishing, London, 345 pp.

Roos, B.M. & B.J. Smits 1983 Rotliegend and Main Buntsandstein gasfields in block K13 – a case history – Geol. Mijnbouw 62: 75–82.

Rijks Geologische Dienst (Geological Survey of The Netherlands): Geologische Atlas van de Diepe Ondergrond van Nederland (English version, Geological Atlas of the Subsurface of The Netherlands), scale 1:250000: Sheet I Vlieland-Terschelling, 1991. Explanation, 79 pp., 16 encls; Sheet II Ameland-Leeuwarden, 1992. Explanation, 86 pp., 15 encls.

Schierbeek, P. 1987 Remarks on the mining legislation and its application. In: Visser, W.A., J.I.S. Zonneveld & A.J. van Loon (eds.) Seventy-five years of geology and mining in The Netherlands (1912–1987). Royal Geol. and Mining Soc. of The Netherlands (KNGMG), The Hague: 33–37.

Van Adrichem Boogaert, H.A. & W.F.J. Burgers 1983 The development of the Zechstein in The Netherlands – Geol. Mijnbouw 62: 83–92.

Van den Bosch, W.J. 1983 The Harlingen Field, the only gas field in the Upper Cretaceous Chalk of The Netherlands – Geol. Mijnbouw 62: 145–156.

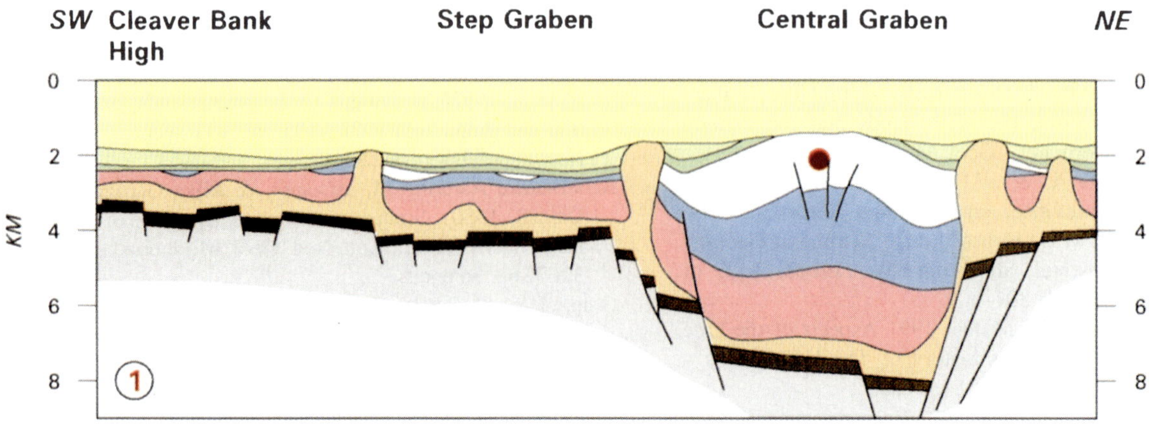
SW
Cleaver Bank
High
Step Graben
Central Graben
NE
0
2
4
6
8
KM
1

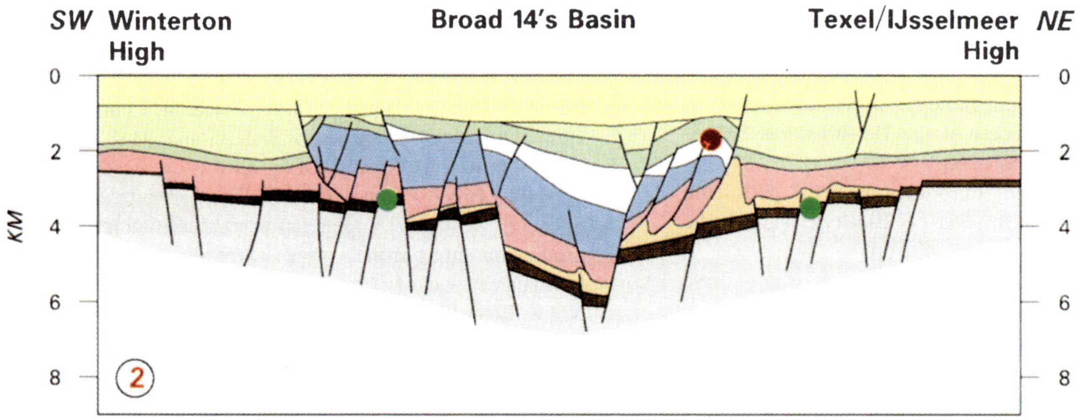
SW
Winterton
High
Broad 14's Basin
Texel/IJsselmeer
High
NE
0
2
4
6
8
KM
2

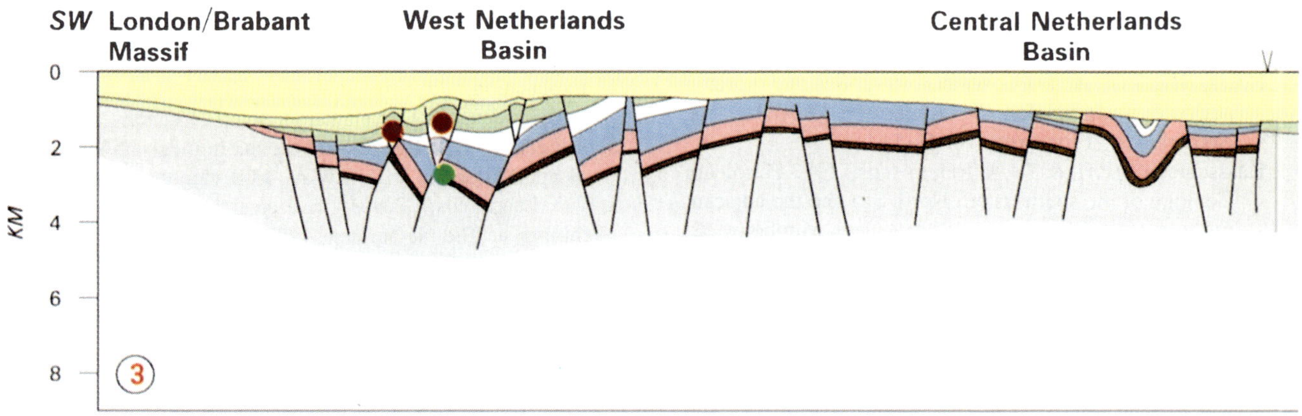
SW
London/Brabant
Massif
West Netherlands
Basin
Central Netherlands
Basin
0
2
4
6
8
KM
3

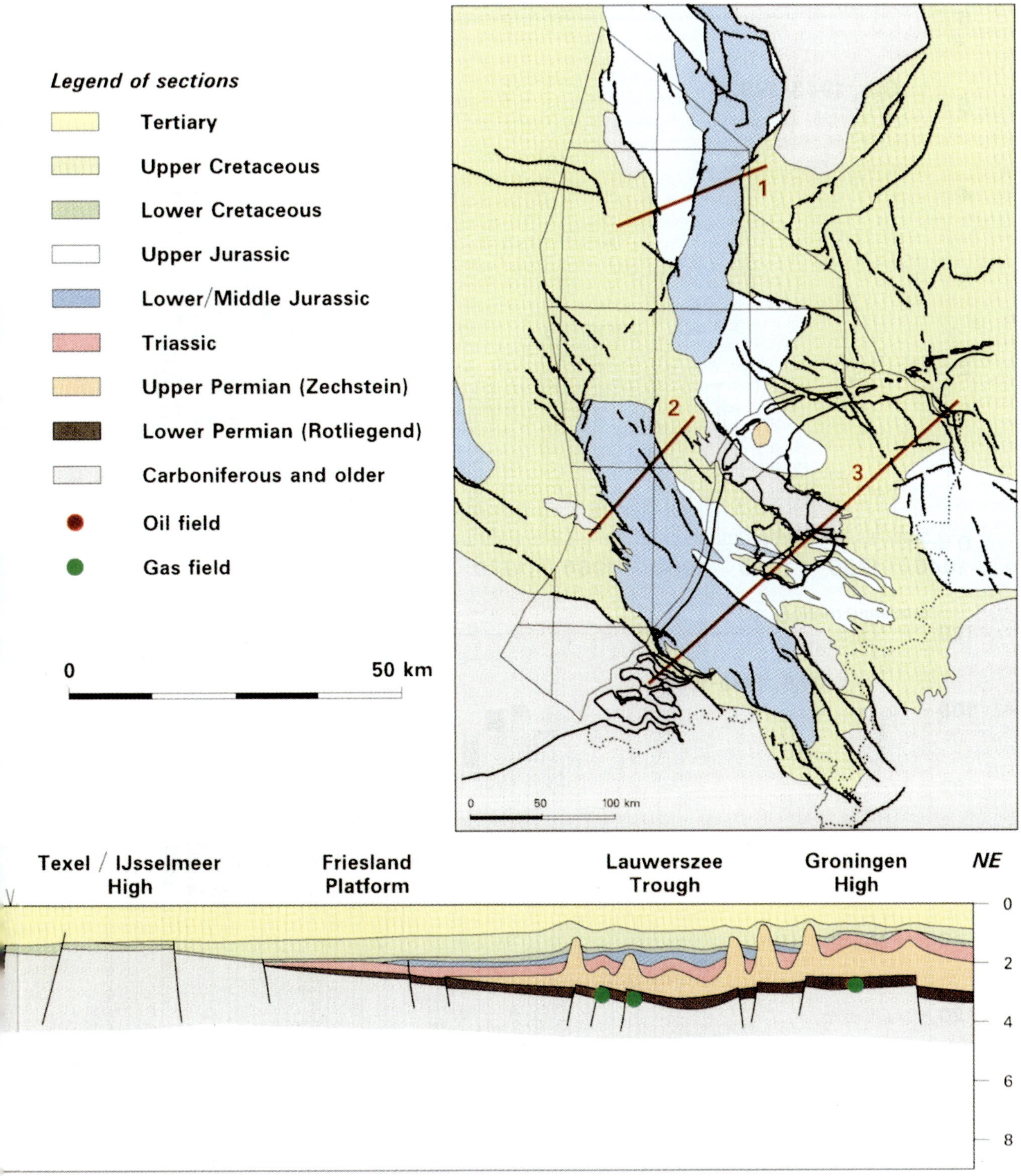

Fig. 11. Regional cross-sections. For legend of map see Figure 10.

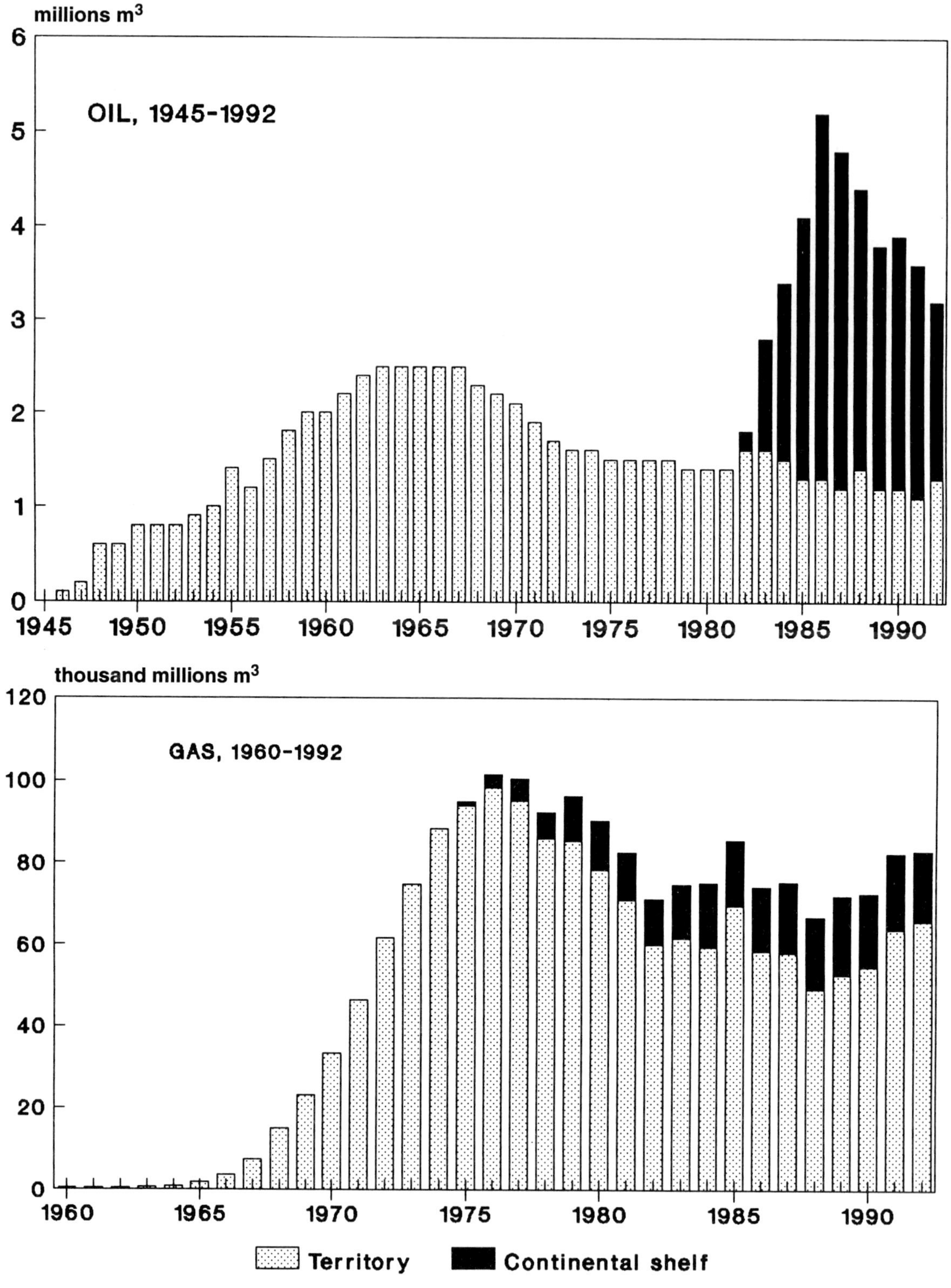

Fig. 4. Annual production of oil and gas.

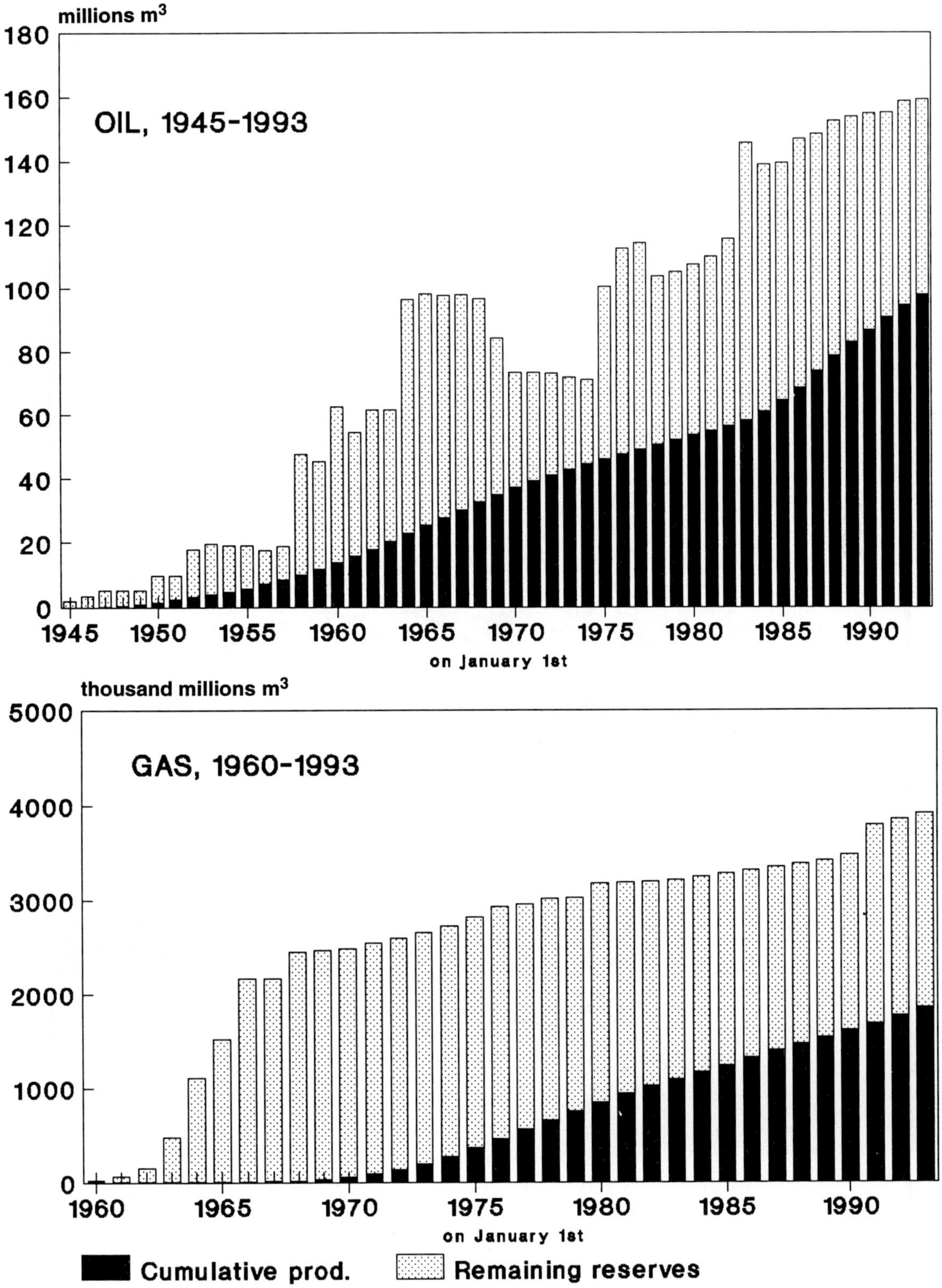

Fig. 5. Cumulative production and remaining reserves of oil and gas, shown by year.

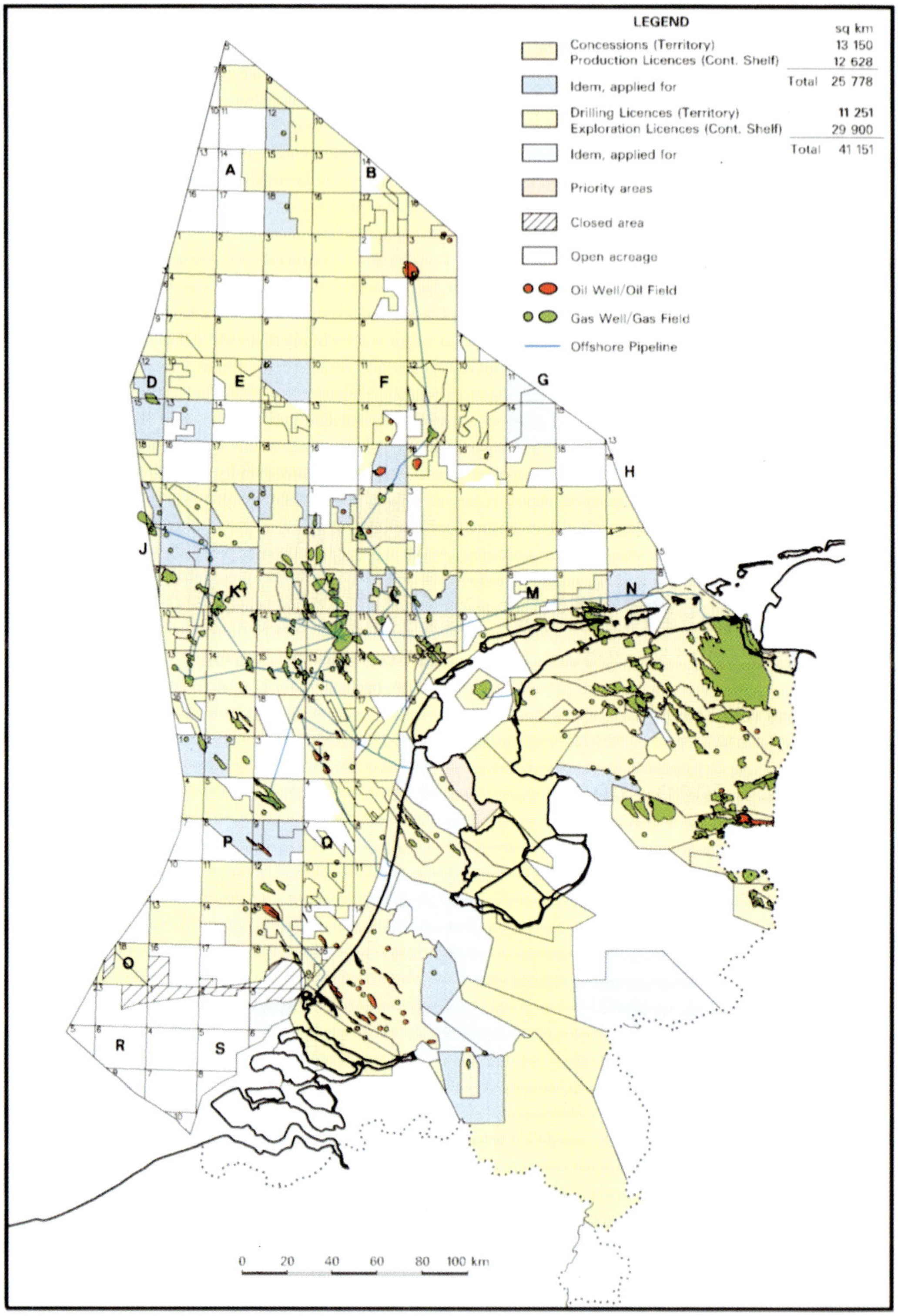

Fig. 6. Licences, fields and offshore pipelines. Licence areas as per April 1st, 1993.

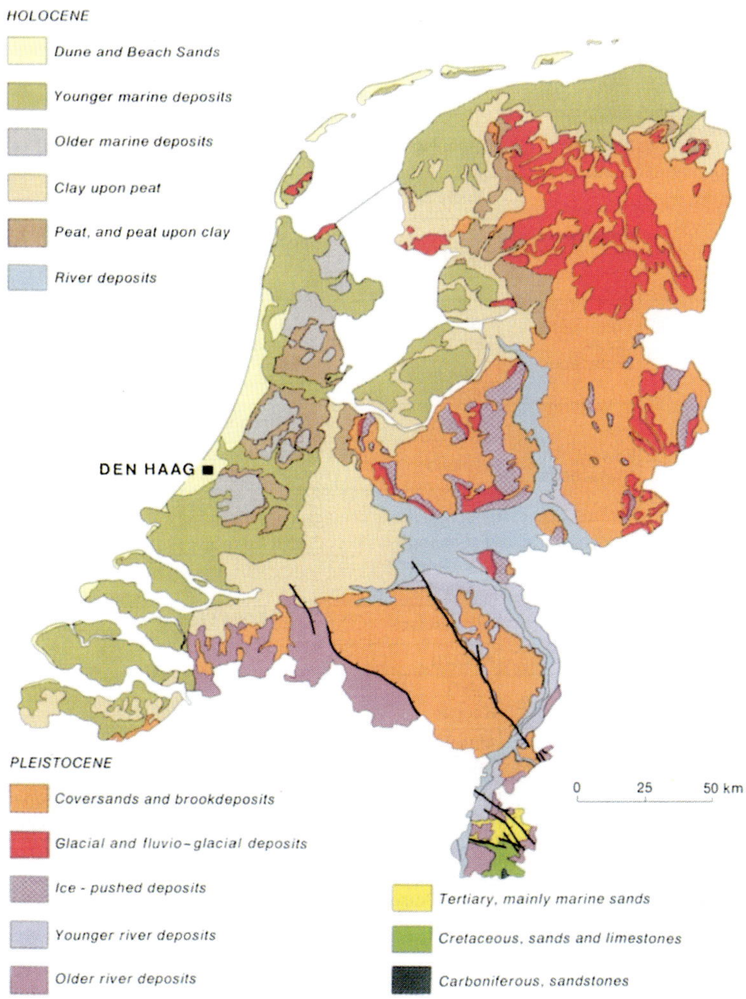

Fig. 7. Surface geology.

Van der Baan, D. 1990 Zechstein reservoirs in The Netherlands. In: Brooks, J. (ed.) Classic Petroleum Provinces. Geol. Soc. Spec. Publ. 50: 379–398.

Van Lith, J.G.J. 1983 Gas fields of Bergen concession, The Netherlands. Geol. Mijnbouw 62: 63–74.

Van Staalduinen, C.J., H.A. van Adrichem Boogaert, M.J.M. Bless, J.W.Chr. Doppert, H.M. Harsveldt, H.M. van Montfrans, E. Oele, R.A. Wermuth & W.H. Zagwijn 1979 The geology of the Netherlands – Meded. Rijks Geol. Dienst, 31: 9–49.

Van Wijhe, D.H. 1987 Structural evolution of inverted basins in the Dutch offshore – Tectonophysics 137: 171–219.

Van Wijhe, D.H., M. Lutz & J.P.H. Kaasschieter 1980 The Rotliegend in the Netherlands and its gas accumulations – Geol. Mijnbouw 59: 3–24.

Wong, Th.E., Th.H.M. van Doorn & B.M. Schroot 1989 "Late Jurassic" petroleum geology of the Dutch Central North Sea Graben – Geol. Rundschau 78: 319–336.

Zagwijn, W.H. 1989 The Netherlands during the Tertiary and the Quaternary: A case history of Coastal Lowland evolution – Geol. Mijnbouw 68: 107–120.

Ziegler, P.A. 1990 Geological Atlas of Western and Central Europe. Shell Int. Petrol. Mij., The Hague, 2nd ed., 239 pp, 56 encls.

Zijp, F.R. 1987 Structural evolution, stratigraphic sequences and subsurface reservoir horizons. In: Visser, W.A., J.I.S. Zonneveld & A.J. van Loon (eds.) Seventy-five years of geology and mining in The Netherlands (1912–1987). Royal Geol. and Mining Soc. of The Netherlands (KNGMG), The Hague: 269–284.

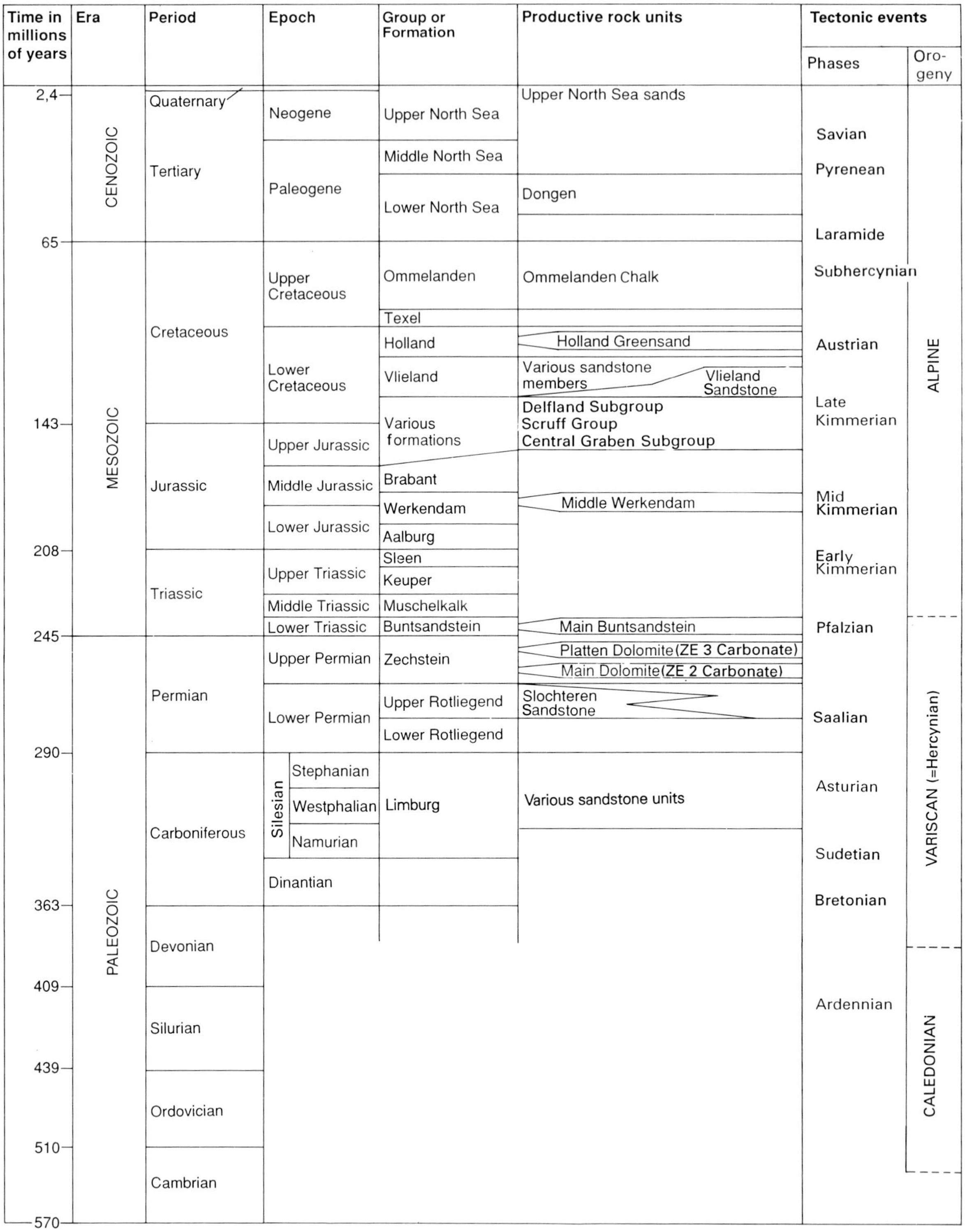

Fig. 8. Summary stratigraphy and tectonic events.

Table 1. Main features of exploration and production licences.

Territory Licence type	Obligations, terms	Duration
Priority declaration	seismic exploration	up to 18 months
Drilling licence	usually up to 4 wells, depending on permit conditions	up to 10 years, depending on permit conditions and number of wells drilled
Concession		
till 1976	various conditions	perpetual
from 1976 to mid 1988	state participation oil 50% gas 50%	perpetual
from mid 1988	state participation oil 50% gas 50%	depending on development plans
Continental shelf **Licence type**	**Obligations, terms**	**Duration**
Reconnaissance licence	seismic exploration	to be specified, usually 6 months
Exploration licence		
till 1976	expenditure obligation per km^2	15 years (50% relinquishment after 10 years)
from 1976	expenditure obligation per km^2 + workprogram	10 years (50% relinquishment after 6 years)
Production licence		
till 1976	state participation oil 0% gas 40%	40 years
from 1976	state participation oil 50% gas 50%	40 years

Table 2. Summary history of oil and gas exploration and production. For localities see Figure 2.

Early days		
1923	First oil shows eastern Netherlands	Corle (near Winterswijk)
1938	First oil shows western Netherlands	The Hague (Mient)
Start-up phase		
1943	Oil discovery eastern Netherlands	Schoonebeek – Bataafsche Petroleum Maatschappij (BPM)
1945	First significant oil production	Schoonebeek – BPM, from 1947 onwards Nederlandse Aardolie Maatschappij (NAM)
1948	Gas discovery eastern Netherlands	Coevorden – NAM
1951	First gas production	Coevorden – NAM
1953	Oil discovery western Netherlands	Rijswijk (near The Hague) – NAM
1954	First oil production western Netherlands	Rijswijk – NAM
Main phase		
1959	Discovery Groningen gas field	Slochteren-1 – NAM
1961	First offshore well	Kijkduin Zee-1 (near The Hague) – NAM
1963	First gas production Groningen field	Groningen – NAM
1964	First gas discoveries northwestern Netherlands	Amoco, Elf Petroland, NAM
1968	Offshore gas discovery	P6 – Mobil
1970	Offshore oil discovery	F18 – Tenneco
1975	First gas production offshore	L10 (discovered 1970) – Placid
1982	First oil production offshore	Q1 (discovered 1979) – Union, now Unocal
1984	Gas discovery western Netherlands	Botlek (near Rotterdam) – NAM
1993	Planned start of oil production northern offshore	F3 (discovered 1974) – NAM

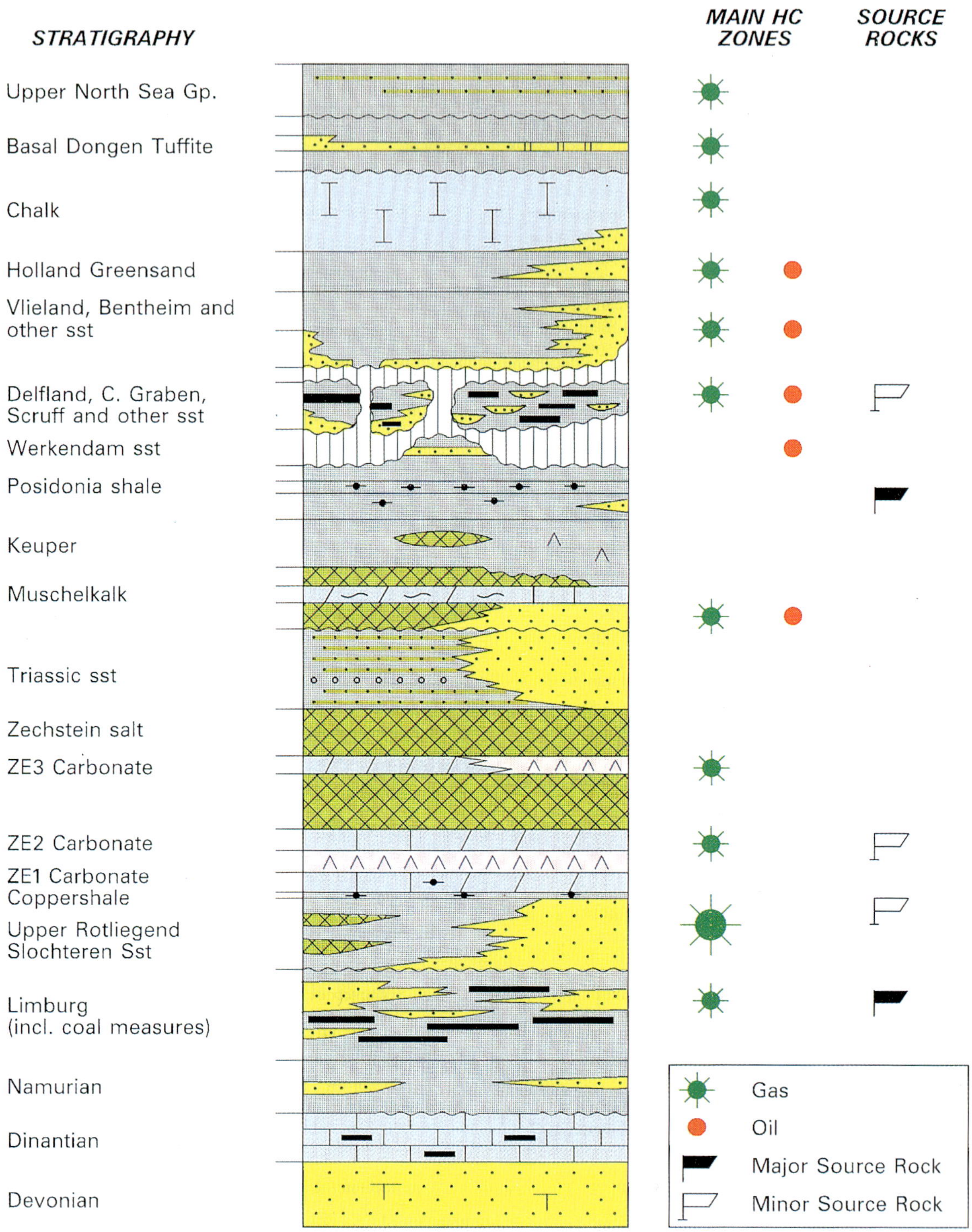

Fig. 9. Hydrocarbon plays.

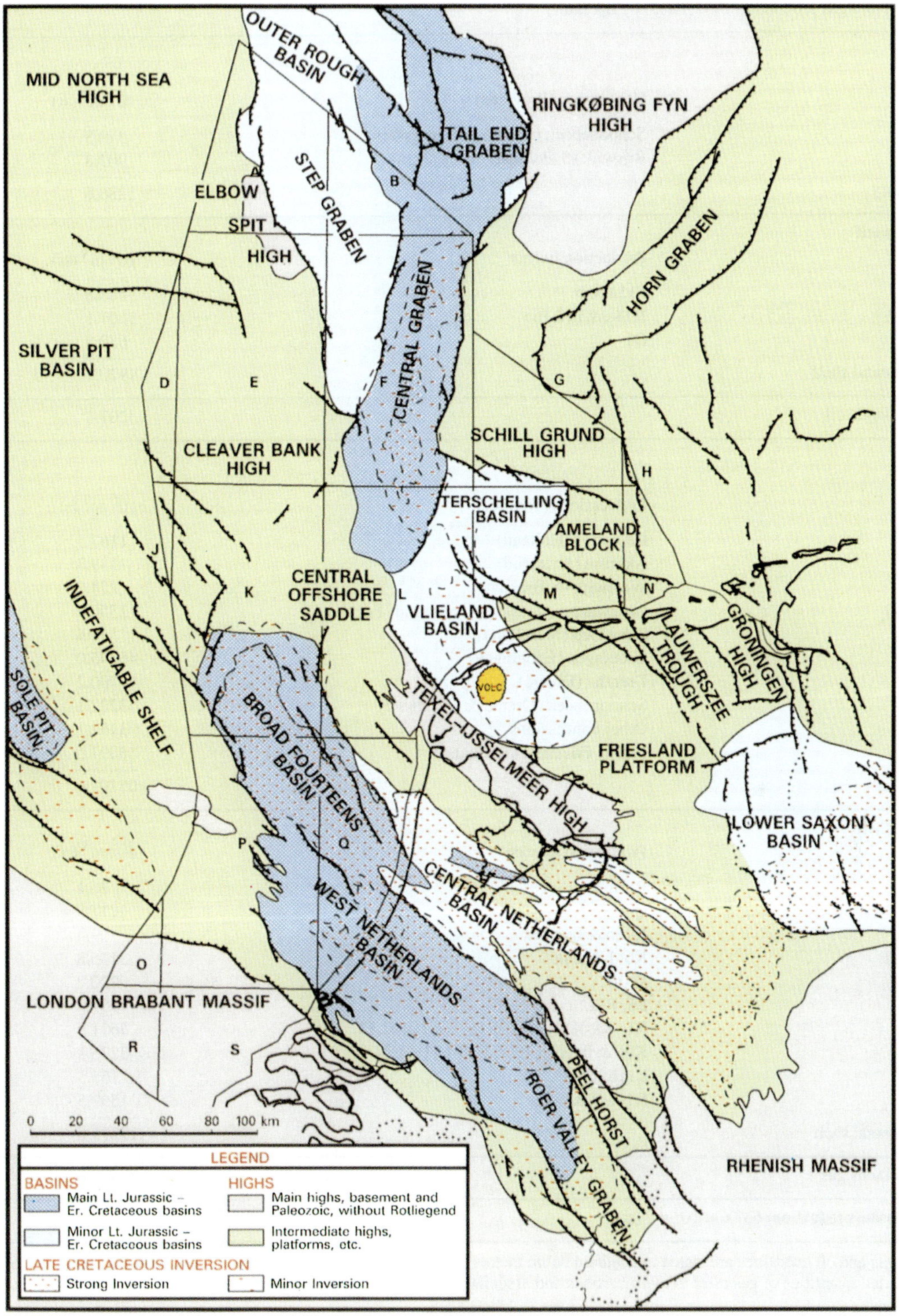

Fig. 10. Mesozoic structural geology.

Table 3. Oil and gas production 1992, listed by operator.

OIL Territory Company	Concession (Province)	10^3 m^3 (st)
NAM	Schoonebeek (Drente & Overijssel)	379.3
NAM	Rijswijk (S Holland)	907.3
Total Territory		1286.6
Continental shelf Company	Production licence	10^3 m^3 (st)
Amoco	P15a & b	188.8
Conoco	K18a & b; L16a	1101.1
Unocal	Q1	630.8
Total Continental shelf		1920.7
Total Netherlands oil		3207.3
GAS Territory Company	Concession (Province)	10^6 m^3 (st)
Amoco	Bergen (N Holland)	1167.5
Chevron	Akkrum (Friesland)	139.3
Clyde	Waalwijk (N Brabant)	274.0
Elf Petroland	Leeuwarden, Zuidwal, Oosterend (Friesland)	2550.4
	Slootdorp (N Holland)	53.4
NAM	Groningen (Groningen)	46345.0
	Drenthe (Drente)	5700.2
	Schoonebeek (Drente & Overijssel)	3222.1
	Other concessions	3152.1
NAM/Mobil	Noord-Friesland (Friesland)	3097.6
Total Territory		65701.6
Continental shelf Company	Production licence	10^6 m^3 (st)
Amoco	P15a & b	26.1
Clyde	Q8	217.3
Conoco	K18a & b; L16a	9.3
Elf Petroland	K6–L7; L4a	3135.8
Lasmo	J3-J6	220.2
Mobil	P6; P12	967.3
NAM	K6; K8-K11; K14; K15; L2; L5a; L13	7691.9
Placid	K9a & b; K9c; K12; L10-L11a; L14	3276.1
Unocal	L11b; Q1	165.5
Wintershall	K10a; L8a	1569.5
Total Continental shelf		17279.0
Total Netherlands gas		82980.6

Total condensate production: 637.6×10^3 m^3 (st)

- Natural gas and oil quantities are stated in standard cubic metres (m^3 (st)) at 1.01325 bar and 15 °C (ISO standard 5024–1976(E)).
- The reported quantities of gas refer to hydrocarbons and associated non-combustible gases.

Rondeel et al. (eds), Geology of gas and oil under the Netherlands, 1–5, 1996.

Resources and resourcefulness *

Mark Moody-Stuart
At the time this paper was presented, Mark Moody-Stuart was Group managing director and Exploration and Production co-ordinator for the Royal Dutch/Shell Group of companies. Since then he relinquished the role of Exploration and Production coordinator but remains a Group managing director.

Let me start by commending your conference theme. Longer-term oil supply will require a balance between 'old' and 'new' oil: between

– increasingly difficult oil from mature areas;
– frontier oil; and
– optimum production from major resource holders.

A key to this balance is technology – under which I include technical, managerial and financial/commercial expertise. In my first two categories – mature areas and frontier provinces – technology can reduce inherently high-cost oil. In the third – the major resource holders – it can help make affordable the increased production that will be needed. In doing so, technology could catalyse greater co-operation in the oil industry: between the resources of the major hydrocarbon holders, and the resourcefulness – if one might call it that – of the private industry.

If I had a criticism of the title of this conference, it would be the absence of the word 'gas'. In a sense, gas *is* 'the new oil', and most of our companies see much of our growth in gas. But there are important differences. Prospects for gas exports depend on developments in the major markets of North America, Europe and the Far East, mainly Japan, each of which behaves in a different way. In general, commercial and environmental factors should favour gas growth, but there is no global gas price, other than its varying link to oil. And although two areas – the former Soviet Union and Iran – hold over half the world total of reserves, the international markets still remain largely regional and bilateral. For many projects, such issues as reliability and pricing will be the deciding factors once an adequate reserve base exists. And in this the challenges of the gas industry are the same as those for oil – ensuring that we, as commercial organisations, can develop and bring the reserves to market at the price levels we might expect.

Consider for a moment the economic background against which our industry must work. For some years post-1986, there were still those who encouraged themselves with the thought that prices would kick up in the early to mid-1990s. By now we have surprisingly uniform agreement on the outlook, if not on the objective – from analysts, oil companies, producers and consumers. That outlook is for oil prices at least to the end of the decade in the range we have seen since 1986. A sombre outlook for producers. But let us bear in mind that when the sages of the world unite to agree that something is inevitable, it often does not happen!

You are all familiar with the arguments. Over the rest of this decade, OECD economies should recover slowly, but oil demand will grow only sluggishly. In the developing countries of Asia, even if economic growth slows somewhat from recent high levels, oil demand will continue to climb. Growth in China in particular is likely to be significant, and possibly in the Indian subcontinent also. Parts of Latin America may also grow rapidly. In the countries of the former Soviet Union, demand is likely to continue to decline through economic restructuring and efficiency improvements, but there should be a return to growth in East Europe.

Although the rate of growth in oil demand varies from scenario to scenario, considerable growth is present in all, driven by the non-OECD countries. Overall, one might envisage an average increase of 1.5 per cent a year in world oil demand, to some 73 million b/d in the year 2000.

Now, supply. OPEC's share of world production fell from 50% in 1973 to 30% in 1985. The increases in oil prices post-1973, together with concerns over stability of supply, stimulated significant volumes of much higher-cost non-OPEC oil. This development of what were then frontier areas, such as the North

* Keynote address to the American Association of Petroleum Geologists' International Conference *'New views on old world oil – Technology leads the way'*, The Hague, 18 October 1993.

Sea and the North Slope, was driven by private sector technology and capital denied an outlet in many OPEC areas. OPEC's share now stands at some 40%. In volume terms, OPEC's increase since 1985 is over 9 million b/d. Again in most scenarios one sees a continued increase in the call on OPEC, with non-OPEC fairly flat.

In spite of this increased call on OPEC, overcapacity seems set to stay – exacerbated by the expansion in hand in almost every OPEC country. Much of this will be in place by 1996, bringing excess OPEC production capacity – even excluding Iraq – to some 4 to 5 million b/d.

At present, oil prices are essentially determined by market perceptions, not by marginal economics or producer power. In spite of the ongoing conditions of over-supply, the market appears to keep prices within a fairly narrow range – a floor of around $ 15 at which there are expectations of increased OPEC cohesion and a ceiling of around $ 20, perhaps based on some perception of an economically sustainable maximum from a consuming country viewpoint.

Many organisations are now making forecasts and plans based on this range – a real-terms Brent price of $ 18 (in 1993 dollars) is fairly common, a change towards conservatism when compared to the last few years, and not a bad fit to history since 1986. But are we being conservative enough? Does the market think in real terms? There is, to say the least, cause for concern. Real-terms prices this year are the lowest since 1973. I believe that much of the upstream industry can live with a future in a world with Brent prices in the region of $ 18 in 1993 dollars, although development in some of the higher-cost areas of the North Sea, for example, may be affected. But if we see a price steady in nominal dollars, the impact of future operating costs, particularly in areas where operating costs are already high, will become critical.

What might change? As we all know, over half the world's proven oil reserves are in just four countries – Saudi Arabia, Iran, Iraq and Kuwait. Four more – Abu Dhabi, Venezuela, Mexico and the CIS – hold another quarter. Into the next century, the increases in production will come mainly from these countries, a fact which may at least begin to affect the market perceptions which govern the oil price.

In contrast, the first category I want to consider today – the mature areas of the USA and Europe – hold a minor – and falling – share. This has inevitably been reflected in production, as easy oil gives way to more difficult, and therefore expensive, reserves. This decline is likely to continue, its speed dictated by costs. Cost pressures in the industry are continually upward, and will not go away. With a flat oil price ceiling, we must reduce the cost base, both capital and operating.

Another wave of creative application of technology, similar to that which followed the oil price collapse in 1986, can prolong production from existing fields, enable new developments to go ahead, and allow access with minimum intrusion to prospective new areas. In turn, governments in mature areas will need to take note of the changing realities of oil economics, and be prepared to adjust fiscal regimes in the interests of maximising recovery. I believe they will do this, since the alternative is rapidly diminishing domestic production.

In addition to increased cost-effectiveness, new technology must deliver environmental and safety benefits. These objectives are not incompatible: slim hole drilling, for example, involves less cost, less waste and less space. Lightweight platforms, often unmanned, mean less materials consumed, fewer people at risk and lower operating costs. Nothing less than the highest environmental commitment and assurance of technical integrity and reliability will be sufficient if governments are to lift restrictions on prospective but environmentally sensitive areas.

Let me now turn to my second category: frontier oil, in which I also include frontier activity in established areas. In recent years, frontier production has been sufficient to offset declines in mature areas. Today's opportunities are perhaps wider than at any time since the end of the 1960s, but location, technical complexity, distance from market and lack of existing infrastructure combine to make many of these difficult and high-cost. The question is: can development be made viable at current prices? Clearly, much will depend on governments setting fiscal structures that can accommodate the costs, and still leave sufficient incentive for companies to justify these high-risk investments. This is not a question of subsidy: it is a matter of maximising value. Negotiations will involve the closest scrutiny – by both parties – of capital requirements, budgets and potential revenues, and ever more emphasis on costs and margins. However, the prospect of developing their own resources should be sufficient spur to encourage agreement.

Unlike mature areas, there is no existing infrastructure for piggy-backing; but there is the compensating advantage that projects can be planned from scratch, using the newest thinking and taking into account current – and future – environmental and safety requirements.

An important area of frontier activity will be deep water. Here, technology has already made great strides. Concrete platforms are being built for water depths of over 300 m, steel platforms are already producing in over 400 m, a subsea well in Brazil produces in 781 m, the Auger tension leg platform in over 800 m in the US Gulf of Mexico is due onstream next year, and Mars will follow in almost 900 m. Such developments are at the leading edge of technology; but perhaps an even greater challenge will be to managerial and business skills, to bring them into production on schedule and within budget. Some of the earlier examples I have quoted could not be built in today's economic climate.

Finally, let us consider the major resource holders. Most of the incremental 10 million b/d call on OPEC by the year 2000 would come from the Middle East members. Of the other major resource holders, Mexico could add over half a million b/d, depending on the availability of domestic investment and the acceptability of that from overseas. Only the CIS shows declining oil output, although with some recovery expected by the end of the decade.

But although the longer-term demand prospect must be encouraging, these countries will need major investment to increase production to meet demand. Into the next century, the five major Gulf producers alone will need to invest up to $ 35 billion a year – comparable with likely expenditure during the 1990s for the whole Upstream industry.

At least the money should be available. Capital investment in the oil industry in the 1990s is likely to be only some 2% of the world total, compared with some 3.5% in the 1980s. Although oil's share may climb again to 3% early next century, this should not cause problems, certainly for the Upstream. However, projects will vary in attractiveness in competing for investment funds. The lesson of the past is that finance and technology flows relentlessly towards opportunities, not to the lowest-cost production. Capital is always scarce for investments likely to offer poor returns.

The position of the major resource holders has changed dramatically since the early 1980s. Many already face severe competition for funds from many pressing development priorities. In some cases, investment merely to maintain the oil infrastructure has been lacking, resulting in some reservoirs not being managed to best effect, and even suffering potentially permanent damage.

Increasing production requires both application of technology in day-to-day efforts to increase production from existing fields, and major long-term projects, with their managerial and technological challenges.

So, where will the capacity be added? In a world of rational economics, capacity would be added in the lowest-cost areas. The bulk would be in the major resource holders of the Gulf OPEC members. But the higher-cost producers both in OPEC and in non-OPEC developing countries will want their share of the cake. These often have the greatest population pressures and severe shortages of funds for major developments. Because of their need both for capital for major developments and technology to reduce the cost of producing their higher-cost resources, they will be more likely to form alliances with the international oil industry.

Such alliances are not new. Indonesia, Nigeria and Gabon, all OPEC members, have long had profitable and stable arrangements with the international oil industry to develop their resources, which tend to be distributed in very large numbers of clastic reservoirs. These are unlike those of the Middle Eastern OPEC members, whose reserves are in highly productive and often massive carbonate reservoirs, with correspondingly lower costs.

This trend is continuing: North African OPEC producers Libya and Algeria, as well as countries such as Venezuela, plus of course the non-OPEC countries of the former Soviet Union – Russia, Kazakhstan, Azerbaijan – all are turning to a greater or lesser extent to the international industry.

In the competition for market share in an oversupplied market, the international suppliers of technology and capital are thus likely to be called upon to redress the balance between the very large low-cost producers, and the smaller but significant producers whose resources are higher-cost. Think of the successful and productive alliances forged over recent years to mutual benefit in Oman, Yemen and Syria.

But the call on the international oil industry may not be limited to the higher-cost producers. Even a major low-cost producer such as Saudi Arabia might choose that route if it were convinced that the cost-reducing technology or increases in ultimate recovery generated through reservoir management techniques were sufficiently attractive to overcome the political barriers to international industry involvement and to remunerate all parties adequately.

What are the barriers? They lie half-buried and almost forgotten, but still there, in many countries and organisations. Historical and emotional barriers, perhaps best summed up by the mirror-image words 'exploitation' and 'expropriation'.

Sovereignty is the key, and I believe has long been fully acknowledged and accepted by the international industry. Likewise, there is full recognition of the need for local industry involvement. However, acceptable international arbitration arrangements are essential for major investments. Production Sharing Contracts, now very often the industry preference, were once a source of concern because the investor had no rights to the assets. I believe that we are no longer concerned about asset ownership, and have long since accepted only limited lifting rights to produced oil. In fact, provided that there is a mechanism for an adequate financial return, rights to oil are not important.

However, on the industry side there are also limits and sensitivities. Experience has shown that a service arrangement based on the supply of skilled manpower or a cost-plus arrangement does not give the company an adequate reward and incentive. To maintain our technological base, we seek opportunities for risk/reward and performance/reward investment. I also believe that few companies would be prepared to accept an arrangement completely disconnected from oil price – so many of our industry costs are indirectly related to the price of oil. But the forms of agreement are very varied. The financial and fiscal arrangements under which Shell companies operate in for example Australia, Brunei, Malaysia, Oman, Syria, Nigeria, the UK and the Netherlands are all completely different, but each can be satisfactory in its own way.

The new arrangements proposed by Venezuela for reactivating fields and related exploration are different again. The company clearly acts as a contractor in producing hydrocarbons, but its reward is ingeniously linked to its technical performance and there is an indexation to oil price. The arrangement shows how agreements can be crafted to acknowledge the realities of legal and political sensitivities.

When seeking technology, to what sort of international company should a major resource holder turn? Is it not possible for a major state company whose parent government has access to sufficient funds to do the job itself, using the off-the-shelf technology available from the service industry? This is certainly one solution, and in some cases it has been very effective. The deep water oil of the Campos Basin, the light hydrocarbons of the El Furrial trend and the light Palaeozoic hydrocarbons of Saudi Arabia are examples of some of the most significant new plays of the past decade discovered and developed, or being developed, by Petrobras, Petroléos de Venezuela SA and Saudi Aramco without involvement of international oil companies.

For the counter argument, I believe one has to turn to the North Sea, or to the North Slope of Alaska, or the Gulf of Mexico. It's hard to imagine that such developments could have been handled by a state company. In these areas one sees a great variety of companies, large and small, competing and co-operating, learning from one another, with a very rapid spread of any successful initiative or technique. The situation is far from static – the successful grow and the laggards wither. The service companies have a vital role to play in this, and are an important mechanism in the diffusion of technology throughout the industry. No single company or organisation has all the answers, and to entrust one's industry to one company alone, state or private, is to run the risk of complacency, stagnation and inefficiency. In providing a benchmark against which to judge performance and productivity, the presence of international competition is a powerful stimulus to domestic industry.

To provide full value, the foreign investor must be able to offer much more: not just a vague promise of 'technical services', or indeed the essential services of the service contracting industry, but an integrated package of financial, technological and managerial skills, specifically targeted at adding value to each individual opportunity on offer. To put it more bluntly: if he cannot increase the size of the cake, no-one is going to offer him a slice.

That is the benefit that the international oil and gas industry brings – wide experience in reservoir development and management in an endless variety of environments, and equally wide experience in developing major international gas projects and the related markets, as well as a system geared to picking up the learning points from one area and transferring them rapidly to another. Within my own organisation, we see people as the key to this. While keeping a basis of common high standards, we are broadening our recruitment of technical professional staff before putting them through a high-quality training and development process.

Part of the contribution of an international company to the economy of a country is the development of a cadre of national staff. In our case, we try to ensure that these staff have had the opportunity to broaden their experience by working in different countries; for example, in our upstream operations almost 900 non-Dutch and non-British staff, of over 40 nationalities, are working outside their own countries. In this way, staff develop confidence in their own abilities on an international scale, while at the same time contributing from their own experience. I believe that companies with truly international staff are more likely to be able to adapt flexibly. Equally important, organisations and companies have an opportunity to refresh themselves and ventilate their thinking.

Rotation of staff internationally is one of the most powerful methods of disseminating technology. Another, of course, is conferences like this. We will no doubt be hearing of many fascinating technological advances and insights over these coming three days. All will have a role to play, later if not sooner, in the identification and assessment of oil and gas opportunities – in enabling the size and value of the cake to be measured and increased. Their full benefit will only come through integration with other technological advances across the exploration and production business. The winners will be those that apply technology, not in terms of isolated breakthroughs, but across a broad, advancing front – holding back the pressure of increasing costs, and helping to ensure the continued supply of world energy.

Rondeel et al. (eds), Geology of gas and oil under the Netherlands, 7–10, 1996.

The role of oil and gas in the Dutch energy policy *

C.W.M. Dessens
Ministry of Economic Affairs, Postbus 20101, 2500 EC Den Haag, the Netherlands

Introduction

Recently I have been told about an article of Dr. Van Waterschoot van der Gracht, one of the founding fathers of the Geological Survey of the Netherlands, in an AAPG Bulletin, dating from 1936, in which he stated: "I believe that Palaeozoic accumulations of petroleum may be expected in Europe" (Van Waterschoot van der Gracht 1936). We can smile about it today, but at the time, it was a daring prediction.

At most places in Europe, the Palaeozoic formations lie at great depths, deeper than anyone had drilled at the time. Dr. Van Waterschoot's 1936 forecast can be compared with someone saying today that large gas reserves are to be found at a depth of five to ten kilometres. Most of us would reject such predictions out of hand. They don't fit with our current scientific insights. Nevertheless, some caution is called for. After all, oil and gas exploration is an activity which regularly produces surprises. Just when we think that we have reached the bottom of the bottle, new discoveries are made. Striking examples of this can be documented from the Netherlands, of which I shall mention a few.

In 1938, oil was discovered near The Hague, entirely by accident. It happened during a public demonstration of how drilling works. Entirely unexpectedly, the drillers struck oil at a depth of 600 m. Not long afterwards, the Schoonebeek oil field was discovered, which is still one of the largest onshore oil fields in western Europe.

The biggest surprise came in 1959, when the Groningen gas field was discovered. Originally, estimates of the reserve were cautious. But gradually it became clear that one of the largest gas fields in the world had been drilled.

At that moment, in the early 1960s, the Dutch government began to realise that its mining policy would have to be drastically revised. A system was designed which, in its main points, is still intact today and has proved extremely successful. To present an idea of the Netherlands' current position:

- The Netherlands is the fourth largest gas producer in the world.
- It supplies more than 40% of the European Community's gas demand.
- Dutch energy production virtually covers domestic consumption.

I will briefly explain how this position has been reached. My address will focus on three points:

1. The main lines of Dutch energy policy.
2. The results of this policy.
3. Possible future developments.

The main lines of energy policy

One of the basic principles of the government's energy policy is that gas reserves should be used in a way that provides most benefits. To achieve this, several measures have been taken. To start with, the gas policy is aimed at close coordination of upstream and downstream activities. Energie Beheer Nederland BV (EBN), a specialised company owned by the State, participates both in gas production and in the main Dutch gas marketing and transport company, NV Nederlandse Gasunie. Furthermore, the private parties (Shell and Esso) involved in the production of the most important Dutch gas field (Groningen), also participate in Gasunie.

State participation through EBN in other exploration and production ventures, operated by private companies, also serves the general interest.

In addition, the Minister of Economic Affairs has been granted a number of decisive powers with respect to Gasunie's activities. His approval is needed for

* Keynote address to the American Association of Petroleum Geologists' International Conference *'New views on old world oil – Technology leads the way'*, The Hague, 18 October 1993.

Gasunie's yearly Marketing Plan, its gas purchases and sales contracts, and its investments in infrastructure.

Finally, the Dutch government at an early stage determined a special pricing principle for gas since gas is a special commodity. In the energy market, gas is always an alternative for other energy sources and it is not easy to process. Therefore the normal bargaining processes cannot apply to gas, and the price mechanism of, say, the cattle market is not valid. In view of this situation, the principle of 'market value' was introduced.

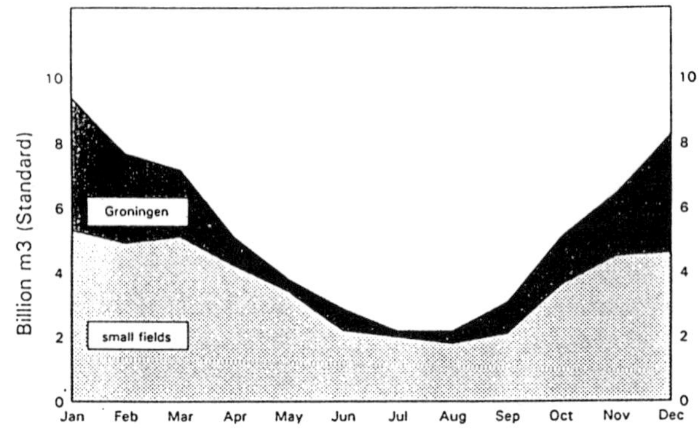

Fig. 1. A typical profile of the yearly gas production in the Netherlands, which illustrates the balancing of gas production by the Groningen field.

Market value principle

In the Netherlands, the price of gas has been linked to a product for which the market mechanism does apply. We use the term 'market value' for this. The principle is that the price of gas should be linked to that of the closest alternative. For the households, that is heating oil, and for large-scale users, fuel oil. In electricity generation a linkage to coal is also possible.

The market value principle is attractive for both consumers and producers. The advantage for the consumer is a price guarantee. Gas is never more expensive than its alternatives. Therefore, the consumer will not easily switch to another fuel. The producers also benefit, as they are given a sales guarantee. In the Netherlands, the costs of transport and distribution are exceptionally low. There is one transport company, which operates on cost base and is therefore not a profit-centre. As a result, most of the natural gas earnings are passed on to the producers. We call this the net-back principle.

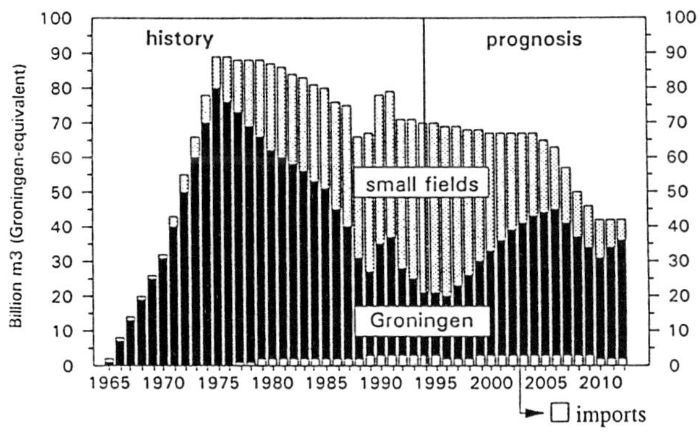

Fig. 2. Gas intake by Gasunie from fields in the Netherlands and abroad.

Small fields policy

Another important element in the Dutch energy policy is that of the small fields. After the first oil crisis in 1973, the country's energy dependence on politically less stable countries was realised. A policy aimed at discovery and development of our national reserves as far as possible has been developed since.

This policy could only be realised if sensible use was made of the Groningen gas field. This field is so large that it can be used to absorb considerable fluctuations in the supply from other fields and in demand. Maximum use is made of the Groningen field in winter, while production is cut back in summer, as shown in Fig. 1. Groningen therefore acts as a balance. Figure 2 illustrates that Groningen also serves as a balancing field over longer periods.

The Groningen field covers the difference between supply and demand over a number of years, as well as within the year. This affords a high degree of freedom in the development of smaller fields, which is extremely attractive for the oil companies.

An important feature of the Dutch energy policy has been to keep the fiscal and financial rules of the game as constant as possible. This offers operators a high degree of security, a significant factor, particularly where large investments are involved. Where has this mining policy led us? This brings me to my second point.

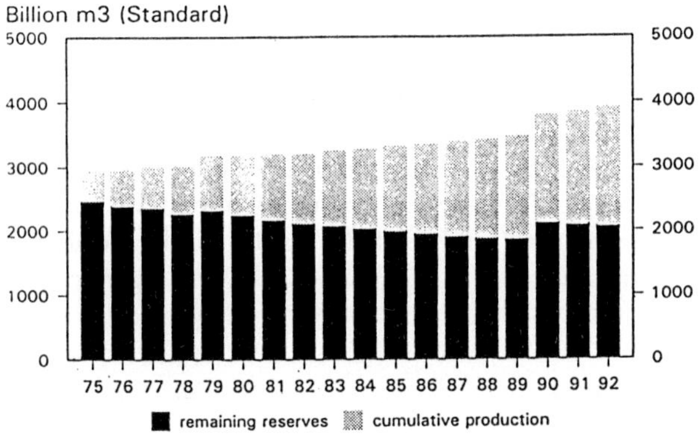

Fig. 3. Development of gas reserves in the Netherlands 1975–1992. Historical data (year end).

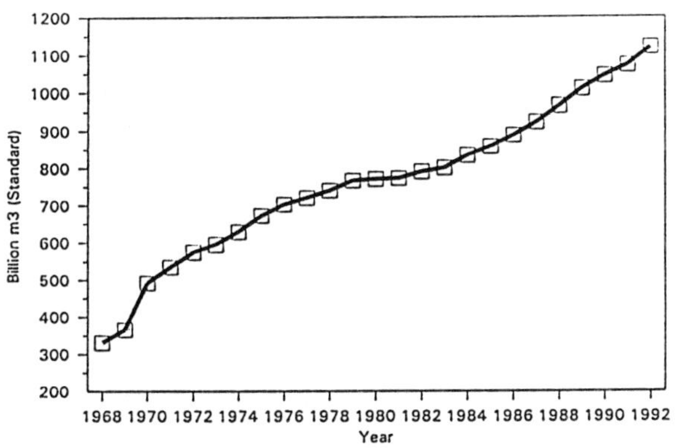

Fig. 4. Newly found, cumulative gas reserves in the Netherlands, 1968–1992, excluding the Groningen gasfield.

Results of energy policy

I should like to illustrate the results of the Dutch mining policy with some graphs. Figure 3 shows the development of Dutch gas reserves. The graph is based on data which are published yearly by the Ministry of Economic Affairs. It shows that reserves of some 3900 billion (10^9) m^3 have now been confirmed. Of this amount, a total of about 1850 billion m^3 has been produced.

Of the 3900 billion m^3, some 2750 billion m^3 lies in the Groningen gas field. The other fields together contain about 1150 billion m^3, which is a substantial volume. The development of the natural gas reserves is expressed in Fig. 4 as a creaming curve. The graph shows little or no creaming in recent years. Each year, additional gas volumes are discovered in the order of some 35 billion m^3.

Three-dimensional seismic makes an important contribution to this success. It is one of the most significant technological developments of the last twenty years. About 35 000 km^2 of the Netherlands, including almost 60% of its total licence area, is now covered with 3D. I suspect that the Netherlands is one of the most intensively recorded countries in the world. As a result, exploration wells have a high success rate. In 1992, one in two exploration wells was successful.

So far, 275 gas fields and 38 oil fields have been discovered. More than half of the gas fields contain reserves of less than 2 billion m^3.

Having outlined what has been achieved, I shall now try to give an idea of what can be expected in the future.

Potential developments

According to the Geological Survey of the Netherlands, 200 to 400 billion m^3 of natural gas is still to be discovered. In this respect there is no need to worry about the near future. However, it will demand a substantial effort to discover and produce this gas. A complicating factor here is that many of the prospects are situated under environmentally sensitive areas such as the Waddensea and the IJsselmeer lake.

Unfortunately, the oil industry still suffers from an image of being a polluting industry. Oil blow-outs and burning gas wells are regularly shown pictures in the media. This creates the idea that the oil industry does not operate safely, an idea which is subsequently used as an argument to keep oil companies out of environmentally sensitive areas.

I have some difficulty with this picture of the industry. Practice has shown that the operators working in the Netherlands satisfy the highest safety and environmental standards. I know that they apply the latest technology to maintain these standards or to improve them still further. I am therefore convinced that they are able to drill in sensitive areas in a responsible fashion.

There is a second factor which affects the exploration and production climate: the continuing low price of oil. Some operators allow themselves to be discouraged by this, and reduce their drilling activities or transfer them to greener pastures. This trend can be seen in the Netherlands, but also in neighbouring countries.

In times of recession, industry tends to point an accusing finger at the government. This is understandable, but is it fair? As I explained at the start of my address, the Dutch government has developed a system which is beneficial to society, but also to industry. I should like to recall the small fields policy, the market value pricing policy and the net-back principle.

The Dutch government has also developed programmes to increase publicity on what lies beneath the surface of the Netherlands, within the limits, of course, of what the Mining Act allows. I am referring specifically to the activities of the Geological Survey, mapping programmes, deep seismic research and gas atlas. The industry can benefit from these studies.

You will not hear me say that things cannot be done better. Particularly for the smallest fields, those with a reserve of less than 3 billion m^3, special incentives will probably be needed. The government is discussing this subject with the organisation of Dutch operators.

However, do not expect miracles from the government. In the first instance, the industry has to work its own miracles. I am referring here to measures such as cost cuts and coordination. I am convinced that it is possible to develop marginal fields at lower costs, although this will require innovative and technological brainwork. This may ask for a kind of cultural change in the industry, but I think the results will be worth the effort.

References

Van Waterschoot van der Gracht, W.A.J.M. 1936 Possibility of oil and gas production from Paleozoic formations in Europe – Amer. Ass. Petroleum Geol. Bull. 20: 1476–1493

Rondeel et al. (eds), Geology of gas and oil under the Netherlands, 11–18, 1996.

New oil and gas – Technology leads the way *

Jacques Halfon†

Executive Vice-President, Exploration and Production, Elf Aquitaine, Tour Elf, Cédex 45, 92078 Paris La Défense, France

Introduction

Following the sharp rise in oil prices in 1973, oil companies endeavoured to increase their reserves and production. This was attempted through the development of unconventional oil from bituminous shales or tar sands and the extensive use of enhanced oil recovery. However, when oil prices dropped in 1985, such projects were no longer viable due to the huge investments required. Subsequently, new reserves were found by applying new techniques to the search for traditional oil and gas accumulations.

Improvements in seismics

In exploration, a substantial improvement in the quality of acquisition and (re)processing of seismic data has taken place over the past decade.

The improvement of the resolution of 3D seismic data ensures a better definition of traps and can result in the discovery of new oil and gas reserves. An example on a field offshore Nigeria clearly demonstrates the extent of improvement achieved between 1975 and 1992. Figure 1 displays a 2D seismic line shot in 1975. Three wells were drilled very close to this line, proving different oil-bearing zones (indicated in thick black). On the early version of this seismic section, the quality of the data was rather poor, especially with respect to fault definition, and was not good enough for locating additional wells. A seismic section taken from a 3D survey recorded in 1987 is shown on Fig. 2. This line is almost superimposable on the previous 2D line. On this line, the fault definition is sharply increased, and markers which previously seemed to be continuous (e.g. the marker shown in dotted line) now appear to be broken by faults. Improved fault definition is particularly critical in deltaic environments in order to accurately delineate the various compartments created by growth faulting.

Nevertheless, the seismic resolution of the producing levels was still not sufficiently reliable to enable a decision to develop the field. Therefore a 3D high-resolution survey was shot in 1992. A line taken from this survey (Fig. 3) clearly indicates a substantial improvement throughout the section, particularly in the producing levels. This is a critical factor for estimating the volume of 'Original Oil in Place'. It has since been decided to go ahead with development.

Because of the increased use of 3D seismic acquisition, a major new challenge has emerged in geophysics: optimisation of the velocity models (where regional geological and geophysical knowledge is crucial), requiring the use of Massively Parallel Processing computers, has become a necessary step in defining new traps. As an example, the seismic definition of pre-salt series can be significantly improved. In the Precaspian Basin (Kazakhstan), a 2D line was processed by Elf using two different methods. In the example shown in Figs 4 and 5, the objective lies within the pre-evaporitic series (discontinuous event at a depth between 3000 and 4000 m). A conventional 2D post-stack depth migration, using the MIGPACK TM software (Fig. 4), failed to restore a satisfactory image of the section below the salt diapir and the flanks of the diapir itself. On a further attempt, the line went through a pre-stack depth migration (Fig. 5). A significant improvement was obtained at the objective level and around the edges of the salt dome.

Pre-stack depth migration is being used more and more often to improve resolution in pre-salt series (e.g. North Sea, Gabon) and to obtain a better image of events within areas of complex tectonics.

---

* Keynote address to the American Association of Petroleum Geologists' International Conference *New views on old world oil – Technology leads the way'*, The Hague, 18 October 1993.

† Deceased December 27, 1994

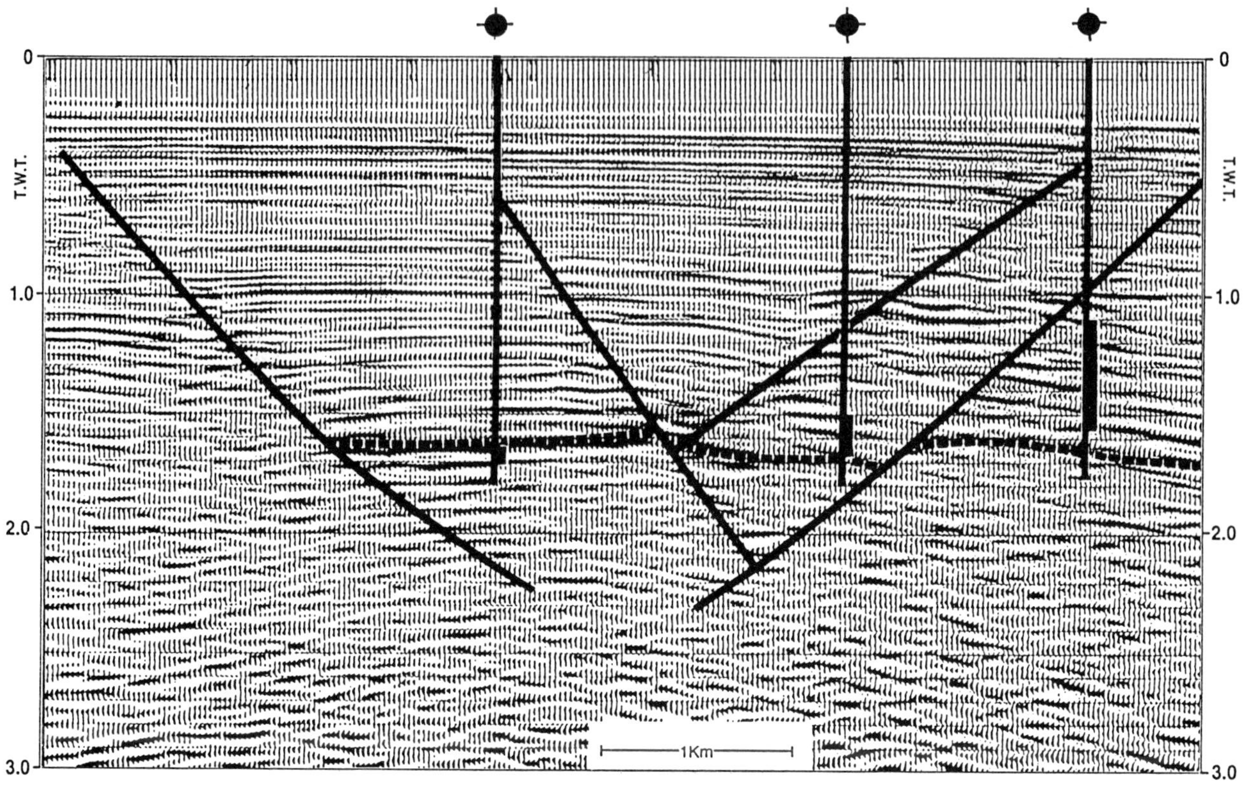

Fig. 1. A 2D seismic line shot in 1975 offshore Nigeria represents a rather poor data quality. There was a real difficulty for locating additional wells. Depth in two-way traveltime (TWT) in seconds.

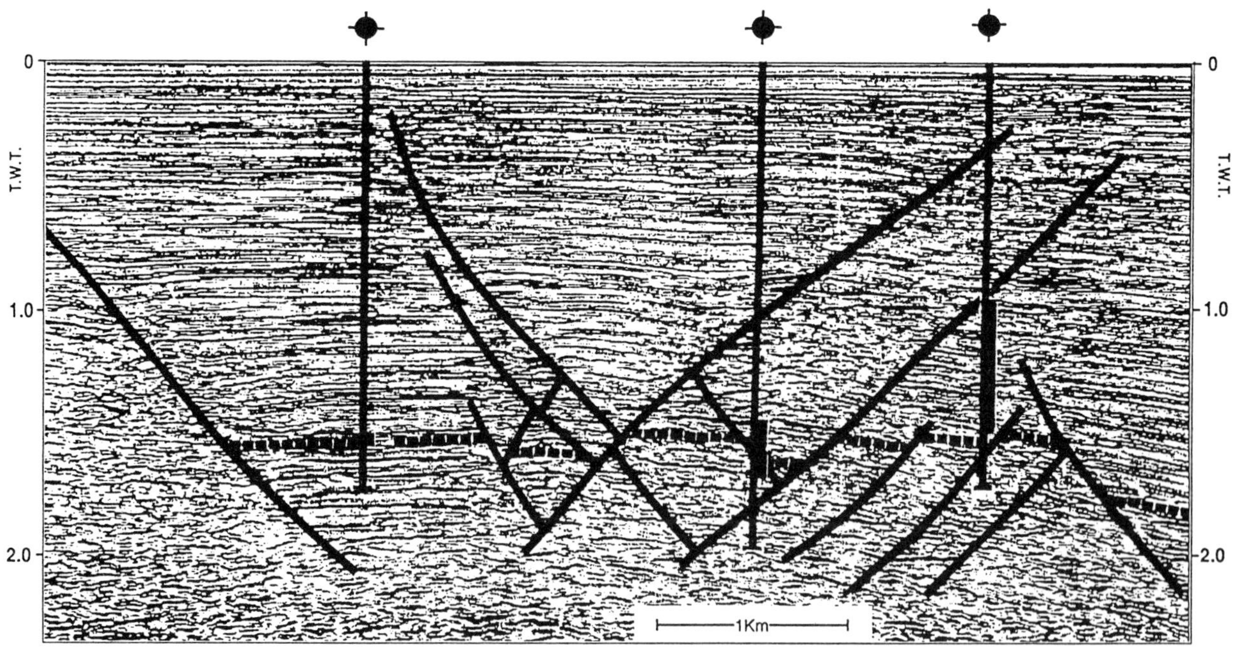

Fig. 2. A standard 3D section, shot in 1987 along the same line as that of Fig. 1, shows a better definition of the fault pattern and an increased accuracy in horizon picking. Depth in two-way traveltime (TWT) in seconds.

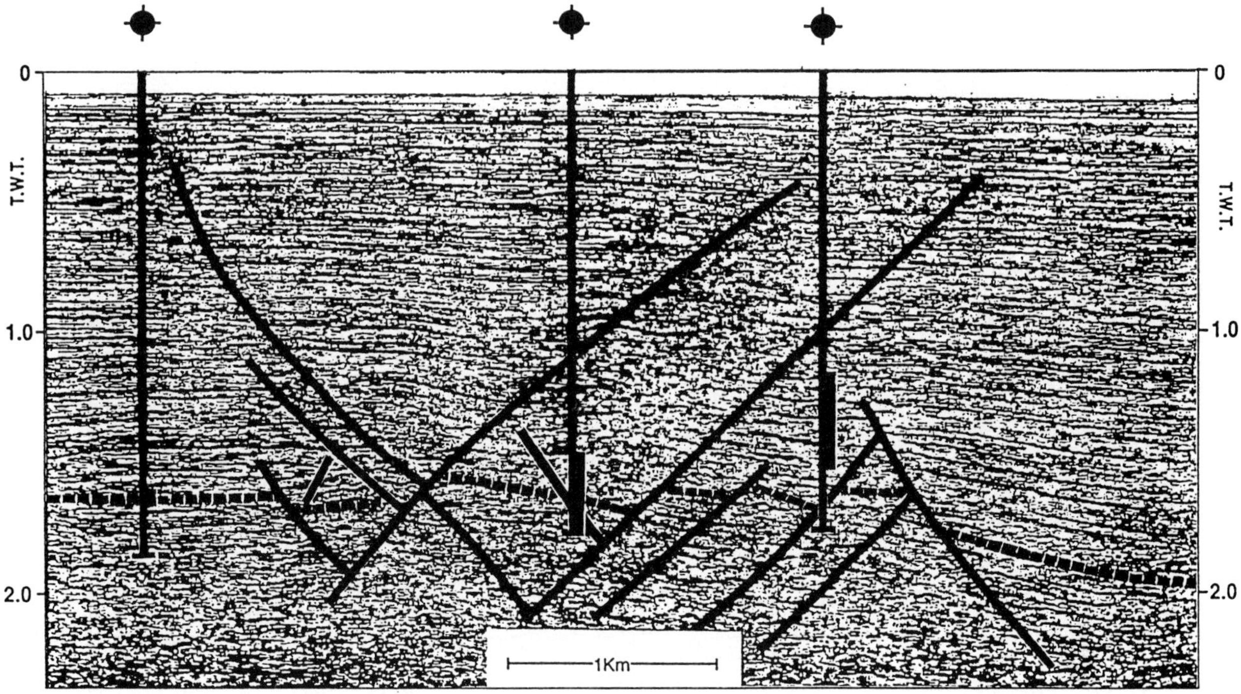

Fig. 3. A 3D high-resolution section, shot in 1992 along the same line as that of Fig. 1, shows a substantial improvement within the production levels. Depth in two-way traveltime (TWT) in seconds.

An additional example of the improvement obtained through sophisticated processing is illustrated by a new seismic tool called Deltastack, a program developed by Elf Aquitaine. Deltastack has been used for some time to predict lithology and to evaluate formation pressures in clastic series (mostly within deltaic environments).

Through the statistical processing of seismic velocities and other attributes, it has become possible to delineate the top and the lateral extension of a particular reservoir. In the example shown (Nigeria, Fig. 6), the oil-bearing interval corresponds to the yellow and pink section (higher velocity interval). It will be noticed that the top of the reservoir has an irregular shape, and that the sand distribution is controlled by growth fault activity. This kind of representation proved to be extremely helpful in visualising the reservoir geometry and optimising the location of additional delineation and development wells. Elf's research efforts are presently focused on new applications of Deltastack in various sedimentary environments.

Deltastack is also used to predict overpressured intervals. On Fig. 7, the undercompacted shales correspond to the blue intervals indicating lower velocities. On the left-hand side of the section, the lower part of the well actually intersected an undercompacted interval down to total depth, thus confirming the interpretation. This type of predictive capability is obviously quite crucial for both explorationists and drillers (e.g. exploration below overpressured zones, choice of casing programme).

Integrated studies (geology, geophysics, reservoir engineering)

We clearly can improve our analysis and results by the use of new technologies in a specific technical area. In what follows it is demonstrated how development projects can benefit from integrated studies and team work.

A comprehensive image of the reservoir can be obtained by integrating all the available field data involving geology and geophysics, and reservoir engineering. This type of integrated approach is of paramount importance for optimising the development of complex fields and in particular for choosing the best drainage mechanism and well pattern (Fig. 8).

A good example is N'Kossa, an offshore field in the Congo. This field was discovered in 1985 and is

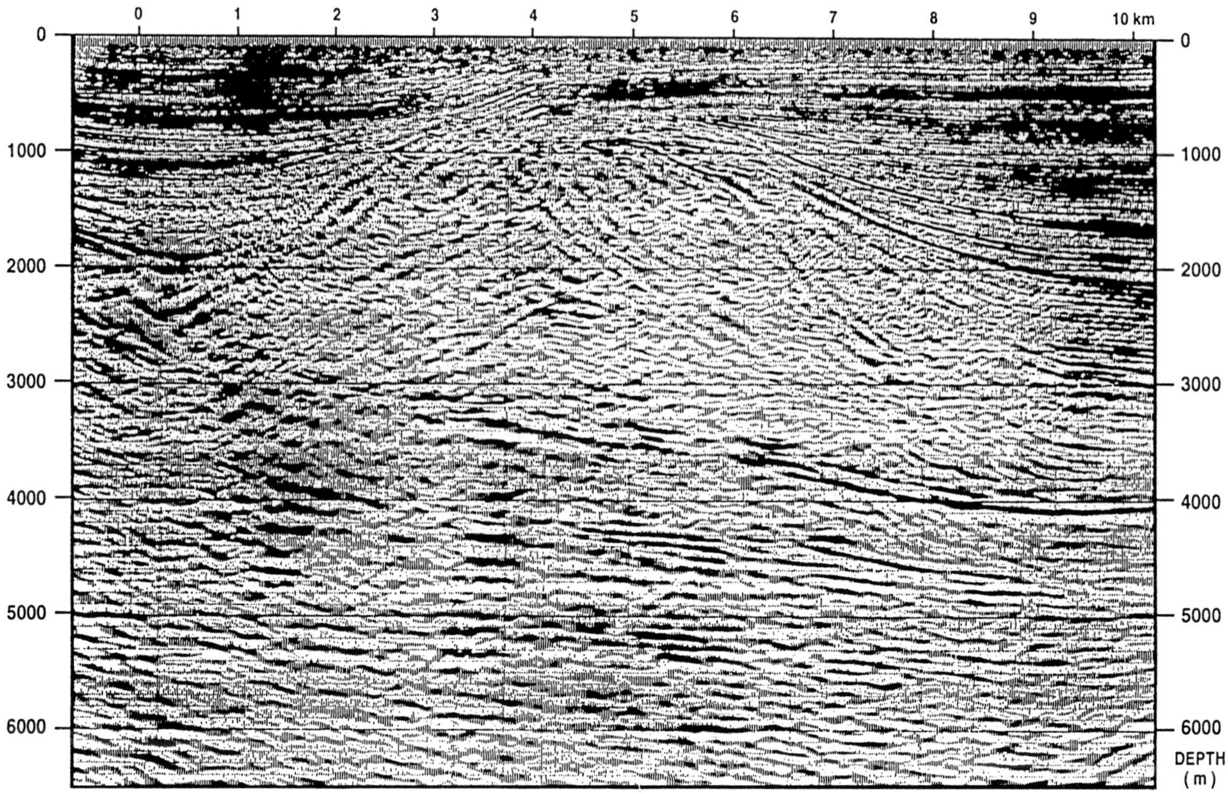

Fig. 4. A conventional 2D post-stack depth migration (with MIGPACK TM software) failed to restore a satisfactory image of the section below the salt diapir (Kazakhstan).

operated by Elf Congo. The water depth ranges from 150 to 300 m and the reservoir lies at a depth of about 3000 m (Fig. 9).

The reservoirs are of Albian age and include dolomites, limestones and sandstones. Field data analysis concluded that permeability was likely to be heterogeneous throughout the reservoir. Moreover the fluid characteristics are unusual: the fluid evolves from a gas condensate at the top to a light oil at the bottom.

Extensive studies of seismic and well data have enabled Elf to build a comprehensive geological model. Sedimentological analysis was the most important step and confidence in the sedimentological model stemmed from the knowledge of the area that Elf had acquired over decades of involvement in West Africa.

In the case of the N'Kossa project where gas and water are re-injected into the reservoir, it was absolutely essential to have an accurate geological model in order to obtain a reliable simulation. The reservoir simulation showed that due to the fluid characteristics, recovery could be increased by maintaining the initial pressure which prevents condensate deposition in the reservoir. It also showed that this pressure maintenance could be obtained by simultaneous gas and water injection and that the use of horizontal gas injection wells would provide efficient drainage while reducing drilling costs.

Innovative developments

Confidence in the results of the N'Kossa reservoir model based on integrated studies led Elf to commit itself to a major project including recovery and export of condensate and of liquefied petroleum gas (LPG) , gas compression and reinjection, and water treatment and reinjection.

For this project, it was necessary to look for innovative solutions, in particular because of the weight of the equipment to be installed (30 000 t), the water depth (200 m) and the lack of big derrick barges in West Africa.

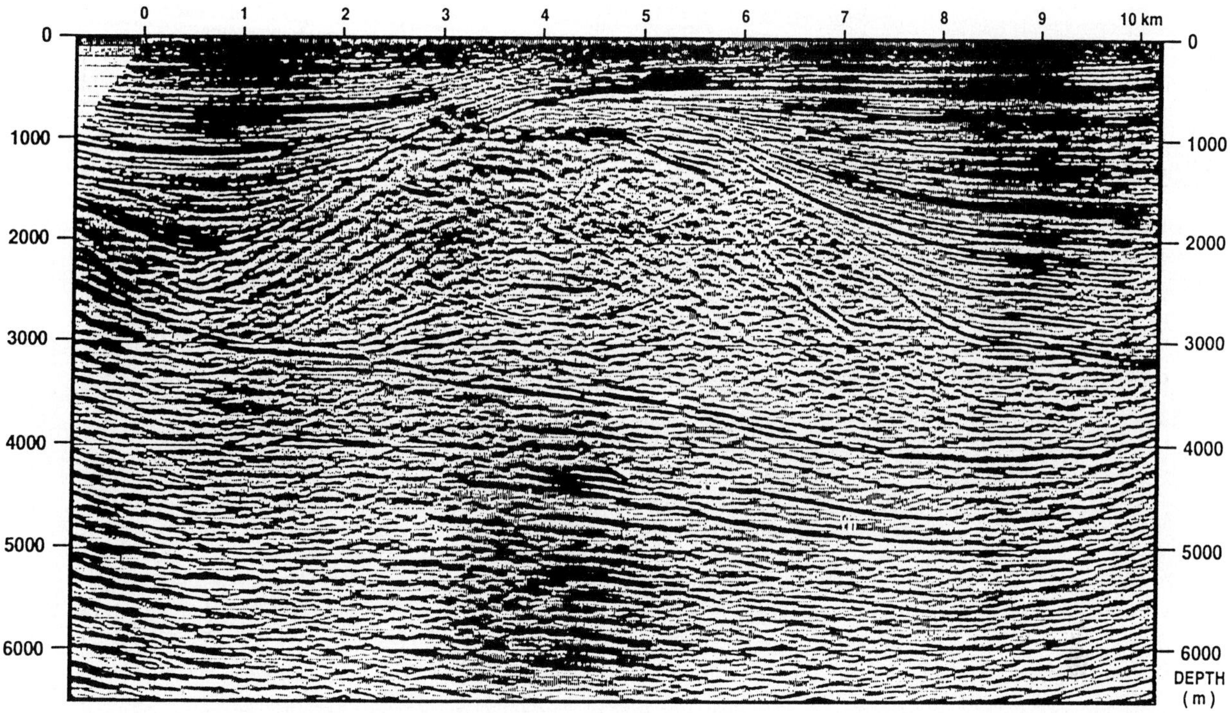

Fig. 5. A pre-stack depth migration (with MIGPACK TM software) induced a significant improvement at the objective level below the salt diapir with respect to the post-stack migration of Fig. 4 (Kazakhstan).

All the equipment will be installed on a floating concrete barge with a life span of more than 30 years, moored alongside one of the two drilling platforms (Fig. 10). The cost of installation of the equipment will be reduced because all the work will be done on the building yard and there will be no expenditure for barge mobilisation and demobilisation.

The future

It is clear that a major oil company must endeavour to stay one step ahead in technical expertise and innovation. Nevertheless, mastering a specific technique does not in the long term guarantee a competitive advantage because most innovations in the oil industry are rapidly available on the market. A good example is horizontal drilling that Elf was the first company to develop and to use industrially in 1980 (Fig. 11).

In any case, the effort to discover and produce oil and gas in traditional areas will reach its limits in the near future and we need to look for new leases. The next technical challenge will be deep offshore exploration and production. West African countries have recently signed contracts with major oil companies. However, there are areas where technical expertise is not the only relevant factor, areas where efficient organisation and financial capacity are essential. For instance, at the present time, this is the case of a number of projects considered in Central Asia. The current projects in Kazakhstan are represented on Fig. 12.

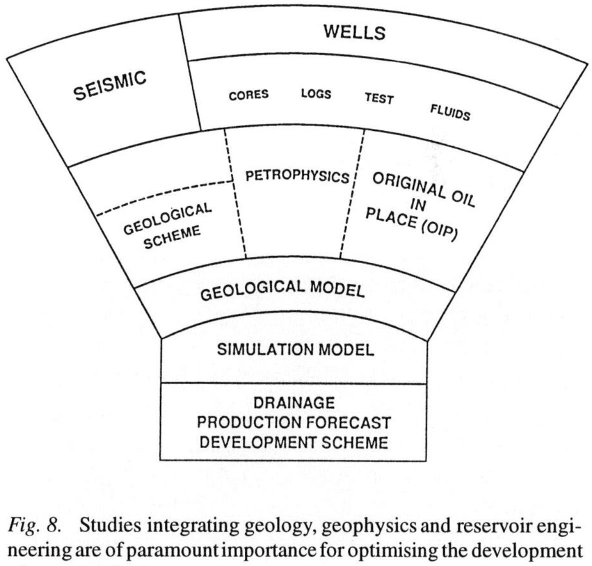

Fig. 8. Studies integrating geology, geophysics and reservoir engineering are of paramount importance for optimising the development of complex fields.

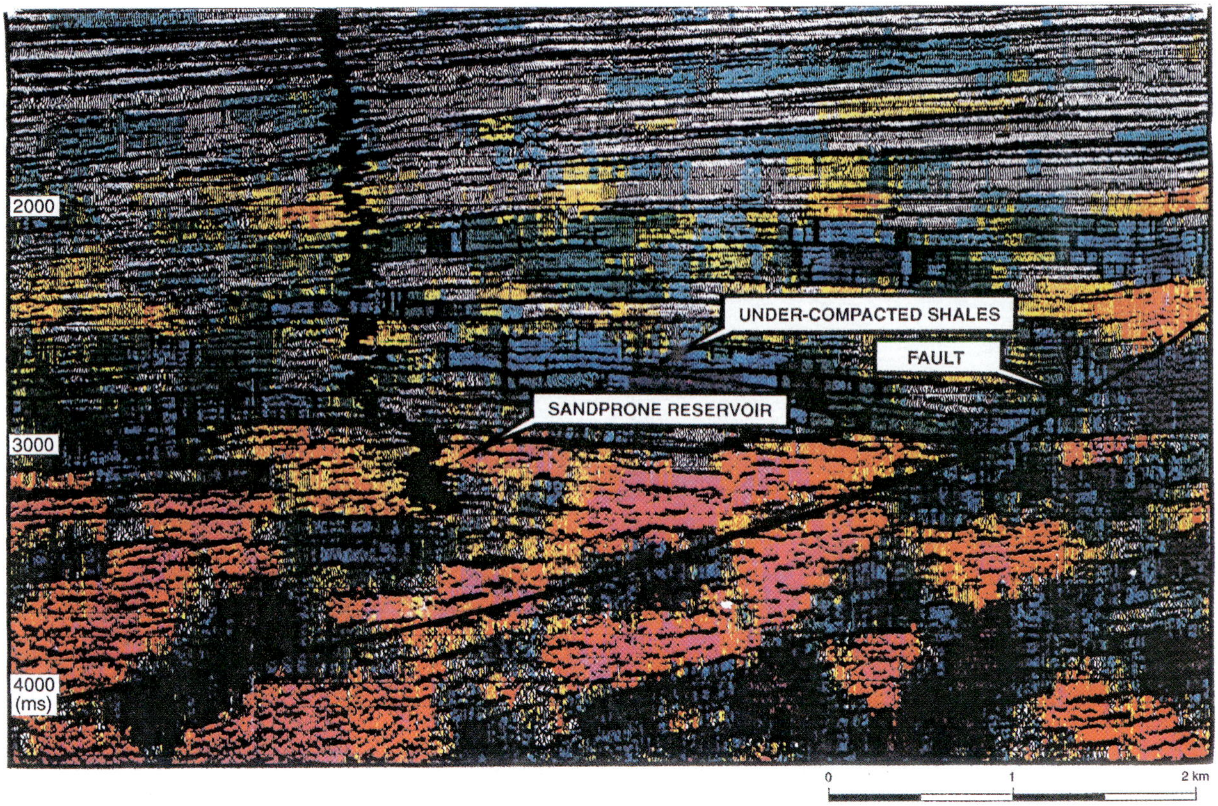
2000
3000
4000
(ms)
UNDER-COMPACTED SHALES
FAULT
SANDPRONE RESERVOIR
0
1
2 km

0
2 km
1000
2000
3000
4000
5000
(ms)
TOP OF THE UNDER-COMPACTED ZONE

← *Fig. 6.* The new seismic tool Deltastack allows to delineate the top and the lateral extension of a particular reservoir (higher-velocity interval in pink and yellow). Depth (TWT) in milliseconds.

← *Fig. 7.* The Deltastack has a second application: prediction of overpressured intervals (= low-velocity intervals in sustained blue).

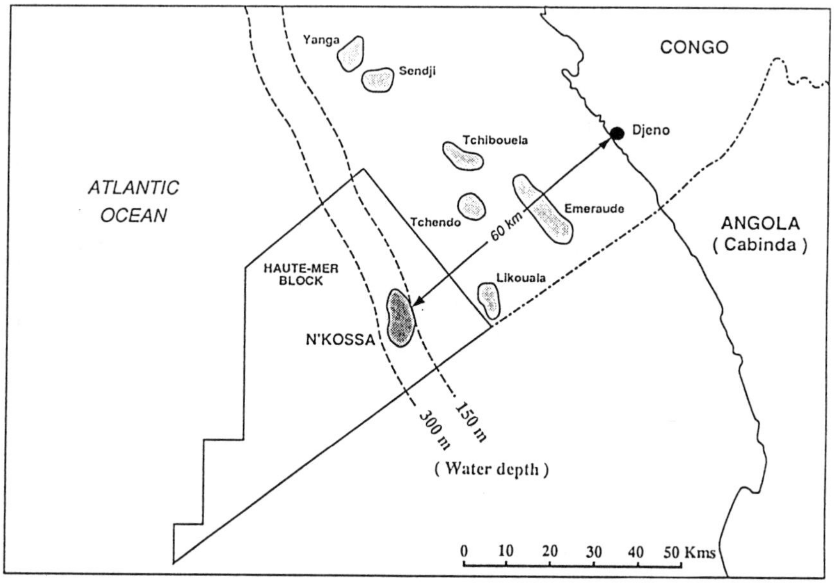

Fig. 9. N'Kossa field, offshore Congo, discovered by Elf Congo in 1985. Water depth ranges from 150 to 300 m.

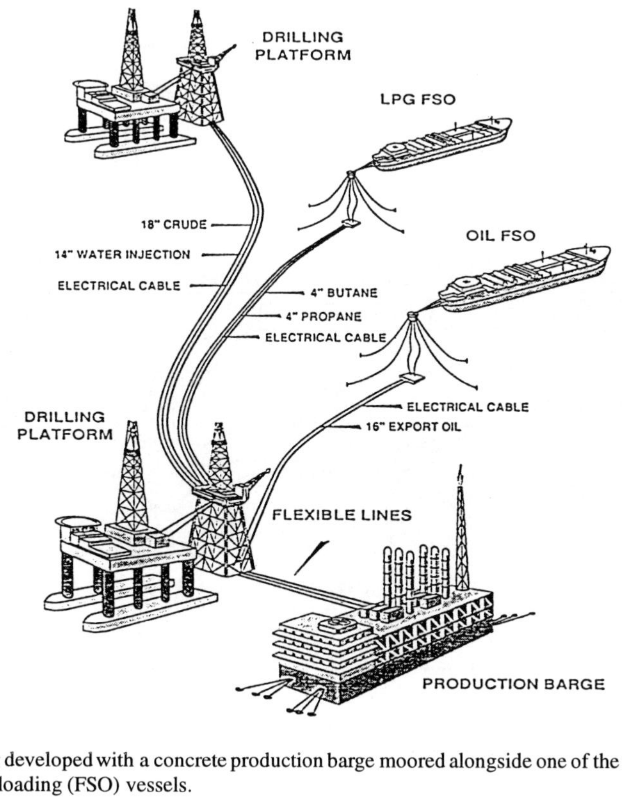

Fig. 10. The N'Kossa field is being developed with a concrete production barge moored alongside one of the two drilling platforms. Production is stored on two floating storage offloading (FSO) vessels.

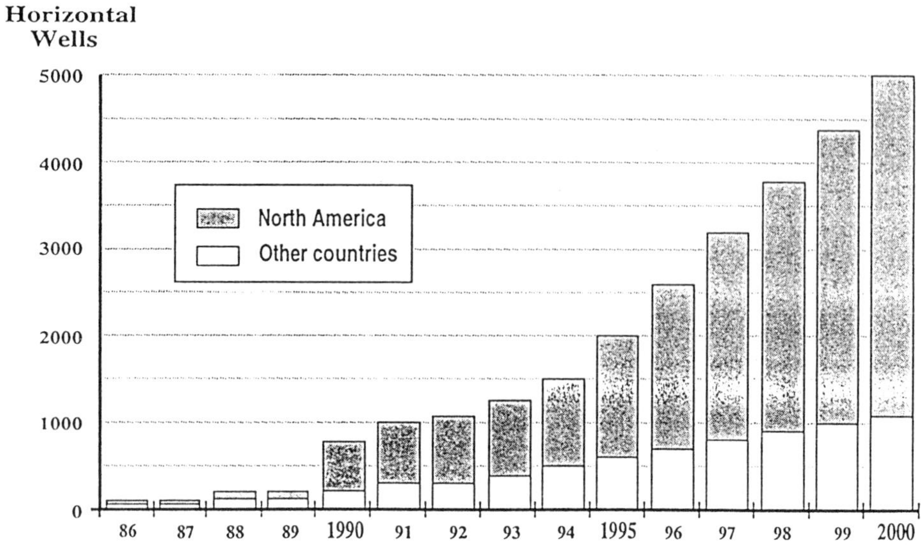

Fig. 11. Number of horizontal wells: past and future. Elf was the first oil company to develop and industrially use horizontal wells in 1980. This technology is now widely used in the industry.

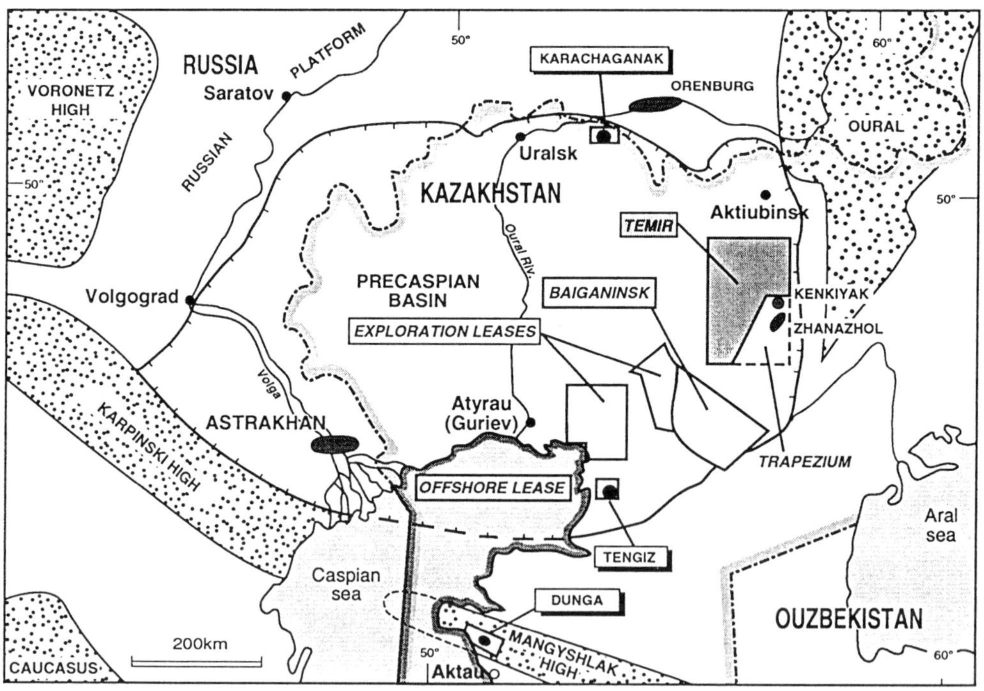

Fig. 12. Major oil and gas fields in the Kazakhstan Precaspian Basin and status of exploration and production leases by 1993.

Rondeel et al. (eds), Geology of gas and oil under the Netherlands, 19–30, 1996.

Natural gas in the Netherlands: exploration and development in historic and future perspective

J.N. Breunese & F.B. Rispens
Geological Survey of the Netherlands, Postbus 157, 2000 AD Haarlem, the Netherlands

Key words: exploration history, gas reserves

Abstract

The discovery in 1959 of the giant Groningen gas field with reserves of 2750 billion (10^9) cubic metres (bcm) triggered a strong revival of the hydrocarbon exploration in the Netherlands. Over the last decades, the country has proven to be a very prolific hydrocarbon province, particularly for natural gas. Supported by the favourable exploration climate, exploration efforts have been generally at a high and sustained level. Well over 250 gas accumulations have been discovered throughout the stratigraphic column in various plays. Field sizes range from medium (of the order of 50 bcm) down to very small (2 bcm or less). The total initial reserves in these fields are around 1150 bcm. A mature stage of exploration has now been reached in many areas. Within a few years virtually all production acreage and large parts of the exploration acreage in the Netherlands will have been covered with 3D seismic. The introduction of 3D seismic has led to an improvement of exploration drilling efficiency and to increased technical success rates. Moreover, the application of 3D seismic has indicated additional prospectivity undetected before. It provides a major opportunity, and challenge, for finding incremental reserves. On the development side, nearly all large and medium-size gas fields have, or shortly will, come on stream. However, the majority of the relatively large number of small and marginal fields is still undeveloped. The challenge here is to reduce economic limits by further expansion of the infrastructure and application of modern marginal-field development technology.

Introduction

The following overview of natural gas exploration and development in the Netherlands is presented from a historic and future perspective. The objective is to show important aspects and trends in exploration and production activities. The history of these activities in the Netherlands will be reviewed in order to illustrate how and under which conditions the present situation has emerged, and to look how it is likely to develop in the future.

The paper will not focus on individual fields but on groups of fields in terms of exploration efforts, the specific plays they are found in, and on their reserves, field sizes and stages of development.

Note that in this paper, all volumes of natural gas, i.e. hydrocarbons and associated non-combustible gases, are specified under standard conditions of 1.01325 bar and 15 °C (ISO 5024–1976(E)).

A detailed account of the exploration and production history up to 1987 for oil and gas in the Netherlands was presented by Knaap & Coenen (1987).

Exploration

Trigger

The first commercial gas discovery in the Netherlands was that of Coevorden in 1948 (Knaap & Coenen 1987). The most important event and trigger for natural gas exploration and development in the country, and even abroad in the United Kingdom and Germany, undoubtedly has been the discovery of the giant Groningen field. The discovery well Slochteren-1, drilled in 1959, demonstrated Permian Rotliegend sandstones to be gas-bearing. After a few other successful exploration wells, it became clear that a huge

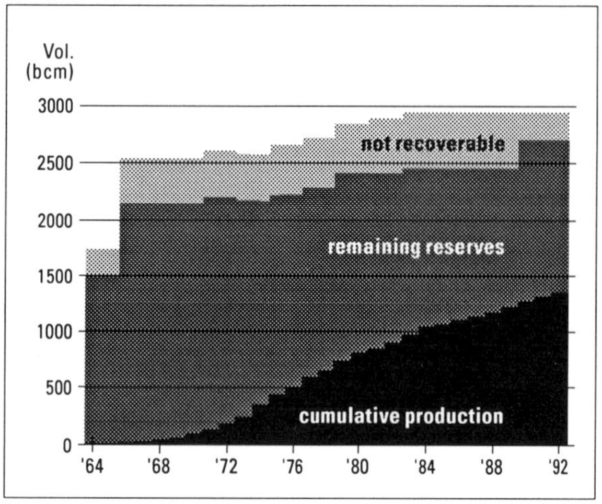

Fig. 1. The Groningen gas field, reserves and production.

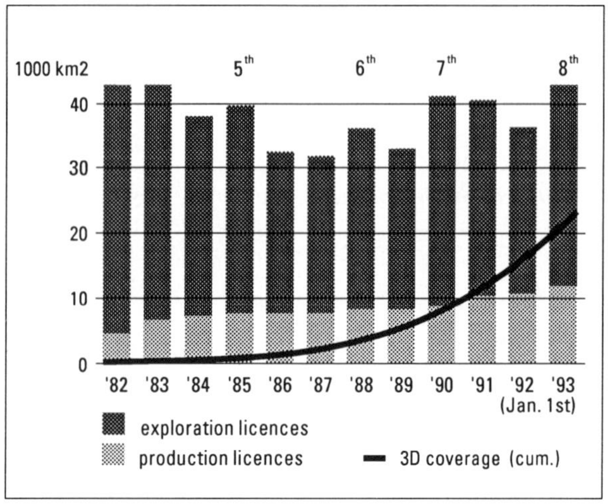

Fig. 2. Licence acreage and 3D seismic coverage offshore (January 1993). Note: 5th through 8th round awards indicated.

gas accumulation had been found in what was then named the Slochteren Formation. The accumulation has been charged from Carboniferous coal measures and is sealed by Zechstein evaporites (Van Wijhe et al. 1980).

The extent of the Groningen field was only recognized a few years after the discovery well. The appreciation of the volume of Gas Initially In Place (GIIP) and of the reserves has been growing over the years (Fig. 1). At present the GIIP estimate of the Groningen field stands at around 2870 billion (10^9) cubic metres (bcm), of which more than 90% is expected to be ultimately recoverable. The Groningen field therewith is the largest gas field in western Europe and ranks amongst the largest fields in the world.

Climate

The large Groningen field has proven to be a very good and flexible producer. These favourable properties have been the basis for the Dutch government's 'small fields policy' developed in the 1970s (Dessens, this volume). The goal of this policy was, and still is, to stimulate the exploration, development and production of small fields and to use the Groningen field as a balancing producer to match demand. The policy has been very successful over the last decades in that many smaller fields have been discovered and developed. Therefore, the production from the Groningen field has been lower than originally anticipated.

Other factors that have favourably contributed to the exploration climate in the Netherlands are stable, well-defined licensing conditions, guaranteed sales of gas to N.V. Nederlandse Gasunie (Dessens, this volume) and shallow water conditions in the offshore.

Efforts

Acreage
Many oil and gas operating companies have shown continuous interest in being active in the Netherlands. Large parts of the available acreage have been licensed over the last 25 years and the present licence map shows that interest has not ceased as yet (PGK 1993: fig. 6; this volume).

As an example we take the offshore, opened up for exploration in 1968 after the Continental Shelf Mining Law was established in 1965. Exploration acreage has been granted in eight rounds up till now. As from 1982, the licensed acreage has been maintained at a high level between 56 and 75% of the total acreage, with a steady growth in production acreage (Fig. 2).

Drilling rate
Exploration drilling has generally been at a high and sustained level as is shown in Figs 3 and 4. Oil price fluctuations appear to have had little effect.

Results

Success rates
Overall, technical success rates have been at a level of around 1 in 3, both onshore and offshore (Figs 3 and

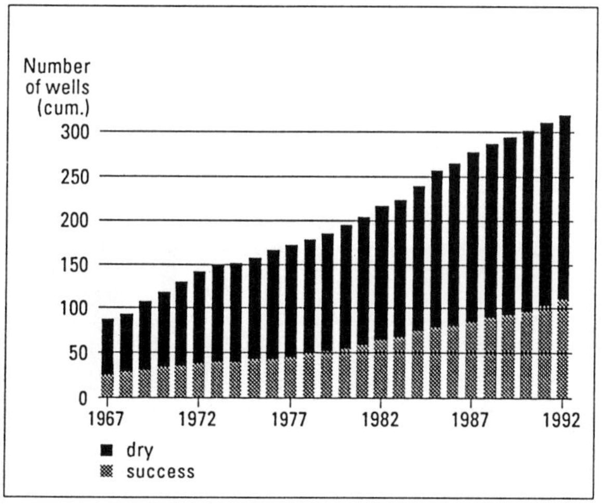

Fig. 3. Exploration drilling rate and success rate onshore.

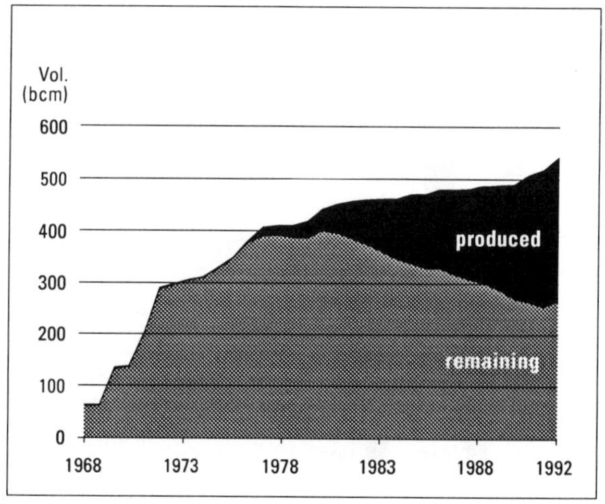

Fig. 5. Gas reserves and production onshore, excluding Groningen field.

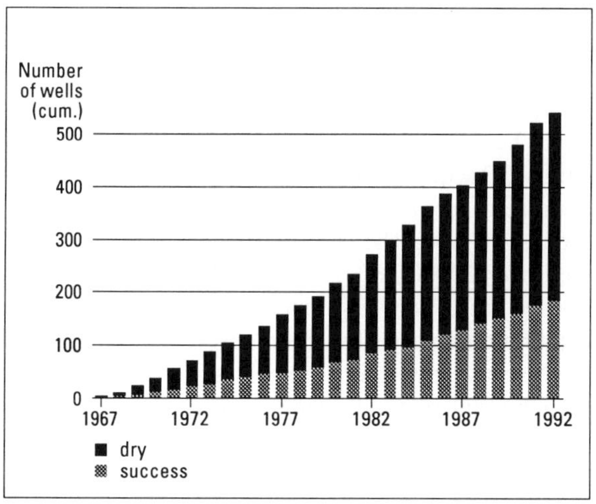

Fig. 4. Exploration drilling rate and success rate offshore.

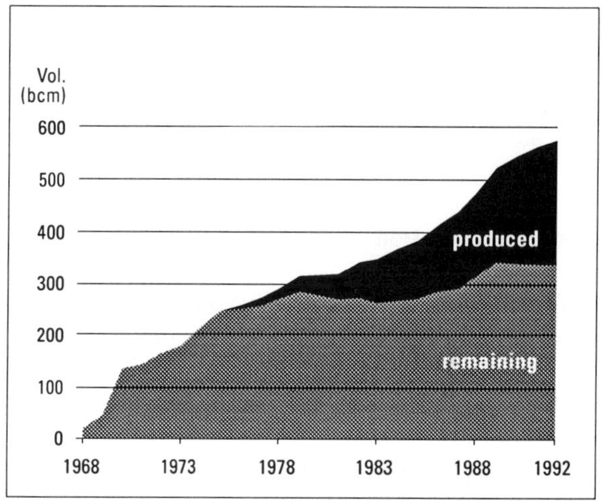

Fig. 6. Gas reserves and production offshore.

4), which is high in comparison with many other areas in the world.

Reserves and production

Up till present, total gas reserves of 1138 bcm, equivalent to some 40% of the initial reserves of the Groningen field, have been discovered in 277 fields outside that field. The cumulative growth of initial gas reserves is shown in Figs 5 and 6 (onshore: 552 bcm in 126 fields; offshore: 586 bcm in 151 fields). Both onshore and offshore, a number of relatively large discoveries with reserves in the order of tens of bcm, were made in the early years of exploration after the discovery of the Groningen field. Up till around 1985, the discovery curve shows a more pronounced creaming-off onshore than offshore. At first glance, the onshore has reached a more mature stage of exploration than the offshore. A review of recent exploration given below will put this observation in a different perspective.

Field sizes

Exploration has resulted in a series of natural gas discoveries, that have tended to become smaller in volume with time. This is clearly illustrated in Fig. 7 by a comparison of the present-day field size distribution with that of 1980, halfway the main exploration phase. It

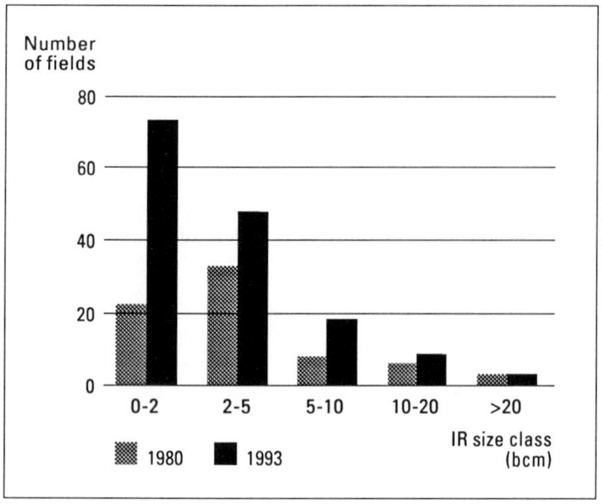

Fig. 7. Gas field size distribution offshore. IR = initial reserves.

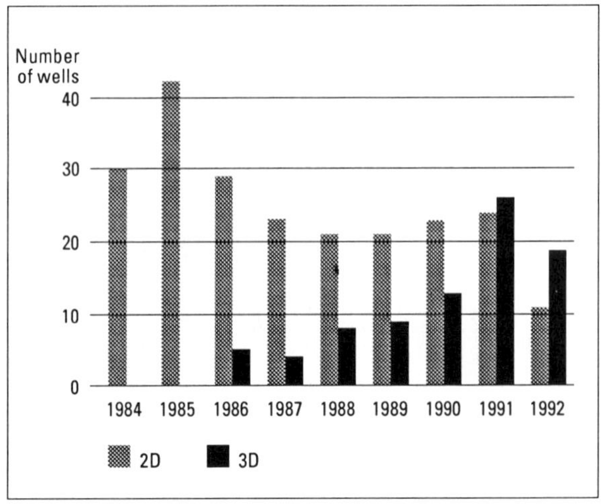

Fig. 9. Exploration drilling rate, 2D versus 3D seismic based.

shows a relatively strong increase in the range of very small fields with reserves below 2 bcm. Considering the steady drilling effort and success rate, this again indicates the maturing of exploration.

Recent exploration

Before trying to see what future exploration may bring, it is useful to have a closer look at recent exploration. Specifically, the introduction of 3D seismic in exploration during the mid 1980s, and the further diversification of exploration over various geological plays, will be shown to be important for the future.

3D Seismic

Coverage

In the Netherlands, the acquisition of 3D seismic surveys started in the early 1980s, both onshore and offshore. At present virtually all fields, certainly the producing ones, have been 3D-covered (Fig. 8). This has provided a wealth of information used for optimizing field appraisal and development. In addition, 3D seismic became important in the exploration stage as will be discussed below, taking the offshore as an example.

The first offshore 3D surveys covered only modest areas ranging from a few to some tens of square kilometres (Fig. 2). They were planned for optimizing development of individual discoveries. Around 1987, however, the capacity of offshore 3D seismic acquisition had been increased to a level at which an area equivalent to that of one offshore block (i.e. 400 km^2) could be covered in a single survey. Since then, the annual 3D acquisition has strongly progressed, peaking at a total of around 5000 km^2, equivalent to 12 offshore blocks, in 1991. The present offshore 3D coverage of well over 20000 km^2 greatly exceeds the production licence area, implying that a significant part of the exploration acreage has been covered (Fig. 2). Over the last six years therefore, 3D seismic has become an important exploration tool, replacing 2D seismic at great pace.

Drilling and success rates (2D versus 3D)

Exploration drilling in the Netherlands has changed from exclusively 2D-based up to 1986 towards mainly 3D-based now within a period of only six years (Fig. 9). Last year (1992), 2D-based drilling indeed strongly declined in comparison with previous years.

In the Netherlands, technical success rates have been significantly higher for 3D than for 2D-based exploration drilling: the rate for 3D averages around 65%, which is almost twice as high as the average for 2D over the same period and before (Fig. 10). It should be noted, that much of the 3D-based exploration drilling has been carried out within acreage that already had been intensively drilled on the basis of 2D seismic. The 3D seismic located additional prospects in 'mature' areas and also led to a high exploration drilling efficiency.

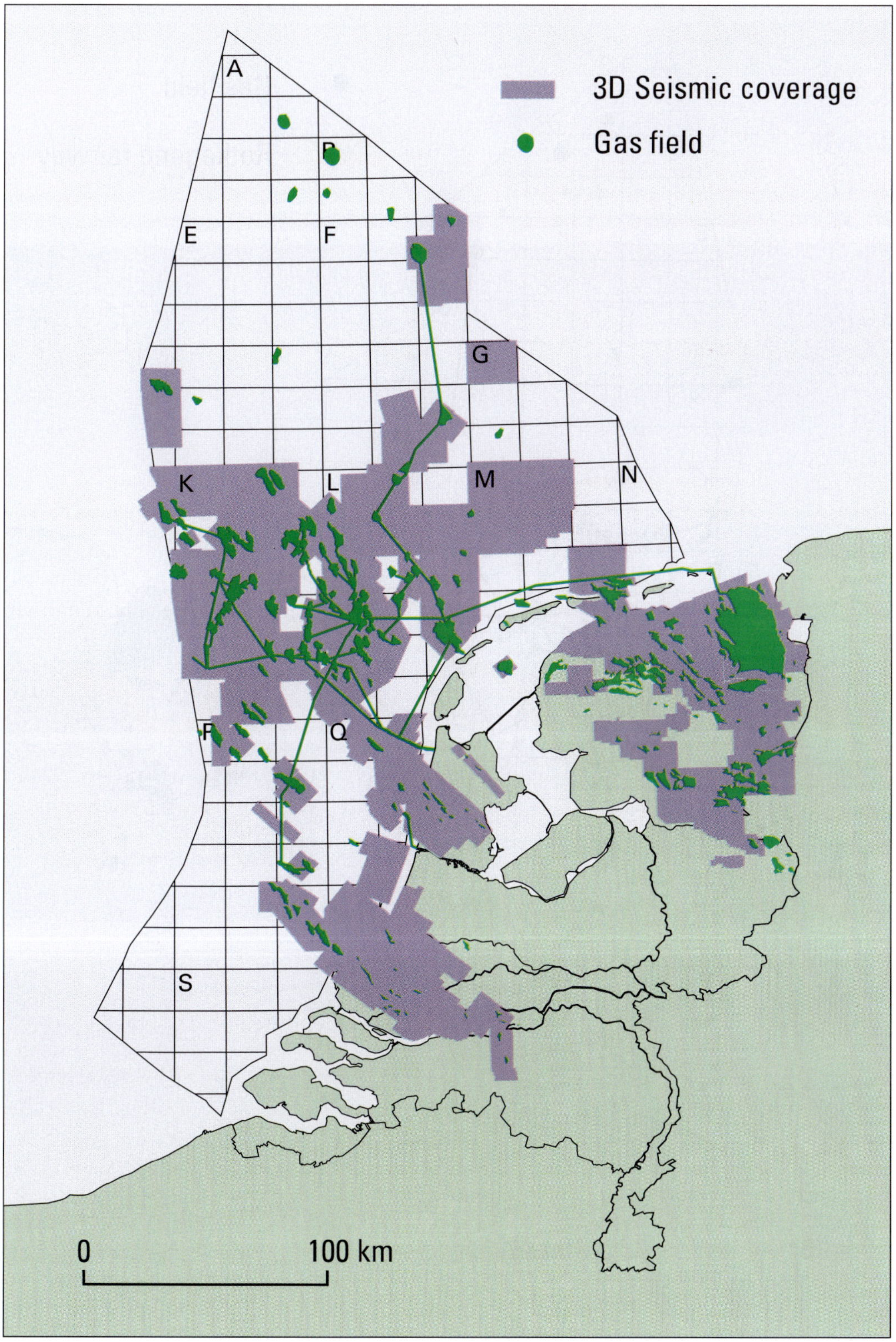

Fig. 8. 3D seismic coverage of gas fields (January 1993).

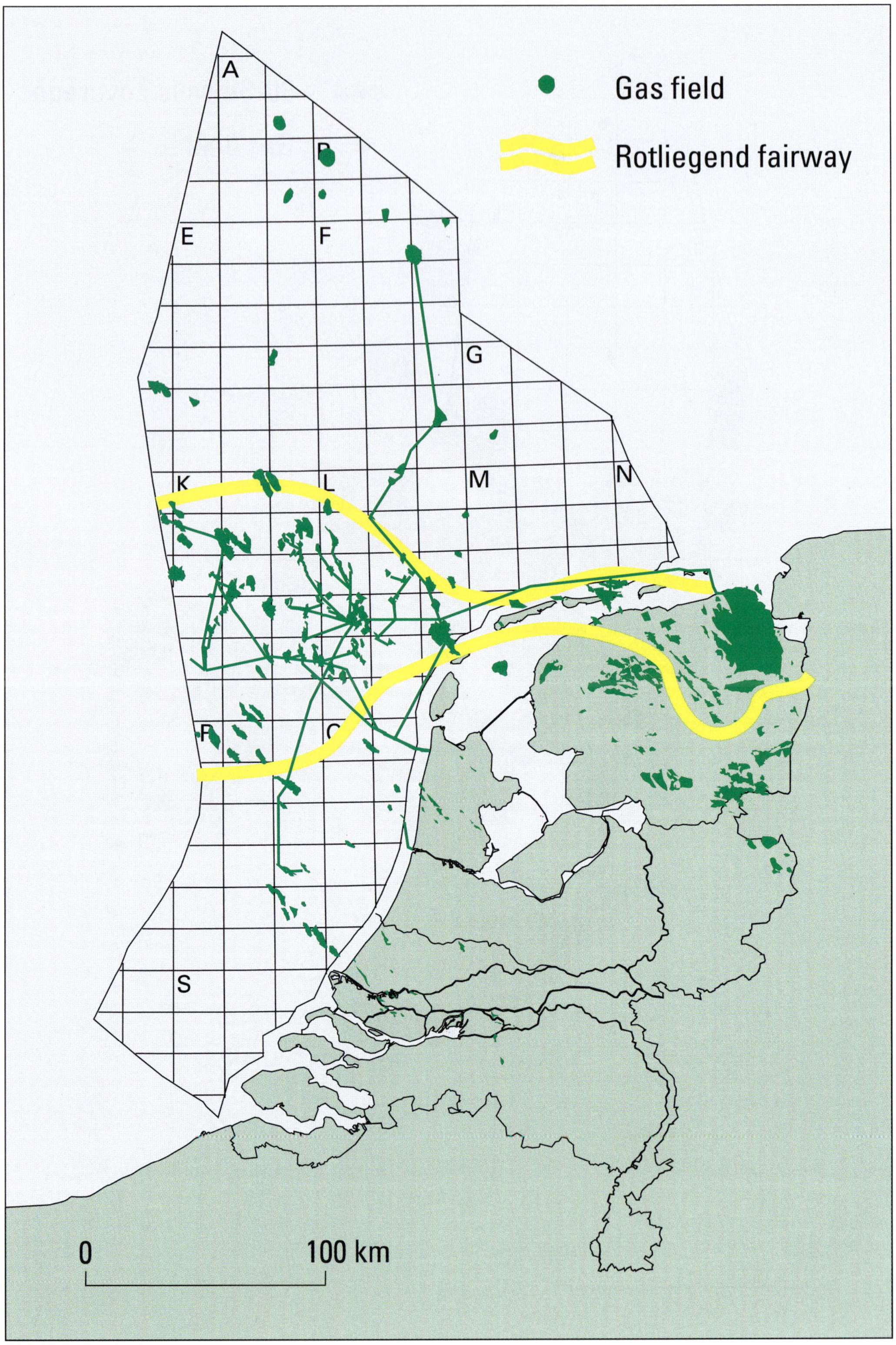

Fig. 11. Rotliegend gas fairway.

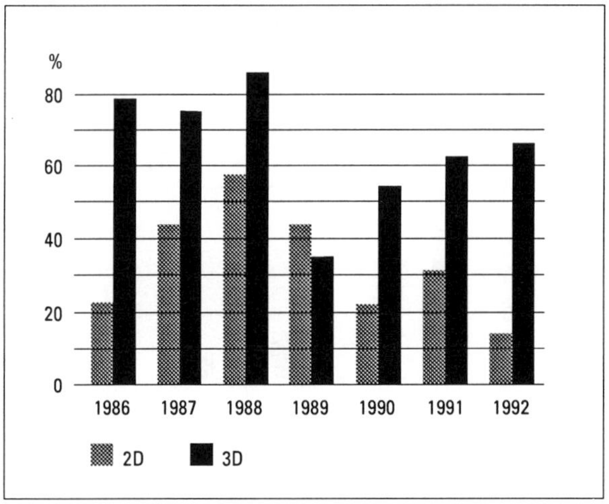

Fig. 10. Exploration success rate, 2D versus 3D seismic based.

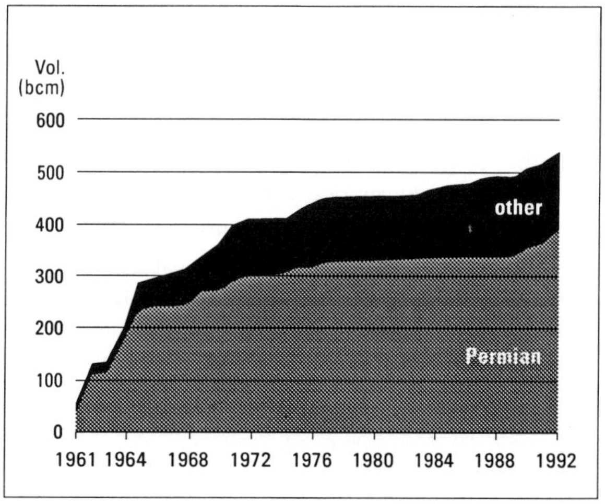

Fig. 12. Growth of initial gas reserves discovered in Permian Rotliegend and non-Permian reservoirs onshore, excluding Groningen field.

Table 1. Initial gas reserves by play (bcm).

	Onshore	Offshore	Total
Post-Triassic	63	32	95
Triassic	60	83	143
Permian*	390	454	844
Pre-Permian	39	17	56
Total	552	586	1138

* Groningen field (2750 bcm) excluded.

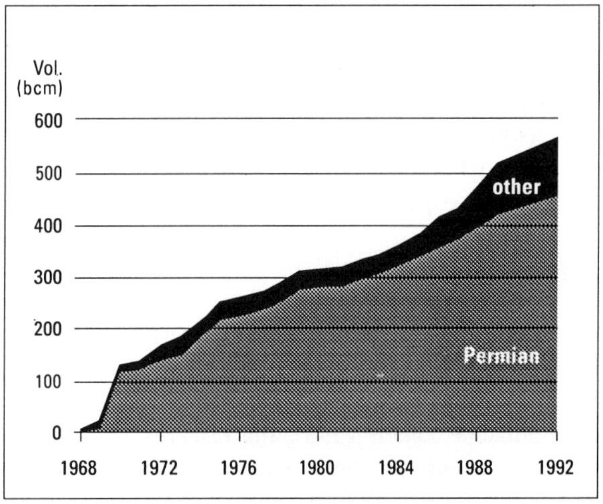

Fig. 13. Growth of initial gas reserves discovered in Permian Rotliegend and non-Permian reservoirs offshore.

Play-wise growth of reserves

Based on the play concept proven in the Groningen field, many other gas discoveries have been made in the Permian Rotliegend fairway in the northern onshore and central offshore (Fig. 11). Less important in number and volume, gas discoveries have also been made in other plays, notably in pre-Permian Carboniferous and in Triassic and post-Triassic reservoirs (Zijp 1987). Generally these discoveries are restricted to smaller basins or areas.

Onshore

The onshore discovery profile (Fig. 12) shows that non-Permian discoveries, some of which predate the Groningen discovery, have contributed significant gas volumes (Table 1). Nevertheless, this profile has been rather flat between 1975 and 1985. Recently, however, a clear upturn is observed due to successful exploration in the Triassic gas play in the southwestern onshore. Moreover, in 1990 the first 3D-based exploration well, targeted to the Rotliegend play in the north (Grijpskerk-1), turned out to be surprisingly successful. This well opened up significant new potential as has already been proven by subsequent drilling in a play that had previously been deemed 'mature'.

Offshore

The growth of initial reserves offshore has been largely determined by discoveries in the Rotliegend (Fig.

13, Table 1). Apart from a few early, relatively large discoveries, the Rotliegend discovery rate from 1970 onwards has been fairly steady at around 15 bcm per year.

Other plays, both in pre- and post-Permian reservoirs, have shown modest early discoveries and subsequently a minor contribution until 1985. From that year onwards the non-Permian plays, in particular the Triassic play, have started to contribute significantly, giving rise to an upturn in the total discovery profile. Note that most of the non-Permian discoveries have been made on the basis of 2D seismic. These discoveries are nowadays followed up by 3D-based exploration drilling.

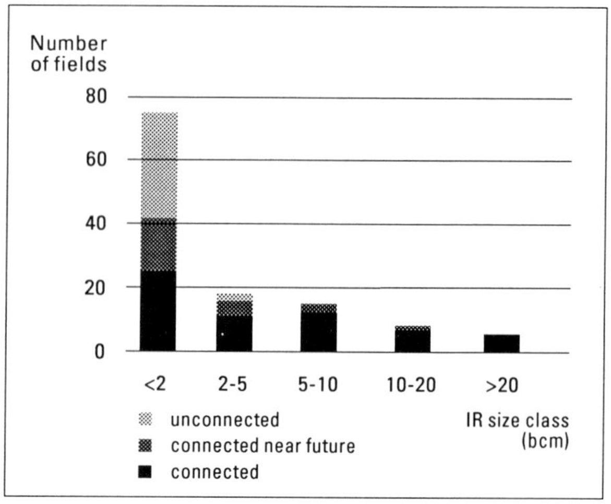

Fig. 15. Field size distribution and degree of development onshore. IR = initial reserves.

Future exploration

Effort

The drilling activity in the Netherlands has strongly declined since 1992, in line with what has been observed elsewhere. Still, it is expected that this activity can be maintained at about the 1992 level. The 3D seismic acquisition still continues over exploration acreage offshore, thus securing new and valuable information on prospectivity.

Future reserves

The Geological Survey of the Netherlands routinely assesses the exploration potential of the country. According to the 1993 estimate (Ministerie van Economische Zaken 1993), the reserves from future discoveries are expected to range between 190 and 410 bcm. To put these numbers in perspective: the most likely value represents about one quarter, and the maximum over one third, of the already discovered non-Groningen gas reserves. Only identified prospects in 'proven' plays have been taken into account in the assessment. This leaves room for speculating about yet unidentified prospects, to be generated from 3D seismic data, or even about yet unproven play concepts.

Development

Introduction

Each individual discovery has its own story with regard to development opportunities: a complex of geological, technical and economic factors eventually will determine which fields will come on stream and when. Here, the relationship will be discussed between field size distributions as derived from exploration results and size distributions of fields that have been or are likely to become connected to the production infrastructure. This will indicate where apparent thresholds for development exist and which part of the discovered initial reserves is likely to be brought on stream under the present conditions.

Evidently, there are significant differences between onshore and offshore regarding the techno-economic conditions for development. It is therefore appropriate to treat these areas separately.

Onshore

The proven onshore gas fields are shown in Fig. 14: connected or producing fields as well as unconnected or non-producing fields. By virtue of the extensive pipeline network, 83% of the discovered onshore reserves at present is in connected fields. Taking into account the likely near-future developments within say the next five years, an additional 77 bcm will become connected (Table 2). From a development point of view, the present discovery population will by then have been strongly creamed off, leaving only very small discoveries undeveloped: of the remaining 37 discoveries with reserves less than 2 bcm, 26 are smaller than 0.5 bcm (Fig. 15). Connection of these very small accumulations into the Gasunie transport net-

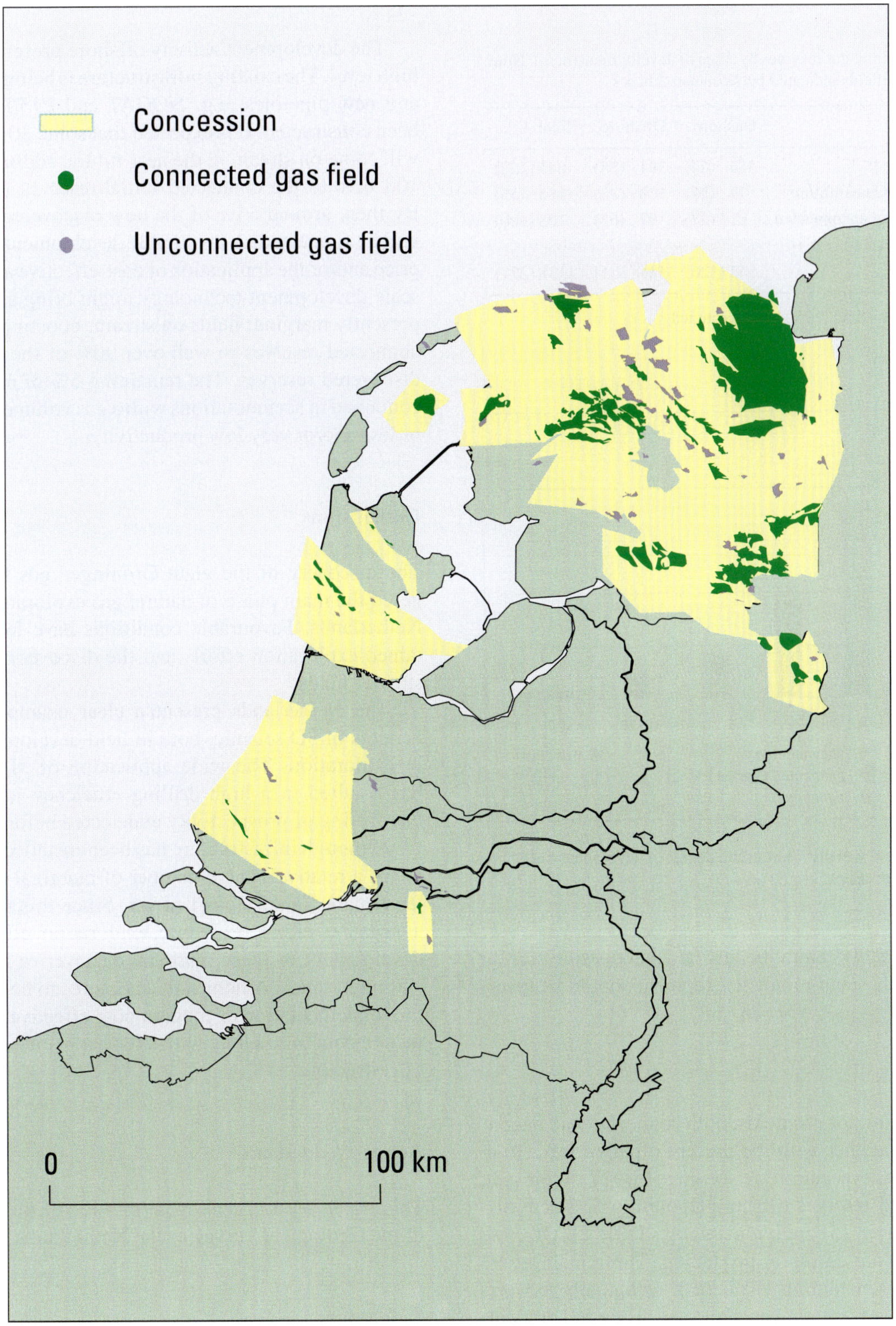

Fig. 14. Connected and unconnected gas fields onshore (January 1993).

Table 2. Initial gas reserves by stage of development (bcm). Note: number of fields indicated between parentheses.

	Onshore	Offshore	Total
Connected*	454 (64)	391 (58)	845 (122)
Connected near future	77 (25)	108 (30)	185 (55)
Remaining unconnected	21 (37)	87 (63)	108 (100)
Total	552 (126)	586 (151)	1138 (277)

* Groningen field (2750 bcm) excluded.

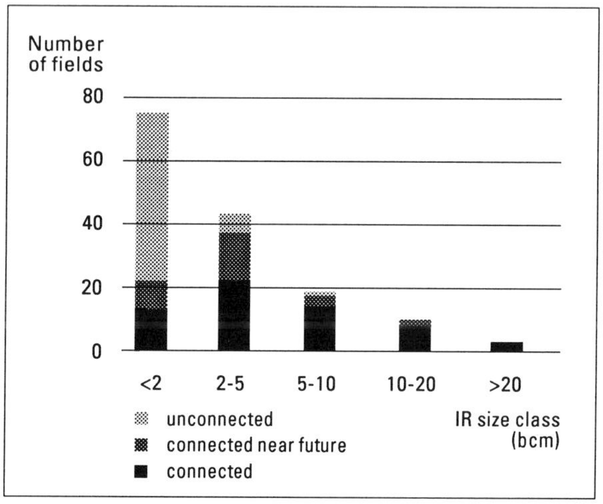

Fig. 16. Field size distribution and degree of development offshore. IR = initial reserves.

work probably cannot be justified and possibilities for small-scale stand-alone development should be investigated.

Offshore

The locations of gas fields, both connected and unconnected, together with the present offshore infrastructure for gas evacuation are shown in Fig. 17. At present, discovered reserves offshore amount to 586 bcm distributed over 151 individual fields (Table 2). Of these fields, 58 are currently producing. The remaining 93 gas fields, containing 195 bcm of technically recoverable reserves, are not yet connected. They fall predominantly in the minor field size categories (Fig. 16). This results from the fact that recent discoveries on average have been smaller and that the larger discoveries have been developed first.

The development activity offshore presently is at a high level. The existing infrastructure is being extended and new pipelines, e.g. NOGAT and P15/P18, have been constructed. It is expected that some 30 gas fields will come on stream in the near future, adding another 108 bcm to the connected initial reserves (Table 2). By then, around 85% of the now discovered reserves will be connected. A favourable development of the oil price and/or the application of cost-effective and small-scale development technology might bring another 24 presently marginal fields on stream, boosting the total connected reserves to well over 90% of the presently discovered reserves. The remaining 6% of reserves is contained in accumulations with a gas volume of 1 bcm or less and/or very low productivity.

Conclusions

The discovery of the giant Groningen gas field triggered the main phase of natural gas exploration in the Netherlands. Favourable conditions have led to sustained exploration efforts and the discovery of many smaller fields.

The Netherlands present a clear example of the benefits of 3D seismic, both in field development and in exploration. The wide application of 3D seismic has resulted in a high drilling efficiency and in the uncovering of prospectivity undetected before.

Although infrastructure has been installed in many areas, a relatively large number of marginal discoveries remains undeveloped as yet. Since this may have a negative effect on future exploration efforts, the development of these marginal discoveries is of crucial importance. Amongst the measures to be taken are the implementation of modern cost-effective development technology, along with a further expansion of the infrastructure.

Acknowledgement

This paper is published by permission of the Director of the Geological Survey of the Netherlands.

References

Dessens, C.W.M. 1995 The role of oil and gas in the Dutch energy policy. In: Rondeel, H.E., D.A.J. Batjes & W.H. Nieuwenhuijs (eds) Geology of gas and oil under the Netherlands – this volume

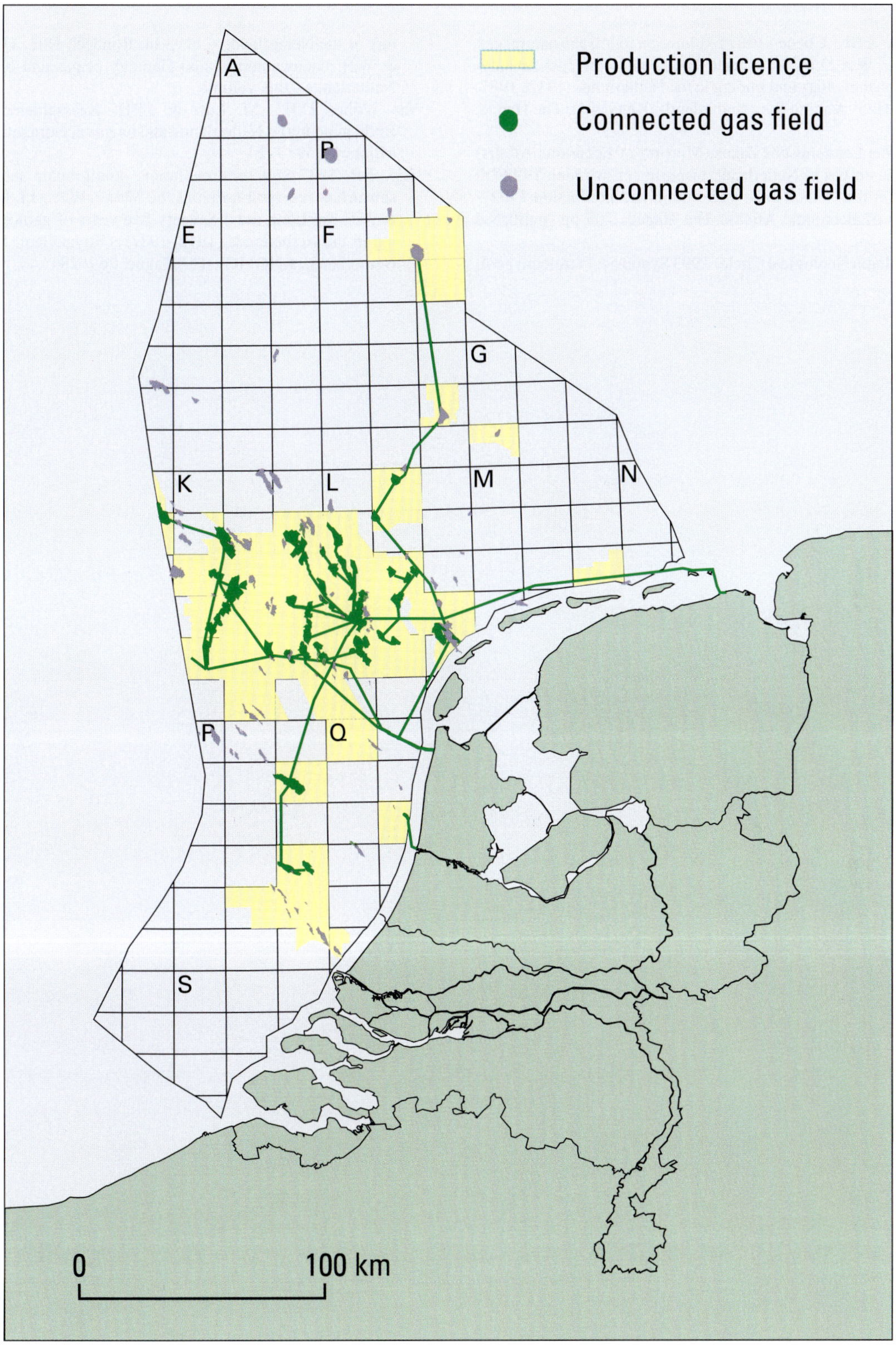

Fig. 17. Connected and unconnected gas fields offshore (January 1993).

Knaap, W.A. & M.J. Coenen 1987 Exploration for oil and natural gas. In: Visser, W.A., J.I.S. Zonneveld & A.J. van Loon (eds) Seventy-five years of geology and mining in the Netherlands (1912–1987) – Royal Geol. Mining Soc. Netherlands (KNGMG), The Hague: 207–230

Ministerie van Economische Zaken (Ministry of Economic Affairs) 1993 Olie en gas in Nederland: opsporing en winning 1992/Oil and gas in the Netherlands: exploration and production 1992 – Ministry of Economic Affairs, The Hague: 105 pp (published yearly)

PGK (Petroleum Geological Circle) 1993 Synopsis: Petroleum geology of the Netherlands – 1993 In: Rondeel, H.E., D.A.J. Batjes & W.H. Nieuwenhuijs (eds) Geology of gas and oil under the Netherlands – this volume

Van Wijhe, D.H., M. Lutz & J.P.H. Kaasschieter 1980 The Rotliegend in the Netherlands and its gas accumulations – Geol. Mijnbouw 59: 3–24

Zijp, F.R. 1987 Structural evolution, stratigraphic sequences and subsurface reservoir horizons. In: Visser, W.A., J.I.S. Zonneveld & A.J. van Loon (eds) Seventy-five years of geology and mining in the Netherlands (1912–1987) – Royal Geol. Mining Soc. Netherlands (KNGMG), The Hague: 269–284

Rondeel et al. (eds), Geology of gas and oil under the Netherlands, 31–43, 1996.

Geothermal history of the Carboniferous in South Limburg, the Netherlands

H. Veld[1,2], W.J.J. Fermont[1], H. Kerp[3] & H. Visscher[2]
[1]*Geological Survey of the Netherlands, Postbus 126, 6400 AC Heerlen, the Netherlands;* [2]*Laboratory of Palaeobotany and Palynology, Utrecht University, Heidelberglaan 2, 3584 CS Utrecht, the Netherlands;* [3]*Abteilung Paläobotanik, Westfälische Wilhelms-Universität, Hindenburgplatz 55–57, D-48143 Münster, Germany*

Key words: basin modelling, coalification trends, heatflow, overburden, Variscan orogeny, vitrinite reflectance

Abstract

The geothermal history of Upper Carboniferous rocks in the former mining district of South Limburg (the Netherlands) is complex. Maturity modelling as calibrated by vitrinite reflectance measurements and volatile matter content shows that the present-day coalification patterns of the Carboniferous were established in Late Westphalian to Early Permian times. Variations in the degree of coalification were mapped using as a reference level the upper Westphalian A coal seam GB 23. Modelled burial and thermal histories for the area indicate that regional variations in overburden, heatflow and in timing of thermal events in front of the Variscan orogen caused the spatial variation in coalification. In the southern part of the area, differences in overburden combined with a generally high heatflow determined the coalification patterns. In the northeastern part of South Limburg an additional thermal event, related to a Permian(?) intrusion, overprinted the earlier coalification pattern. The coalification data are consistent with basement heatflow values during the Carboniferous in the order of 46 to 71 mW/m^2 and erosion following the Variscan orogeny of some 2900 m of sediments around Sittard and as much as 5400 m in the southeasternmost part of South Limburg.

Introduction

Despite the fact that much research has been carried out on coal and coal-bearing strata in the former mining district of South Limburg (the Netherlands), an overview of coalification patterns and trends does not exist. Only a few scattered data (e.g. Bless et al. 1981) and isovol maps of parts of the district (Patijn 1963) have been published. Although all coal mines were abandoned in the early seventies, coal research has continued and several reconnaissance wells were drilled in the early eighties north of the former mining district. This paper includes part of the results of the coal-petrological investigations carried out within a coal inventory project by the Geological Survey of the Netherlands (Fermont 1986). The objective of this study is to integrate old and new coalification data with the geological and subsidence history of the South Limburg area and to relate them to the thermal setting of the Late Palaeozoic Variscan foredeep.

Progressive burial of organic matter enclosed in sedimentary strata results in maturation, or coalification when referring specifically to coal. Coalification is the progressive transformation of peat through brown coal and (sub)bituminous coal to anthracite. A variety of physical and chemical properties changes with coalification, including calorific value, and the percentages of fixed carbon and volatile matter. The most universally used measure of maturity is vitrinite reflectance. Vitrinite reflectance is an optical parameter defined as the percentage of vertically incident white light reflected from a highly polished surface of the vitrinite maceral telocollinite. Vitrinite is frequently the most abundant constituent in coals but is also well represented as dispersed particles in other sediments. Reflectance values range from 0.2% at the peat stage to well over 2.4% at the anthracite rank. The maturation process is primarily a function of time and temperature. Temperature increases with depth along the thermal gradient which is controlled by the physical properties of the sediments and by the heatflow history of the basin. The maturity level of organic matter can be related to the thermal history of a basin by simulating the transformation process of organic matter with chemical kinetic models (Tissot & Espitalié 1975, Burnham & Sweeney 1989). In the present study the thermal history of the Upper Carboniferous sediments in South Limburg has been evaluated by comparing calculated vitrinite reflectance equivalents with measured vitrinite reflectance values.

Geological outline of South Limburg

The geological framework of South Limburg has been discussed by various authors (e.g. Van Waterschoot van der Gracht 1938, Patijn 1963, Kimpe et al. 1978). The Upper Palaeozoic sediments of the area have been deposited in the Variscan foreland basin which extended from Poland to Ireland. The area is located in front of the Aachen Overthrust that is part of the Variscan thrust zone that runs from northern France through Belgium into Germany. A structural map of the pre-Permian sub-crop is presented in Fig. 1. The area is bordered to the southwest by the Brabant Massif. The Feldbiss Fault and its continuation in the First NE Main Fault separate the area from the Roer Valley Graben to the northeast.

Several WSW – ENE striking faults in the southeastern part of the area are considered to have formed during the Late Carboniferous Variscan orogeny.

The most important fold structures are the Puth and Waubach anticlines (Fig. 1). The Puth Anticline has a steep eastern flank. It forms the northern end of an anticlinal structure with a curved foldaxis. On the basis of seismic data, Dusar & Langenaeker (1992) have interpreted the Puth Anticline as a 'positive flower structure' related to dextral strike-slip movements on the Bordière Fault. The Waubach Anticline has a SW – NE axis, parallel to the local Variscan strike. The pattern of the tectonic structures has been influenced by the presence of the rigid Brabant Massif. The origin of the N – S striking folds and the rotation of the Variscan strike around the eastern margin of the massif has been interpreted as reflecting local variations in stress field orientations at its southern and eastern edges (Wrede 1985).

Denudation and erosion took place after the Variscan orogeny. The unconformity between unfolded Mesozoic and Cenozoic sediments and faulted, and locally folded, Carboniferous rocks gently dips towards the northwest. The variable thickness of the Dinantian limestones is possibly related to the early formation of graben-like structures (Bless et al. 1981). It has been estimated to be at least 900 m. These limestones are unconformably overlain by Namurian sandstones and siltstones with local intercalations of thin coal beds. The thickness of the Namurian sediments is known only partially. The Namurian has been eroded where it subcrops against its unconformable cover in the south. Where it is conformably overlain by Westphalian deposits, wells have never reached its base. The thickness of the Namurian in the study area is approximately 900 m (H.J.M. Pagnier, pers. comm. 1994). Thicknesses of the Westphalian A and B are well established in the former mining area and measure 900 m and 800 m, respectively. The Westphalian C is not completely preserved; its upper part is missing. The thickness of the Westphalian C is estimated at approximately 900 m (Van de Laar & Fermont 1989). The Westphalian A to C deposits consist of sandstones, shales, and up to 4% coal. Westphalian D and Stephanian deposits are not known from the area. Westphalian D sediments have been reported, however, from wells in the adjacent Belgian Campine Basin (Kimpe et al. 1978). Here, the Westphalian D Neeroeteren Sandstone consists predominantly of sandstones and shales, and reaches a thickness of approximately 800 m (Bless et al. 1977).

Along the Dutch-Belgian border in the south, the post-Carboniferous cover has a maximum thickness of approximately 100 m. Towards the north it increases rapidly in thickness (up to 690 m near Sittard). No Permian deposits are known from South Limburg. In the Belgian Campine District, however, 30 m of Zechstein nearshore deposits have been recognized. In the north of the study area red-coloured Buntsandstein deposits unconformably overlie the Carboniferous. Locally they can reach a thickness of several hundreds of metres, up to approximately 470 m near Sittard. Jurassic sediments are not recorded in the study area. In the Roer Valley Graben, however, Lias bituminous shales are present. The position of the depositional edge of the Late Jurassic to Early Cretaceous basin indicates that South Limburg was an area of non-deposition or erosion during this period (Van Staalduinen et al. 1979). Approximately 300 m of Upper Cretaceous is present in most of the area. The lower part consists predominantly of fine-grained sands and clays, the upper part consists of chalk (Romein 1963).

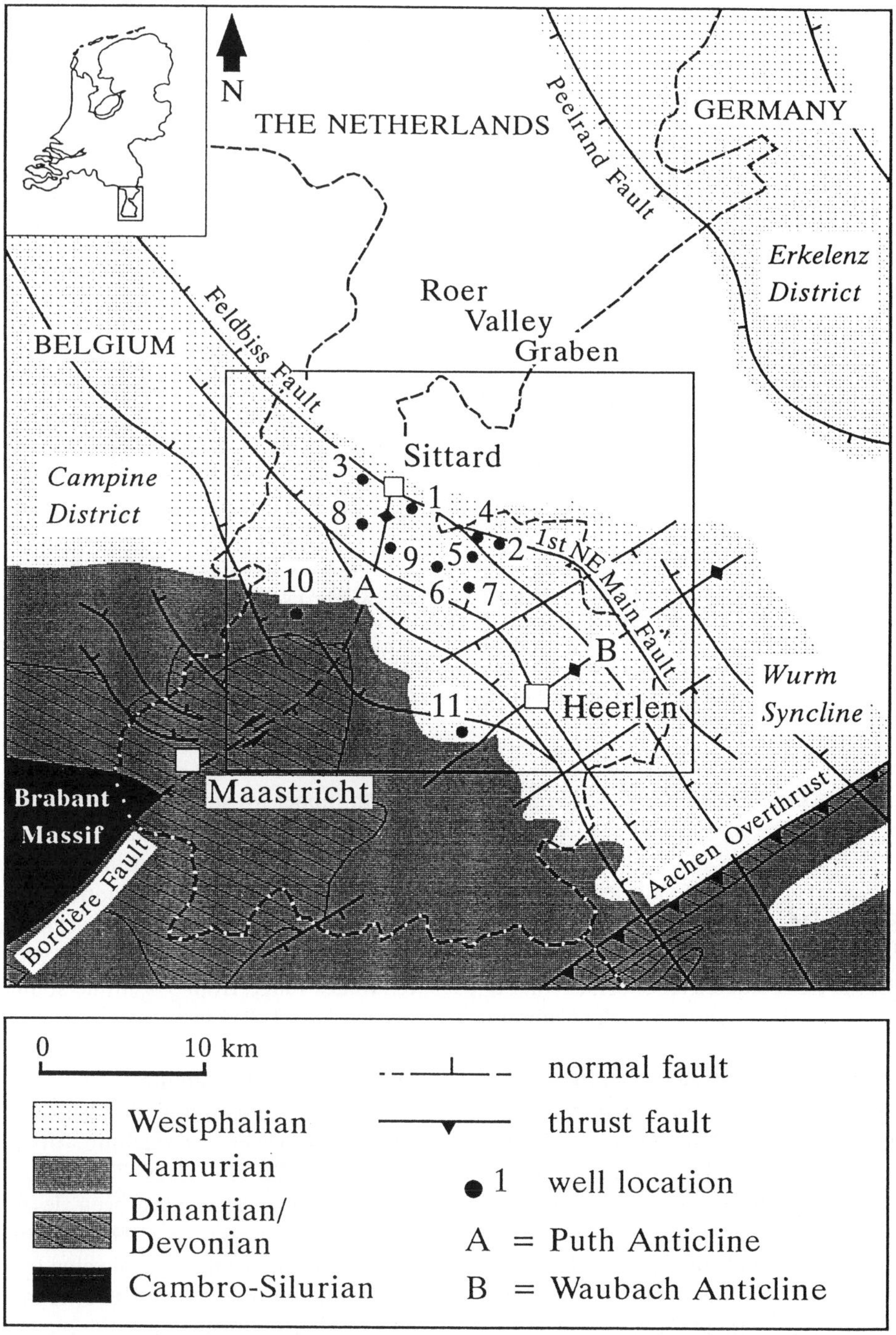

Fig. 1. Structural map at the eroded top Paleozoic in South Limburg, with locations of the wells studied. 1 = KPK-1; 2 = XLV; 3 = LBR-1; 4 = RAA-1; 5 = XL; 6 = LI; 7 = XIX; 8 = XIV; 9 = LV; 10 = GVK-1; 11 = GB113. For abbreviations see Table 1.

The maximum total thickness of the Permian, Triassic and Jurassic deposits north of study area is approximately 1200 m (M.C. Geluk, pers. comm. 1994). For the Belgian Campine Basin the cumulative thickness of these deposits has been estimated at approximately 1400 m (De Craen & Swennen 1992).

The thickness of the Upper Cretaceous to Quaternary deposits southwest of the Roer Valley Graben

Table 1. Data summary of the wells and the modelled heatflow and overburden values.

No	Well	Location	TC	TD	M	VR	N	HF	OV
1	KPK-1	Kemperkoul	489	1665	–	0.67–1.26	93	62	2700
2	XLV	Jabeek	789	1355	182	0.79–0.95	28	66	2900
3	LBR-1	Limbricht	828	1074	211	0.71–0.84	23	60	2600
4	RAA-1	Raath	864	1203	272	0.80–0.91	38	61	2650
5	XL	Wiggelraderhof	348	1002	516	0.82–1.19	48	59	2450
6	LI	Oirsbeek	348	566	626	0.82–0.95	22	60	2600
7	XIX	Douvergenhout	350	1383	769	0.91–1.68	27	55	2300
8	XIV	Geleen	607	1044	1004	0.88–1.04	19	48	1800
9	LV	Munstergeleen	430	775	1119	0.92–1.04	15	46	1800
10	GVK-1	Geverik	325	1687	2683	2.39–2.92	8	54	1850
11	GB 113	Winthagen	154	700	2144	2.59–3.14	16	65	3100
	Seam GB 23		var	–	1718	1.40	16	50	1800
						2.40		71	3400

TC = top of the Carboniferous (in metres below rotary table); TD = total depth (m below rotary table); M = the amount of eroded Carboniferous sediment (m) relative to the top of the Westphalian C in well Kemperkoul-1; VR = range of mean random vitrinite reflectance (% Rr); N = number of samples; HF = calculated basement heatflow (mW/m^2) at a constant overburden of 2700 m; OV = calculated amount of overburden (m) on top of the Westphalian C at a constant heatflow of 62 mW/m^2. The numbers refer to the well locations on Fig. 1.

most likely never exceeded 800 m (Van Staalduinen et al. 1979, Geluk 1990).

Material and methods

The locations of the wells studied are indicated on Fig. 1. A summary of the depth intervals of the Carboniferous is given in Table 1. The wells have been drilled for mining exploration purposes, for stratigraphical or economic reconnaissance studies, and as part of a coal inventory project carried out by the Geological Survey of the Netherlands in 1982–1986. Some of the organic petrological results of these wells have been published earlier (Veld et al. 1993). The penetrated Carboniferous strata were fully cored and individual wells yielded up to 93 coal samples, each from a different coal seam. The thicknesses of the penetrated Carboniferous intervals range from 218 to 1176 m. Well Geverik-1 has been drilled in 1987 by the Geological Survey for a different purpose, an underground pump accumulation station (OPAC project).

Coalification data from the former mining district were obtained from literature, mining companies and Geological Survey records. Most of the seams exploited were restricted to a few coal mines. Seam GB 23 (Furth, upper Westphalian A), however, was exploited all over the district. From this seam a considerable number of volatile matter data was available, collected over several decades of exploration activity. This seam was therefore selected to construct a coalification map based on volatile matter values (% VM). Volatile matter values range from 9% in the southeast to 24% in the northwest. These values were converted to vitrinite reflectance values using the correlation diagram of Teichmüller & Teichmüller (1984).

For the evaluation of the thermal maturity in the study area a 1D basin modelling software package was used (IFP/BEICIP – GENEX). The basic concepts of this computer model have been described in several papers (e.g. Ungerer et al. 1990). The time-temperature path from which a theoretical value for the maturity of an organic particle is calculated, is intrinsically related to its burial history. For the reconstruction of the burial history and basin subsidence the program uses the backstripping method. The necessary input consists of stratigraphical time-depth information (thicknesses, ages, lithology and depositional water depths). The decompaction model is based on the algorithms described by Perrier & Quiblier (1974). Petrophysical parameters of a formation, such as porosity, matrix thermal conductivity and matrix heat capacity, are cal-

Table 2. Input of stratigraphical and lithological data for modelling of temperature and burial history for well Kemperkoul-1.

Base of Unit or *Event*	Age (Ma)	Depth (m)	Missing (m)	Eroded (m)	Lithology (%)					Cond.	Heat cap.
					Sst	Sh	Co	Lst	Dol		
Quaternary	1.6	110	–	–	90	5	–	5	–	3.33	2.86
Miocene	23.3	290	–	–	85	10	5	–	–	2.41	2.74
Oligocene	36.6	430	–	–	50	45	–	5	–	2.52	2.62
Erosion	40.0										
Palaeocene	63.6	489	–	–	10	–	–	90	–	3.04	2.72
Erosion	64.0			300							
Maastrichtian	74.0	–	300	–	–	10	–	90	–	2.84	2.66
Erosion	144.0			100							
Erosion	150.0			1100							
Middle Jurassic	178.0	–	50	–	30	70	–	–	–	2.20	2.48
Lower Jurassic	208.0	–	400	–	15	85	–	–	–	2.04	2.39
M – U Triassic	235.0	–	150	–	50	50	–	–	–	2.46	2.60
Buntsandstein	245.0	–	600	–	90	10	–	–	–	3.23	2.84
Erosion	251.0			30							
Zechstein	256.1	–	30	–	70	30	–	–	–	2.79	2.72
Erosion	300.0			2700							
Stephanian(?) + Westphalian D	309.0	–	2538	–	80	20	–	–	–	2.99	2.78
U Westphalian C	309.5	–	162	–	40	59	1	–	–	2.23	2.53
L – M Westphalian C	311.0	1227	–	–	25	71	4	–	–	1.85	2.40
U Westphalian B	312.0	1598	–	–	22	75	3	–	–	1.89	2.39
L Westphalian B	313.0	2050	–	–	13	85	2	–	–	1.88	2.35
Westphalian A	315.0	2950	–	–	14	85	1	–	–	1.96	2.37
Namurian	326.0	3850	–	–	20	74	1	–	5	2.06	2.43
Dinantian	360.0	4750	–	–	5	5	–	80	10	2.96	2.69

Sst = sandstone; Sh = shale; Co = coal; Lst = limestone; Dol = dolomite; Cond. = matrix conductivity (W/m/K); Heat cap. = matrix heat capacity ($MJ/m^3/K$).

culated on the basis of the program default values of ten pure lithologies and their relative volume proportions in that formation (Table 2).

For the determination of the temperature history the model considers the vertical heat transfer from the basement towards the surface. Temperatures are calculated using a transient heatflow model to account for thermal inertia of rocks in situations of rapid heating or cooling (high sedimentation or erosion rates). The equation calculates the total heat transfer as the summation of heatflow due to conduction, the heat transported by moving fluids and the heat generated or stored by the rocks (Ungerer et al. 1990).

One of the options to define the vertical thermal history is by postulating a heatflow model at the basement. The calculated temperature history of a specific sediment interval can be calibrated with measured maturity data. In GENEX, vitrinite reflectance equivalents are obtained through a correlation curve between vitrinite reflectance and the transformation ratio of the standard type IV organic matter. This transformation ratio is calculated with the primary cracking model of Espitalié et al. (1988), which applies first-order Arrhenius reaction kinetics with a distribution of activation energies. Program default values have been used for the description of activation energy distributions and for the translation of tranformation ratios into vitrinite reflectance equivalents.

Coalification patterns

Eleven wells have been studied in detail (Fig. 1, Table 1). The vitrinite reflectance trends with depth

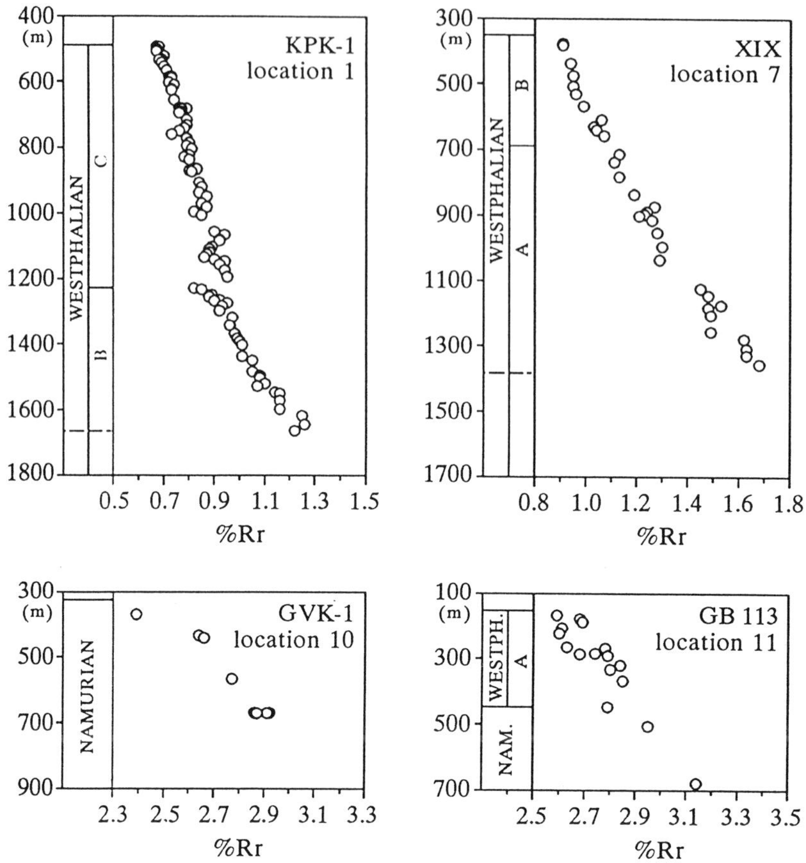

Fig. 2. Vitrinite reflectance (% Rr) plotted against depth (m below rotary table) for four selected wells. The location of the wells is shown in Fig. 1.

for four selected wells are presented in Fig. 2. Well Kemperkoul-1 (KPK-1, no. 1 on Fig. 1) shows an irregular increase from approximately 0.67% Rr at the top of the Westphalian to 1.26% Rr at total depth. The coalification gradient increases with depth. In the uppermost part of the sequence it amounts to 0.35% Rr/km and in the Westphalian B it has increased to 0.80% Rr/km. Around the Aegir marine band (Westphalian B-C boundary) at about 1227 m depth, a significant drop of the reflectance values occurs, the meaning of which has been discussed elsewhere (Veld & Fermont 1990). The vitrinite reflectance values of well XIX (Fig. 1, no. 7) range from 0.91 at the top to 1.68% Rr at total depth. Here too an increase of the coalification gradient with increasing depth is observed, from 0.60 to 1.06% Rr/km. Fewer data points are available from the southern deep wells. For well GB 113 (Fig. 1, no. 11) reflectance values for 23 samples show a relatively large scatter. These high-rank coals range from 2.59 to 3.14% Rr. The wide scatter is presumably due to the bireflectance of anthracites, which increases the variance of randomly measured vitrinite reflectance values. For well GVK-1 (Fig. 1, no. 10) only eight samples from the Namurian are available. The vitrinite reflectance values range from 2.39 to 2.92% Rr.

In order to exclude the effect of stratigraphical differences, a single-seam coalification map was constructed for the Westphalian seam GB 23. The iso-rank lines in Fig. 3 show two different areal trends. From southeast to the northwest the reflectance decreases from over 2.4 to less than 1.4% Rr. Towards the northeast, however, the degree of coalification increases again. Here the vitrinite reflectance values increase to well over 2.1% Rr. One of the irregularities in the coalification pattern is in the Puth Anticline, south of Sittard, where the core of the anticline shows a depression of the coalification values to under 1.6% Rr. This may indicate that folding took place before the final coalification pattern was established.

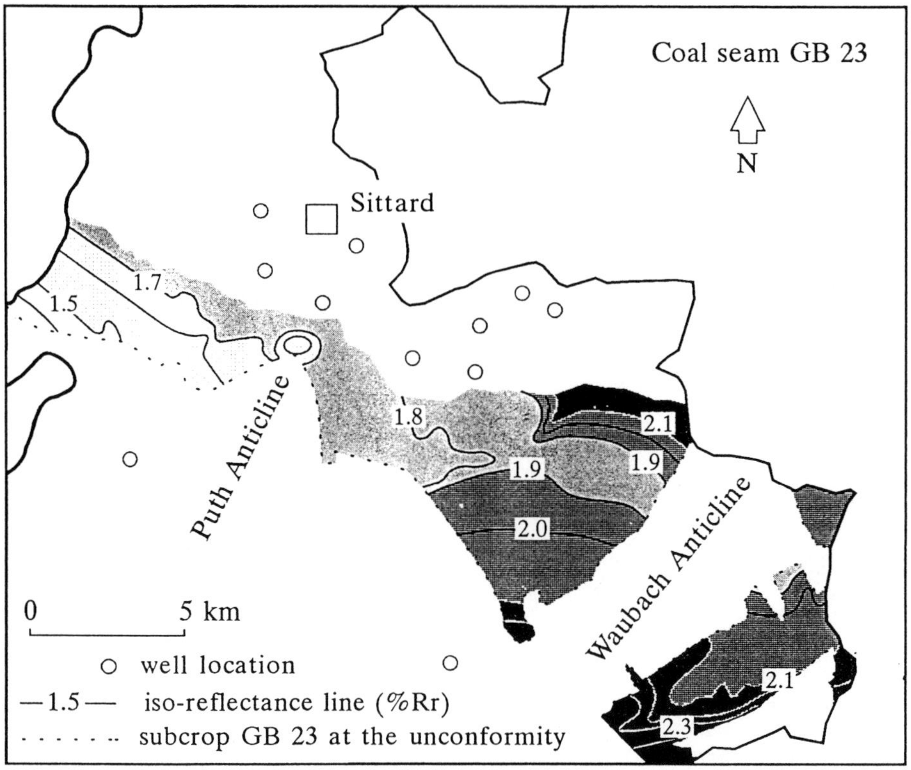

Fig. 3. Lateral variation in coalification of coal seam GB 23 (Furth, upper Westphalian A), calculated from volatile matter using the correlation diagram published by Teichmüller & Teichmüller (1984).

Burial and geothermal modelling

The present depth of the Carboniferous in South Limburg is too shallow to account for the observed coalification. Due to insufficient information on the post-Carboniferous development, the burial history of the area and the timing of coalification are difficult to determine reliably. Vitrinite reflectance measurements of sediments unconformably overlying the Carboniferous clearly show a much lower maturity than the underlying Carboniferous. Wolf & Bless (1987) reported reflectance values of 0.32 to 0.46% Rr in Upper Cretaceous rocks directly overlying the highly mature Dinantian. These data indicate a pre-Cretaceous coalification for the Carboniferous. No maturity data are available for the Triassic sediments in the area. The regional picture suggests that the maximum amount of Triassic to Early Jurassic overburden was less than 1200 m. This is not sufficient to account for the high degree of coalification of the Carboniferous strata. Therefore a Middle to Late Jurassic coalification for these strata is here ruled out. Information from the nearby Ruhr Basin in Germany clearly indicates a pre-Zechstein origin of the coalification patterns in the Carboniferous of that area (Teichmüller 1987).

Despite the limited information on the post-Carboniferous subsidence history of South Limburg, it is inferred to be relatively straightforward. Rapid Late Carboniferous subsidence followed by rapid Early Permian uplift are assumed to have controlled coalification patterns. This is in agreement with the generally accepted hypothesis that Late Carboniferous sedimentation and subsidence in the Variscan foredeep terminated as a consequence of the Asturian tectonic phase of the Variscan orogeny (Robert 1989).

Well KPK-1 (Figs. 1, 2) was choosen as a reference for the burial and geothermal modelling. This well yielded a large number of reflectance data, enabling the establishment of a detailed coalification trend. For an extrapolation of the results it is assumed that several characteristics, such as stratigraphical and lithological parameters and thicknesses, remain constant over the whole area. The absolute ages are based on Harland et al. (1990) and for the Late Carboniferous on Lippolt et

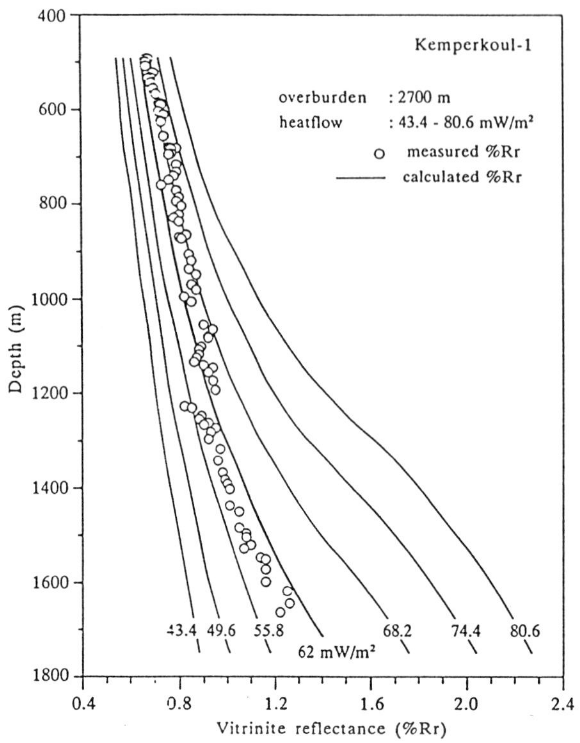

Fig. 4. Kinetically modelled vitrinite reflectance profiles, obtained using a constant overburden of 2700 m on the eroded top of the Westphalian C (489 m below rotary table) and variable basement heatflow values, and measured vitrinite reflectance to illustrate the fit between modelled and measured maturity profiles for well KPK-1. The best fit is obtained with an overburden of 2700 m and a heatflow of 62 mW/m^2. The 10% intervals relative to 62 mW/m^2 refer to the sensitivity matrix of Table 3.

al. (1984). The crustal thickness has been estimated to be approximately 34 km (Betz et al. 1988), for which the program calculated a radiogenic heatflow of 18.90 mW/m^2. The present-day surface heatflow of the South Limburg area ranges between approximately 60 and 70 mW/m^2 (Ramaekers 1991). This requires a present basement heatflow of 34.0 mW/m^2.

Temperature is the most sensitive parameter determining the state of organic maturation. Palaeotemperatures strongly depend on palaeo-overburden and palaeo-heatflow, neither of which can be measured directly. By varying these two parameters in the model, measured vitrinite reflectance values can be compared with calculated values, and in case of a large number of reflectance measurements, as in well KPK-1, only limited combinations of basement heatflow values and of depths of burial remain possible. The palaeosurface temperature during the Carboniferous was set at 23 °C. Palaeo-waterdepths were considered to be less than 100 m. Heatflow values were varied from 10 to 100 mW/m^2. The maximum overburden thicknesses on top of the Westphalian C were varied from 500 to 5000 m. A best curve fit between calculated and measured reflectance data was achieved with a basement heatflow of 62 mW/m^2 and an overburden of 2700 m at the site of well KPK-1 (Fig. 4).

The input data for modelling the sedimentation history of KPK-1 are given in Table 2. The burial history diagram on the basis of these model parameters is represented in Fig. 5a for the top Westphalian C, seam GB 23 and top Dinantian. These curves indicate a maximum burial at 300 Ma of approximately 2850 m for the top Westphalian C and 6360 m for the top Dinantian. The (imposed) basement heatflow history and the related surface heatflow pattern are shown in Fig. 5b. The transient effects of high sedimentation and erosion rates on the surface heatflow history are here clearly demonstrated. The temperature histories for the Late Dinantian to late Westphalian C are presented by the three curves in Fig. 5c. The highest temperatures are reached at the time of maximum burial at 300 Ma, or slightly later. For the top Dinantian the temperature of approximately 252 °C results in a vitrinite reflectance of 4.16% Rr. Uppermost Westphalian C strata reached a maximum temperature of 118 °C with a corresponding vitrinite reflectance of 0.69% Rr. The maximum temperature reached by coal seam GB 23 in the KPK-1 model is determined at 183 °C for which a vitrinite reflectance of 2.06% Rr has been calculated. This maximum temperature for GB 23 has been determined for each well and for 16 separate data points. These data have been used to construct the palaeotemperature map for GB 23 of Fig. 6.

The maximum temperature that has been reached by a sediment particle can be calculated by several methods, for instance by adjusting the amount of overburden and/or the basement heatflow. In order to gain a quantitative insight in the possible variation of overburden and basement heatflow for the study area two model studies have been made.

In the first model the heatflow was kept at a constant value. By applying a basement heatflow of 62 mW/m^2, like in KPK-1, the variation in overburden was calculated. The overburden consists of two parts. The first is the part eroded relative to the top of the Westphalian C in KPK-1. This part has been estimated from the regional stratigraphical framework and is given in Table 1 in column M. The second part consists of the overburden thickness that needs to be assumed to

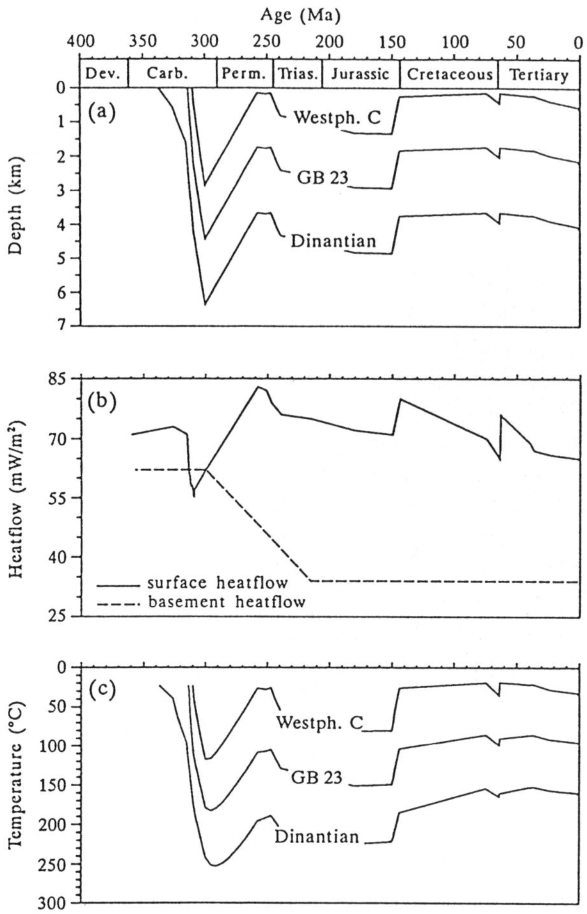

Fig. 5. Results of the burial and geothermal modelling of well KPK-1 with a basement heatflow of 62 mW/m^2 and an overburden of 2700 m at the end of the Carboniferous. (a) Burial history of top Westphalian C, coal seam GB 23 and the top Dinantian; (b) Heatflow history, showing the imposed basement heatflow of 62 mW/m^2 during the Carboniferous and the resulting surface heatflow. Note the transient thermal effect of high sedimentation and erosion rates on surface heatflow values; (c) The resulting temperature history for the three levels of Fig. 5a indicates maximum temperatures at or immediately after deepest burial.

fit the local maturity level. In column OV of Table 1 the total calculated overburden on top of the Westphalian C is given. It ranges from 1800 m in the northwest to 3400 m in the south of the area.

In the second model the Late Carboniferous – Early Permian overburden on top of the Westphalian C has been kept identical throughout the area. The heatflow pattern necessary to induce the present-day coalification pattern was evaluated. For the calculation of the heatflow variation, the modelling results of well KPK-1 were used, i.e. an overburden of 2700 m on top of the Westphalian C. For the other wells the eroded amount of Westphalian and older strata was added and subsequently the heatflow variation was calculated to fit the measured vitrinite reflectance data. The output of this modelling is given in Table 1, column HF. The imposed basement heatflow at the identical overburden of 2700 m has been calculated to range between 46 and 71 mW/m^2.

The geographical variation in maximum temperature of seam GB 23 is shown in Fig. 6 (cf. Fig. 5c). The iso-temperature lines show trends similar to the iso-rank lines (Fig. 3). Palaeotemperatures range from less than 170 °C in the northwest to more than 200 °C in the southeastern part of the area.

Sensitivity analysis and errors

The influence on modelling of variations in both the basement heatflow and the amount of overburden is important for assessing the offset of the measured vitrinite reflectance profiles. The modelling results of well KPK-1 have been used as a reference for a sensitivity analysis. The sensitivity matrix presented in Table 3 shows the combined effects of varying basement heatflow and overburden on the modelled vitrinite reflectance values for coal seam GB 23. This matrix covers an 1890 to 3510 m overburden range and a 43.4 to 80.6 mW/m^2 heatflow range at 300 Ma. With an overburden of 2700 m and a heatflow of 62 mW/m^2 ('best fit') the modelled vitrinite reflectance for coal seam GB 23 is 2.06% Rr. The matrix demonstrates that modelled vitrinite reflectance values are more sensitive to changes in heatflow than to proportional changes in the amount of overburden. With a heatflow of 62 mW/m^2 a reduction of 10% in the amount of overburden reduces vitrinite reflectance from 2.06 to 1.90% Rr. A 10% reduction in heatflow at an overburden of 2700 m results in a lowering of the reflectance value for GB 23 to 1.73% Rr. As an example, the effect of heatflow variations on calculated vitrinite reflectance values at an overburden thickness of 2700 m is given in Fig. 4.

Because all other input parameters have been kept constant, the heatflow and the overburden model will both result in a similar palaeotemperature map. Discrepancies between the maximum temperatures from both models derive from the variability in the 'best fit' between measured and calculated reflectance values. The maximum difference between the two sets of calculated vitrinite reflectance values for GB 23 was less than 0.20% Rr.

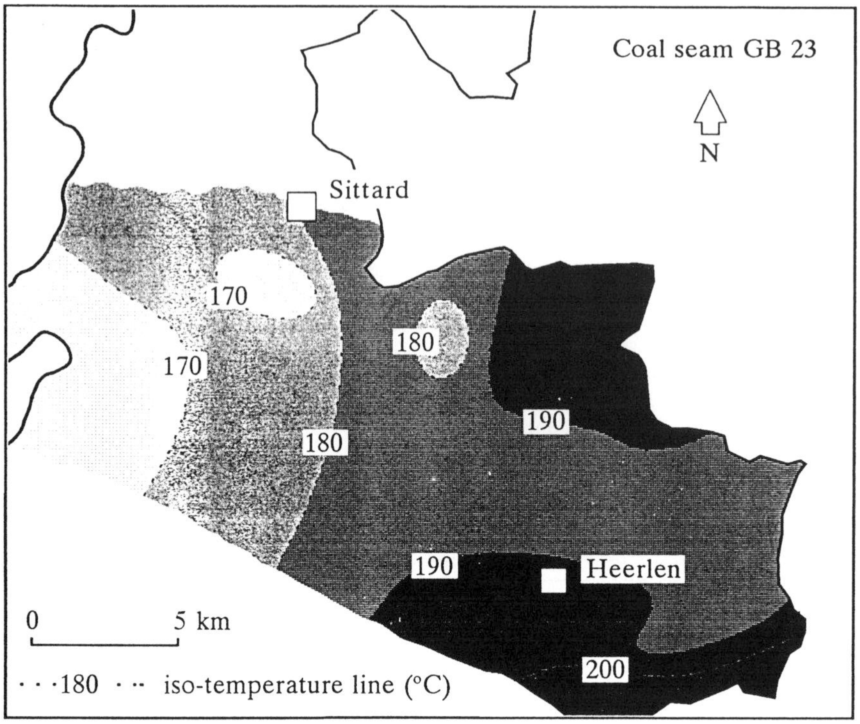

Fig. 6. Palaeotemperature map of seam GB 23 (upper Westphalian A) at the time of deepest burial (300 Ma), calculated using the modelling results of well KPK-1 as a reference for the whole area (see Fig. 5c).

Table 3. Sensitivity matrix showing the effects of varying heatflow (HF) and overburden (OV) on the calculated vitrinite reflectance values for coal seam GB 23 for the locality of well Kemperkoul-1.

	OV (m)	*1890*	*2160*	*2430*	*2700*	*2970*	*3240*	*3510*
HF (mW/m^2)		70%	80%	90%	100%	110%	120%	130%
43.4	70%	1.04	1.05	1.09	1.14	1.22	1.34	1.50
49.6	80%	1.16	1.20	1.27	1.37	1.58	1.72	1.88
55.8	90%	1.34	1.44	1.59	1.73	1.92	2.07	2.19
62.0	100%	1.61	1.74	1.90	2.06	2.20	2.34	2.50
68.2	110%	1.87	2.02	2.16	2.31	2.48	2.64	2.83
74.4	120%	2.11	2.25	2.42	2.59	2.79	2.98	3.16
80.6	130%	2.32	2.50	2.68	2.90	3.10	3.36	3.67

The conversion of volatile matter values into vitrinite reflectance equivalents also gives a maximum error of approximately 0.20% Rr for individual data points. A change of 0.20% Rr in the vitrinite reflectance value of GB 23 is in our model correlated with a change in palaeotemperature of approximately 8 °C. A maximum error for the temperature map (Fig. 6) is therefore less than 10 °C, but will generally be less than 5 °C.

Discussion

Detailed coalification studies have provided an approximate record of the geothermal setting of the Variscan foredeep of western Europe (Robert 1989). Palaeogeothermal gradients during the Late Palaeozoic usually ranged from 40 to 80 °C/km (Teichmüller & Teichmüller 1986, Von Winterfeld 1994). The surface heatflow values during this period ranged from 75 to 130 mW/m^2 (Ritter 1986, Teichmüller 1987). The modelled basement heatflow values of South Limburg ranged from 46 to 71 mW/m^2. Due to the transient thermal effects, the highest surface heatflow values in our model were reached during the Permian (Fig. 5b). During the Carboniferous, surface heatflow values ranged from 61 to 74 mW/m^2. Although comparison of literature data is hampered by the fact that different geothermal concepts were applied, the results from the present area indicate that the geothermal history is in general agreement with that of other localities within the Carboniferous foredeep basin (Robert 1989).

The temperature map of seam GB 23 can be considered representative for the temperature distribution at the end of the Carboniferous, although absolute values will differ at other stratigraphical levels. The temperature distribution (and thus the coalification pattern) is complex and difficult to explain using simple assumptions of identical overburden or heatflow throughout the area. Such assumptions result in extreme differences in both modelled heatflow and overburden over very short distances.

The coalification pattern in the Wurm Syncline directly to the southeast of, and parallel to the Waubach Anticline has been the subject of detailed research. The coalification pattern here of the Westphalian A coal seam Großlangenberg closely corresponds to the pattern of seam GB 23 in the South Limburg area (Teichmüller & Teichmüller 1979). Increasing coalification in seam Großlangenberg towards the southwest is evident, with reflectance values up to 2.3% Rr (10% VM) directly in front of the Aachen Overthrust. The Großlangenberg seam in the southwest of the Wurm Syncline shows an almost similar degree of coalification as GB 23 south of Heerlen which has calculated maximum palaeotemperatures of over 200 °C (Fig. 6). For this area south of Heerlen (Fig. 6) an additional overburden of 3400 m on top of the Westphalian C would be needed to explain the observed level of coalification. However, for well LV (Fig. 1), approximately 1800 m of overburden has been calculated (Table 1). This would imply differences in overburden of 1600 m within 20 km distance. Alternatively, if the coalification pattern (and temperature pattern) is exclusively attributed to differences in heatflow this would result in a difference in basement heatflow of 25 mW/m^2. The present data and modelling results are not conclusive as to whether the high levels of coalification in the south of the study area and in the Wurm Syncline can be attributed either to a high overburden thickness or to high heatflow values. In the light of the spatial trends it may be a combination of both. Teichmüller & Teichmüller (1979) interpreted the increase in coalification of seam Großlangenberg towards the southwest as the result of increasing heatflow in that direction and to a possibly shallower Moho. An alternative explanation for the differences in coalification may be the common occurrence of hydrothermal mineralizations as reported from the mining area south of the Waubach Anticline (Kimpe et al. 1978). Von Winterfeld (1994) explained the high levels of coalification in front of the Variscan overthrust by means of 'tectonic heating' during orogeny. Nappe formation and tectonic stacking would induce a flow of high-temperature fluids towards the foreland basin. The available thermal history model, however, is not capable of modelling lateral heatflows.

The observed trend of decreasing coalification values towards the west, away from the Variscan thrust-front, as established for seam GB 23, continues in the neighbouring Campine Basin (Langenaeker 1992). This implies that the differences in degree of coalification between the Campine and South Limburg have a common cause. This may be either a decrease in heatflow or a decreasing overburden towards the northwest. In addition to this trend, Langenaeker (1992) established an increase of coalification towards the northeast within the Campine, similar to the southeastward increase for seam GB 23 south of Sittard, which he explained as reflecting increasing burial depth.

The coalification pattern around the Puth Anticline suggests that this structure already existed prior to the main phase of coalification. Lines of equal rank in the Wurm Syncline, however, are parallel to the folded sedimentary sequence, in support of a mainly pre-orogenic phase of coalification (Teichmüller & Teichmüller 1979). These two different observations imply either two phases of coalification or two phases of folding. The Wurm Syncline and the Waubach and Puth anticlines are believed to have formed more or less simultaneously during the Variscan orogeny, although their structure may have been modified later (Wrede 1985, Dusar & Langenaeker 1992). Therefore, two phases of coalification, separated by the formation of the Puth Anticline, seem to be more likely to explain the coalification pattern in the northern part of the area, as previously suggested by Fermont (1986).

Across the German border, approximately 25 km east of Sittard, an area with relatively high-rank coals around the Sophia Jacoba Mine, in the Erkelenz mining district, has been related to the presence of a Permian(?) intrusion that influenced the coal seams in a wide area (Teichmüller & Teichmüller 1966). Calculated temperature variations around this mine indicate a maximum temperature of approximately 310 °C for a stratigraphical level comparable to seam GB 23 (Erren & Bredewout 1991). Extrapolated into the northeast of the South Limburg area this would result in maximum temperatures of 200 to 225 °C for seam GB 23. This corresponds with the temperature we calculated (Fig. 6). Therefore, an influence of the Erkelenz intrusion on the coalification pattern in South Limburg is likely. A coalification high in the Meeuwen-Bree area, in the north-east of the Campine Basin, has also been ascribed to the Erkelenz intrusion (Dusar 1982).

Although details are not known, the thickness of the eroded section of Carboniferous rocks of post-Westphalian C age certainly was of the order of 1500–3000 m. This implies that the thickness of eroded Late Carboniferous sediments was approximately 2900 m around Sittard, and in the south, towards the Belgian border, at least as much as 5400 m. This volume of rock has been a main source for the Permian and Triassic deposits in the north. In addition, these volumes indicate that either the Variscan nappes extended much further to the north than can be deduced from the preserved structures, or alternatively, that in front of the Variscan orogen a molasse basin evolved which was subsequently completely eroded.

Acknowledgements

We gratefully acknowledge sample preparations and vitrinite reflectance measurements by L. Jegers. The Director of the Geological Survey of the Netherlands (RGD) is thanked for giving permission to publish this paper. The manuscript was greatly improved by the constructive comments of M.P. Ormerod and anonymous reviewers. The work was supported by the Netherlands Foundation for Earth Science Research (AWON) with financial aid from the Netherlands Organization for Scientific Reseach (NWO). This is paper nr. 941105 of the Netherlands Research School of Sedimentary Geology (NSG).

References

Betz, D., H. Durst & T. Gundlach 1988 Deep structural seismic reflection investigations across the northeastern Stavelot-Venn Massif – Ann. Soc. Géol. Belgique 111: 217–228

Bless, M.J.M., J. Bouckaert, M.A. Calver, J.M. Graulich & E. Paproth 1977 Palaeogeography of Upper Westphalian deposits in NW Europe with reference to the Westphalian C of the mobile Variscan belt – Meded. Rijks Geol. Dienst 28: 101–147

Bless, M.J.M., P. Boonen, J. Bouckaert, C. Brauckmann, R. Conil, M. Dusar, P.J. Felder, H. Gökdag, F. Kockel, M. Laloux, H.R. Langguth, C.G. van der Meer Mohr, J.P.M.Th. Meessen, F. op het Veld, E. Paproth, H. Pietzner, J. Plum, E. Poty, A. Scherp, R. Schulz, M. Streel, J. Thorez, P. van Rooijen, M. Vanguestaine, J.L. Vieslet, D.J. Wiersma, C.F. Winkler-Prins & M. Wolf 1981 Preliminary report on Lower Tertiary – Upper Cretaceous and Dinantian – Famennian rocks in the boreholes Heugem-1/1a and Kastanjelaan-2 (Maastricht, the Netherlands) – Meded. Rijks Geol. Dienst 35: 333–415

Burnham, A.K. & J.J. Sweeney 1989 A chemical kinetic model of vitrinite maturation and reflectance – Geochim. Cosmochim. Acta 53: 2649–2657

De Craen, M. & R. Swennen 1992 Sedimentology and diagenesis of the ankeritized basal Zechstein conglomerate in the Campine Basin (Bree borehole, NE Belgium) – Geol. Mijnbouw 71: 145–160

Dusar, M. 1982 Exploration for coal in the Belgian Campine – Publ. Natuurhist. Gen. Limburg 32: 27–39

Dusar, M. & V. Langenaeker 1992 De oostrand van het massief van Brabant, met beschrijving van de geologische verkenningsboring te Martensheide – Belg. Geol. Dienst, Prof. Paper 255: 22 pp

Erren, H. & J.W. Bredewout 1991 Model calculations on intrusive cooling and related coalification of the Peel-Erkelenz coalfield (the Netherlands and Germany) – Geol. Mijnbouw 70: 243–252

Espitalié, J., P. Ungerer, I. Irwin & F. Marquis 1988 Primary cracking of kerogens. Experimenting and modeling C_1, C_2-C_5, C_6-C_{15} and C_{15+} classes of hydrocarbons formed – Org. Geochem. 13: 893–899

Fermont, W.J.J. 1986 Evaluatie van het koolpetrografisch onderzoek in Limburg en de Achterhoek/Twenthe. In: Krans, Th.F. et al. (eds) 'Eindrapport Project Inventarisatieonderzoek Nederlandse Kolenvoorkomens' – Internal Report Geol. Survey of the Netherlands, Geol. Bureau. Heerlen. Rapp. nr. GB2107. Appendix I: 62 pp

Geluk, M. 1990 The Cenozoic Roer Valley Graben, southern Netherlands – Meded. Rijks Geol. Dienst 44: 65–72

Harland, W.B., R.L. Armstrong, A.V. Cox, L.E. Craig, A.G. Smith & H.G. Smith 1990 A Geological Time Scale 1989. Cambridge University Press, Cambridge: 263 pp

Kimpe, W.F.M., M.J.M. Bless, J. Bouckaert, R. Conil, E. Groessens, J.P.M.Th. Meessen, E. Poty, M. Streel, J. Thorez & M. Vanguestaine 1978 Paleozoic deposits east of the Brabant Massif in Belgium and the Netherlands – Meded. Rijks Geol. Dienst 30: 37–103

Langenaeker, V. 1992 Coalification maps for the Westphalian of the Campine coal basin – Belg. Geol. Dienst, Prof. Paper 256: 38 pp

Lippolt, H.J., J.C. Hess & K. Burger 1984 Isotopische Alter pyroklastischer Sanidinen aus Kaolin-Kohlentonsteinen als Korrelationsmarken für das mitteleuropäische Oberkarbon – Fortschr. Geol. Rheinl. u. Westf. 32: 119–150

Patijn, R.J.H. 1963 Tektonik von Limburg und Umgebung – Verh. Kon. Ned. Mijnb. Gen., Geol. Serie 21: 9–24

Perrier, R. & S. Quiblier 1974 Thickness changes in sedimentary layers during compaction history; methods for quantitative evaluation – Am. Ass. Petroleum Geol. Bull. 58: 507–520

Ramaekers, J.J.F. 1991 The Netherlands. In: Hurtig, E., V. Čermák, R. Haenel & V. Zui (eds) Geothermal Atlas of Europe – Hermann Haack, Gotha: 81–83

Ritter, U. 1986 Heat flow during the Carboniferous and Mesozoic of the Northwest German Basin – Geol. Rundschau 75: 293–300

Robert, P. 1989 The thermal setting of the Carboniferous basins in relation to the Variscan orogeny in Central and Western Europe – Int. J. Coal Geol. 13: 171–206

Romein, B.J. 1963 Present knowledge of the stratigraphy of the Upper Cretaceous (Campanian-Maastrichtian) and Lower Tertiary (Danian-Montian) calcareous sediments in southern Limburg – Verh. Kon. Ned. Mijnb. Gen., Geol. Serie 21: 93–104

Teichmüller, M. 1987 Recent advances in coalification studies and their application to geology. In: Scott, A.C. (ed) Coal and coal-bearing strata: Recent advances – Geol. Soc. Spec. Publ. 32: 127–169

Teichmüller, M. & R. Teichmüller 1966 Die Inkohlung im Saar-Lothringer Karbon, verglichen mit der im Ruhrkarbon – Z. dt. geol. Ges. 117: 243–279

Teichmüller, M. & R. Teichmüller 1979 Ein Inkohlungsprofil entlang der linksrheinischen Geotraverse von Schleiden nach Aachen und die Inkohlung in der Nord-Süd-Zone der Eifel – Fortschr. Geol. Rheinl. u. Westf. 27: 323–355

Teichmüller, M. & R. Teichmüller 1984 Verbreitung und Eigenschaften tiefliegender Steinkohlen in der Bundesrepublik Deutschland – Glückauf-Forschungsh. 45: 140–153

Teichmüller, M. & R. Teichmüller 1986 Relations between coalification and palaeogeothermics in Variscan and Alpidic foredeeps of western Europe. In: Buntebarth, G. & L. Stegena (eds) Paleogeothermics. Evaluation of geothermal conditions in the geological past – Lecture Notes in Earth Sciences 5: 53–79

Tissot, B.P. & J. Espitalié 1975 L'évolution de la matière organique des sédiments: application d'une simulation mathématique – Rev. Inst. Fr. Pétr. 30: 743–777

Ungerer, P., J. Burrus, B. Doligez, P.Y. Chénet & F. Bessis 1990 Basin evaluation by integrated two-dimensional modeling of heat transfer, fluid flow, hydrocarbon generation and migration – Am. Ass. Petroleum Geol. Bull. 74: 309–335

Van de Laar, J.G.M. & W.J.J. Fermont 1989 On-shore Carboniferous palynology of the Netherlands – Meded. Rijks Geol. Dienst 43: 35–73

Van Staalduinen, C.J., H.A. van Adrichem Boogaert, M.J.M. Bless, J.W.Chr. Doppert, H.M. Harsveldt, H.M. van Montfrans, E. Oele, R.A. Wermuth & W.H. Zagwijn 1979 The geology of the Netherlands – Meded. Rijks Geol. Dienst 31: 9–49

Van Waterschoot van der Gracht, W.A.J.M. 1938 A structural outline of the variscan front and its foreland from south-central England to eastern Westphalia and Hessen – C.R. 2me Congr. Strat. Carb., Heerlen, 1935: 1487–1565

Veld, H. & W.J.J. Fermont 1990 The effect of a marine transgression on vitrinite reflectance values – Meded. Rijks Geol. Dienst 45: 151–169

Veld, H., W.J.J. Fermont & L.J.F. Jegers 1993 Organic petrological characterization of Westphalian coals from the Netherlands: correlation between Tmax, vitrinite reflectance and hydrogen index – Org. Geochem. 20: 659–675

Von Winterfeld, C.H. 1994 Variszische Deckentektonik und devonische Beckengeometrie der Nordeifel – Ein quantitatives Modell – Aachener Geowissenschaftliche Beiträge 2: 319 pp.

Wolf, M. & M.J.M. Bless 1987 Coal-petrographic investigations on samples from the boreholes Thermae 2000 and Thermae 2002 (Valkenburg a/d Geul, the Netherlands) – Ann. Soc. géol. Belgique 110: 77–84

Wrede, V. 1985 Tiefentektonik des Aachen-Erkelenzer Steinkohlengebietes. In: Beiträge zur Tiefentektonik westdeutscher Steinkohlenlagerstätten. Geol. Landesamt, Nordrhein-Westfalen, Krefeld: 9–103

Rondeel et al. (eds), Geology of gas and oil under the Netherlands, 45–56, 1996.

The Rotliegend sedimentation history of the southern North Sea and adjacent countries

J.P. Verdier
Exxon Exploration Company, P.O. Box 4778, Houston, Texas 77210, USA

Key words: aeolian, beach, desert, dune, fluvial, lake deposits, NW Europe, Permian, sabkha

Abstract

In this study more than 900 well logs and 125 core descriptions from the UK, the Netherlands and Germany are analyzed and integrated into a refined tectono-stratigraphic subdivision. Three periods of Rotliegend deposition (Schneverdingen, Slochteren and Hannover) correspond to distinct phases of basin extension. A climatic cyclicity of drying upward sequences is superimposed upon the tectonically-driven phases of sediment accumulation, resulting in a fine-scale cyclicity that can be correlated across a variety of facies in the Southern Permian Basin. The fluvial, aeolian, and sabkha and lake environments are represented by conglomerates, sandstones and claystones respectively. The aeolian facies identified in this study include a variety of deposits found in a modern desert environment and range from sheet sand, minor dune, barkhan dune, transverse dune and star dune to longitudinal dune. The main controlling factors that were responsible for the sedimentological differences across the basin are the evolution of the tectonic elements, the increasingly arid climate and the westward prograding extension of the basin. These distinctive features are retained and enhanced during post-Rotliegend basin evolution. Prevailing winds, blowing from the north-east, explain in part the distribution of the various aeolian facies across the basin.

Introduction

Since the discovery of the giant Groningen gas field in 1959, followed in the mid-sixties by many smaller accumulations in the offshore and onshore areas, the Southern Permian Basin, has seen a lot of industry activity. More than 150 trillion cubic feet of gas have been discovered to date. The Southern Permian Basin covers an area which extends from the eastern coast of England to eastern Poland. This represents a distance of approximately 1600 km from west to east and 450 km at its maximum width from north to south.

This study focuses on the western portion of the basin from the British shores of the North Sea to the former west-east German border. The principal tectonic elements of the basin during Early Permian times are: the London – Brabant Massif to the west and south-west, the Mid-North Sea and Ringkøbing – Fyn Highs to the north (that separate the Northern from the Southern Permian Basin) and the Variscan Mountains to the south.

The objectives of the study were:

1. to develop a properly time-constrained basin evolution scenario and to establish the facies distribution within the selected time slices;
2. to provide answers to questions relating to sedimentological differences noticed in the basin and to their hydrocarbon implications; and
3. to design a stratigraphic model that is applicable to either more easterly areas or to other regions which are less explored but geologically similar.

It should be noted that the data base is less complete onshore than offshore for reasons of industry activity and accessibility (Table 1).

Hydrocarbon occurrence: chronology and distribution

The discovery in 1959 of the giant Groningen gas field in the Netherlands onshore triggered a now 45 years long successful search for hydrocarbons in the stud-

Table 1. Composition of the Rotliegend data base used in the study.

	Number of wells	Wells used for facies and time slice maps	Wells used for isopach maps	Wells with cores
UK offshore	573	386	379	86
NL offshore	210	197	80	43
NL onshore	184	150	–	–
Germany offshore	30	22	–	–
Germany onshore	172	150	–	–
TOTAL	1169	905	459	129

ied area of the Southern Permian Basin (Encl. 1). The Groningen gas field was estimated to contain at the time of discovery an approximate 58 trillion cubic feet of recoverable raw gas; today this amount has been revised to 100 trillion cubic feet of raw gas. The hydrocarbons, contained in reservoirs of Rotliegend age, are almost exclusively gas which is sourced from the underlying Carboniferous source rock.

Following the Groningen discovery, Annerveen and Ameland Oost were found in 1962 and 1964 respectively in the Netherlands onshore. Exploration then moved into the UK sector of the North Sea and in 1965 West Sole was discovered and was quickly followed by discoveries at Indefatigable (1966), Leman (1966), Viking (1969; Gray 1975), Galleon (1969) and Barque (1971). In the early eighties the exploration interest shifted to the Silverpit Basin where gas accumulations were discovered in Carboniferous reservoirs in an area where the Rotliegend has no reservoir potential. The most recent largest fields discovered in the UK offshore (containing between 1 and 2 trillion cubic feet of raw gas) are Vulcan (1983), Ravenspurn North (1984) and North Valiant (1985). A total of 110 Rotliegend gas fields were discovered over a 30-year exploration period in the UK offshore.

Exploration in the Netherlands offshore began in 1968 in the K and L blocks. K08-FA (1970), K15-FB (1975), L15-FA (1978), L13-FC (1984) and K15-FG (1988) are among the largest fields (Oele et al. 1981). More than sixty fields have been discovered to date in the Netherlands offshore. Recently significant onshore Rotliegend accumulations at Grijpskerk (1990) and Anjum (1992) have renewed the exploration interest in the northeast Netherlands.

Significant gas discoveries in the studied area of Germany are: Groothusen (1965), Wustrow-Salzwedel (1966–1968), Dehtlingen (1971), Söhlingen (1980), Taaken (1982), Bötersen (1984), and Walsrode (1990) (Bender & Hedemann 1983). Gas contained in these German fields occurs only in reservoirs of Hannover age, except in the Söhlingen field where gas also occurs in a reservoir of Schneverdingen age (Drong et al. 1982). Exploration in the German offshore has revealed to date only non-commercial accumulations of hydrocarbon gas generally associated with large quantities of nitrogen gas (up to 66%).

Stratigraphic subdivision

The Rotliegend sediments were deposited during distinctive prograding cycles. As tectonic extension moved westward, the Rotliegend deposits progressively onlapped the edge of the expanding basin. Traditional formation subdivisions such as Leman or Slochteren Sandstones and Ten Boer Claystone or Silverpit Formation are not synchronous depending upon their location in the basin. However, the cessation of Rotliegend deposition was nearly synchronous across the maximum extent of the basin, allowing a more meaningful stratigraphic subdivision of the interval to be defined from top to bottom. The stratigraphic subdivision defined in the deepest and eastward sector uses halite cyclicity which provides a time element for a refined stratigraphy (Gast 1991; Fig. 1). Correlation of this stratigraphy from the center of the saline lake into the sandstones and conglomerates at the basin edge works reasonably well despite the shaling out of the halite (Encl. 2a). The challenge is to untangle

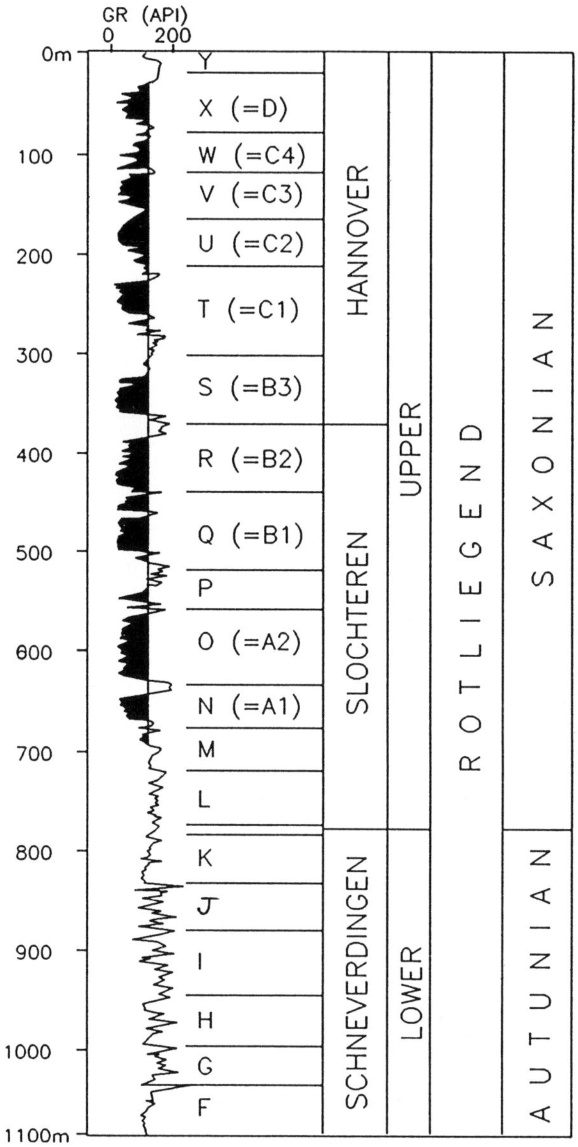

Fig. 1. German lithostratigraphic subdivision of the non-volcanic Rotliegend in the German North Sea Consortium B2 well, offshore Germany; location on Encl. 1 (Gast 1991).

the inevitable intricacies of lithostratigraphy and time stratigraphy using primarily gamma ray logs, keeping in mind the similarity of the gamma ray expressions of the aeolian and fluvial deposits that are the reservoirs for the gas accumulations in the basin. This approach to stratigraphic subdivision based on halite and claystone deposits, correlated using gamma ray logs, has obvious limitations when these two lithologies become rare or absent as in the sand-prone west and south sectors of the basin. A lithostratigraphic subdivision addressing the sand-dominant facies was then developed as industry activity progressed and discovered gas accumulations in reservoirs deposited in aeolian and fluvial environments (Glennie et al. 1978). The accuracy of this subdivision is quite satisfactory at field scale or for short-distance correlations but does not offer a reliable tool for regional time correlations through the entire spectrum of lithologies found in the basin.

A methodology based on climatic cyclicity was developed by Clemmensen (1987) and Clemmensen et al. (1989) in Scotland and the United States of America, and was applied by the industry to the Southern Permian Basin by GAPS Geological Consultants in 1990 (unpublished report).

GAPS divided the Rotliegend in the western part of the basin into five units of climatically-driven drying upward sequences (Encl. 2b). Each drying upward cycle can usually be divided into a lower 'wet' phase followed by an upper 'dry' phase. One cycle may involve only aeolian facies (damp interdune, dry interdune, dune core) or be composed of mixed facies associations (fluvial fan, fluvial sheetflood to aeolian deposits; Encl. 2c). Definition and interpretation of this subdivision require a complete suite of logs : gamma ray, sonic, neutron and density, supplemented by and tied to core descriptions (Fig. 2).

This subdivision works particularly well in fluvial and aeolian environments, but less well in sabkha and lake deposits (except when halite deposits are present) due to the difficulties to define the 'wet' and 'dry' phases in fine-grain deposits. At the margins of the basin, synsedimentary tectonics can overprint the climatic control of sedimentation. The 'wet' phase is often absent or reduced due to reworking during the dry phase, especially in the interior desert environment.

The refined stratigraphic subdivision proposed in this study integrates the GAPS scheme and prior subdivisions (Fig. 3) and can be described as follows: the Rotliegend deposition occurred during the Autunian and Saxonian and is subdivided into three periods: Schneverdingen, Slochteren and Hannover (Gast 1991). The Autunian stage contains not only volcanic flows, but also a sedimentary package that is assigned to the Schneverdingen Formation. The Schneverdingen sedimentation period is only recorded in the eastern part of the studied area related to the beginning of tectonic extension.

Each of the three periods of sedimentation involves a full suite of environments from fluvial to aeolian, sabkha and lake (Encl. 2c). The subdivision can be

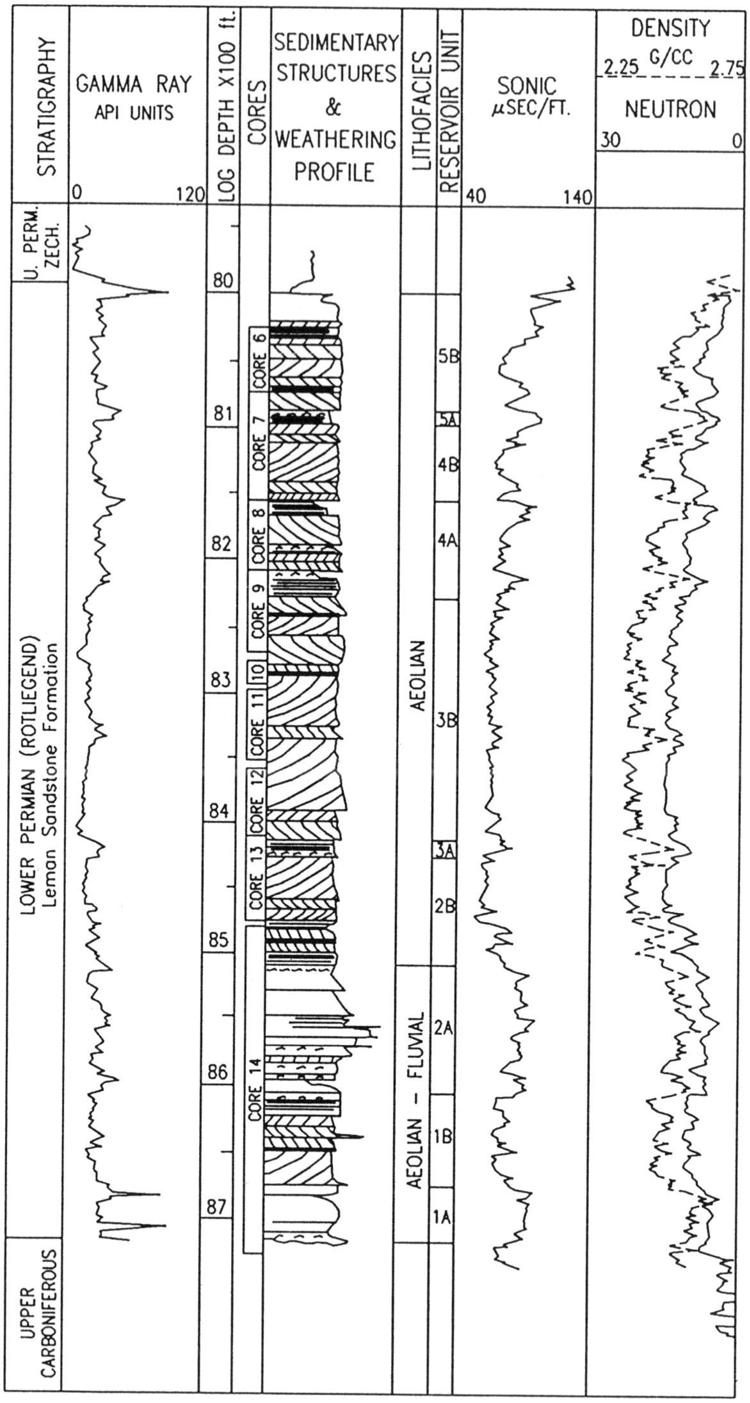

Fig. 2. Example of the GAPS subdivision of the Rotliegend in the Shell/Esso well 48/25–1 in the Sole Pit area of the UK sector of the North Sea; location on Encl. 1.

applied to the eastern part of the Southern Permian Basin and probably also to the little explored Northern Permian Basin and to other Permian basins in north-western Europe where absence of fossils prevents the application of the conventional time stratigraphic subdivision. The new proposed subdivision, despite its limitations, is a valuable addition to the understanding of the basin's sedimentary history.

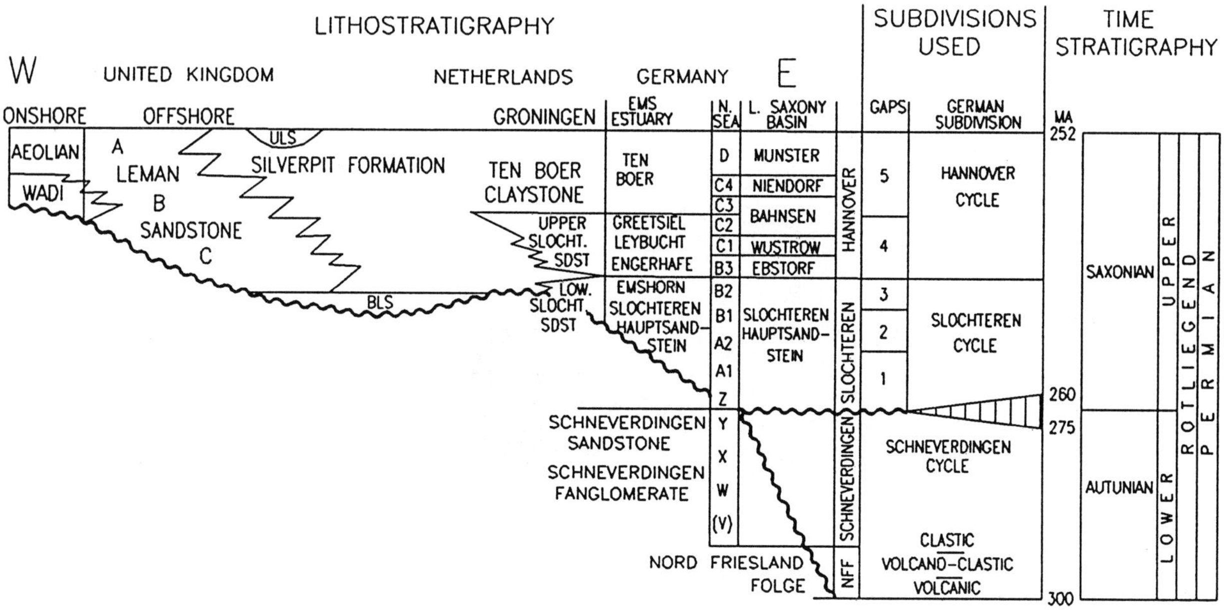

Fig. 3. Stratigraphic subdivisions of the Rotliegend in the UK, the Netherlands and Germany (ULS = upper Leman sandstone, BLS = lower Leman sandstone).

Tectonic framework

Late Carboniferous–Rotliegend tectonism in the Southern Permian Basin can be explained by the reactivation (by compression or extension) of lineaments inherited from earlier tectonic events such as the Variscan, Caledonian and Cadomian orogenies. Moreover, the Southern Permian Basin, during the Permian, echoes the tectonic events of great magnitude taking place far away and clearly records them.

The right megashear couple, which cuts across NW Europe during the Permian and corresponds to the final consolidation of Pangea through the emplacement of the Mauretanides and Uralids orogens, establishes the E – W orientation of the basin which was already present during Stephanian times (Ziegler 1982). The Northern and Southern Permian Basins are parallel to the Variscan Deformation Front and to the Mid-North Sea and Ringkøbing-Fyn Highs which separate them.

The NW – SE lineaments created by the emplacement of the Polish – German Caledonides and the closure of the Tornquist Ocean began to affect sedimentation during the Early Carboniferous (Encl. 3). Those lineaments are expressed by the Texel – IJsselmeer and Zandvoort – Maasbommel – Krefeld Ridges in the Netherlands and by the Dowsing Fault Zone in the British offshore. These elements were also active during the Rotliegend as they experienced reactivation through either compressional or extensional deformation.

The final pulse of the northward displacement of the Variscan Deformation Front was accompanied by a westward shift in the final consolidation of Gondwana – Laurussia (Ziegler 1982). This triggered inversion and erosion in the western part of the future Southern Permian Basin while the eastern sector continued to subside. This is recorded by thick Stephanian sediments in the eastern part of the basin (Plein 1993) while thin or no deposition occurred to the west (Encl. 4). This situation continued during Autunian times when Pangea started to break up and extension of greater magnitude began.

The structural grain of the British Caledonides with its NE – SW orientation is recorded mostly in the northwest of the basin where it was already apparent during Early Carboniferous times through the 'block and basin' (horst and graben) topography. Expression of this grain is seen in the Cleveland Hills area (England) which is perpendicular to the future Sole Pit Basin (Encl. 1).

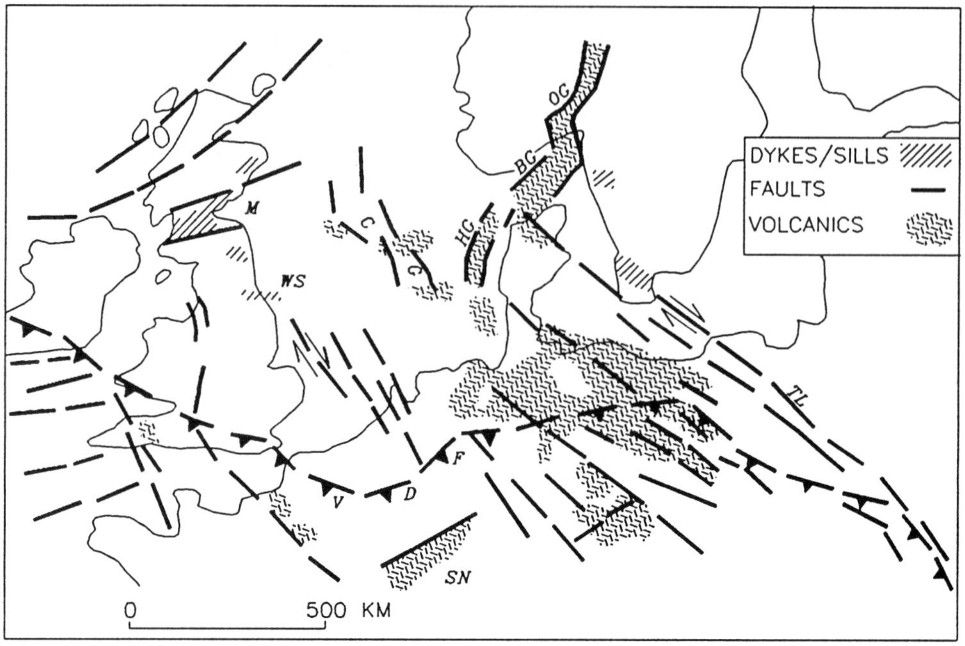

Fig. 4. Distribution of Early Rotliegend volcanics, sills and dykes in north-west Europe. M: Midland Valley, CG: Central Graben, HG: Horn Graben, BG: Bamble Graben, OG: Oslo Graben, WS: Whin Sill, VDF: Variscan Deformation Front, TL: Tornquist-Teisseyre Line, SN: Saar-Nahe Basin (after Glennie 1986).

The inversion occurring in the west was counterbalanced by volcanic activity taking place in the east in the German – Polish Caledonides region and elsewhere in N – S oriented grabens (Fig. 4). It occurred mostly during Stephanian and Autunian times (Plein 1993). Dykes and sills, such as the Whin Sill in the British offshore and onshore, are the only igneous manifestations present to the west and reflect the low degree of extension in this part of the basin at that time (Fitch & Miller 1967).

The greater and earlier extension which affected the east continued throughout the Saxonian as is clearly shown by the isopach map of the total Rotliegend (excluding volcanics; Encl. 5). This map depicts two distinct areas:

1. an area which includes the Netherlands and German offshore plus the northern part of Germany onshore with thicknesses ranging from 1500 to 6000 feet; and
2. an area which comprises the British offshore, the Netherlands onshore and the southern part of the area of the future Lower Saxony Basin with thicknesses less than 1500 feet. The orientations of the main Rotliegend sediment packages (NW – SE, NE – SW and N – S) within the overall east – west oriented basin express basin physiographies that existed during Dinantian through Stephanian times.

The V-shape of the area of the future Lower Saxony Basin was controlled by the Rhenish Massif and the Harz Mountains which are made of stacked Variscan thrust sheets that inhibited the pre-Carboniferous grain from influencing basin extension to the same extent as it did elsewhere in the basin (Gast 1991).

Wind direction and climate evolution

Dune orientation can be best determined using dipmeter data and core observations. Northeasterlies are the predominant winds in the Southern Permian Basin (Glennie 1990). An hypothesized decrease in the wind velocity from east to west in the basin could explain the distribution of the dunes the size of which appears to increase in a westward direction from small dunes to star dunes. However a change in their overall orientation is also noted from NE – SW in Germany and the Netherlands to predominantly NW – SE in the offshore UK. This change of orientation and magnitude of the dunes in the western part of the basin could be due to the combined effect of southeasterly winds with northeasterlies but could also be caused by topographic features like the Inde and the Winterton Pediments. These pediments are oriented NW – SE and could have slowed down the prevailing winds, created an entrapment for the blown sands, and facilitated the development of dunes of great height along the same orientation.

An overall increase in temperature in the area of the Southern Permian Basin from Carboniferous to Permian times was caused by the displacement of this region of Pangea towards latitudes similar to those of the present-day North-African and Middle-East deserts (Glennie 1990). The climate also became more arid on the northern side of the Variscan Mountains due to the reduced effect of their topography on precipitation, caused by ongoing erosion. The expansion of the Rotliegend lake through time was the result of a more efficient and greater connection with the proto-Atlantic to the north which resulted in the inundation of the entire Southern Permian Basin at the beginning of Zechstein times (Smith 1989).

Environments of deposition

Three types of environments (fluvial, aeolian and sabkha) are differentiated and mapped across the studied part of the basin. Net isopach maps were prepared for each of the three facies of the Rotliegend and demonstrate basin evolution.

The map of the fluvial deposits, composed of sandstones and conglomerates, shows the paucity of such sediments on the northern side of the basin (Encl. 6). This is interpreted as evidence of the lack of topographic relief in the Caledonian uplands which were the main sediment source during the Carboniferous. The main sources of sediment during Rotliegend times were the Variscan Mountains to the south, and to a much lesser degree the British portion of the London – Brabant Massif. Thick accumulations of fluvial deposits also surround the Texel – IJsselmeer Ridge in the area of the future Broad Fourteens basin and in the L quadrant in the Netherlands offshore. The relatively thin fluvial deposits onshore in Germany may indicate recycling of fluvial sediments into aeolian deposits during the dry phases of deposition.

The Rotliegend contains a variety of depositional facies recognized in modern aeolian environments. These sand-rich facies include: minor dunes, barkhan dunes, transverse dunes, star dunes, sheet sands and shore belt deposits accumulated along the southeastern shores of the lake (Encl. 7). Aeolian deposition was strongly controlled by the NW – SE structural grain in the British offshore. The future Broad Fourteens basin and the L quadrant are the site of the thickest pile of aeolian deposits in the Netherlands offshore and adjacent onshore. A region located between the Groningen High area and the Texel-IJsselmeer Ridge contains more than 400 net feet of aeolian deposits. The German offshore contains aeolian deposits accumulated in grabens during the early Schneverdingen extension phase, and the onshore shows the E – W orientation of a shore belt which includes aeolian and beach sediments deposited during later Rotliegend times (Drong et al. 1982).

Sabkha deposits are traditionally referred to as 'transitional' or 'waste zone' sediments due to their poor reservoir or seal qualities and are found mostly on the shores of the lake (Encl. 8). They consist mainly of siltstone and argillaceous sandstone which accumulated as adhesion ripples. Low aeolian dunes less than 2 feet high can also be found. They have poor reservoir qualities due to the low connectedness and porosity of the reservoir bodies, except when early chloritization inhibits later diagenesis (R. Gast, pers. comm. 1992).

Sedimentation history

Schneverdingen

The Schneverdingen sedimentation was restricted to the eastern part of the studied area where extension was initiated during the Autunian (Encl. 9). The Schneverdingen sediments onlap the area north of the Groningen High and the Oldenburg High, largely mimicking the western limit of the Stephanian Basin (Plein 1993).

The Southern Permian Basin was shallow and cut by N – S oriented grabens at its southern edge (Drong et al. 1982). The horsts are topped by volcanics, and the grabens contain fluvial and aeolian sediments. In the offshore, the horst and graben topography was covered by mid-Schneverdingen when rapidly onlapping lake deposits of shale, silt and halite became prevalent in the northern half of the basin. The German onshore depicts a N – S oriented horst and graben topography contained in the V-shaped basin bounded by the Rhenish Massif and the Harz Mountains.

In the aeolian and fluvial sandstones, relatively favorable reservoir characteristics with porosity values approaching 10% are not uncommon and are due to the coarseness and sorting of these deposits (Glennie 1990). The coarseness softened an otherwise damaging diagenesis due to deep burial. Aeolian sandstone deposits are particularly porous in the Söhlingen area where gas accumulations are present.

Early Slochteren

During the early Slochteren the basin experienced significant expansion to the west, but at the same time became more constrained in a N – S direction in its eastern part (Encl. 10). This reflected the ongoing stretching of the basin that arose due to the right megashear cutting across western Europe (Ziegler 1982). This tectonic influence is recorded in Germany by the deposition of a conglomerate at the base of the Slochteren. In the British area, a large pediment extended between the Sole Pit and the Silverpit subbasins, linking Quadrants K-49 and -53 to the hinterland. At the end of Slochteren sedimentaion, the basin was approximately half the size it eventually achieved by the end of Rotliegend deposition.

Fluvial deposits are scarce and limited to the area near the Dowsing Fault Zone and reflect the low topography of the source area. This situation remained unchanged during the entire Slochteren sedimentation until the southwestward migrating Dowsing Fault Zone reached the London – Brabant Massif.

In the Netherlands offshore and onshore the early Slochteren deposition was influenced by the unstable and topographically higher Variscan Mountains. The greater depositional gradient allowed deposition of larger amounts of fluvial sediments.

Middle Slochteren

During the middle Slochteren, the basin continued to expand westward, but the subsidence rate slowed down (Encl. 11). The Dowsing Fault Zone migrated further in a southwestward direction, but the Inde Pediment retained the same areal extent. Clastic input continued to dominate sedimentation in the Netherlands offshore while the onshore was still barely affected by extension and subsidence. In Germany, the offshore area was dominated by lake and evaporite deposition while onshore only a slight southward expansion of the basin occurred.

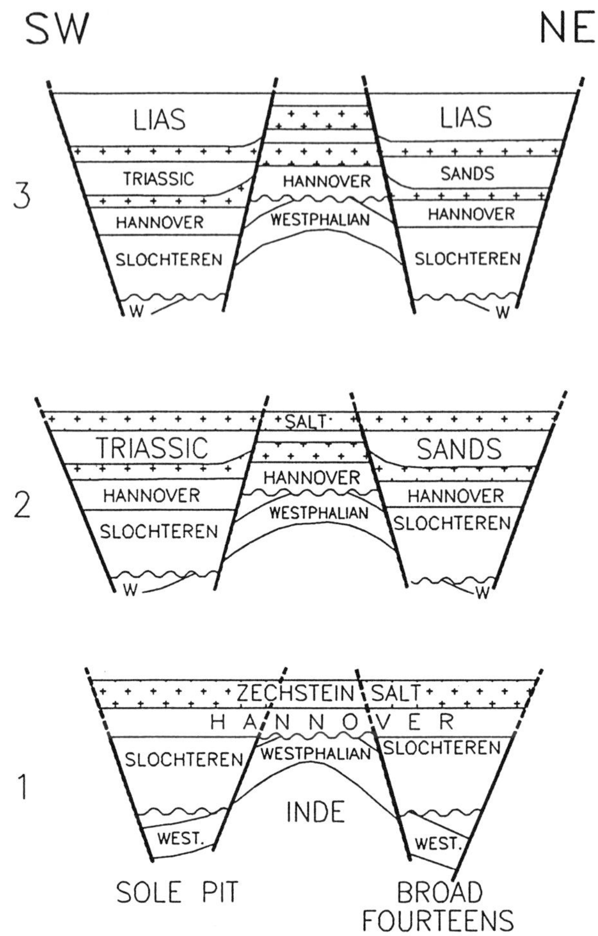

Fig. 5. Model explaining the effect on the isolation of the Rotliegend reservoirs in the Inde region ('High and dry') during the gas generation and migration process from Jurassic times onwards.

Late Slochteren

At the close of the Slochteren period, the basin continued to expand to the west but retained the same limits in the east of the studied area (Encl. 12). The East Midlands Pediment was reduced to a narrow salient and the Inde Pediment had almost entirely disappeared due to onlap.

The absence of deposition during most of the Slochteren period in the area of the Inde and Winterton Pediments created a barrier between the Cleaver Bank, Silverpit and Sole Pit subbasins. The Inde – Winterton positive topographic feature was enhanced during

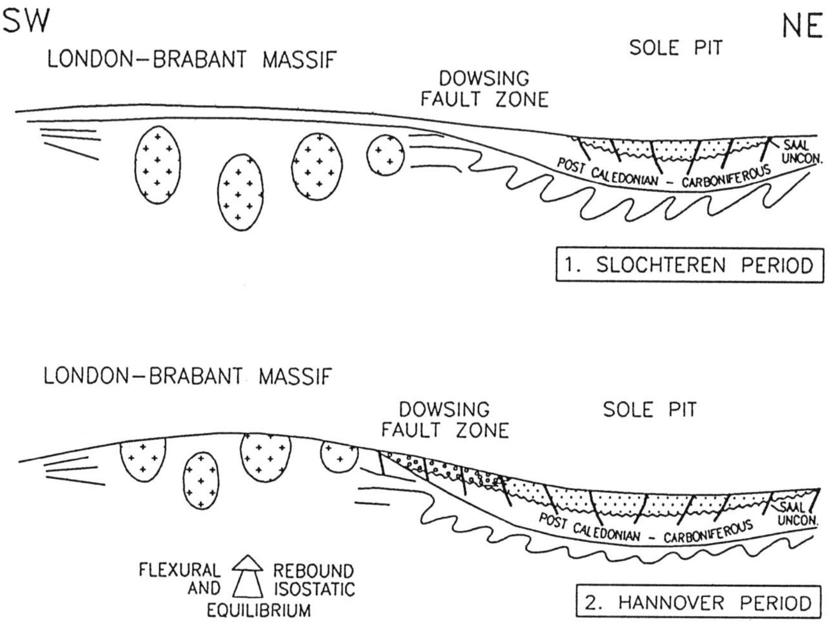

Fig. 6. Distribution mechanism of the fluvial (circles) and aeolian (dots) deposits in the offshore UK controlled by flexural rebound of the London – Brabant Massif during the Rotliegend back-stepping of the Dowsing Fault Zone. Crosses mark granitic plutons. SAAL. UNCON. = Saalian unconformity at base of Rotliegend.

the Mesozoic and Cenozoic times by salt movements, subsidence and inversion phases. This prevented easy hydrocarbon migration from the Sole Pit and the Broad Fourteens kitchen areas into the former Inde and Winterton areas from the Jurassic onwards and explains in part the underfilled character of the gas reservoirs, in the misnamed 'high and dry' area, deposited mostly during Hannover times over these former pediments (Fig. 5).

Continued uplift of the hinterland occurred in the Netherlands, as evidenced by a large amount of fluvial material in Quadrants K, L, P and Q, where deposits up to 400 feet are present. The Texel – IJsselmeer Ridge, however, was no longer a predominant source area for sediments, as is shown by a decrease in fluvial facies at its margins. The areas of the Ems Low and the Oldenburg and Hamwiede Highs remained as active sources of sediments; at the same time the lake was transgressing to the south, thus limiting the sabkha and the aeolian deposits to a narrow belt along its shores.

Early Hannover

During the Hannover period three major changes occurred:

1. the basin reached its maximum extent to the west and south;
2. the lake shore sand belt had its maximum development onshore Germany; and
3. the London – Brabant Massif became a more important area of clastic sediment supply due to the unroofing of granitic plutons (Danto & Megson 1990). The relatively thin-crusted London – Brabant Massif underwent extension due to the back-stepping Dowsing Fault Zone which resulted in reduced coherence which triggered its uplift. The platform achieved isostatic balance by unroofing of its granitic plutons which were then eroded and deposited as Hannover fluvial and aeolian deposits (Fig. 6).

The NW – SE structural grain, so apparent thus far in the basin fill history, became subdued to the west as the back-stepping Dowsing Fault Zone reached the more rigid and shallower lower Paleozoic of the London – Brabant Massif (Encl. 13). Despite the fluvial influxes at this end of the basin, aeolian deposition remained predominant in this area.

The Texel – IJsselmeer Ridge had totally disappeared. The first indication of a future connection with the Northern Permian Basin appeared west of the Schill

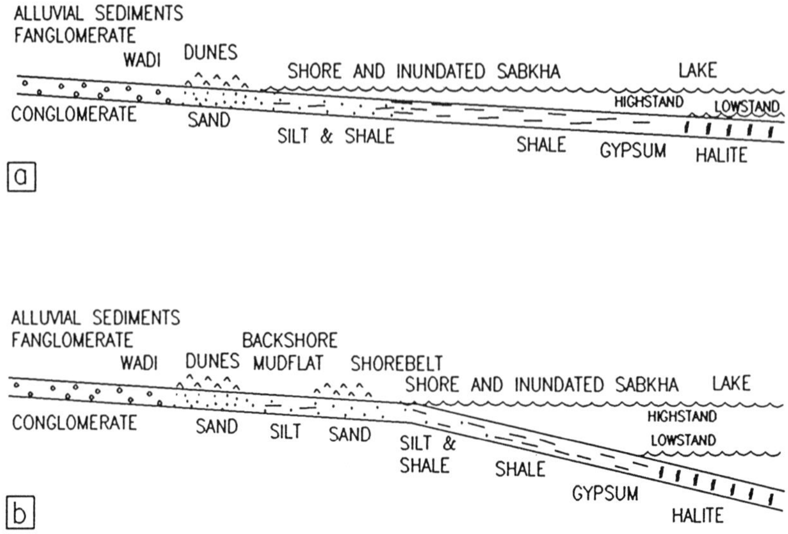

Fig. 7. Effect of shoreface gradient on the development of a shoreline sand belt. (a) sedimentary sequence across a shore with a gentle gradient, (b) sedimentary sequence across a shore with a steep gradient.

Grund High. The area of the Ems Low remained a predominant source region, with continuing deposition of coarse clastics into the Groningen area.

Beach sands were deposited during this time along the southern shores of the lake in Germany (Gast 1991). These southward migrating stacked sandstones are distributed in an E – W oriented belt fashion. They interfinger lakeward with playa flat deposits and landward they merge into sabkha deposits. They are best developed where the pediment is topographically high and the shoreface has a steep gradient (Fig. 7). These conditions were present particularly north of the Oldenburg and Hamwiede Highs.

Late Hannover

Finally, during late Hannover, the lake attained its maximum development and was linked with the Northern Permian Basin (Smith & Taylor 1989; Encl. 14). Alluvial fans activated during the preceding period continued to shed sediments into the British sector of the North Sea in Quadrants 47 and 48, largely sourced by the East Anglian Granite of the London – Brabant Massif. The East Midland Spur was entirely peneplained during late Hannover with the exception of an isolated butte, north of the Market Weighton Granite. During late Hannover deposition, the subsidence rate was high and dune fields such as the aeolian Durham Yellow Sands in onshore England were predominant (Yardley 1984). Halite precipitation ended in the British sector of the North Sea and a veneer of sheet sands was deposited during the dry phase of this last cycle.

The alluvial fans in the Netherlands were still active and continued to reach the lake across the sabkha. The connection with the Northern Permian Basin was now well established during late Hannover and lake as well as fluvial, aeolian and sabkha deposits are recorded in Quadrants A and B in the German offshore. Halite deposition decreased and was entirely limited to the German offshore while the area of the Ems Low continued to act as a source of sediments. In Germany, the southward expansion of the basin reached its maximum. The Hamwiede High was no more than a small isolated flat-topped hill and the expression of the Oldenburg High was limited to a narrow ridge. The sand belt in this area was replaced gradually by lake and sabkha deposits during this period.

Conclusions

The stratigraphic subdivision introduced here has considerably improved the understanding of the paleogeographic evolution of the Southern Permian Basin during Rotliegend time. The main controlling factors that were responsible for the sedimentologic differences across the basin are the evolution of the tectonic elements, the increasingly arid climate and the westward-prograding extension of the basin.

Late Variscan and older tectonic lineaments are important controls of Rotliegend deposition. Heterogeneity of the hinterland surrounding the basin explains the differences in facies distribution:

(i) the Caledonian uplands were inactive;

(ii) the London – Brabant Massif was peneplained; its rigid character and thin crust prevented uplift and erosion during most of Rotliegend times; and

(iii) the Variscan hinterland was unstable, due to its thick crust, and prone to rise as soon as extension started. The Variscan Mountains offered a constantly rejuvenated source of sediments and this explains the predominance of fluvial over aeolian deposits along the former Variscan Deformation Front.

Acknowledgements

During the preparation of this compilation, the author was able to use well files and in-house reports prepared by Esso Exploration and Production UK, Shell Exploration and Production UK, Nederlandse Aardolie Maatschappij, BEB Erdgas und Erdöl, and GAPS Geological Consultants. Their contribution to this work and permission to publish it are gratefully acknowledged.

Special thanks are extended to GAPS Geological Consultants who developed the stratigraphic subdivision presented here and allowed its publication.

Finally, I am indebted to Exxon Exploration Company for having given me the opportunity to present this work at the AAPG conference in The Hague in October 1993 and letting me release it for publication. Special thanks go to Q.L. Ballard, P. Bourland, D.A. Leary and T.E. McReynolds for helping me to prepare the text and the illustrations.

Enclosures (in pocket)

1. Rotliegend gas fields and physiographic features.
2. Stratigraphic profiles of the Rotliegend.
3. Rotliegend tectonic elements map.
4. Pre-Permian subcrop map.
5. Total Rotliegend sediment isopach map.
6. Total Rotliegend fluvial deposit isopach map.
7. Total Rotliegend aeolian and shore belt sands isopach map.
8. Total Rotliegend sabkha deposit isopach map.
9. Schneverdingen facies distribution map.
10. Early Slochteren facies distribution map.
11. Middle Slochteren facies distribution map.
12. Late Slochteren facies distribution map.
13. Early Hannover facies distribution map.
14. Late Hannover facies distribution map.

References

All Exxon, Shell, NAM, BEB and GAPS proprietary reports used in this study are omitted from the following list of references. Sources listed which are not referred to in the text have been consulted for the preparation of the Enclosures.

Arthur, T.J., D. Pilling, D. Bush & L. Macchi 1986 The Leman sandstone formation in UK Block 49/28 – Sedimentation, diagenesis and burial history. In: Brooks, J., J.C. Goff & B. van Hoorn (eds) Habitat of Paleozoic gas in NW Europe – Geol. Soc. London Spec. Publ. 23: 251–266

Bender, F. & H.A. Hedemann 1983 Zwanzig jahre erfolgreiche Rotliegend-Exploration in Nordwestdeutschland – weitere Aussichten auch im Präperm – Erdöl-Erdgas Zeitschr. 99: 39–49

Brouwer, G.C. 1972 The Rotliegend in the Netherlands. In: Falke, H. (ed) Rotliegend essays on European Lower Permian – Brill, Leiden: 34–42

Clemmensen, L.B. 1987 Complex star dunes and associated aeolian bedforms, Hopeman Sandstone (Permo-Triassic), Moray Firth Basin, Scotland. In: Frostick, L.E. & I. Reid (eds) Desert sediments – Ancient and Modern – Geol. Soc. London Spec. Publ. 35: 213–231

Clemmensen, L.B., H. Oslen & R.C. Blakely 1989 Erg Margin deposits in the Lower Jurassic Moenave Formation and Wingate Sandstone, Southern Utah – Geol. Soc. Am. Bull. 101: 759–773

Danto, J.A. & J.B. Megson 1990 A buried granite beneath the East Midland shelf of the Southern North Sea Basin – J. Geol. Soc. London 147: 133–140

Drong, J.J., E. Plein, D. Sanneman, M.A. Schuepbach & J. Zimdars 1982 Der Schneverdingen-Sandstein des Rotliegenden – eine äolische Sedimentfüllung alter Grabenstrukturen – Z. dt. geol. Ges. 133: 699–725

Fitch, F.J. & J.A. Miller 1967 The age of the Whin sill – Geological J. 5: 233–249

Gast, R.E. 1991 The Perennial Rotliegend saline lake in NW Germany – Geol. Jb. 119: 25–59

Glennie, K.W. 1970 Desert sedimentary environments. Developments in sedimentology 14 – Elsevier, Amsterdam: 222 pp

Glennie, K.W. 1972 Permian Rotliegendes of Northwest Europe interpreted in the light of modern desert sedimentation studies – Am. Ass. Petroleum Geol. Bull. 56: 1048–1071

Glennie, K.W. 1986 Development of NW Europe's Southern Permian gas Basin. In: Brooks, J., J.C. Goff & B. van Hoorn (eds) Habitat of Paleozoic gas in NW Europe – Geol. Soc. London Spec. Publ. 23: 3–22

Glennie, K.W. 1990 Lower Permian – Rotliegend. In: Glennie, K.W. (ed.) Introduction to the petroleum geology of the North Sea – Blackwell Scientific Publ., London: 120–152

Glennie, K.W. & P.L.E. Boegner 1981 Sole Pit Inversion Tectonics. In: Illing, L.V. & G.D. Hobson (eds) Petroleum geology of the continental shelf of NW Europe – Heyden, London: 110–120

Glennie, K.W. & A.T. Buller 1983 The Permian Weissliegend of NW Europe; the partial deformation of aeolian dune sands caused by the Zechstein transgression – Sediment. Geol. 35: 43–81

Glennie, K.W., G.C. Mudd & P.J.C. Nagtegaal 1978 Depositional environment and diagenesis of Permian Rotliegendes sandstones in Leman Bank and Sole Pit areas of the UK Southern North Sea – J. Geol. Soc. Lond. 135: 23–34

Glennie, K.W. & D.M.J. Provan 1990 Lower Permian Rotliegend reservoir of the Southern North Sea gas province. In: Brooks, J. (ed) Classic Petroleum Provinces – Geol. Soc. London Spec. Publ. 50: 399–416

Gray, I. 1975 Viking Gas Field. In: Woodland, A.W. (ed) Petroleum and the continental shelf of Northwest Europe – Applied Science Publ. Barking, Essex: 241–248

Hancock, N.J. 1978 Possible cause of Rotliegend sandstone diagenesis in Northern West Germany – J. Geol. Soc. Lond. 135: 33–40

Kettel, D. 1983 The East Groningen massif – detection of an intrusive body by means of coalification – Geol. Mijnbouw 62: 203–210

Kocurek, G. 1981 The significance of interdune deposits and bounding surfaces in aeolian dune sands – Sedimentology 28: 753–780

Lee, M., J.L. Aronson & S.M. Savin 1989 Timing and conditions of Permian Rotliegend Sandstone Diagenesis, Southern North Sea: K/Ar and Oxygen isotopic data – Am. Ass. Petroleum Geol. Bull. 73: 195–215

Lutz, M., J.P.H. Kaasschieter & D.H. Van Wijhe 1975 Geological factors controlling Rotliegend gas accumulation in the Mid-European Basin. In: Proc. 9th World Petroleum Congr., Tokyo, 2: 93–103

Magraw, D. 1978 New boreholes in the Permian beds of Northumberland and Durham – Proc. Yorks. Geol. Soc. 42: 157–183

Marie, J.P.H. 1975 Rotliegendes stratigraphy and diagenesis. In: Woodland, A.W. (ed) Petroleum and geology of the continental shelf of NW Europe – Applied Science Publ. Barking, Essex: 205–211

Nagtegaal, P.J.C. 1979 Relationships of facies and reservoir quality in Rotliegendes desert sandstones, Southern North Sea region – J. Petroleum Geol. 2: 145–158

Oele, J.A., A.C.P.J. Hol & J. Tiemans 1981 Some Rotliegendes gas fields of the K and L blocks, Netherlands offshore (1968–1978) – a case history. In: Illing, L.V. & G.D.Hobson (eds) Petroleum geology of the continental shelf of NW Europe – Heyden, London: 289–300

Plein, E. 1978 Rotliegend-Ablagerungen im Norddeutschen Becken – Z. dt. geol. Ges. 129: 71–97

Plein, E. 1993 Bemerkungen zum Ablauf der paläogeographischen Entwicklung im Stefan und Rotliegend des Norddeutschen Beckens – Geol. Jb. 99–116

Plumhoff, F. 1966 Marines Ober-Rotliegendes (Perm) im Zentrum des Nordwestdeutschen Rotliegend-Beckens – Erdöl Kohle 19: 713–720

Randall, B.A.O. 1980 The great Whin sill and its associated dyke suite. In: Robson, D.A. (ed) The geology of NE England – The Natural Society of Northumbria Spec. Publ.: 67–75

Smith, D.B. 1989 The Late Permian paleogeography of North-East England – Proc. Yorks. Geol. Soc. 47: 285–312

Smith, D.B. & J.C.M. Taylor 1989 A 'North West Passage' to the Southern Zechstein Basin of the UK North Sea – Proc. Yorks. Geol. Soc. 47: 313–320

Sørensen, S. & B.B. Martinsen 1987 A paleogeographic reconstruction of the Rotliegendes deposits in the Northeastern Permian Basin. In: Brooks, J. & K.W. Glennie (eds) Petroleum Geology of NW Europe – Graham & Trotman, London: 497–508

Stäuble, A.J. & G. Milius 1970 Geology of the Groningen Gas Field, Netherlands – Am. Ass. Petroleum Geol. Mem. 14: 359–369

Van Adrichem Boogaert, H.A. 1974 Outline of the Rotliegend (Lower Permian) in the Netherlands. In: Falke, H. (ed) The continental Permian in Central, West and South Europe – Reidel, Dordrecht: 21–37

Van Lith, J.G.J. 1983 Gas fields of the Bergen concession, the Netherlands. – Geol. Mijnbouw 62: 63–74

Van Veen, F.R. 1975 Geology of the Leman gas field. In: Woodland, A.W. (ed) Petroleum and geology of the continental shelf of NW Europe – Applied Science Publ., Barking, Essex: 223–231

Van Wijhe, D.H. 1987 Structural evolution of inverted basins in the Dutch offshore – Tectonophysics 137: 171–219

Van Wijhe, D.H., M. Lutz & J.P.H. Kaasschieter 1980 The Rotliegend in the Netherlands and its gas accumulations – Geol. Mijnbouw 59: 3–24

Walker, M. & W.G. Cooper 1987 The structural and stratigraphic evolution of the Northeast margin of the Sole Pit Basin. In: Brooks, J. & K.W. Glennie (eds) Petroleum Geology of NW Europe, Graham & Trotman, London: 262–275

Yardley, M.W. 1984 Cross-bedding in the Permian Yellow sands of County Durham – Proc. Yorks. Geol. Soc. 45: 11–18

Ziegler, P.A. 1982 Geological atlas of Western and Central Europe – Elsevier, Amsterdam: 130 pp

Rondeel et al. (eds), Geology of gas and oil under the Netherlands, 57–78, 1996.

Development of the Permo-Triassic succession in the basin fringe area, southern Netherlands

M.C. Geluk, A. Plomp & Th.H.M. van Doorn
Geological Survey of the Netherlands, Postbus 157, 2000 AD Haarlem, the Netherlands

Key words: hydrocarbons, paleogeography, reservoir rocks, Southern Permian Basin

Abstract

During the Permian, the fringe area of the Southern Permian Basin formed a platform attached to the London-Brabant Massif. During the Triassic the platform was transformed into basins with differentiated subsidence. Upper Rotliegend sandstones have a much wider distribution than previously recognized. The Rotliegend deposits are capped by sandstones, claystones and carbonates of the Late Permian Zechstein Group. An important feeder system of clastics from the London-Brabant Massif was active in the present-day offshore area of the Netherlands. In the course of time, the location of major sandstone deposition shifted westward. Deposition of the Triassic Buntsandstein was governed by the development of a large feeder system which transported clastics from the Vosges in NE France northward through the Roer Valley Graben and West Netherlands Basin into the Off Holland Low. Influx of coarse clastics ceased during the Middle Triassic. The feeder system was responsible for the deposition of the gas-bearing sheet sandstones of the Volpriehausen, Detfurth and Hardegsen Formations. The importance of the Triassic play is demonstrated by the over 100×10^9 m^3 Gas Initially In Place (GIIP) that were discovered since 1968. Subtle unconformities characterize the base of the Volpriehausen and especially the Detfurth Formation. These formations are separated by a regional unconformity from the overlying sandstones and claystones of the Solling Formation, and the claystones, evaporites and sandstones of the Röt Formation. The Muschelkalk Formation consists of shallow marine carbonates and claystones which are interbedded with evaporites. Rock salt was deposited in the Central Netherlands Basin and locally in other areas. Extension during the Late Triassic led to uplift and erosion. After the extension ceased, the upper part of the Keuper Formation was deposited unconformably upon sediments of the Muschelkalk or Röt Formations.

Introduction

The aim of this paper is to present the regional Permo-Triassic lithostratigraphic and paleogeographic framework of the basin fringe area based on a set of isopach maps, highlighting the depositional trends. We will focus on clastic potential reservoir units. The regional development of these units, which are of importance for hydrocarbon exploration, will also be illustrated by means of log correlations.

Geological studies of the Permo-Triassic succession in the southern Netherlands were initiated as part of the recent revision of the stratigraphic nomenclature of NAM & RGD (1980). Industry representatives (Amoco, Nederlandse Aardolie Maatschappij (NAM) and Elf Petroland), together with the staff of the Geological Survey of the Netherlands (RGD), revised the Permo-Triassic part of this nomenclature. A separate nomenclature had previously been designed for the basin fringe deposits as, owing to their overall sandy development, they fell completely outside the basinal lithostratigraphic framework (NAM & RGD 1980). Several studies carried out since have indicated that these basin fringe deposits can be fitted into the already existing basinal stratigraphy. This has been further elaborated by the industry and the RGD (Van Adrichem Boogaert & Kouwe 1993–1995, Sections D, E). The new lithostratigraphy meets the demands of the industry, since an enhanced mappability of reservoirs and seals is thus obtained.

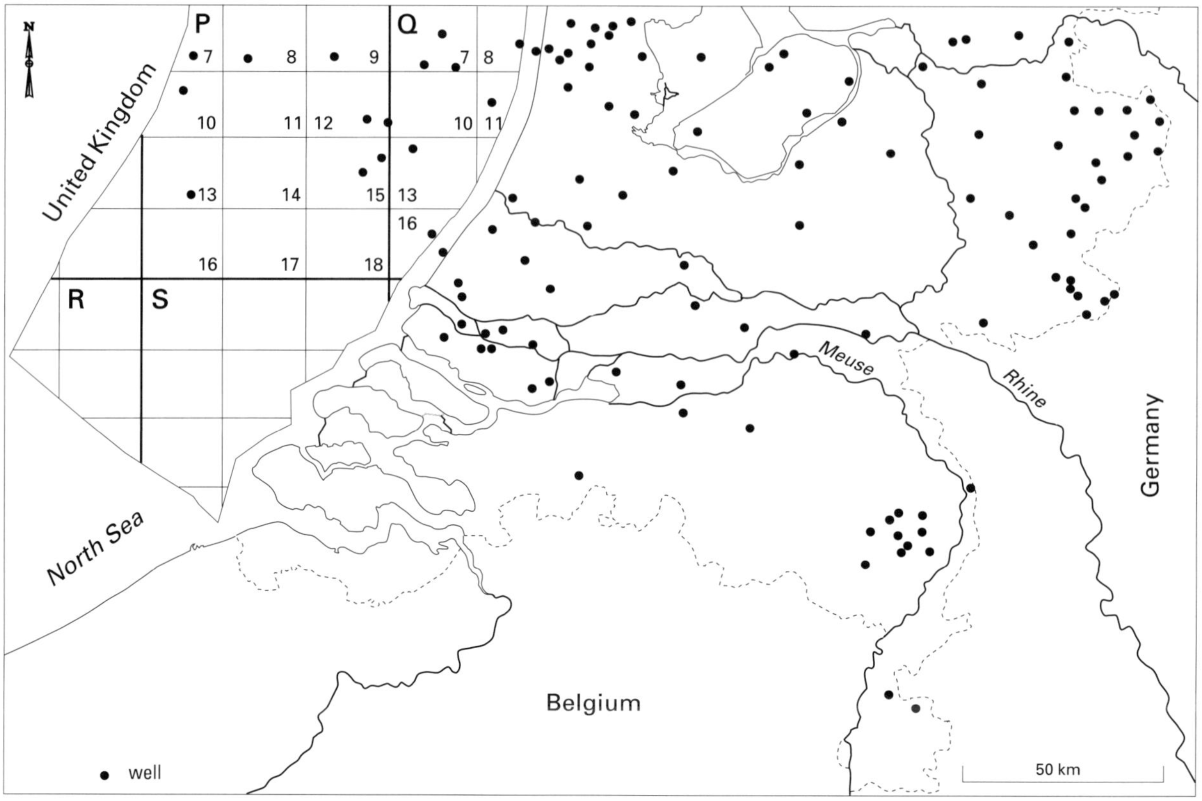

Fig. 1. Location of wells used in this study.

The studied area covers the southernmost part of the offshore P and Q quadrants and the central and southern Netherlands onshore area (Fig. 1). The study is based on log correlation of released offshore well data, older onshore wells, core studies and literature. Offshore well data are released in the Netherlands ten years after completion; onshore well data are not released. There is an agreement between RGD and onshore operators allowing publication of older exploration and production data. The data-base of this study is presented in Fig. 1 and the Appendix.

Previous work

Publications on the development of the Permo-Triassic succession in the southern Netherlands are scarce. A number of papers present regional information, for example Haanstra (1963), Heybroek (1974), Van Adrichem Boogaert (1976), Van Adrichem Boogaert & Burgers (1983), Van Staalduinen et al. (1979), Van Waterschoot van der Gracht (1909, 1918), Van Wijhe et al. (1980), and Van Wijhe (1987). Detailed information concerning specific areas in the Netherlands or adjacent parts of neighbouring countries is supplied by Ames & Farfan (1995; this volume), Boigk (1956), Brennand (1975), Brueren (1959), Demyttenaere (1989), Haanstra (1963), Hilden (1988), Muchez et al. (1992), RGD (1993), Roos & Smits (1983), Van Lith (1983), Winstanley (1993), and Wolburg (1956, 1957, 1961, 1968, 1969).

Structural framework

The subsurface of the southern Netherlands can be divided into three NW – SE striking Jurassic – Early Cretaceous basins, namely the West Netherlands Basin, Central Netherlands Basin and Roer Valley Graben (Fig. 2). The development of these basins dates back to Permian or Triassic times. The basins are bordered to the south by the London – Brabant Massif and to the southeast by the Rhenish Massif. These massifs are part of the northern rim of the Variscan

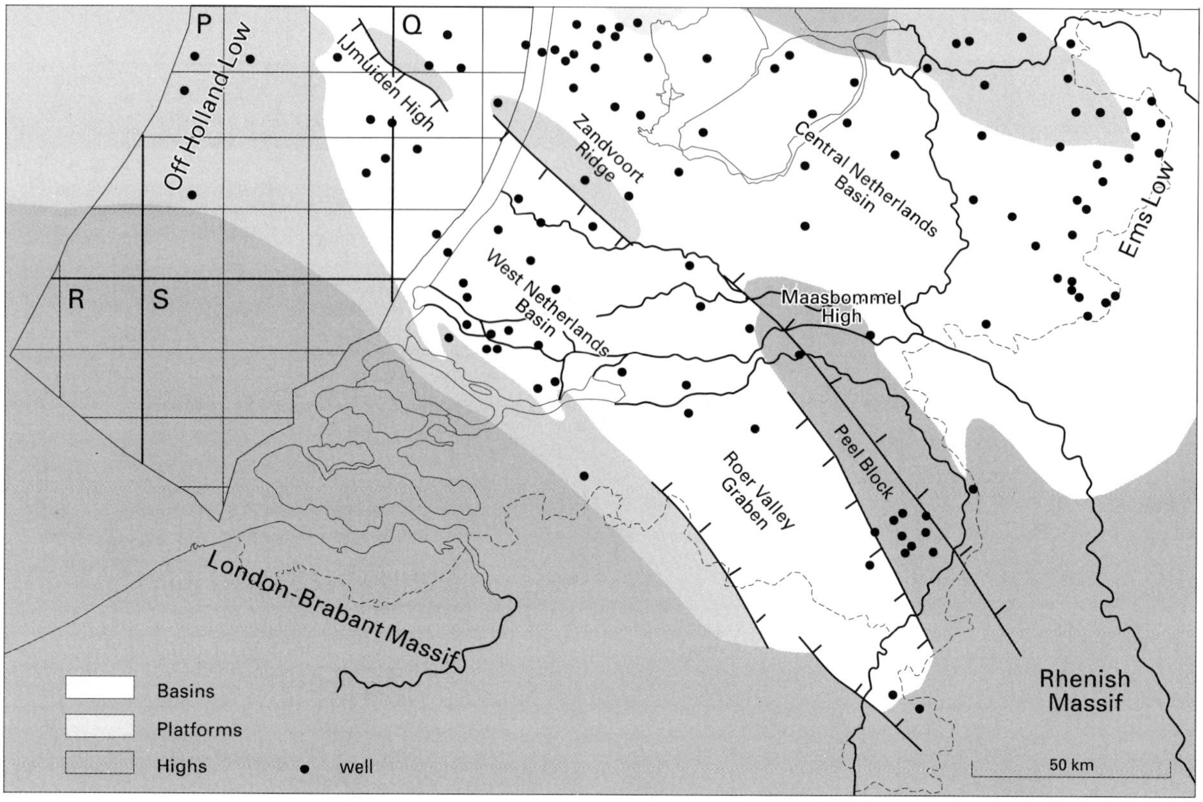

Fig. 2. Structural framework of the study area during the Late Jurassic to Early Cretaceous. Two main Permo-Triassic structural elements, the Off Holland Low and the Ems Low are also indicated (adapted from Van Adrichem Boogaert & Kouwe 1993–1995).

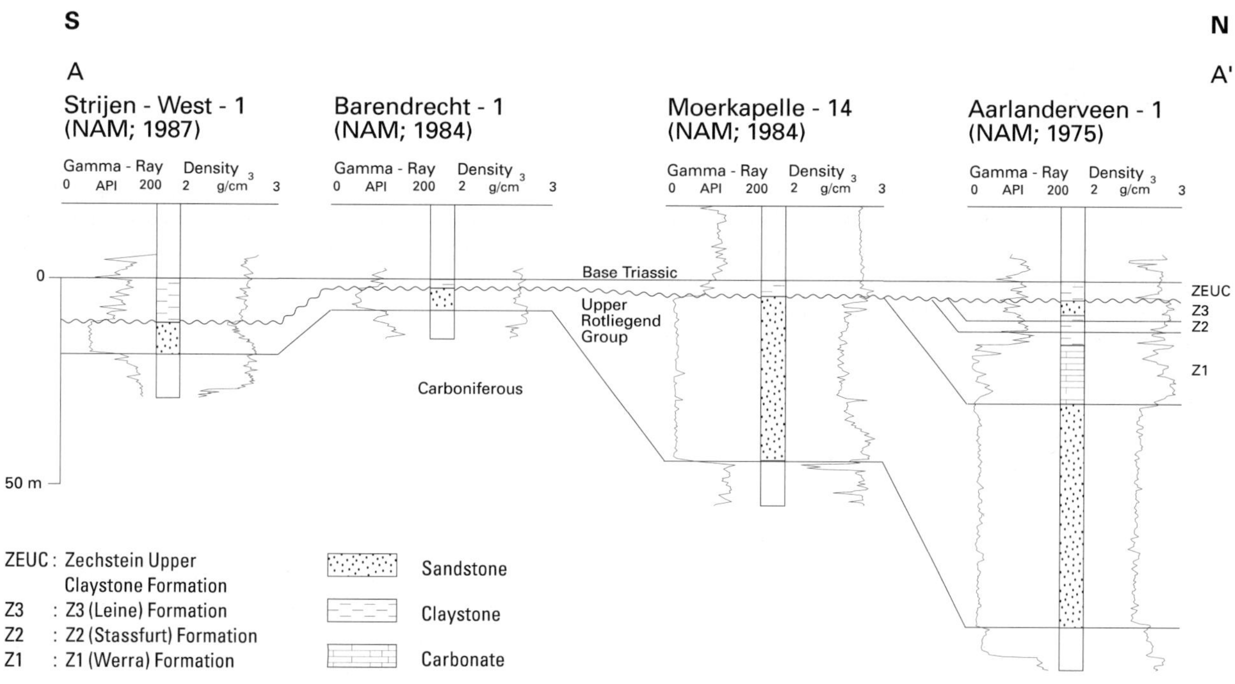

Fig. 3. Lithostratigraphic correlation of the Permian sequence in the West Netherlands Basin. For location see Fig. 4.

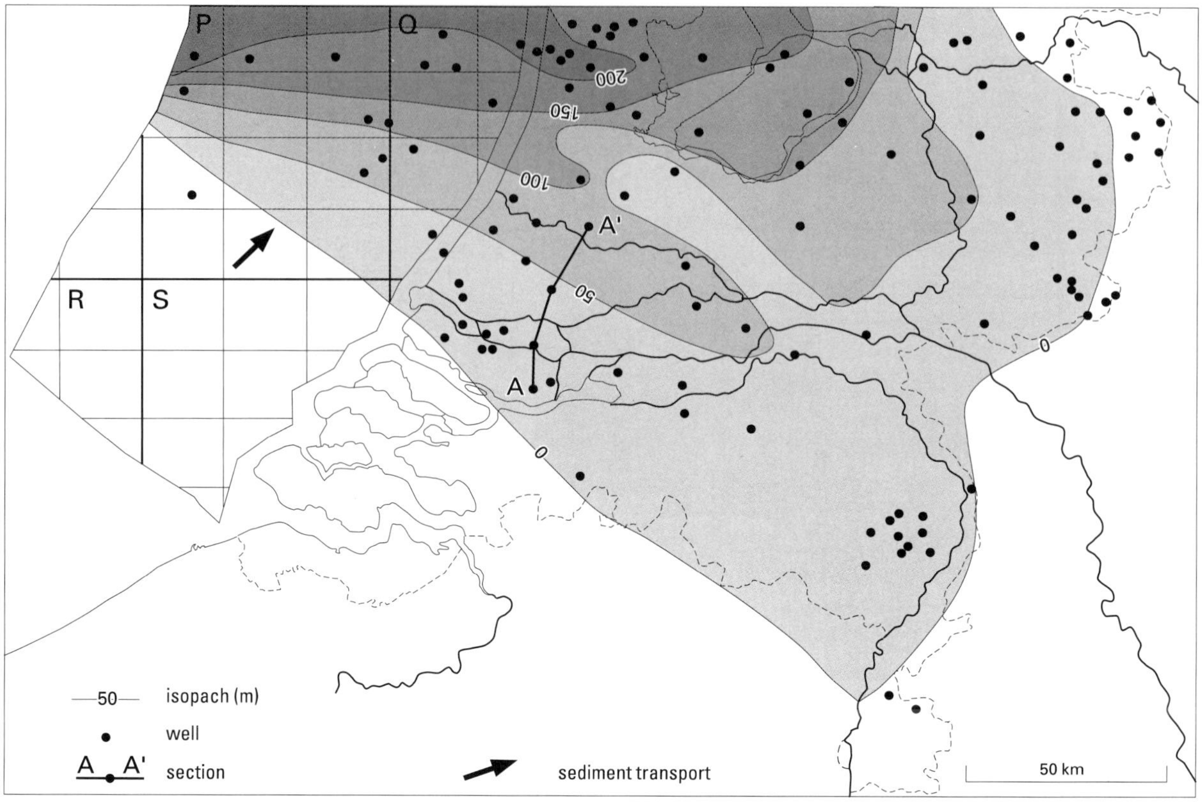

Fig. 4. Isopach map of the Upper Rotliegend Group. Onshore, the main subsidence occurred in the incipient West Netherlands and Central Netherlands Basins; the Maasbommel High and Zandvoort Ridge acted as a single high. Correlation section A-A′ is presented in Fig. 3.

orogenic belt. The West Netherlands Basin is separated from the Central Netherlands Basin by a fault-bounded ridge, the Mid Netherlands Fault Zone (Van Adrichem Boogaert & Kouwe 1993–1995). This fault zone comprises from SE to NW the Peel Block, Maasbommel High, Zandvoort Ridge and IJmuiden High. The Mesozoic geological history of the basins has been summarized by Van Adrichem Boogaert & Kouwe (1993–1995), Van Staalduinen et al. (1979), Van Wijhe (1987) and Ziegler (1990).

Geological history

The Permo-Triassic succession is subdivided into four major lithostratigraphic groups: the Permian Upper Rotliegend and Zechstein Groups and the Triassic Lower and Upper Germanic Trias Groups (Van Adrichem Boogaert & Kouwe 1993–1995). The boundary between the Zechstein Group and the Lower Germanic Trias Group is well-recognizable on logs in the entire area. While realizing that the chronostratigraphic Permo-Triassic boundary is situated within the Lower Buntsandstein Formation (Van der Zwan & Spaak 1992; Van Adrichem Boogaert & Kouwe 1993–1995), we will use the lithostratigraphic boundary as the boundary between the Permian and Triassic.

Permian

The Permian succession is subdivided into the Upper Rotliegend Group and the Zechstein Group. Permian deposits were previously recognized only in the northern part of the studied area (see Van Wijhe et al. 1980: fig. 9, Ziegler 1990). Recent work, however, has shown that Permian rocks occur in the southern part as well (A. Speksnijder, pers. comm.). Only in a small area in the southeastern Netherlands the Permian was found absent below the Triassic (wells Raath-1 and Limbricht-1).

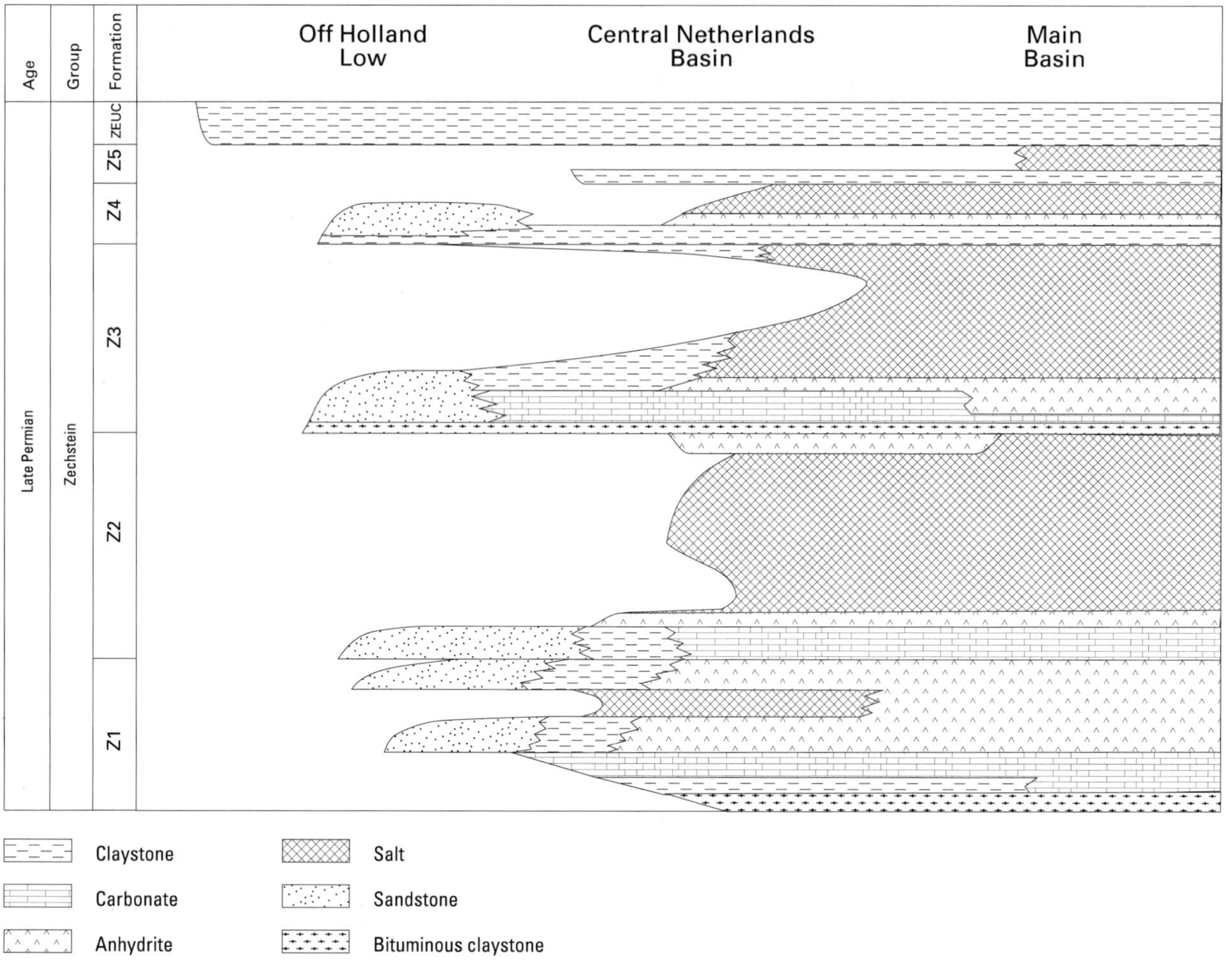

Fig. 5. Stratigraphic diagram of the Zechstein Group (Late Permian) from basin fringe to basin centre. RO: Upper Rotliegend Group; Z1: Z1 (Werra) Formation; Z2: Z2 (Staßfurt) Formation; Z3: Z3 (Leine) Formation; Z4: Z4 (Aller) Formation; Z5: Z5 (Ohre) Formation; ZEUC: Zechstein Upper Claystone Formation; RB: Lower Germanic Trias Group.

Detailed log correlation in the West Netherlands Basin has revealed the occurrence below the Triassic of a sandstone-claystone succession which was originally assigned to the Carboniferous (Fig. 3: wells Strijen-West-1 and Barendrecht-1). Palynological evidence has indicated a Permian age for this succession in the offshore NAM well Q16–7 (Van Adrichem Boogaert & Kouwe 1993–1995). A further argument is the good well-to-well correlation of this succession, which would not be the case for the Carboniferous that is truncated by a major unconformity while its paralic deposits are, moreover, laterally highly variable in facies.

Upper Rotliegend Group (Early?-Late Permian)

The distribution of the Upper Rotliegend Group is shown in Fig. 4. The group consists of porous, light-coloured sandstones and conglomerates (Slochteren Formation) which frequently coarsen upward. Deposition was governed by subsidence of the incipient West Netherlands and Central Netherlands Basins. The fluvial clastic influx into the basin mainly occurred through the Off Holland Low (see Fig. 2). The sediments were partly reworked into aeolian deposits. A high occupied the eastern Netherlands, and the area of sedimentation was delineated by the London – Bra-

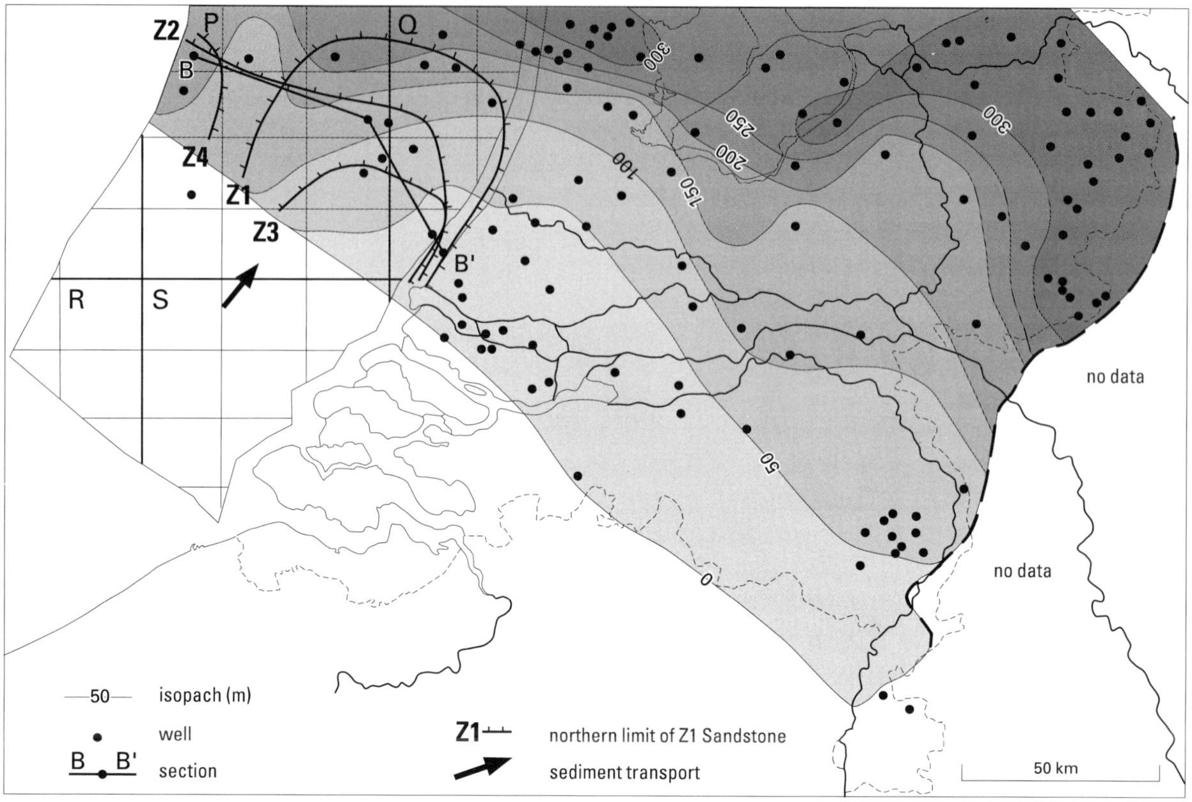

Fig. 6. Isopach map of the Zechstein Group. Note the strong subsidence in the Ems Low. The distribution limits of the Z1, Z2, Z3 and Z4 Sandstones in the offshore are shown. Correlation section B-B′ is presented in Fig. 7.

bant Massif in the south and the Rhenish Massif in the southeast. Two wells on the Zandvoort Ridge, Weesp-1 and Waverveen-1, penetrated a thin Upper Rotliegend section, less than 10 m, representing only the youngest part; this could have been caused either by synsedimentary movements of the ridge or by faulting. Seismic data did not provide a solution to this uncertainty. We favour synsedimentary movements in view of similar thinning of part of the overlying Triassic sequence in both wells.

The age of the deposits is still a matter of debate; in Germany paleomagnetic data indicate a Late Permian age (Menning et al. 1988). No data are available from the Netherlands, but based on the conformable contact between the Upper Rotliegend and the youngest deposits of the overlying Zechstein Group (Fig. 3: wells Barendrecht-1 and Moerkapelle-14), there is presumed to be no major hiatus between these groups. A Late Permian age for at least part of the Upper Rotliegend Group is therefore likely.

Zechstein Group (Late Permian)

The Zechstein Group is composed of five evaporite cycles (Z1–Z5) and an overlying claystone unit (Fig. 5). A typical evaporite cycle is composed of claystone – carbonate – anhydrite – rock-salt – anhydrite. The first three cycles were deposited in a marine environment, the higher cycles in a continental, playa-type setting. In the offshore and southern onshore areas, clastics occur, but the cyclicity can still be demonstrated in these deposits. Especially the basal claystones of the cycles are excellent correlation markers. The distribution of potential reservoir sandstones is limited to the offshore area where they occur in the Z1 to Z4 Formations (Fig. 6).

During deposition of the Zechstein Group, pronounced subsidence of the Central Netherlands Basin occurred, which connected the Ems Low and the Off Holland Low (Fig. 6). Carbonates, anhydrites and rock salt were deposited in this basin; at its southern flank fine-grained clastics intercalate with evaporites. The

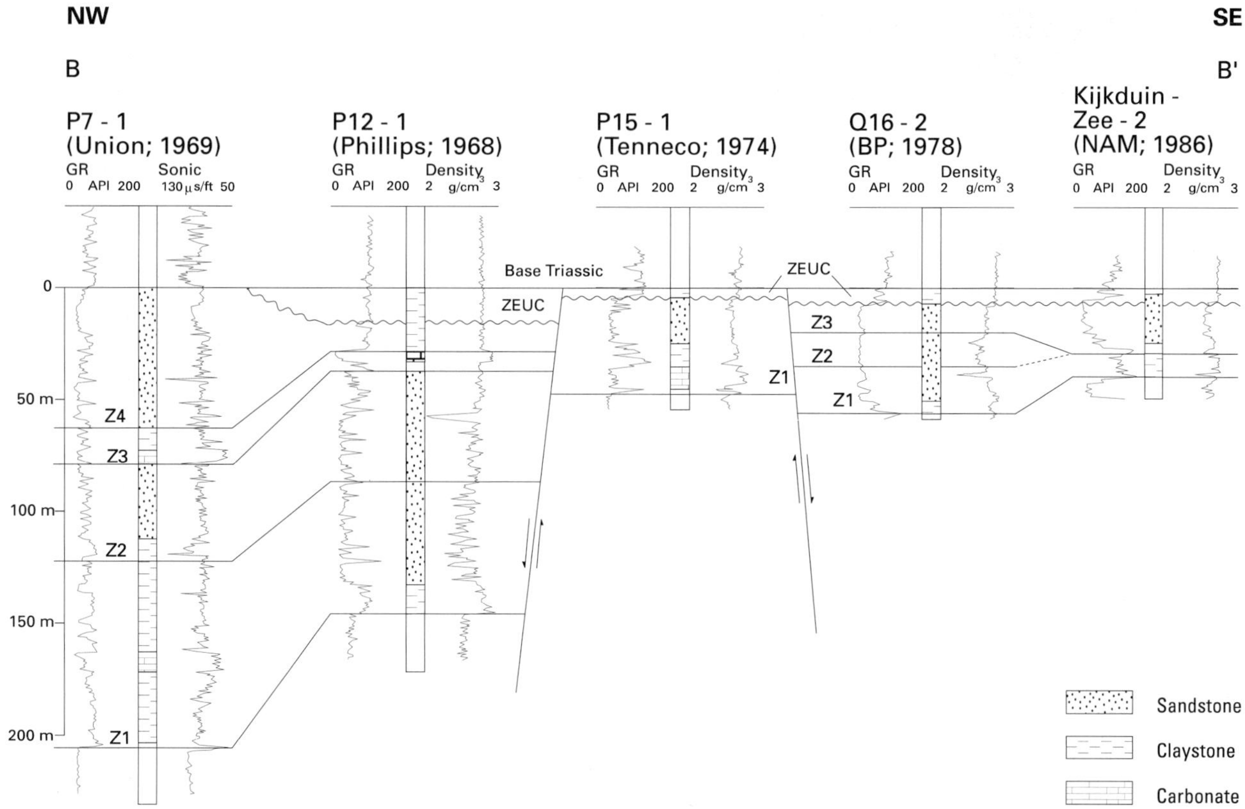

Fig. 7. Lithostratigraphic correlation of the Zechstein Group between P7–1 and Kijkduin-Zee-2. For location see Fig. 6.

group is less than 50 m thick in the southern onshore (an area here referred to as the Southern Netherlands platform), attaining a thickness of over 500 m in the Central Netherlands Basin. The subsiding area extended over previous highs, for example in the eastern Netherlands. The influx of coarser clastics was limited to the offshore. In the course of time, sandstone deposition shifted westward (Fig. 6). Extension led to the development of half-grabens in the area; it ceased already during deposition of the Zechstein Group. The Z1 and Z2 Formations still show considerable influence by synsedimentary faulting, but by the time the Z3 Formation was deposited faulting had ceased and, as a consequence, lateral facies variations decreased. At the end of the Zechstein, tectonic movements occurred again.

The Z1 and Z4 Sandstones are the lateral equivalent of anhydrites and rock salt, and represent in sequence stratigraphic terms lowstand deposits (Tucker 1991). The Z2 and Z3 Sandstones grade laterally into carbonates and were deposited during highstands. The Z4 Sandstone is considered to be the equivalent of the Hewett Sandstone as described in the UK sector of the southern North Sea. The sandstones are often tightly cemented with carbonate, dolomite or anhydrite cement. Only one core is available from the sandstones (Q16–2). Here light-coloured sandstones of the Z3 Formation display a regular interbedding with claystone laminae. The pattern of thickening and thinning of the sandstone sets and the occurrence of double clay-drapes and reactivation surfaces are very similar to descriptions for Holocene tidal deposits by Visser (1980).

The relation to sea-level has important implications for the distribution of these sandstones. Figure 6 shows that they were shed further into the basin during lowstand (e.g. Z1, Z4 Sandstone) than during highstand (Z2, Z3 Sandstone). In particular the Z3 Sandstone is constrained to a narrow zone along the basin margin. The southeastern limits of the Z1 up to Z3 Sandstones occur within a narrow zone. The coincidence of these limits was caused mainly by the intra-Zechstein uplift and erosion of the Southern Netherlands platform. A simplified NW – SE log correlation clearly displays the

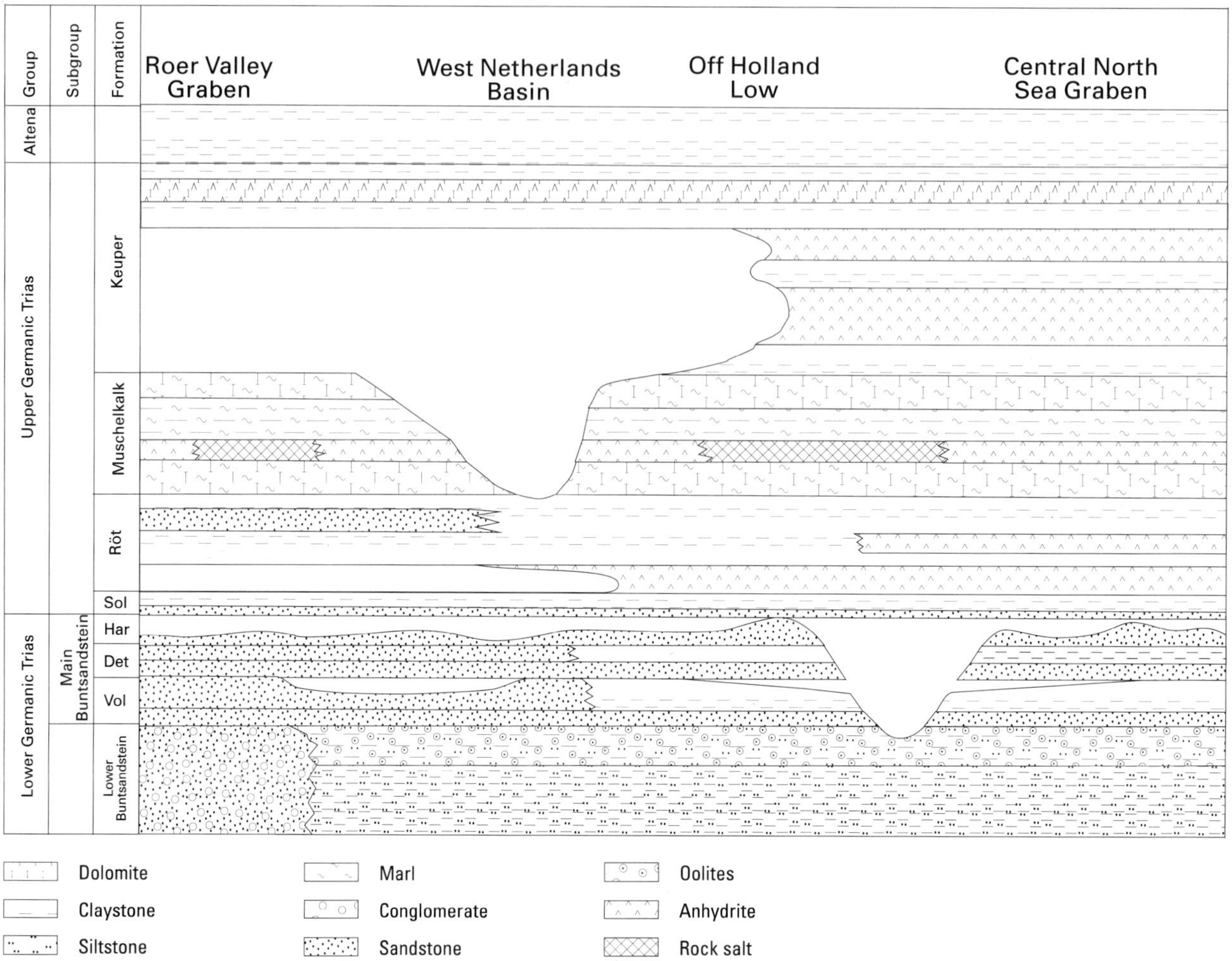

Fig. 8. Stratigraphic diagram of the Triassic. The age of the Altena Group is Rhaetian to Middle Jurassic. Vol = Volpriehausen Formation; Det = Detfurth Formation; Har = Hardegsen Formation; Sol = Solling Formation.

various sandstone bodies (Fig. 7). Note the eastward onlap of the lowermost sequence of the Z1 Formation onto the Southern Netherlands platform and the disconformable contact of the Zechstein Upper Claystone Formation with the underlying deposits. Towards the platform this formation progressively truncates underlying sediments (Figs 3, 5). On the platform, the Zechstein Upper Claystone overlies the lowermost claystones of the Z1 Formation and in places even the Upper Rotliegend Group (Fig. 3). We assume that during deposition of the Zechstein Group this area already formed a platform. This platform was further uplifted near the end of the Permian. Besides this overall uplift, individual fault blocks were uplifted. This was demonstrated in the area of the P15–1 well where the Zechstein Upper Claystone Formation overlies the Z1 Sandstone Member (Fig. 7).

Triassic

A pronounced paleogeographic change marked the onset of the Triassic: the Central Netherlands Basin became inactive and after an initial period of tectonic quiescence and rapid thermal subsidence (Ziegler 1990), extension followed that was accommodated by the NNE – SSW oriented Ems Low and Off Holland Low and by the NW – SE oriented Roer Valley Graben and West Netherlands Basin. The Roer Valley Graben formed the main feeder system of sediment during the Early to Middle Triassic and was subsequently part of a major seaway connection with the Paris Basin, at least until Middle Jurassic times (Ziegler 1990). The importance of the feeder system of Permian age, that originated on the London – Brabant Massif in the offshore had diminished, although it remained active during the Early Triassic.

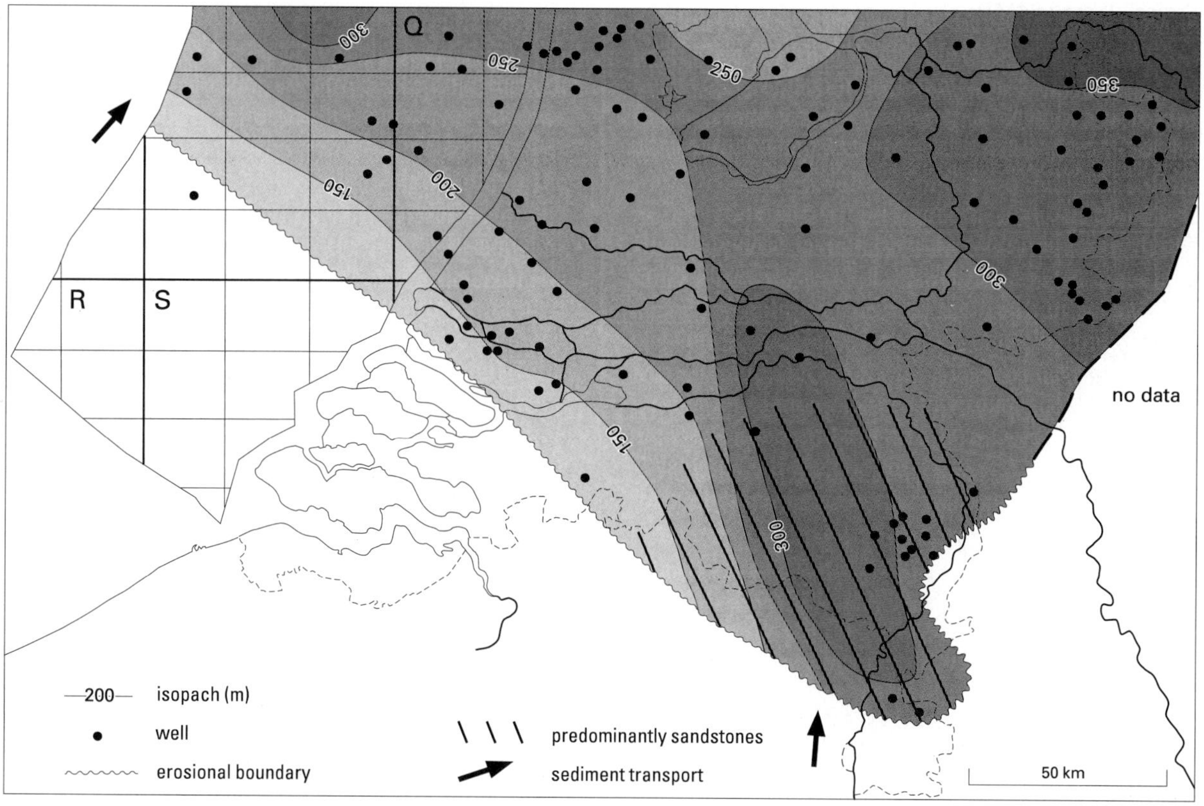

Fig. 9. Isopach map of the Lower Buntsandstein Formation. Strongest subsidence occurred in the Ems Low, Roer Valley Graben and Off Holland Low. The distribution of coarse clastic deposits in the formation is limited to the Roer Valley Graben.

Within the Triassic deposits, two main lithological groups can be distinguished (Fig. 8), namely the Lower Germanic Trias Group, composed of lacustrine claystones and fluvial sandstones, and the Upper Germanic Trias Group, composed of lacustrine to marine claystones, marine carbonates and evaporites, with minor intercalations of fluvio-lacustrine sandstones. During the Early Triassic coarse clastics first built out from the southeast far into the basin, but became afterwards limited again to the basin margin. During the Middle Triassic the supply of clastics from the southeast completely ceased.

The Triassic deposits display a pronounced layer-cake character (Fig. 8), not only evident on a regional scale but also on reservoir scale (e.g. Winstanley 1993). Most Triassic units can therefore be regarded as time slices. The main variations within the Triassic are caused by two phases of extension: (1) the Hardegsen phase which affected the distribution of the sandy Main Buntsandstein, and (2) the Early Kimmerian phase as a result of which the Muschelkalk and Keuper Formations were uplifted and eroded.

Lower Germanic Trias Group

Lower Buntsandstein Formation (Late Permian – Scythian)

The lowermost formation of the group is composed mainly of a cyclic alternation of lacustrine fine-grained sandstones and clayey siltstones (Fig. 8). The areas of strongest subsidence were the Roer Valley Graben, Ems Low and Off Holland Low (Fig. 9). Deposition of the Buntsandstein was governed by the development of a large fluviatile feeder system which transported clastics from the Vosges in NE France northwestward through the Roer Valley Graben and West Netherlands Basin into the Off Holland Low. A branch of this system continued northward into the Ems Low. This system caused deposition of up to 200 m of massive sandstones and conglomerates in the Roer Valley Graben and the adjacent Peel Block; these became covered

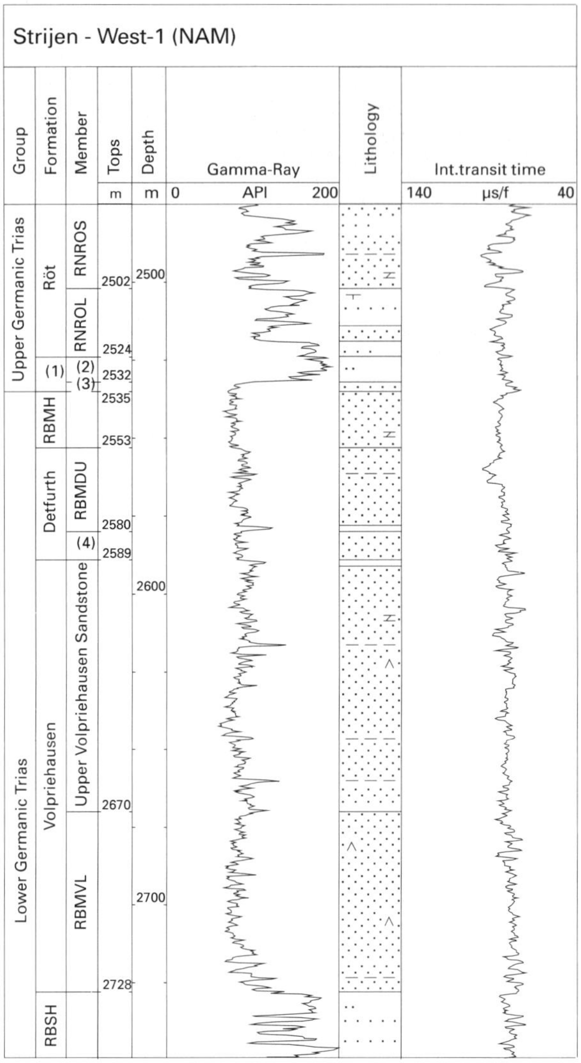

RNROS Röt Sandstone Mb.
RNROL Lower Röt Claystone Mb.
(1) Solling Fm.
(2) Solling Claystone Mb.
(3) Basal Solling Sandstone Mb.
RBMH Hardegsen Fm.
RBMDU Upper Detfurth Sandstone Mb.
(4) Lower Detfurth Sandstone Mb.
RBMVL Lower Volpriehausen Sandstone Mb.
RBSH Lower Buntsandstein Fm.

Fig. 10. Development of the Main Buntsandstein Subgroup (Volpriehausen, Detfurth and Hardegsen Formations) in the West Netherlands Basin examplified by the well Strijen-West-1. The well location is shown in Fig. 11.

by laterally persistent claystone beds. A minor feeder system was active near the P10 block. In the West Netherlands Basin, the formation is gradually thinning and onlapping towards the Brabant Massif, indicative of its positive relief at the time.

Main Buntsandstein Subgroup (Scythian)

The Main Buntsandstein Subgroup displays a cyclic alternation of (sub-)arkosic sandstones and clayey siltstones. The subgroup is composed of the Volpriehausen, Detfurth and Hardegsen Formations. The deposits are mainly of fluvial origin, but in the northern P and Q quadrants aeolian deposits occur as well (Ames & Farfan 1995; this volume). The subdivision of the subgroup for the well Strijen-West-1 is shown in Fig. 10.

The thickness and facies distribution of the individual formations has strongly been governed by the extensional tectonics of the Hardegsen phase. A subcrop map at the base of the overlying Solling Formation is shown in Fig. 11. Subsidence was strongest in the Off Holland Low, and nearly as much in the Roer Valley Graben and Ems Low (Fig. 12). The subsidence during deposition of the various Triassic units, however, was rather spasmodic in time and place (cf. Figs 14, 16). During deposition of the Volpriehausen Formation the strongest subsidence occurred in the Off Holland Low, whereas major subsidence during the deposition of the Detfurth Formation took place in the Ems Low. The outline of the Netherlands Swell is indicated by the 50 m isopach of the Main Buntsandstein (Fig. 12).

An important characteristic of this subgroup is the occurrence of subtle unconformities at the bases of the Volpriehausen and Detfurth Formations. The incision of the Volpriehausen Unconformity, however, is rather limited; only occasionally on the Netherlands Swell have several tens of metres of Lower Buntsandstein Formation been eroded (Geluk & Röhling, in prep.). The Detfurth Unconformity is the most prominent unconformity in the lows. This is illustrated in Fig. 13 for the Roer Valley Graben. This fact has not been recognized before in the Netherlands, although it had long been realized in Germany (Trusheim 1961, 1963). If the Detfurth Unconformity is not identified, it is likely that reservoir sandstones will be correlated with non-reservoir sandstones. Because of the good reservoir characteristics of the Lower Detfurth Sandstone in the basin fringe area, its proper identification could be of economic importance.

The *Volpriehausen Formation* is composed of a Lower Volpriehausen Sandstone Member, a clean sandstone with a blocky appearance on wireline logs, and an Upper Volpriehausen Sandstone Member which contains several claystone intercalations. Towards the

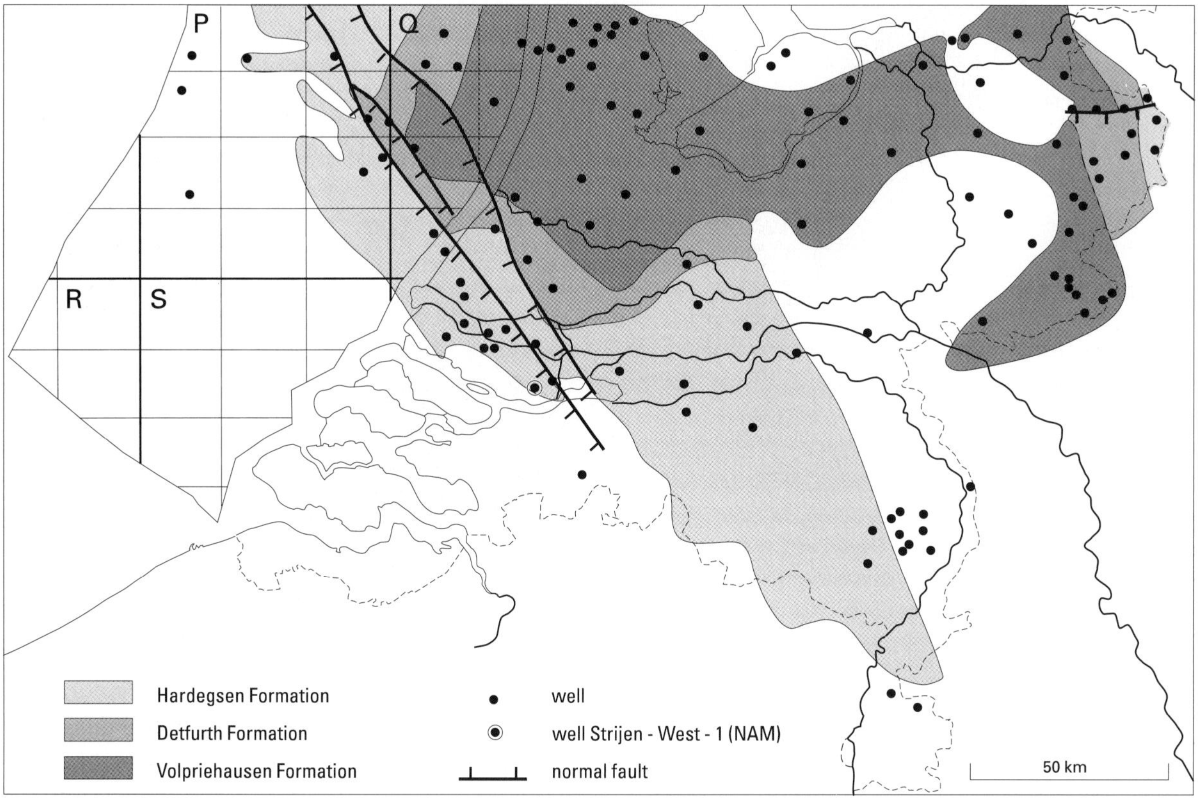

Fig. 11. Subcrop map at the base of the Solling Formation (cf. Fig. 17). The area where the latter formation overlies the Volpriehausen Formation, is named the Netherlands Swell. After Van Adrichem Boogaert & Kouwe 1993–1995.

Netherlands Swell, the Upper Volpriehausen Sandstone grades into the Volpriehausen Clay-Siltstone. The Lower Volpriehausen Sandstone has the best reservoir properties but the Upper Volpriehausen Sandstone is considered a waste zone owing to cementation.

The Lower Volpriehausen Sandstone has a quartz percentage slightly below 50% and high percentages of calcite and dolomite cement in the lowermost part. The average porosity amounts up to 10% in the West Netherlands Basin and up to 15% in the Ems Low. During deposition of the Lower Volpriehausen Sandstone the largest subsidence occurred in the Off Holland Low, where the member reaches a thickness of over 75 m (Fig. 14). On the Netherlands Swell, it shales out. Limited accomodation occurred in the Ems Low, where only 15 m of sandstone is present. This picture is, however, influenced by lateral facies variations, as the lowest beds of the Volpriehausen Clay-Siltstone in the Ems Low pass into the Lower Volpriehausen Sandstone in the Off Holland Low. Removing the effect of this lateral transition would still result in a thickness in excess of 50 m in the Off Holland Low.

The Upper Volpriehausen Sandstone is a reddish-brown, silty sandstone with dolomite, anhydrite and ankerite cements. It is composed of stacked fining-upwards cycles. The thickness of the member is strongly reduced by the Detfurth and Solling Unconformities, decreasing from well over 150 m in the Roer Valley Graben to less than 50 m on the Netherlands Swell (Fig. 15). The occurrence of the uppermost sandy part of the member is limited to the northern part of the Off Holland Low and the Roer Valley Graben.

The *Detfurth Formation* is composed of a Lower and an Upper Detfurth Sandstone Member. Towards the north, the Upper Detfurth Sandstone Member grades into the Detfurth Claystone Member. The Lower Detfurth Sandstone Member is often marked by low gamma-ray values and low velocities (cf. Figs 10, 13). It is one of the best reservoir intervals within the Main Buntsandstein. The unit has a high quartz grain content (up to 60%) and is quartz-cemented. Average porosities

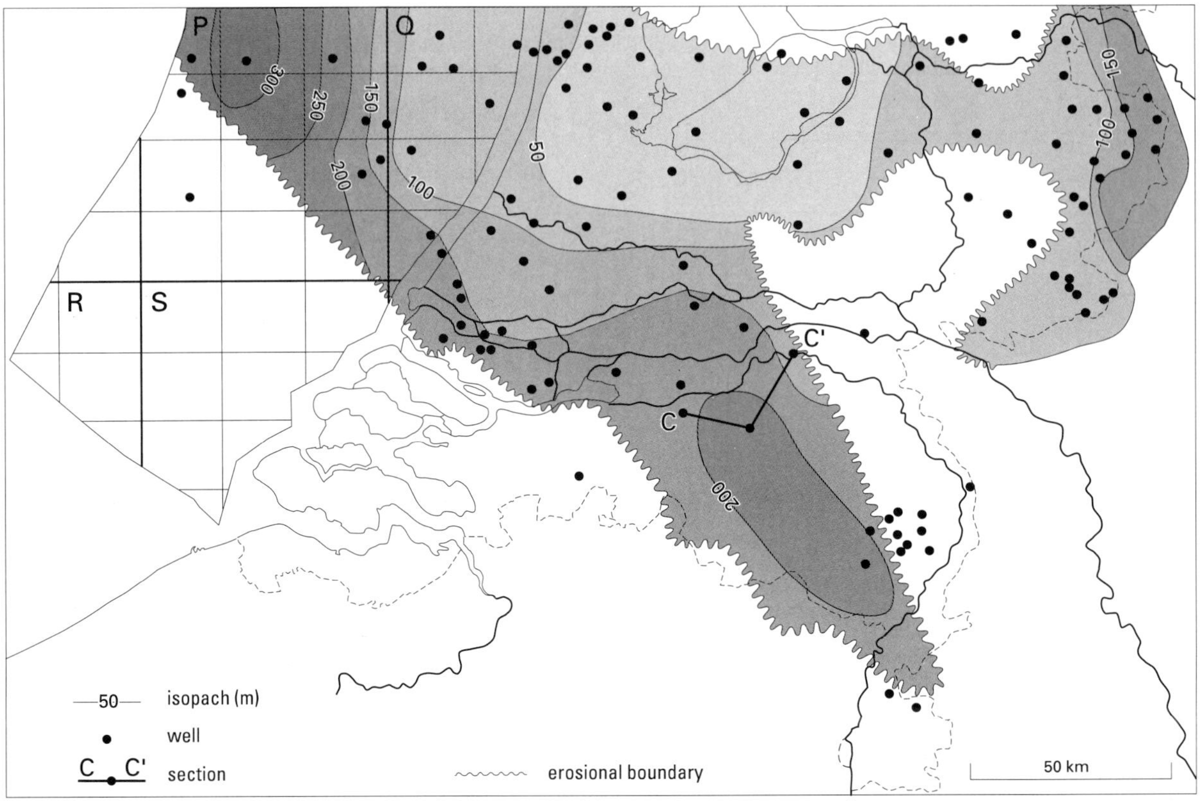

Fig. 12. Isopach map of the Main Buntsandstein Subgroup. Note the match between this map and that of the base Solling subcrop (Fig. 11). Correlation section C-C′ is presented in Fig. 13.

range from 15 to 20% in the West Netherlands Basin to over 20% in the Ems Low. During deposition the greatest subsidence occurred in the Ems Low, where this sandstone reaches local thicknesses in excess of 50 m, compared to 10–22 m in the Roer Valley Graben and Off Holland Low (Fig. 16). The Upper Detfurth Sandstone contains two claystone intervals. The thickness of this member varies between 20 and 30 m in the West Netherlands Basin; the equivalent Detfurth Claystone Member reaches 50 m in the Ems Low.

The transition to the *Hardegsen Formation* is often accompanied by a sudden increase of the sonic transit time (Fig. 10). The formation is characterized by a rapid alternation of sandstones and claystones, but in the southern part of the Off Holland Low and the West Netherlands Basin more massive sandstones occur. Altered feldspars witness leaching. Porosities of up to 20% were reported in the West Netherlands Basin. The occurrence of this formation is limited to the lows (Fig. 11). Its thickness varies strongly as a result of erosion below the Solling Unconformity.

Upper Germanic Trias Group

Solling Formation (Late Scythian; Spathian)

After the tectonic movements of the Hardegsen phase ceased, the Solling Formation was deposited during a major transgression. It onlaps onto different formations of the Main Buntsandstein Subgroup (Fig. 11). In the basin fringe area it is the first laterally extensive claystone above the potential reservoir rocks of the Main Buntsandstein. The formation is thickest in the Ems Low where it reaches over 100 m (Fig. 17). The Ems Low clearly continued to subside, as to a lesser extent, did the Central Netherlands Basin. Previously subsiding areas as the Roer Valley Graben, West Netherlands Basin and Off Holland Low, became platform areas. In these areas, the formation is thinner than 25 m, but it is usually present. In the West Netherlands Basin and the southern Off Holland Low the formation is reduced in thickness below the Röt Formation. The distribution of the Basal Solling Sandstone is indicated in Fig.

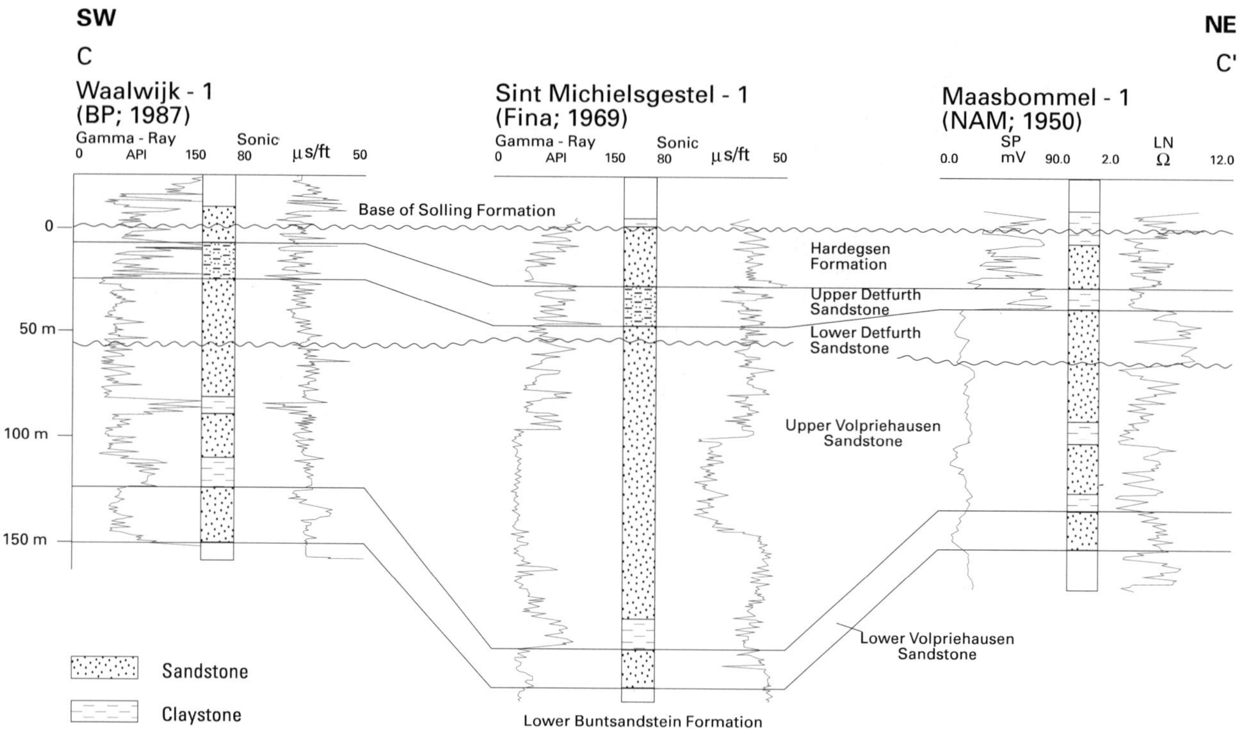

Fig. 13. Lithostratigraphic log correlation of the Main Buntsandstein Subgroup in the Roer Valley Graben. For location see Fig. 12. SP: Spontaneous Potential; LN: Long Normal.

17. Sandstones also occur in the middle part of the formation in the Off Holland Low (blocks P8, Q7).

Röt Formation (Early Anisian)

The Röt Formation comprises an evaporitic lower part and a clastic upper part. The evaporitic part contains anhydrite, halite and claystones; the clastic part is composed mainly of claystones, in the southern onshore area alternating with sandstones. During deposition of the Röt Formation, the strongest overall subsidence was located in the Roer Valley Graben, Central Netherlands Basin and Ems Low (Fig. 18). The formation reaches its greatest thickness in the Ems Low, where it exceeds 250 m. The West Netherlands Basin and the Off Holland Low, both areas of major subsidence during deposition of the Main Buntsandstein Subgroup, continue to show the characteristics of a platform area. Here the Röt Formation is in places even less than 50 m thick.

In the Roer Valley Graben the evaporitic and the clastic part of the formation are both present. In the West Netherlands Basin and the Off Holland Low, however, the clastic part onlaps onto the Solling Formation and the evaporitic part is missing (Fig. 19). At places in the Off Holland Low, the Upper Röt Fringe Claystone Member even rests on a reduced section of the Solling Formation.

The sandstones in the upper part of the formation, known as the Röt Fringe Sandstone Member, form important reservoir rocks in the southern onshore part of the Netherlands. They have excellent porosities (up to 15%) and permeabilities and in some fields they form the best producing unit. Their occurrence is limited to the West Netherlands Basin and the Roer Valley Graben (Fig. 18). The Röt Formation records the last Triassic clastic influx through this graben. Towards the north the sandstones grade rapidly into siltstones, and show an increasing cementation. Their overall thickness decreases northwards from over 50 m in the Roer Valley Graben.

Muschelkalk Formation (Middle Anisian – Early Ladinian)

The Muschelkalk Formation is subdivided into a lower part composed of limestones, dolomites and claystones, an evaporitic middle part, and an upper part of

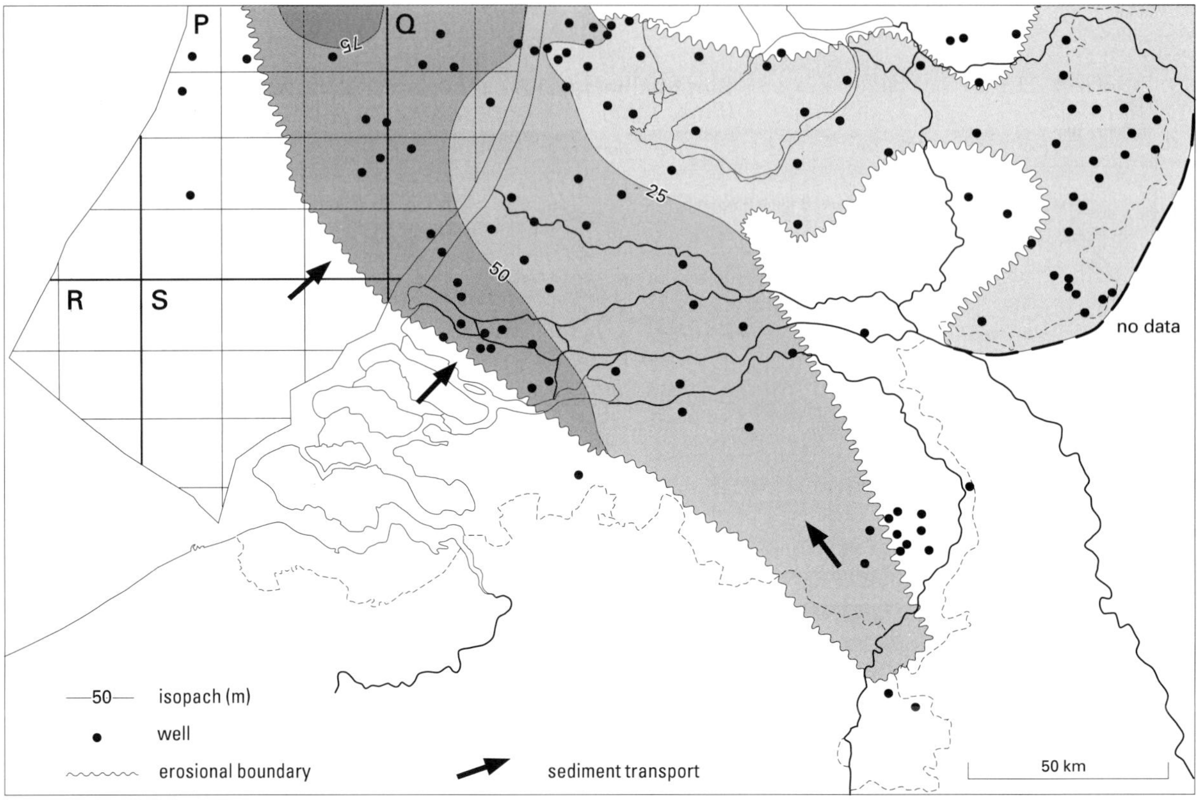

Fig. 14. Isopach map of the Lower Volpriehausen Sandstone Member. The strongest subsidence occurred in the Off Holland Low; the Ems Low was not a low during deposition of this unit.

claystones and dolomites. During its deposition, the strongest subsidence occurred in the Central Netherlands Basin, the Ems Low and the Roer Valley Graben (Fig. 20). In all three basins the formation is halite-bearing, although the salt occurrence in the Roer Valley Graben is a local phenomenon (M. Ferguson, pers. comm.). In the entire southern Netherlands the top part of the formation was removed by Late Triassic erosion. Much of the formation or locally all of it was eroded in the eastern Netherlands, on the Mid Netherlands Fault Zone and at the southern margin of the West Netherlands Basin (Fig. 20).

Keuper Formation (Late Ladinian-Norian)

During the Late Triassic, the area was affected by the Early Kimmerian rifting phase. Uplift prevailed in the study area; rifting occurred to the north. The uplift had a regional character and the amount of erosion varies only gradually. Strongest uplift occurred in the eastern Netherlands, on the Mid Netherlands Fault Zone and the southern margin of the West Netherlands Basin, where erosion cut deeply into the Muschelkalk or locally even the Röt Formation. The Off Holland Low and the northwestern part of the Central Netherlands Basin were affected to a lesser extent; here erosion cut into the lower part of the Keuper Formation.

After these movements ceased, sedimentation resumed with deposition of red-coloured claystones and light-coloured dolomites and carbonates of the Red Keuper Claystone and Dolomitic Keuper (the German 'Steinmergelkeuper') of Norian age; in the eastern Netherlands sedimentation only resumed during the Rhaetian. The hiatus of this Early Kimmerian unconformity spans 30 to over 60% of the duration of the Triassic Period within much of the study area.

Hydrocarbon occurrence

The first successful exploration well in the study area was Mobil's P6–1 which struck gas in the Main Buntsandstein in 1968. The resulting P6-field was the

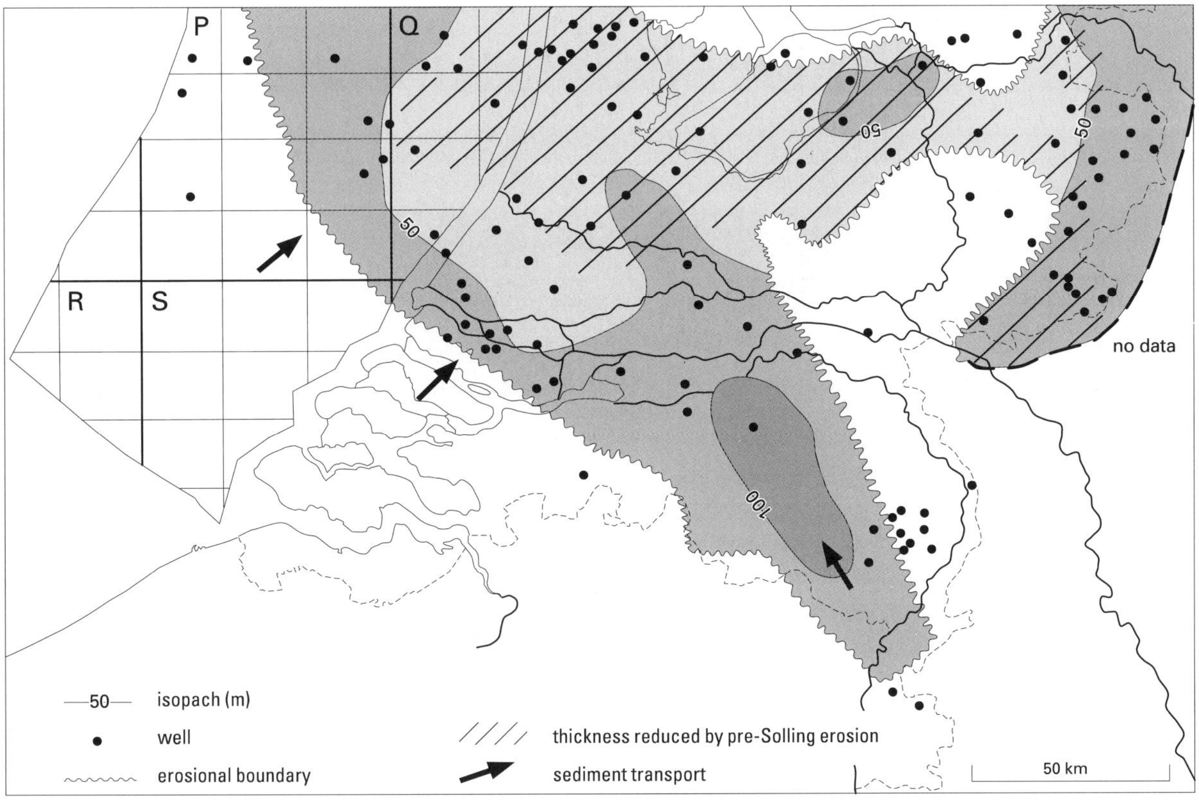

Fig. 15. Isopach map of the Upper Volpriehausen Sandstone and Volpriehausen Clay-Siltstone Members. The area of pre-Solling erosion (Fig. 11) is indicated; outside this area the thickness variations were caused by the Detfurth Unconformity.

first in a series of gas fields to be discovered in Triassic sediments (Fig. 21). The fields are confined to the Main Buntsandstein Subgroup and the Solling and Röt Formations. They occur where the Zechstein salt is thin or absent; this created a migration path from the Carboniferous source rocks to the Triassic reservoir rocks.

Offshore

After the Mobil discovery, wells were drilled by Phillips (P12–1; 1968, dry hole), Tenneco (P15–1; 1974, gas shows) and others. From 1982 on new successes were recorded in P12 by Pennzoil (1982) and Mobil (1986). Amoco explored blocks P15 and P18, awarded in 1980/1983. The first major discovery was the P18–1 field in 1987. This field straddles the border of the P15 and P18 blocks. Further successes within these blocks and in P14 by Wintershall led to a development plan for the eight Triassic fields of Amoco in blocks P15 and P18 (Amoco 1992). Gas had come on stream by the end of 1993.

Onshore

Exploration of the Main Buntsandstein also took place on land. Wells were drilled in the Rijswijk Concession (NAM) and the Eindhoven drilling licence (BP, now Clyde). Gas discoveries were reported by NAM in the so-called Gaag-Monster-Botlek trend from 1982 to 1984 and by BP in the Waalwijk area in 1987 (Winstanley 1993). At the moment more fields have been discovered in these Triassic sandstones in the study area (Fig. 21).

Source rocks

The gas in the studied area is considered to have been generated from the Upper Carboniferous Coal Measures. It migrated through the overlying Rotliegend and Zechstein clastic deposits into the Triassic reservoir

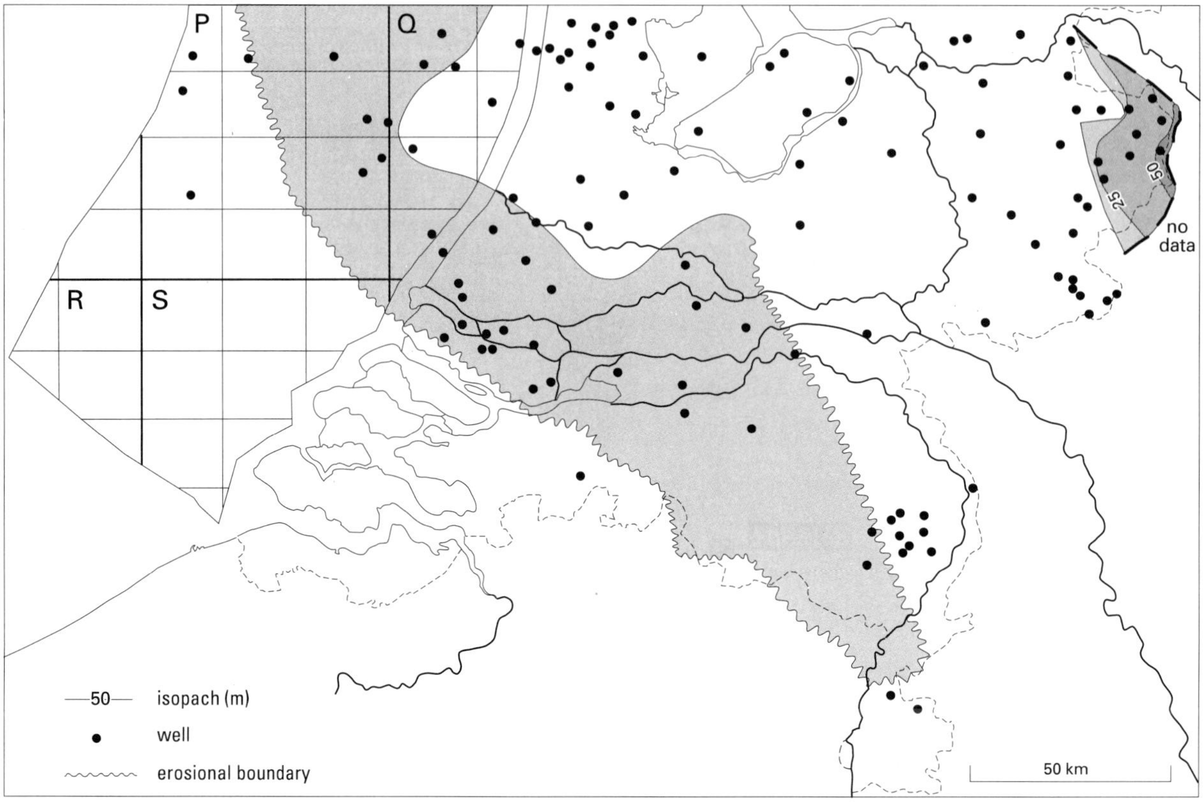

Fig. 16. Isopach map of the Lower Detfurth Sandstone Member. Note the differences in thickness between the Ems Low in the east and the Off Holland Low in the west.

rocks. Recent studies show that the Posidonia Shale Formation of Jurassic age also generated gas and not only oil as previously thought (De Jager et al. 1995). Locally these hydrocarbons were able to migrate into juxtaposed Triassic beds.

Reservoirs and seals

In the Permo-Triassic of the study area the following potential reservoirs have been identified:

The sandstones of the *Upper Rotliegend Group* have been identified as potential reservoirs in the entire study area. The group is sealed by claystones or evaporites of the Zechstein Group in the onshore part of this area. Best prospectivity is expected in the Central Netherlands Basin (Van Lith 1983), where evaporites occur as top-seal. In the southern part of the Off Holland Low, top-seals are absent.

In the *Zechstein Group* potential clastic reservoir rocks were identified only in the Off Holland Low; cementation has locally destroyed the porosity. Top-seals are the claystones within the Zechstein Group or the overlying Lower Triassic.

In the *Lower Germanic Trias Group* a number of potential reservoirs have been identified. In the case of the Lower Buntsandstein Formation these occur only in the southeastern onshore. In the Roer Valley Graben an intra-Triassic top-seal is present. On the Peel Block the play concept would depend on the sealing capacities of the overlying Cretaceous succession.

Potential reservoir rocks within the Main Buntsandstein Subgroup are widespread and occur predominantly within the Lower Volpriehausen Sandstone, Lower Detfurth Sandstone and Hardegsen Formation in the Roer Valley Graben, the West Netherlands Basin and Central Netherlands Basin, and in the Off Holland Low. These reservoirs are top-sealed by claystones, evaporites and tight sandstones in the Solling, Röt, Muschelkalk or Keuper Formations.

Reservoir rocks in the *Upper Germanic Trias Group* are limited to the Röt Fringe Sandstone Member in the southern onshore. The excellent reservoir sand-

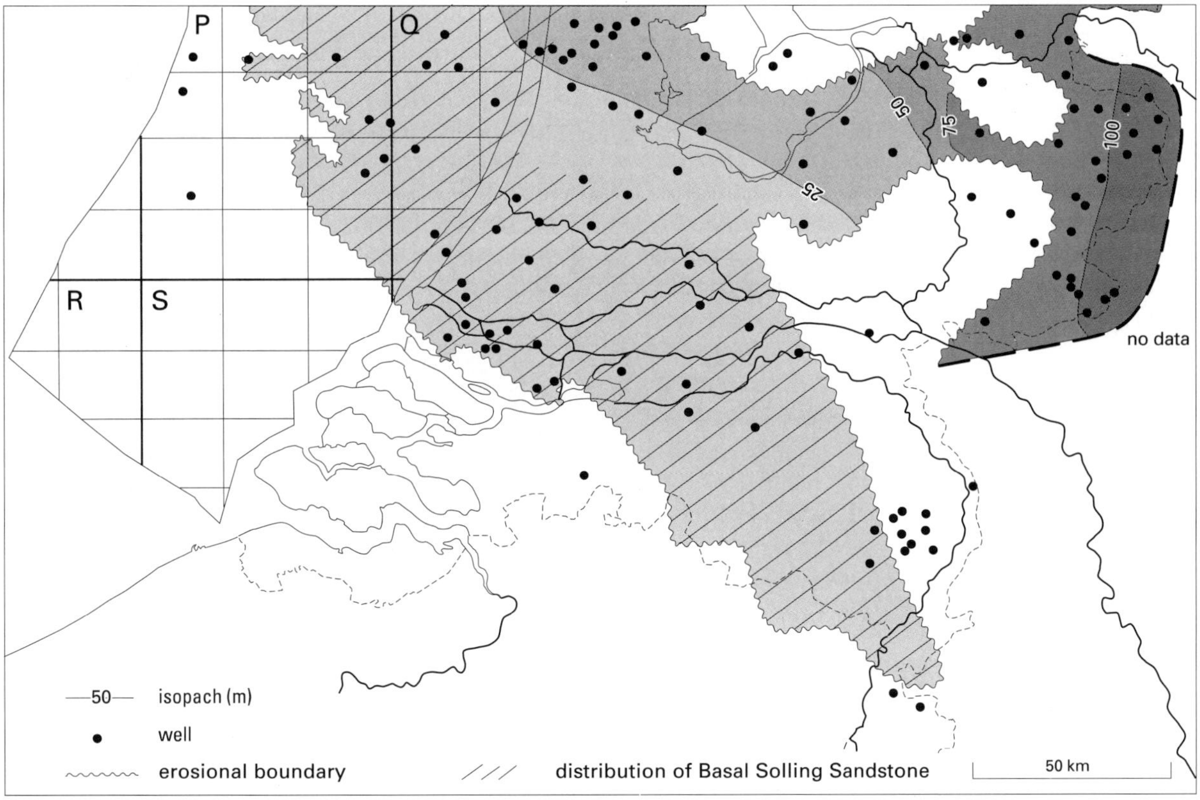

Fig. 17. Isopach map of the Solling Formation. The distribution of the Basal Solling Sandstone Member is indicated.

stones of this member are top-sealed by claystones and carbonates of the Röt, Muschelkalk or Keuper Formations.

In general, Triassic reservoir quality varies strongly and seems dependent on both burial history and the timing of gas generation and entrapment. Porosities range from less than 6 to almost 20%, while gas-saturations range from 30 to 80%.

The importance of the Triassic play in the basin fringe area is demonstrated by the over $100 \times 10^9 m^3$ GIIP (Gas Initially In Place) that were discovered since the P6–1 well of 1968.

Conclusions

1. Several potential clastic reservoir units are present offshore and onshore. Offshore, minor potential reservoirs occur in the Upper Rotliegend Group and Zechstein Group and major reservoirs in the Main Buntsandstein Subgroup. Onshore, potential reservoir rocks occur in the Upper Rotliegend Group, the Main Buntsandstein Subgroup and the Röt Formation.
2. In the study area there are marked differences in basin development in time. During the Permian, deposition was dominated by the subsiding Off Holland Low, Ems Low and Central Netherlands Basin; during the Triassic deposition shows a more differentiated development with, in addition, subsidence of the Roer Valley Graben and West Netherlands Basin. As a consequence of the basinal development the clastic influx in the basin shifted from the Off Holland Low in Permian times to the Roer Valley Graben in Triassic times.
3. Extension faulting occurred during three periods: the Late Permian, the Early Triassic and the Late Triassic. Contemporaneously with the Early Triassic Hardegsen phase, sandy sediments were deposited which present good reservoir rocks.
4. Most of the Permian and Triassic in the basin fringe area is represented by hiatuses. The Early Permian hiatuses are mainly caused by non-deposition, those of the Late Triassic by a combination of erosion and non-deposition.

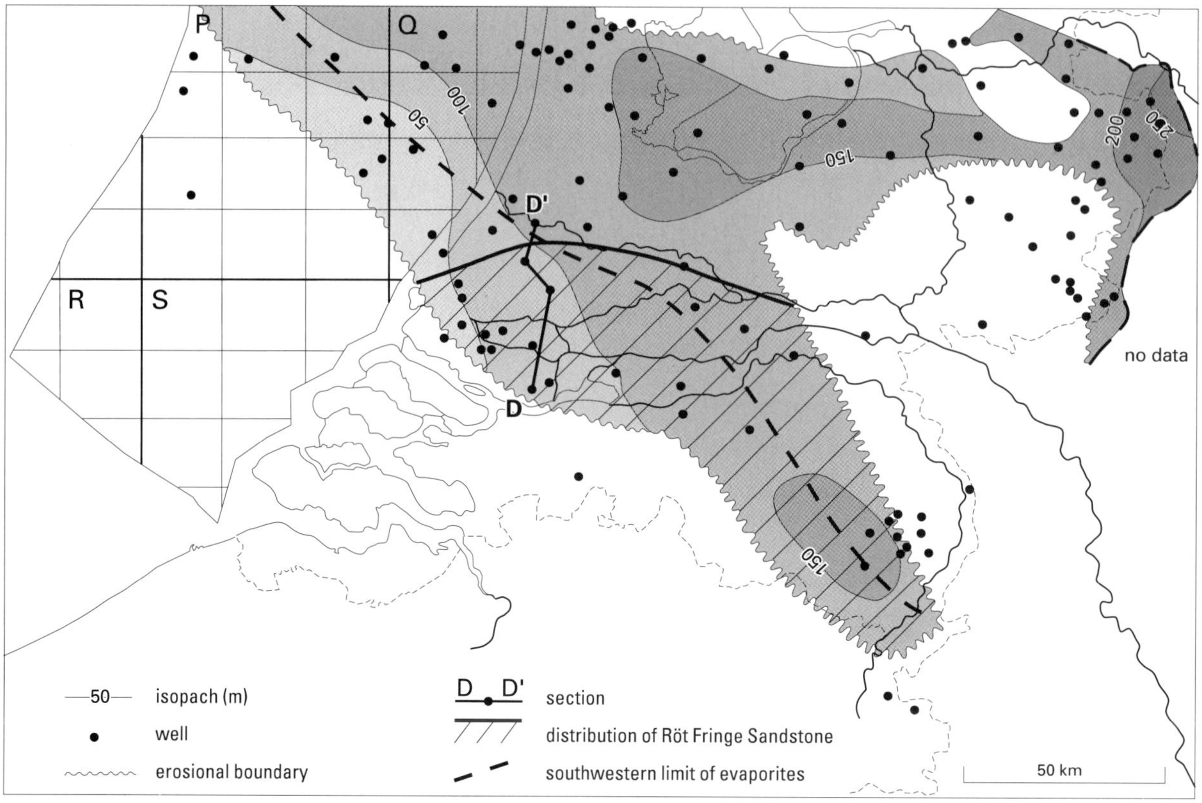

Fig. 18. Isopach map of the Röt Formation. Note the renewed subsidence of the Central Netherlands Basin. Correlation section D-D′ is presented in Fig. 19.

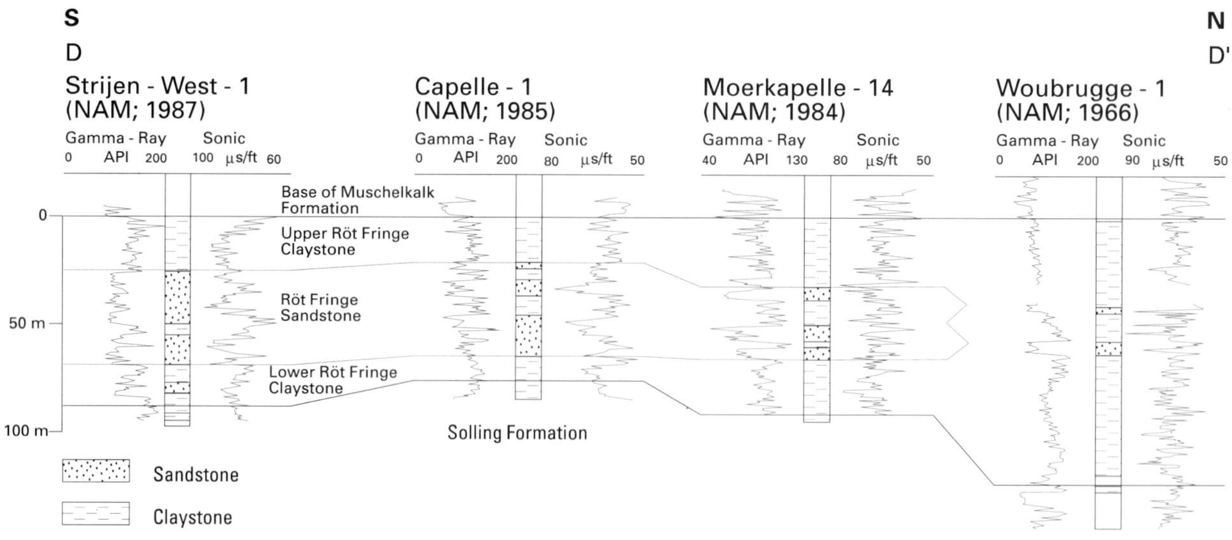

Fig. 19. Lithostratigraphic log correlation of the Röt Formation in the West Netherlands Basin. Note the southward onlap of this formation onto the Solling Formation and the sheetlike character of the Röt Fringe Sandstone. For location see Fig. 18.

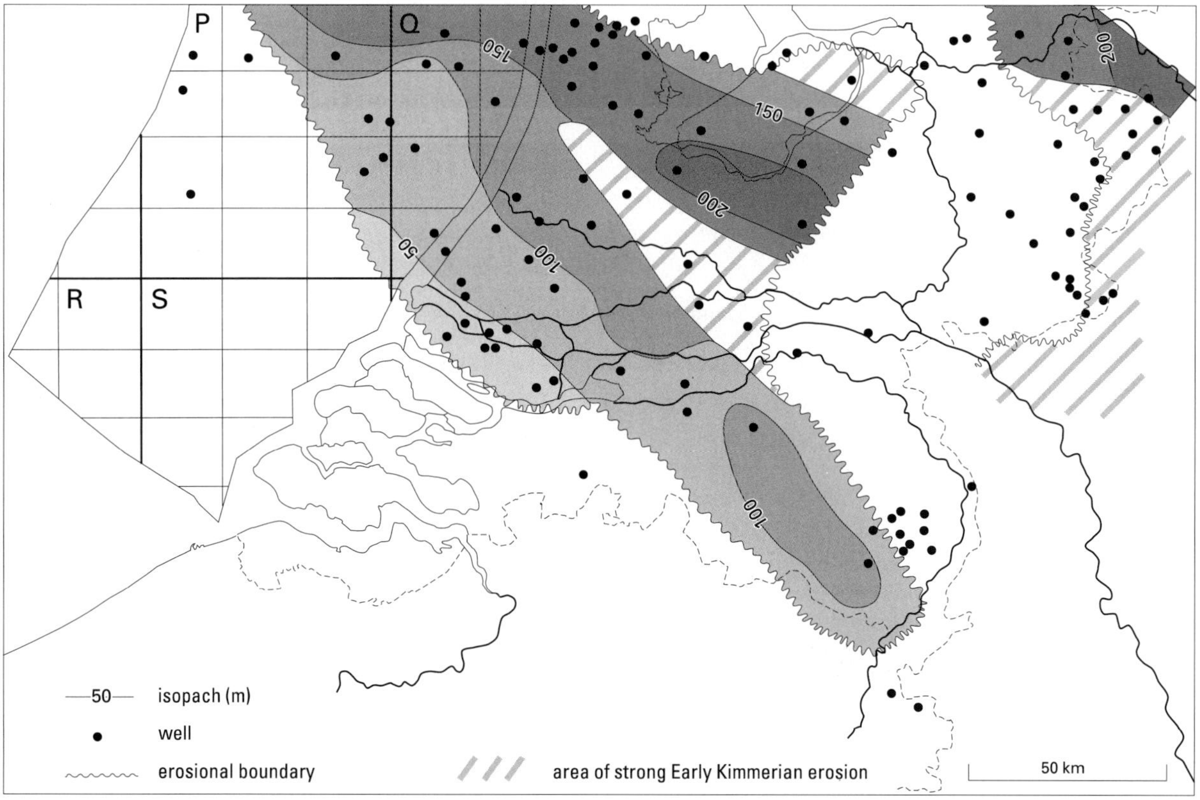

Fig. 20. Isopach map of the Muschelkalk Formation. The formation has been uplifted and eroded during the Early Kimmerian tectonic phase on the Maasbommel High, Zandvoort Ridge and in the eastern Netherlands.

Acknowledgements

The authors are indebted to the Director of the Rijks Geologische Dienst, Chr. Staudt, for permission to present and publish this paper. Thanks are due to the reviewers, A. Speksnijder, D.A.J. Batjes and H.E. Rondeel for many valuable suggestions and constructive comments, which led to improvements in the manuscript. Furthermore the authors wish to thank many colleagues for their contributions and especially A. Speksnijder and K. van Ojik, NAM representatives in the Working group on Permian, Triassic, and Lower and Middle Jurassic. The drawings were prepared by A. Koers and M. Kerlen; the text was revised by F.R. Bianchi and V.A. Verkaik-Drew (Foundation Pangea).

References

Ames, R. & P.F. Farfan 1995 The environments of deposition of the Triassic Main Buntsandstein Formation in the P and Q quadrants, offshore the Netherlands. In: Rondeel, H.E., D.A.J. Batjes & W.H. Nieuwenhuijs (eds) Geology of gas and oil under the Netherlands, this volume

Amoco 1992 P15/P18 gas development project. Amoco Netherlands Petroleum Co., The Hague: 28 pp

Boigk, H. 1956 Ausbildung und Paläogeographie des Buntsandsteins im nördlichen Teil der Niederrheinischen Bucht und seine Beziehung zu benachbarten Gebieten – Geol. Jb. 72: 347–366

Brennand, T.P. 1975 The Triassic of the North Sea. In: Woodland, A.W. (ed.) Petroleum geology and the continental shelf of Northwest Europe, Inst. Petr., Appl. Sci. Publ., Barking: 295–311

Brueren, J.W.R. 1959 The stratigraphy of the Upper Permian 'Zechstein' formation in the Eastern Netherlands. In: I Giacimenti Gassiferri dell' Europa Occidentale, Atti del Conv. Milano, 1957, I: 243–274

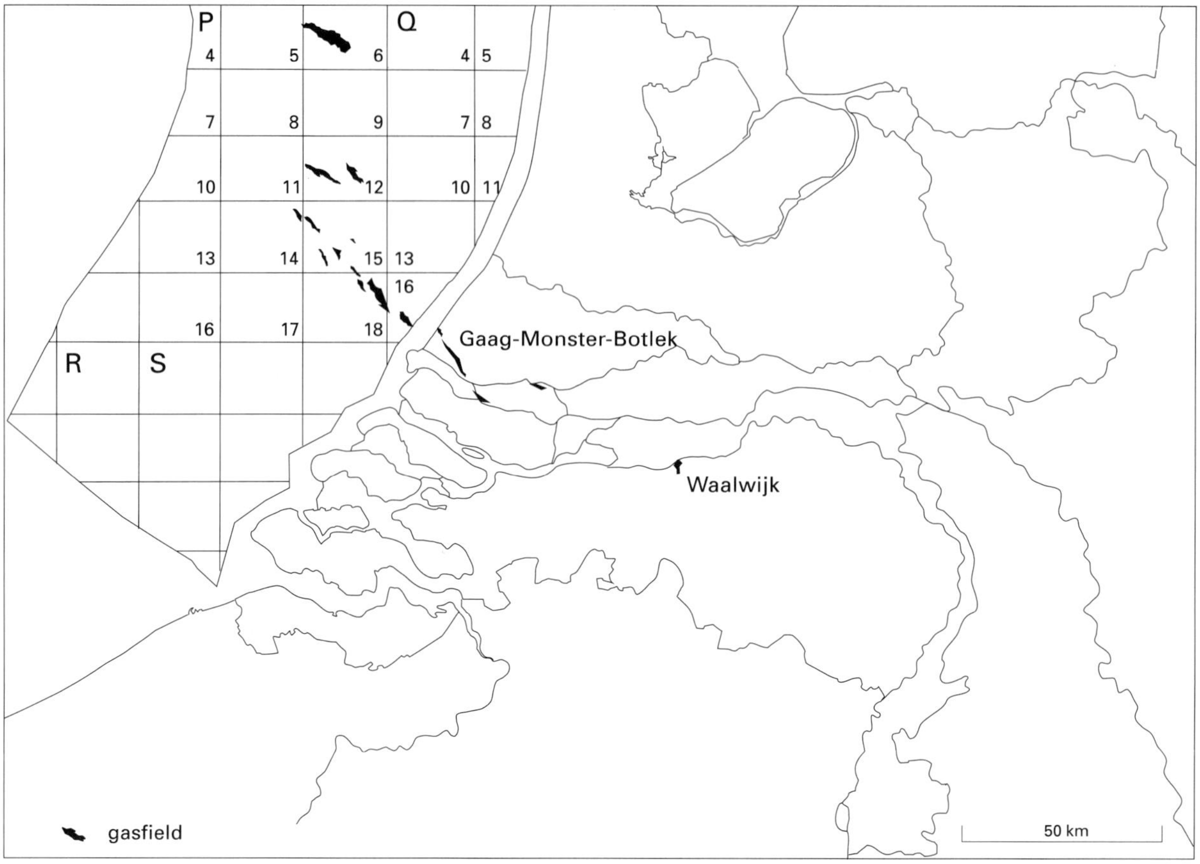

Fig. 21. Map showing gas fields with Permo-Triassic reservoirs.

De Jager, J., M.A. Doyle, P.J. Grantham & J.E. Mabillard 1995 Hydrocarbon habitat of the West Netherlands Basin. In: Rondeel, H.E., D.A.J. Batjes & W.H. Nieuwenhuijs (eds) Geology of gas and oil under the Netherlands, this volume

Demyttenaere, R. 1989 The post-Paleozoic geological history of north-eastern Belgium – Meded. Kon. Acad. Wetensch., Lett. Schone Kunsten België 51: 51–81

Haanstra, U. 1963 A review of Mesozoic geological history of The Netherlands – Verh. Kon. Ned. Geol. Mijnb. Gen. 21(2): 35–55

Heybroek, P. 1974 Explanation to tectonic maps of The Netherlands – Geol. Mijnbouw 53: 43–50, 2 encls

Hilden, H.D. (ed.) 1988 Geologie am Niederrhein – Geologisches Landesamt Nordrhein-Westfalen, Krefeld: 60 pp

Menning, M., G. Katzung & H. Lützner 1988 Magnetostratigraphic investigations in the Rotliegendes of Central Europe – Z. geol. Wiss. 16: 1045–1053

Muchez, Ph., W. Vianne & M. Dusar 1992 Diagenetic control on secondary porosity in flood plain deposits: an example of the Lower Triassic of northeastern Belgium – Sedimentary Geology 78: 285–298

NAM & RGD (Nederlandse Aardolie Maatschappij & Rijks Geologische Dienst) 1980 Stratigraphic nomenclature of The Netherlands – Verh. Kon. Ned. Geol. Mijnb. Gen. 32: 77 pp, 36 encls

RGD (Rijks Geologische Dienst) 1993 Geological atlas of the subsurface of the Netherlands 1:250 000: sheet IV Texel – Purmerend – Rijks Geologische Dienst, Haarlem: 124 pp

Roos, B.M. & B.J. Smits 1983 Rotliegend and Main Buntsandstein gas fields in block K 13 – a case history – Geol. Mijnbouw 62: 75–82

Trusheim, F. 1961 Über Diskordanzen im Mittleren Buntsandstein Norddeutschlands zwischen Ems und Weser – Erdöl Zeitschr. 77: 361–367

Trusheim, F. 1963 Zur Gliederung des Buntsandsteins – Erdöl Z. 79: 277–292

Tucker, M.E. 1991 Sequence stratigraphy of carbonate – evaporite basins: models and applications to the Upper Permian (Zechstein) of northeast England and adjoining North Sea – J. Geol. Soc. 148: 1019–1036

Van Adrichem Boogaert, H.A. 1976 Outline of the Rotliegend (Lower Permian) in The Netherlands. In: Falke, H. (ed.) The continental Permian in Central, West and Southern Europe, Proc. NATO Advanced Study Inst., Series C, 22 – Reidel Publ. Co., Dordrecht: 23–27

Van Adrichem Boogaert, H.A. & W.F.J. Burgers 1983 The development of the Zechstein in The Netherlands – Geol. Mijnbouw 62: 83–92

Van Adrichem Boogaert, H.A. & W.F.P. Kouwe 1993–1995 Stratigraphic nomenclature of the Netherlands; revision and update by RGD and NOGEPA – Meded. Rijks Geol. Dienst 50

Van der Zwan, C.J. & P. Spaak 1992 Lower and Middle Triassic sequence stratigraphy and climatology of the Netherlands, a model – Palaeogeography, Palaeoclimatology, Palaeoecology 91: 277–290

Van Lith, J.G.J. 1983 Gas fields of the Bergen concession, The Netherlands – Geol. Mijnbouw 62: 63–74

Van Staalduinen, C.J., H.A. Van Adrichem Boogaert, M.J.M. Bless, J.W.Chr. Doppert, H.M. Harsveldt, H.M. Van Montfrans, E. Oele, R.A. Wermuth & W.H. Zagwijn 1979 The geology of The Netherlands – Med. Rijks Geol. Dienst 31–2: 9–49

Van Waterschoot van der Gracht, W.A.J.M. 1909 The deeper geology of the Netherlands and adjacent regions, with special reference to the latest borings in the Netherlands, Belgium and Westphalia – Mem. Govt. Inst. Geol. Expl. Netherl. 2 (R.O.V.D.), The Hague: 437 pp

Van Waterschoot van der Gracht, W.A.J.M. 1918 Eindverslag over de onderzoekingen en uitkomsten van de Dienst der Rijks-opsporing van Delfstoffen in Nederland, Amsterdam: 664 pp

Van Wijhe, D.H. 1987 Structural evolution of inverted basins in the Dutch offshore – Tectonophysics 137: 171–219

Van Wijhe, D.H., M. Lutz & J.P.H. Kaasschieter 1980 The Rotliegend in The Netherlands and its gas accumulations – Geol. Mijnbouw 59: 3–24

Visser, M.J. 1980 Neap-spring cycles reflected in Holocene subtidal large-scale bedform deposits: a preliminary note – Geology 8: 543–546

Winstanley, A.M. 1993 A review of the Triassic play in the Roer Valley Graben, SE onshore Netherlands. In: Parker, J.R. (ed.), Petroleum geology of Northwest Europe, Proc. of the 4th Conf., Geol. Soc., London: 595–607

Wolburg, J. 1956 Das Profil der Trias im Raum zwischen Ems und Niederrhein – N. Jb. Geol. Paläont.: 305–330

Wolburg, J. 1957 Ein Querschnitt durch den Nordteil des Niederrheinischen Zechsteinbeckens – Geol. Jb. 73: 7–38

Wolburg, J. 1961 Sedimentations Zyklen und Stratigraphie des Buntsandsteins in NW-Deutschland – Geotekt. Forsch. 14: 7–74

Wolburg, J. 1968 Vom zyklischen Aufbaus des Buntsandsteins – N. Jb. Geol. Paläont. Mh. 9: 535–559

Wolburg, J. 1969 Die epirogenetischen Phasen der Muschelkalk und Keuper-Entwicklung Nordwest-Deutschlands, mit einem Rückblick auf den Buntsandstein – Geotekt. Forsch. 32: 1–65

Ziegler, P.A. 1990 Geological Atlas of Western and Central Europe, 2nd ed. – Shell Internationale Petroleum Maatschappij, The Hague/Geol. Soc. Publ. House, Bath: 239 pp, 56 encls

Appendix

Well database

Well name	Operator	Year
Aarlanderveen-1	NAM	1975
Almere-1	Amoco	1976
America-11	ROVD	1909
Arcen-1	Klein Vink	1987
Asten-1	NAM	1953
Bakkum-Castricum-1	NAM	1964
Barendrecht-1	NAM	1984
Barneveld-1	Amoco	1971
Bergen-1	Amoco	1965
Bergermeer-1	Amoco	1969
Borne-1	KNZ	1982
Botlek-1	NAM	1984
Buurmalsen-1	NAM	1970
Capelle-1	NAM	1985
Corle-1	ROVD	1921
De Lier-45	NAM	1981
De Lutte-4	NAM	1944
Denekamp-1	NAM	1951
Deurningen-Weerselo-4	NAM	1955
Doornspijk-1	NAM	1964
Doornspijk-2	NAM	1965
Dronten-1	Conoco	1965
Egmond-Zee-1	NAM	1984
Eibergen-1	ROVD	1902
Epe-1	Chevron	1971
Ermelo-1	NAM	1969
Everdingen-1	NAM	1965
Gaag-2	NAM	1971
Geesteren-1	NAM	1971
Gelria-Lichtenvoorde	M&H	1924
Haaksbergen-1	NAM	1950
Heiloo-1	Amoco	1965
Heemskerk-1	NAM	1965
Helenaveen-5a	ROVD	1906
Helenaveen-7	ROVD	1906
Hengevelde-1	RGD	1985
Hengelo-1	KNZ	1966
Hellevoetsluis-1	NAM	1969
IJsselmeer-1	Superior	1965
Joppe-1	RGD	1985
Jutphaas-1	NAM	1968
Kijkduin-Zee-2	NAM	1986
Kloosterhaar-1	NAM	1980
Landsmeer-1	NAM	1977
Langenholte-1	NAM	1977
Lelystad-1	Conoco	1970
Lichtenvoorde-2	NAM	1952
Liessel-22	ROVD	1915
Lievelde-1	NAM	1975
Limbricht-1	RGD	1984
Lochem-1	BPM	1943
Lochem-2	BPM	1943
Maasbommel-1	NAM	1950
Maasbree-13	ROVD	1910
Maris-18	ROVD	1914
Meijel-8	ROVD	1907
Middelie-101	NAM	1964
Moerkapelle-14	NAM	1984
Nederweert-1	Fina	1964
Nijmegen-Valburg-1	NAM	1968
Noordwijk-2	NAM	1983
Oost-Flevoland-1	NAM	1965
Oostzaan-1	NAM	1950
Oud-Beyerland-1	NAM	1985
Pernis-West-1	NAM	1987
Plantengaarde-1	ROVD	1908
Punthorst-1	NAM	1984
Raalte-2	BP	1983
Raath-1	RGD	1983
Rammelbeek-2	NAM	1970

Well name	Operator	Year
Ratum-1	ROVD	1911
Rijsbergen-1	NAM	1969
Rossum-Weerselo-3	NAM	1968
Rotterdam-Schulpweg-1	NAM	1984
Rozenburg-1	NAM	1987
Rustenburg-1	NAM	1977
Ruurlo-1	RGD	1984
Schermer-1-Deep	Amoco	1976
Schiphol-1	Amoco	1970
Schiphol-2	Amoco	1970
Sevenum-19	ROVD	1913
Sint-Michielsgestel-1	Fina	1969
Spierdijk-1	Petroland	1984
Spijkenisse-1	NAM	1960
Staphorst-1	NAM	1950
Starnmeer-1	Amoco	1974
Strijen-1	Amoseas	1964
Strijen-West-1	NAM	1987
Tubbergen-5	NAM	1952
Tubbergen-Mander-1	NAM	1967
Ursem-1	NAM	1965
Waalwijk-1	BP	1987
Wassenaar-23B	NAM	1988
Waverveen-1	Amoco	1971
Weesp-1	Amoco	1970
Werkendam-2	NAM	1965
Wijk-Aalburg-1	NAM	1972
Wimmenum-Egmond-1	NAM	1964
Winterswijk-1	NAM	1977
Woubrugge-1	NAM	1966
Zeddam-1	NAM	1954
Zeewolde-1	Conoco	1965
Zuidwolde-1	NAM	1954
P7–1	Unocal	1969
P8–4	Mobil	1984
P9–1-A	Amoco	1969
P10–1	Pennzoil	1969
P12–1	Phillips	1968
P12-A-1	Pennzoil	1978
P13–1	BP	1978
P15–1	Tenneco	1974
P15–2	Amoco	1977
Q7–1	NAM	1973
Q7–2	NAM	1978
Q7–4	NAM	1982
Q8–1	BP	1976
Q11–1	NAM	1969
Q13–3	Amoco	1982
Q16–2	BP	1978

BP = British Petroleum; BPM = Bataafse Petroleum Maatschappij; KNZ = Koninklijke Nederlandse Zoutindustrie (Royal Dutch Salt Industry); M&H = Mees & Hope; NAM = Nederlandse Aardolie Maatschappij; RGD = Rijks Geologische Dienst (Geological Survey of the Netherlands); ROVD = Rijksopsporing van Delfstoffen (Government Institute for the Geological Exploration of the Netherlands).

Rondeel et al. (eds), Geology of gas and oil under the Netherlands, 79–92, 1996.

Geological aspects of the Annerveen gas field, the Netherlands

Epeüs N. Veenhof
Nederlandse Aardolie Maatschappij (Business Unit Groningen), De Vosholen 66, 9611 TD Sappemeer, the Netherlands

Key words: Rotliegend, aquifer influx, case study, fault seal, VSP

Abstract

The Rotliegend Annerveen field was discovered in 1962. Commercial gas production started in 1973. The GIIP (Gas Initially In Place) is estimated at 75.6 × 10^9 m^3. At the time of writing, some 59 × 10^9 m^3 has been produced. The objective of this paper is to give, as a case study, a concise overview of all geological aspects pertaining to the Annerveen field. The field history, stratigraphy, reservoir development and structure are described. Furthermore, particular attention is focused on the problems associated with the structural relationship of Annerveen with the Groningen field and the possibility of aquifer influx into the Annerveen structure. It is concluded that semi-transmissive and sealing faults at the eastern flank as well as aquifer influx at the western flank could cause a loss of gas recovery towards the end of the field's lifetime.

Introduction and field history

The Annerveen field (Fig. 1) was discovered by exploration well Annerveen-Anloo-1 in 1962, just three years after the discovery of the giant Groningen gas field in 1959. After Groningen, Annerveen is the largest gas field in the Netherlands. The field straddles the boundary between the Groningen and Drenthe concessions of the Nederlandse Aardolie Maatschappij (NAM) and is operated by NAM on behalf of the Maatschap Groningen and Drenthe (Energie Beheer Nederland, EBN: 40%, NAM: 60%).

Commercial gas production by NAM started in 1973 via three clusters: the Annerveen (ANN), Wildervank (WDV) and Zuidlaarderveen (ZLV) clusters (Fig. 2). From these clusters a total of 17 wells were drilled. The clusters also comprise facilities to bring the gas to sales specifications after which the gas is delivered to NV Nederlandse Gasunie for further transporation and distribution. Wells Zuidlaren-1 (ZLN-1) and Westerdiep-1B (WTD-1B) are tied in to the Zuidlaarderveen and Wildervank clusters, respectively. The Annerveen-Veendam (ANV), Eexterveen (ETV) and Annerveen-Schuilingsoord (ANS) locations are used for observation. The GIIP (Gas Initially In Place) is currently estimated at 75.6 × 10^9 m^3 and to date some 59 × 10^9 m^3 has been produced. Annerveen can therefore be classified as a mature field. The gas has a high calorific value, the Wobbe index (a measure of the amount of energy released from the gas on combustion) being around 51 MJ/m^3.

Stratigraphy and reservoir development

A brief description of the stratigraphy in the Annerveen area is given below. The Permian, Upper Rotliegend, Slochteren Sandstone is more fully described as it constitutes the productive formation. The stratigraphic development is illustrated by a generalised stratigraphic column (Fig. 3) and the overall structural and stratigraphic setting is displayed by means of a seismic line from a 3D dataset (Fig. 4).

Carboniferous (source rock)

The oldest drilled rocks underlying the Annerveen gas field, are thick Carboniferous sequences (at least 2500 m, Van Wijhe et al. 1980). The Carboniferous is separated from the Rotliegend by an angular unconformity,

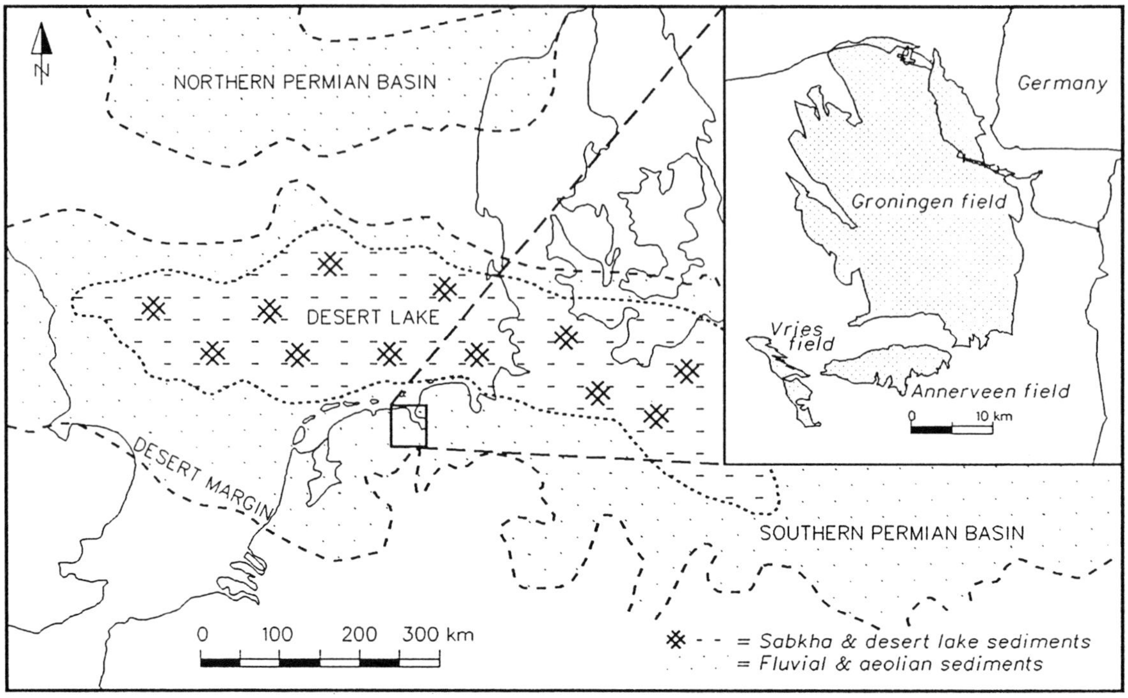

Fig. 1. Early Permian (Rotliegend) palaeogeographical map (after Ziegler 1990) showing the location of the Groningen and Annerveen gas fields.

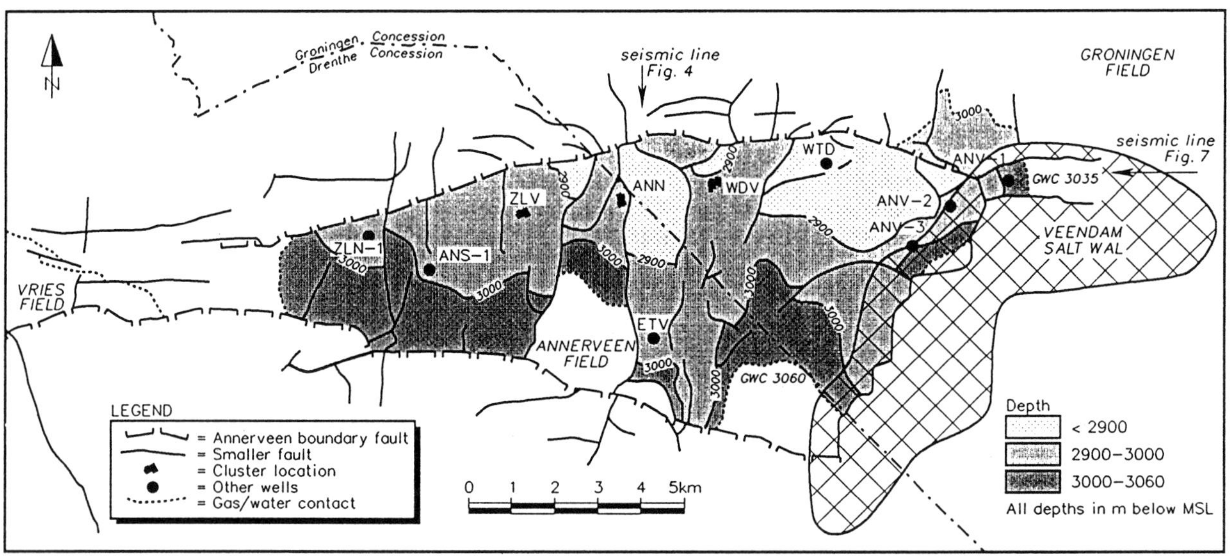

Fig. 2. Generalised depth map of the top Rotliegend Group in the Annerveen field.

the Saalian Unconformity (Fig. 4). Rocks subcropping at this unconformity, are dated as Westphalian C (top of the 'Coal Measures') and Westphalian D (Van Wijhe & Bless 1974). In the Groningen area, the subcropping facies consist of deltaic shales, sandstones and coals with a tendency to a more continental sedimentary environment upwards (Stäuble & Milius 1970). In Annerveen, well penetrations are sparse and incomplete. It is generally accepted that the Westphalian is the source rock for most of the gas in the north-east

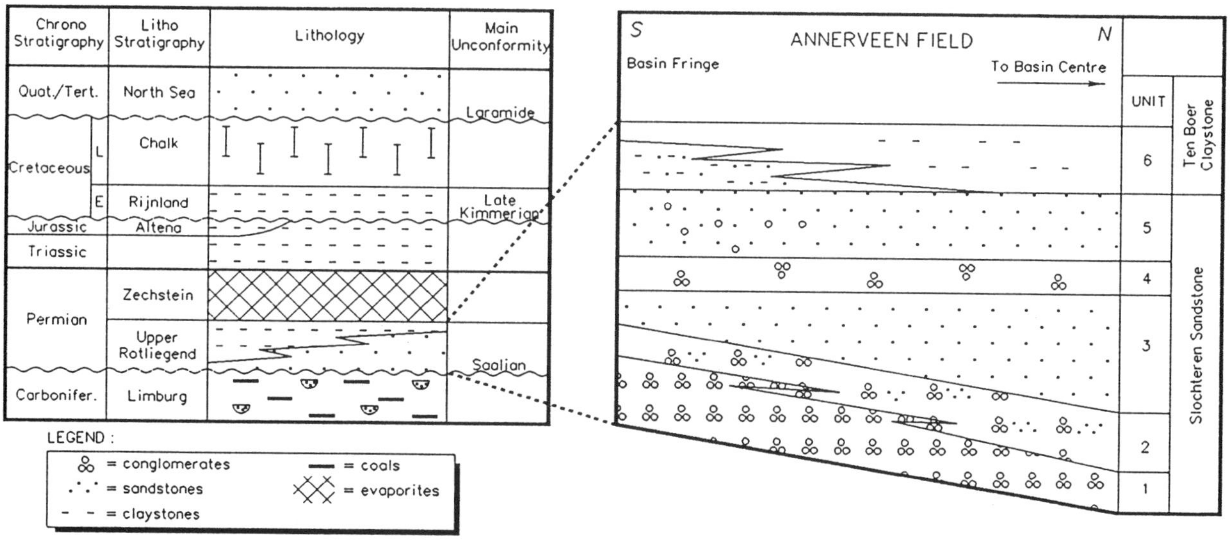

Fig. 3. Generalised stratigraphic column of the Annerveen area, with Annerveen reservoir subdivision (units 1–6).

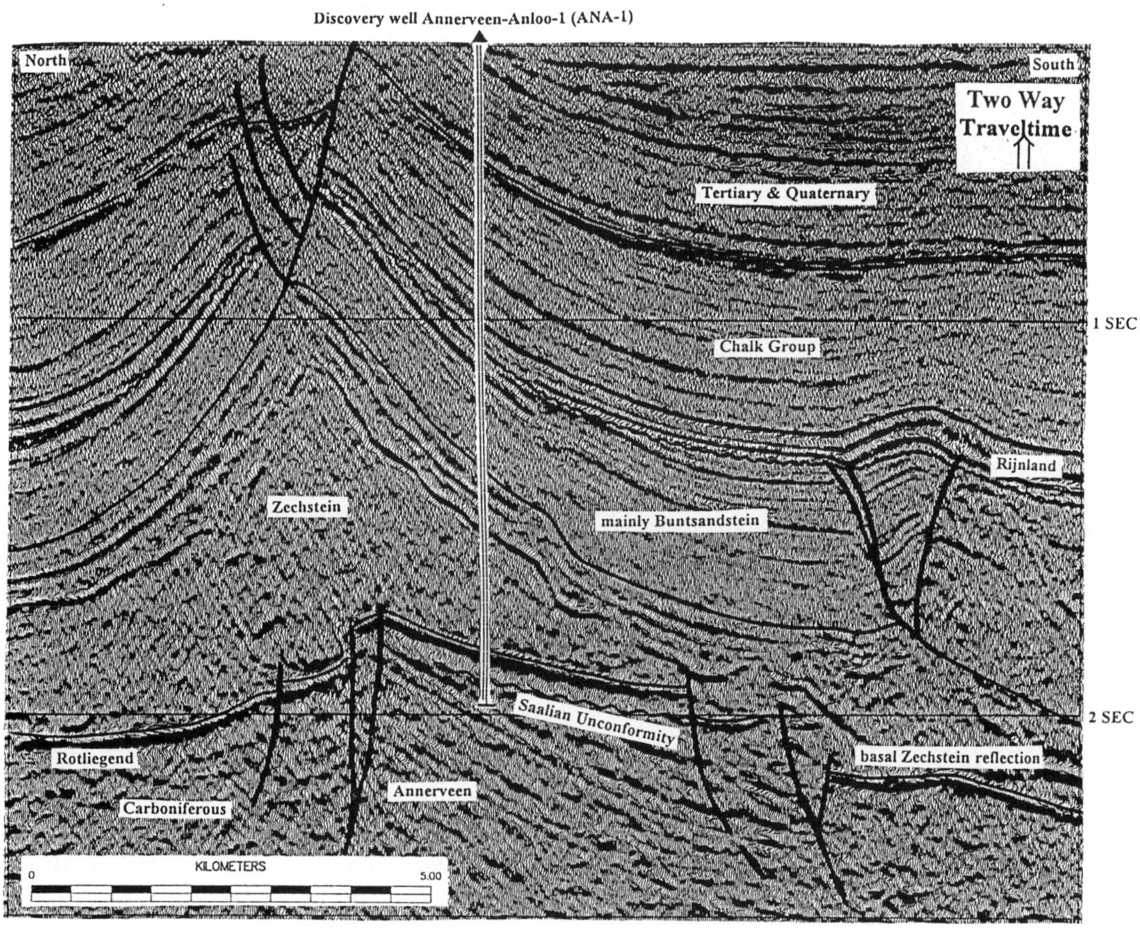

Fig. 4. Seismic profile (north – south) through discovery well Annerveen-Anloo-1 (For location see Fig. 2). Note the angular nature of the Saalian Unconformity and the Rotliegend fault terminations in the Zechstein.

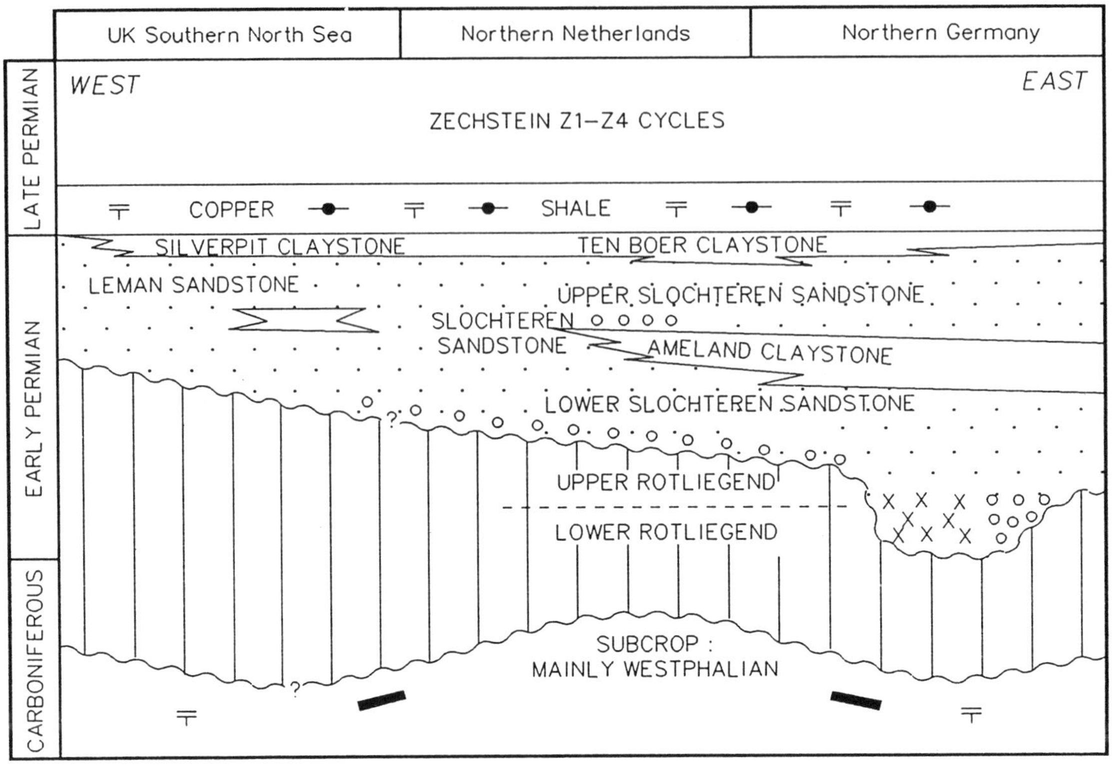

Fig. 5. Schematic stratigraphic correlation diagram along the southern margin of the Southern Permian Basin (after Van Veen 1975, Van Wijhe et al. 1980, Glennie 1990).

Netherlands (e.g. Van Wijhe et al. 1980). Some thin Westphalian sandstones may subcrop at the base of the Rotliegend and contribute to gas production.

Permian: Rotliegend (reservoir)

Lower Permian Rotliegend deposits contain the gas of the Annerveen field. Throughout this paper the name Rotliegend is used to describe sediments belonging to the Upper Rotliegend Group. The Lower Rotliegend Group is absent in the Annerveen area (Fig. 5). For this reason the Rotliegend in this area is thought to be of Saxonian age (Van Wijhe et al. 1980).

In general, the Upper Rotliegend is interpreted to have been deposited in a mixed fluvial, aeolian and lacustrine environment under arid climatic conditions (Glennie 1972). In Annerveen all those facies are recognised (Fig. 6). Overall, a fining upwards trend can be recognised. This coincides with a shift in sedimentary environment from coarse alluvial braid plain clastics via finer fluvial and aeolian clastics to lacustrine deposits. 'Weissliegend' sediments (reworked, structureless, uncoloured sandstones, locally found near the top of the Rotliegend, Glennie 1990) are absent in the Annerveen area where the Zechstein transgressed mainly fine-grained lacustrine clastics.

Reservoir characteristics

Reservoir data for the Rotliegend in the Annerveen field can be summarised as follows:

Thickness	100–150 m
Average porosity	11%
Average net reservoir sands	75%
Average gas saturation	67%
Average horizontal permeability	40 mD
Average well productivity	1–2 × 10^6 m^3/day
Gross rock volume	5.3 × 10^9 m^3
Gas initially in place	75.6 × 10^9 m^3
Estimated recovery factor	
(at a 30 bar abandonment pressure)	93%

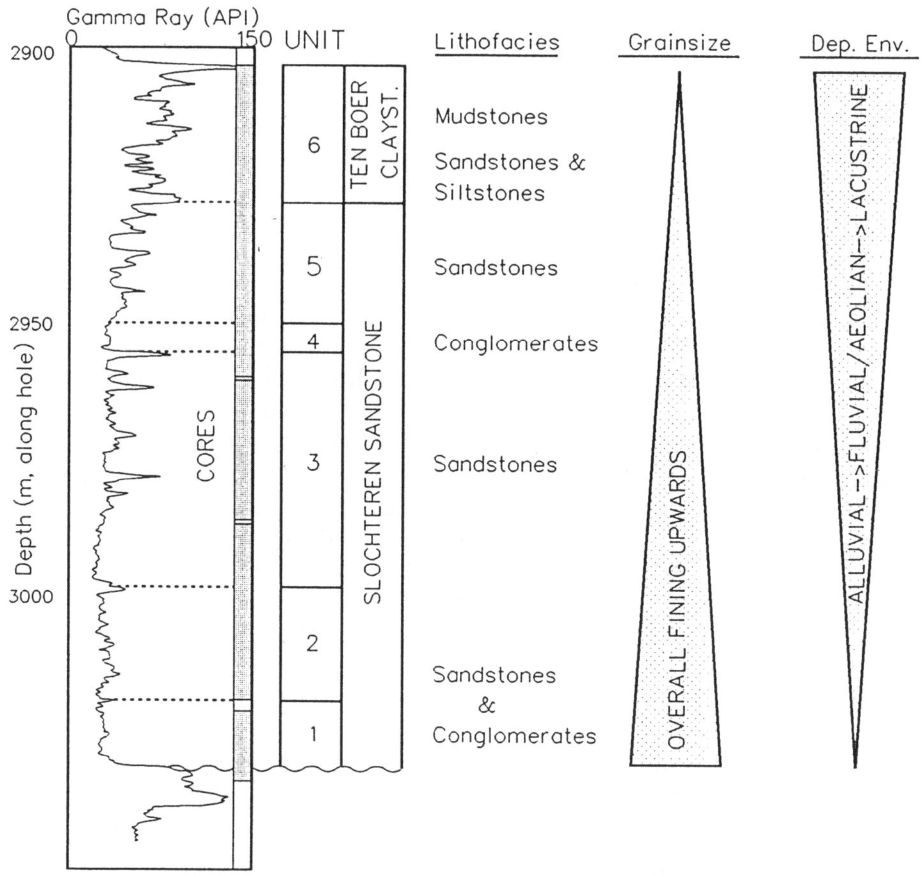

Fig. 6. Rotliegend type-log of the Annerveen field: well Annerveen-1.

Diagenesis

Much of the porosity and permeability is primary depositional. Mild diagenetic overprints are observed such as some feldspar leaching and growth of small amounts of authigenic quartz, kaolinite and illite. Illite growth in the north-eastern Netherlands took place during the Late Jurassic to Early Cretaceous, upon deep burial and prior to Late Kimmerian uplift (Lee et al. 1989).

The relatively minor influence of diagenesis on reservoir potential in Annerveen contrasts with the Rotliegend reservoirs found in the North Sea and in north-west Germany. Early (shallow burial) diagenesis has been observed in the UK sector of the North Sea (Glennie et al. 1978). Early diagenetic cements are generally found in sandstones deposited close to the water table, along sabkha margins of desert lakes or in interdune areas. Most early cements are removed upon burial, resulting in some secondary porosity. Late (deep burial) diagenesis is found both in the Dutch offshore and in Germany. In the Dutch offshore K and L blocks porosities similar to those found in Annerveen have been reported (Oele et al. 1981). Locally in the K and L blocks, deep Mesozoic palaeo-burial, deeper than 4 km, caused growth of fibrous illite which deteriorates permeability (Seemann 1979). Likewise, in Germany, severe illite growth has been observed in reservoirs buried between 4 and 6 km (Hancock 1978). Here, illite growth is attributed to the dissolution of volcanic detrital grains within the Rotliegend.

It can be concluded that primary porosity and permeability has been preserved in Annerveen because:

1. Annerveen is situated on the southern margin of the Southern Permian Basin, therefore relatively far from areas of fine clastic deposition and relatively high above the water table of the desert lake (Fig. 1). This precludes early cementation and development of secondary porosity.

2. The Annerveen – Groningen area has probably never been buried deeper than 4 km (Lee et al. 1989: their fig. 6) which largely precludes deep burial diagenesis.

Reservoir subdivision
The Upper Rotliegend Group can been subdivided into two formations, the Slochteren Sandstone Formation and the Silverpit Claystone Formation (NAM & RGD 1980). In Annerveen, the upper part of the Rotliegend comprises the Ten Boer Claystone Member, which can be correlated with the upper part of the Silverpit Claystone of the UK sector of the southern North Sea (Fig. 5). The Slochteren Sandstone is partly equivalent to the Leman Sandstone in this sector (Van Veen 1975). The intra-Slochteren Ameland Claystone Member is absent over the Annerveen field.

The Slochteren Sandstone Formation of Annerveen has been subdivided into units (Figs 3, 6). This subdivision was made based on core evaluation and log correlation of wells in the Annerveen and Groningen fields. The subdivision attempts to optimally describe the reservoir and serves as a base for multi-layer dynamic reservoir modelling. The subdivision is tabulated below and the members will be briefly described.

Unit	Informal name	Stratigraphic name
Unit 6	Ten Boer Muds and Sands	Ten Boer Claystone
Unit 5	Upper Sands	Slochteren Sandstone
Unit 4	Upper Conglomerates	Slochteren Sandstone
Unit 3	Main Sands	Slochteren Sandstone
Unit 2	Lower Pebbly Sands	Slochteren Sandstone
Unit 1	Basal Conglomerates	Slochteren Sandstone

Unit 1: Basal Conglomerates. The unit consists of conglomerates and pebbly sandstones deposited in a braided stream to alluvial fan environment. It thickens towards the south relative to the overlying unit 2; taken together the thickness remains more or less constant (30–40 m). Well ETV-1 in the south of the field (Fig. 2) penetrates alluvial fan sequences in the lower part of the Slochteren, whereas wells further north have penetrated laterally equivalent braided stream sequences. Reservoir quality is likely to vary considerably over short distances within an alluvial fan setting. Porosities vary from 3–7% and horizontal permeabilities are low: 0.5–10 mD.

Unit 2: Lower Pebbly Sands. Pebbly sandstones alternating with coarse- to fine-grained sandstones form this unit. They alternate with conglomerate stringers and a few shale layers, and are interpreted as fluviatile braid plain sediments with alluvial fan incursions. Porosities and horizontal permeabilities are low in this poorly sorted unit: 6–10% and 3–30 mD respectively.

Unit 3: Main Sands. The unit comprises mixed fluviatile and aeolian, medium- to fine-grained sands deposited in a desert plain and low-energy braided stream environment. Thin conglomerate and shale beds are found. The thickness increases rather regularly towards the north-west and ranges from 20–50 m. This unit constitutes the best reservoir of Annerveen. Porosities range from 10–15% and horizontal permeabilities from 30–300 mD. The correlation between reservoir quality and depositional environment for the Rotliegend has been noted earlier by Nagtegaal (1979). He concluded that, in the absence of diagenesis, porosities (and permeabilities) show a strong dependence on sorting. As aeolian sandstones are in general better sorted than fluviatile sandstones, they show better porosity and permeability trends. This probably explains the better reservoir potential of unit 3 (and 5) at those places where aeolian sediments occur.

Unit 4: Upper Conglomerates. A thin but continuous ‘band’ of conglomerates and pebbly sandstones separates units 3 and 5. A high-energy braided stream environment is envisaged, possibly with storm-induced sheet floods. Porosities are moderate (5–11%) while horizontal permeabilities are low (3–30 mD). The unit is characterised by the absence of intercalated shale beds and is commonly directly underlain by a mudstone bed. This horizon therefore could represent a partial barrier or baffle to vertical gas and water transmissibility.

Unit 5: Upper Sands. This unit is the most prolific reservoir after unit 3. Fine- to medium-grained sand layers alternate with some shale beds and thin layers of conglomerate and pebbly sandstone. An interfingering of fluviatile, aeolian and lacustrine depositional environments is assumed. Thicknesses increase gradually to the north (10–24 m). The porosity varies from 8–14% and increases towards the north and northeast. Horizontal permeabilities are 10–100 mD. The western part of Annerveen (ZLN-1) exhibits the lowest values.

Unit 6: Ten Boer Muds and Sands. In Annerveen, the Ten Boer Claystone Member consists of continuous shale layers with intercalations of fine sandstones, silts and shaly sands. The unit represents the southernmost incursion of the Rotliegend desert lake (Fig. 1). In the Groningen field the Ten Boer Claystone comprises mostly non-reservoir shales and silts and acts as a regional shaly seal towards the north. In Annerveen, being located closer to the Rotliegend basin margin, there is a clear trend from coarse to fine clastics from southeast to northwest. The lower sequences of the Ten Boer contain reservoir quality sands (porosities from 7–13%) and are in communication with the Slochteren Formation. Several wells in Annerveen produce gas from the lower Ten Boer. The upper intervals are considered to comprise largely waste-rock, as evidenced by the fact that pressure delay has been observed in some of the upper sand layers. The thickness of the Ten Boer increases gradually from east to west.

Permian: Zechstein (cap rock)

Rotliegend sedimentation ended in the Late Permian when continental deposits were rather abruptly flooded by a marine transgression (Glennie 1983). Evaporites and carbonates were deposited in the Southern Permian Basin and they also blanketed the Annerveen and Groningen area. In Annerveen, all major Zechstein cycles (Glennie 1990) are present and similar to the cycles described in Stäuble & Milius 1970. Zechstein halokinesis has caused large thickness variations, notably at the eastern flank of the Annerveen field near the Veendam salt wall (Fig. 2). Here, Zechstein salts penetrate into the Tertiary and the salt is mined by solution at depths from 100–200 m below the surface. As the impermeable Zechstein strata form a continuous cover over the Rotliegend, they are interpreted to form the seal for the Annerveen gas accumulation.

Post-Permian

On top of the Permian, most wells in the Annerveen area encounter Triassic, Lower Buntsandstein Formation shales and Röt Formation evaporites and fine clastics, as a subcrop to the Base Cretaceous Unconformity. During the Triassic and Jurassic the Annerveen area was part of the southern rim of the regional Groningen 'palaeo-high' (Stäuble & Milius 1970). Therefore, most of the Upper Germanic Trias Group (Röt, Muschelkalk and Keuper Formations) has been removed due to Late Kimmerian truncation. As Late Kimmerian uplift is considered to have been rather mild in this area (Van Wijhe et al. 1980), the latter formations may have had a reduced depositional thickness. An exception to this is formed by the rim synclines of the Veendam salt wall, covering the eastern part of the Annerveen field (Fig. 7). Here, sedimentation continued and younger deposits, in particular thick Jurassic Altena Group shales, have been preserved as a result of halokinesis during the Triassic and/or Jurassic.

Overlying the truncated Triassic to Jurassic sequence, Early Cretaceous marine shales of the Rijnland Group onlap the Late Kimmerian erosional surface in the area. During the Late Cretaceous, Chalk Group sediments uniformly covered the area. Reduced Chalk Group thicknesses are observed above salt swells.

The Annerveen area continued to subside during the Tertiary. Thick sequences comprising marine shales and glauconitic sands of the North Sea Group were deposited. Sedimentation was affected by continuing halokenitic movements of the Zechstein. The Veendam salt wall pierced through the Cretaceous and Tertiary and is now covered by Late Tertiary and Quaternary sediments only.

Structure

The Annerveen field is covered by 3D seismic reflection data. An early 3D survey (1984) was acquired over the eastern part of the field to resolve structural uncertainties pertaining to the area below the Veendam salt wall. Later on, NAM adopted a 'blanket' approach to 3D acquisition and the western half of the Annerveen field was covered as well (1992).

Most peripheral wells and at least one well per cluster have checkshot surveys and therefore adequate velocity information is present to map and depth-convert the Annerveen structure. The objective seismic horizon is formed by the basal Zechstein reflection (Figs 4, 7). This strong negative loop is caused by an acoustic impedance contrast of basal Zechstein anhydrites with overlying halite. When depth-converted, the top of the gas-bearing reservoir (top Rotliegend Group) can be mapped by adding an isochore of some 50 m to this horizon. Seismic horizon attributes have been used to map the fault pattern in great detail (Hoetz & Watters 1992).

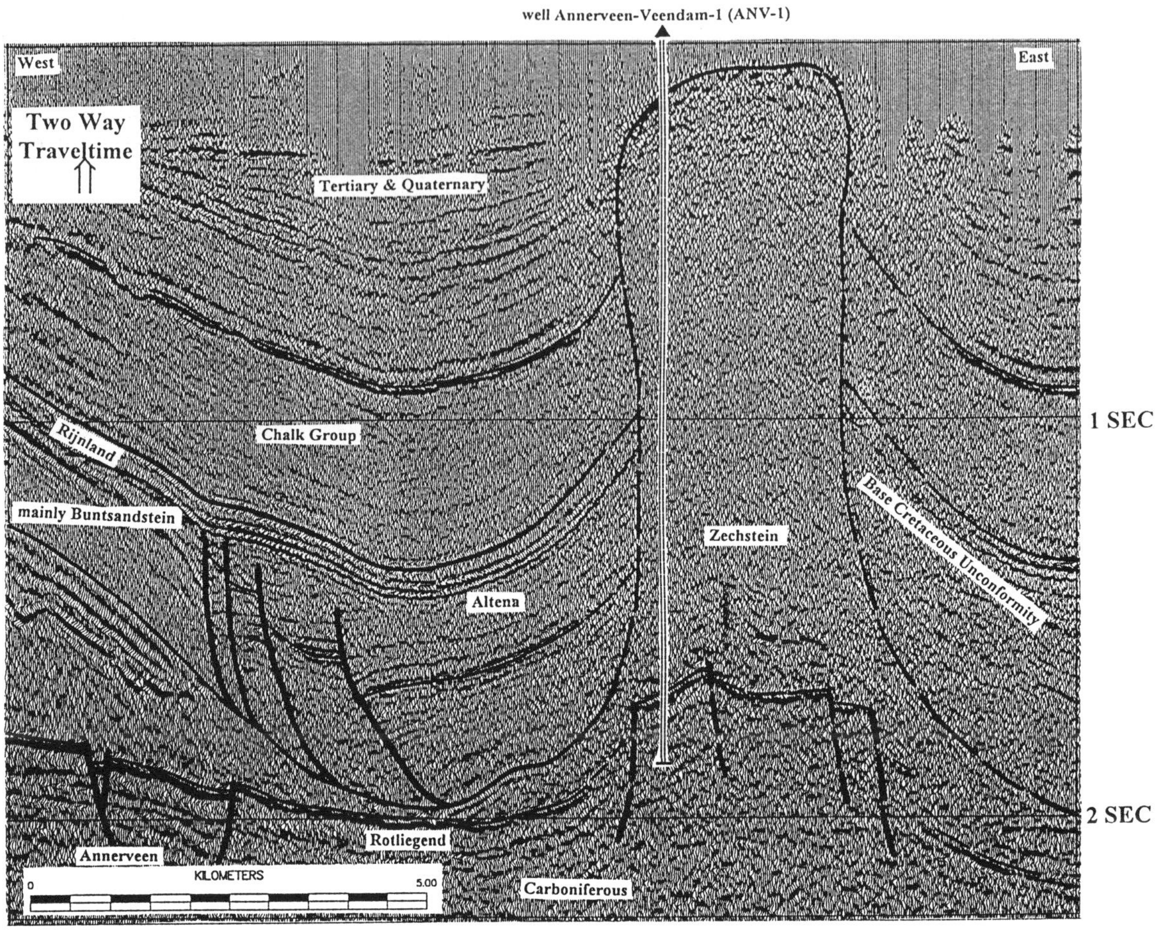

Fig. 7. Seismic profile (east – west) through well ANV-1 (Fig. 2), showing piercement of the Veendam salt wall and Jurassic sedimentation in its rim synclines.

Regional geological setting

The Southern Permian Basin

The Annerveen field is located at the southern rim of the Southern Permian Basin (fig. 1). The basin extends from the English onshore to Poland and probably originated during the Autunian. It subsided as a result of post-Variscan thermal contraction of the lithosphere and was filled with clastics mainly during the Saxonian by northward directed drainage systems, which progressively unroofed the Variscan hinterland (Ziegler 1990). In this basin, palaeotopographic highs existed that control facies and depositional thicknesses (Gast 1988).

The Annerveen area is interpreted to form the southern rim of the regional Groningen intra-basinal palaeohigh (Stäuble & Milius 1970). This palaeohigh probably originated during Variscan times as evidenced by the pre-Permian subcrop (Van Wijhe et al. 1980: fig. 1; see also Fig. 4). The Groningen and Annerveen area continued to be relatively high during the Mesozoic (Stäuble & Milius 1970).

Gas charge and quality

On burial the Westphalian coaly source rocks generated gas which migrated directly into the overlying Rotliegend reservoirs. Jurassic and Tertiary to Quaternary charge phases, before and after Late Kimmerian uplift respectively, were modelled for the Groningen area (Van Wijhe et al. 1980) and are also applicable to the Annerveen area. The major difference between Groningen and Annerveen is found in the gas quality. Groningen gas (low-calorific) has some 14 mole % nitrogen while Annerveen gas (high-calorific) contains only 4 mole % nitrogen. This compositional difference is difficult to explain as it demands knowledge of geological factors controlling nitrogen generation and distribution. Influencing factors may be: source rock facies (primary nitrogen distribution), source rock maturity (fractionation of nitrogen upon generation and expulsion) and structural position (migration and spill of gas derived from different palaeo-drainage areas).

Structural description of Annerveen

The Annerveen gas accumulation is contained in an east – west trending elongate horst block. It is bounded to the north and south by two main boundary fault zones; the field is dip-closed to the west (Fig. 2). Overall, the horst block is tilted towards the south. At its eastern side, the Annerveen structure is seismically masked by the overlying Veendam salt wall.

Annerveen is cut by numerous north – south trending faults none of which has a large enough throw to completely offset the reservoir. The faults as mapped at base Zechstein level are extrapolated vertically to top Rotliegend. They can be compared to intra-reservoir faults (with throws < 10 m) which are detected in the Annerveen wells by means of log correlation. This comparison indicates that faults cannot be detected on the 3D seismic surveys if the throw is less than some 10–20 m (see also Hoetz & Watters 1992).

Three major fault classes are recognised:

1. E – W oblique-slip faults with a largely normal component. On Fig. 4, a change in dip below the Saalian Unconformity indicates that the northern Annerveen boundary fault is a re-activated pre-Saalian structural lineament. A wrench component might be deduced from associated WNW – ESE Riedel-type faults (e.g. north of the ZLV cluster).
2. N – S extensional faults. These faults are, at the boundaries of the field, offset by the E – W trending boundary faults and may therefore also represent older, reactivated Carboniferous lineations.
3. NW – SE faults of extensional nature, trending towards Groningen. They are interpreted to be of post-Permian age (probably Late Kimmerian, Stäuble & Milius 1970).

Timing of faulting

As all faults, offsetting the Rotliegend and basal Zechstein, disappear in the more ductile Zechstein evaporites immediately above the basal Zechstein reflection (Figs 4, 7), it is difficult to arrive at firm conclusions on the timing of generation of the faults. Existing pre-Saalian palaeohighs were probably bounded by Variscan structural elements. These might have bounded the southern rim of the Groningen – Annerveen palaeohigh and were oriented in a N – S and E – W pattern. Subsequently, these structural elements were re-activated during the Late Kimmerian orogenic event in the Late Jurassic to Early Cretaceous. Apart from this re-activation a NW – SE fault trend developed which had a minor impact on Annerveen, but became the dominant trend in Groningen.

Fault sealing

The pressure decline in the Annerveen field indicates that most of the field is well drained. Consequently, the numerous faults which are mapped in the main field do not form transmissibility barriers. Exceptions are formed by peripheral blocks which are found juxtaposed at the hanging wall of the Annerveen northern boundary fault (see below). No obvious relation can be demonstrated between the type of fault (or orientation) and the sealing capacity.

Relationship with the Groningen field: the ANV-1 area

The area between Groningen and Annerveen is seismically masked by the Veendam salt wall and therefore structural resolution is low. Below, an attempt is made to integrate all data from wells, seismic, pressures, etc., in order to arrive at meaningful conclusions on the relationship of Annerveen with the Groningen field.

Well data

Appraisal well ANV-1, drilled in 1969 close to the edge of the Annerveen field, encountered gas of a quality similar to Annerveen (high-calorific, 3.8% nitrogen). It recorded an initial pressure of 345 bara at 2900 m below MSL (Mean Sea Level). The well was converted to an observation well in 1977 when a pressure decline to 273 bara was measured. A gas-water contact of 3035 m below MSL was calculated from the logs which is some 25 m shallower than the average gas-water contact in the Annerveen field (3060 m below MSL).

VSP results

To increase the structural resolution, walk-away Vertical Seismic Profiles (VSP) were acquired in two wells.

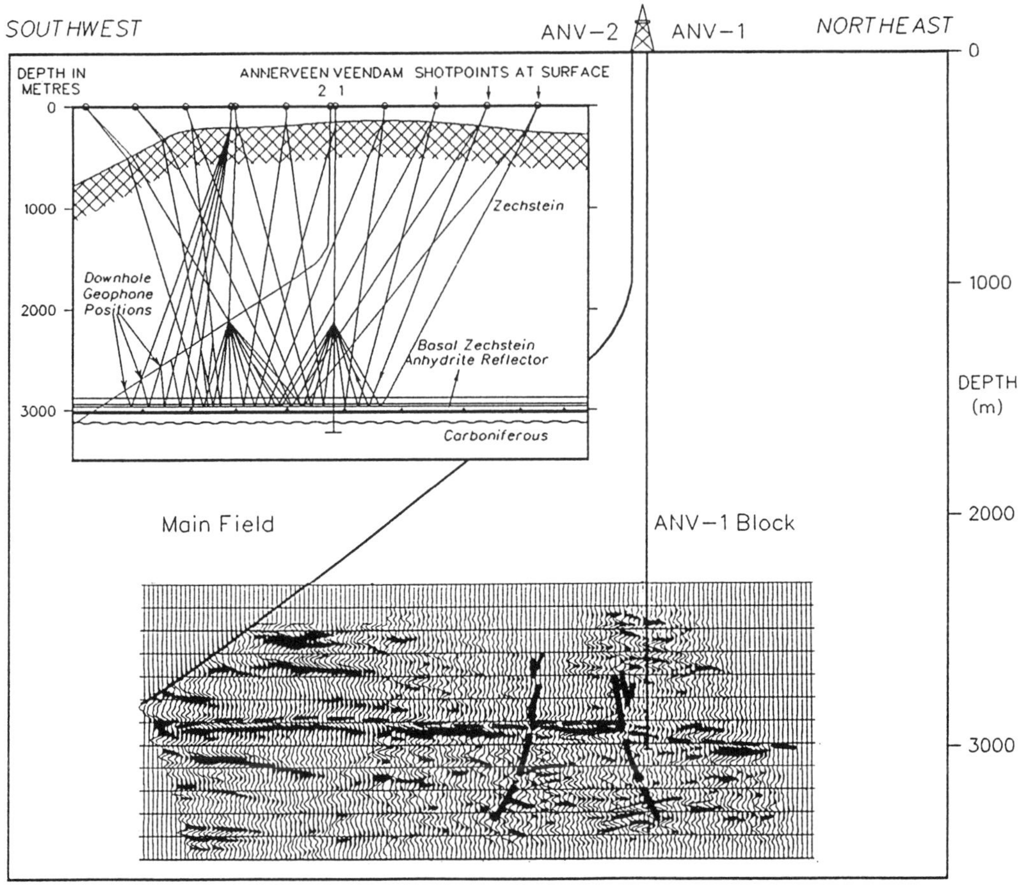

Fig. 8. Combined VSP results of wells ANV-1 and 2. Inset: model showing raypaths of reflected acoustic energy.

One VSP was acquired in well ANV-1 and illustrates the area immediately around the well at reservoir level. Another VSP was recorded in well ANV-2, drilled from the same surface location as the previous well, but then deviated into the main Annerveen field. This VSP shows the area below the well trajectory. Shotpoints were placed at the surface while geophones were lowered in the wells (Fig. 8, inset). Short offsets were used resulting in raypaths mostly through salt. This precluded severe raybending which would have been recorded with surface seismic, using larger offsets. Results of both VSPs have been spliced together and are displayed with a depth scale on Fig. 8. The effect of the travel time or velocity pull-up, caused by the Veendam salt wall (Fig. 7), has been compensated for. Although difficult to interpret, the results seem to indicate that ANV-1 is separated from the main field by a (small) normal fault.

Pressure data

After well ANV-1 was converted to an observation well, regular pressure measurements were taken. The first data are from 1977 when the pressure in the block appeared to be similar to the Annerveen main field (Fig. 9). Later measurements showed that the ANV-1 pressure decline did not follow the relatively rapid pressure decrease of Annerveen but rather followed the pressure trend of ZWD-1, the nearest well in the Groningen field, albeit with a 30 bar lag. Well ZWD-1, a well with Groningen quality gas, itself follows the main Groningen field trend with a small (< 10 bar) lag. Thus the ANV-1 block has Annerveen quality gas, yet its pressure decline follows that of the southern Groningen margin. Currently the fault, or fault zone, between ANV-1 and the nearby part of the Annerveen field is holding a pressure differential of more than 100 bar. As the quality of the seal is difficult to estimate, the risk exists that the fault could be breached once the

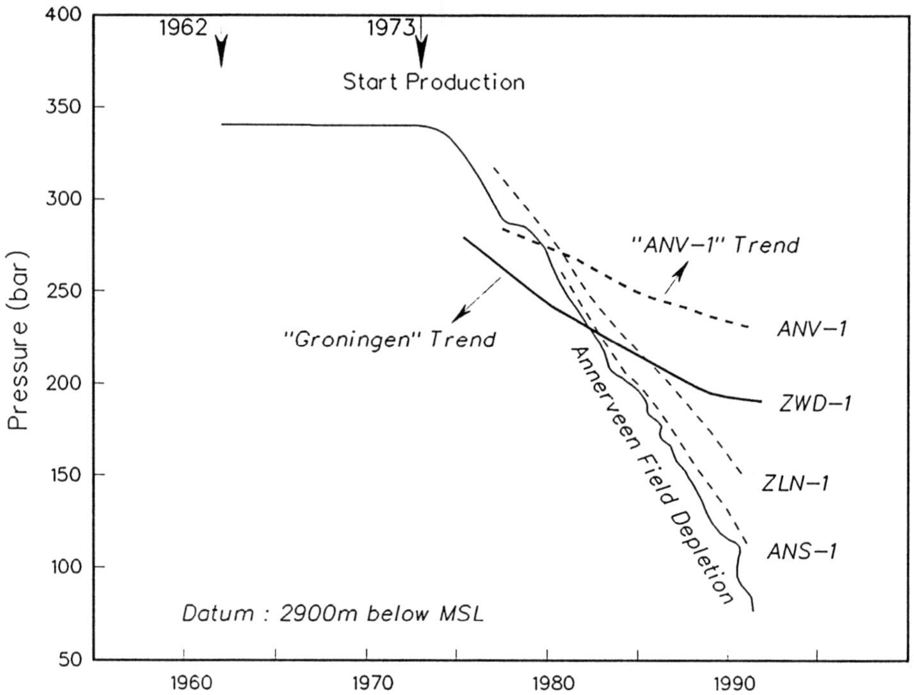

Fig. 9. Reservoir pressure versus time for the Annerveen field and the southern margin of the Groningen field.

pressure differential exceeds the fault's sealing capacity. Subsequently, the ANV-1 gas would flow to the Annerveen main field.

Structural interpretation of the ANV-1 area

In view of the pressure, gas-water contact and gas quality of ANV-1, the fault between the ANV-1 block and the Annerveen main field is interpreted to be sealing. The fault probably belongs to the Annerveen northern boundary fault zone. Just west of ANV-1, there is an area, the 'Oosterdiep area', where throw across the fault is small. A fault plane diagram, in which both foot wall and hanging wall depths have been plotted, is shown on Fig. 10a. It shows that gas-legs across the fault are juxtaposed. The depth of this area (around 3000 m below MSL) coincides with the gas-water contact of southern Groningen and is somewhat higher than the gas-water contact of the Annerveen main field (around 3060 m below MSL). It may be inferred that also this part of the boundary fault is sealing.

Summary

1. The gas quality in ANV-1 is similar to the Annerveen main field; therefore its gas-leg cannot be in direct communication with the Groningen gas.
2. The gas-water contact of ANV-1 is some 25 m shallower than the Annerveen main field and pressures are considerably higher; therefore the gas-legs of Annerveen and ANV-1 are unlikely to communicate.
3. The ANV-1 gas-leg pressures deplete 'like' the southern rim of the Groningen field.

ANV-1 is for these reasons interpreted to have penetrated an isolated, small fault-block juxtaposed to the Annerveen field. The map around ANV-1 has been drawn manually using VSP results and dipmeter data (giving a 14° dip to the east in the Ten Boer Claystone). At the edges of the Veendam salt wall, seismic shadow zones, from where no reflections are received by the recording instruments, are present, obscuring the basal Zechstein reflections. The small ANV-1 block, although separated from the main Annerveen field, is interpreted to be in communication with the southern part of the Groningen field via the aquifer. As Annerveen depletes more rapidly than the Groningen field a pressure differential is building up between ANV-1 and the Annerveen field.

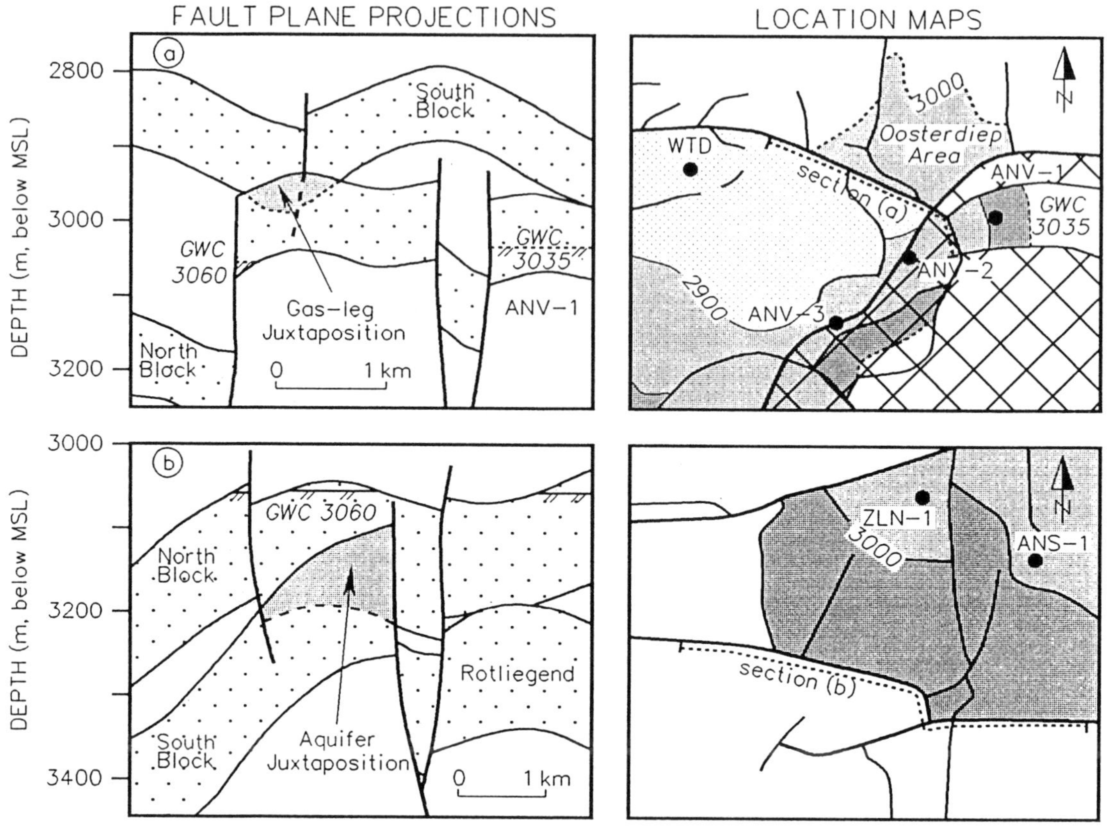

Fig. 10. Fault plane projections of: (a) the 'Oosterdiep area' (at the eastern flank of the Annerveen field); and (b) the ANS-1 area (at the western flank of the Annerveen field).

The aquifers of Annerveen

In a field which is in an advanced stage of production, like Annerveen, there is the risk that water might invade the gas-leg and thus 'close off' gas reserves from the producing wells. In the Annerveen area, two main aquifers are connected to the gas-leg:

1. To the west there is an open Slochteren-Slochteren connection via the aquifer with the neighbouring Vries field. This aquifer is bounded by the east – west running southern boundary fault of Annerveen. Some pressure decline has been measured in the Vries field which is probably related to depletion in the Annerveen field.
2. There is juxtaposition of an aquifer to the south of Annerveen and the ANS-1 area (Fig. 10b).

Pressure data (Fig. 9) show that, apart from well ANV-1, two wells deviate from the line of main field depletion: ANS-1 and ZLN-1. These wells have in common that they are situated at the western flank of Annerveen where the aquifers are connected. From 1980 onwards regular TDT (Thermal Decay Time) logs have been recorded. These logs essentially measure water saturation and can be compared to previous, similar measurements. On Fig. 11 seven TDT logs, taken every two years, have been displayed together with a 1975 base log. When no water saturation increase is present, the curves should overlap. It can be seen that over the years there is a gradual increase in water saturation in the sandstones with curve deflections to the right, i.e. the sandstones with highest porosity. In ANS-1, gas saturation losses have occurred in a 43 m zone above the original gas-water contact. In ZLN-1, this zone is only 11 m thick.

It may be concluded that some aquifer activity is taking place in the western half of the Annerveen field, since full communication and juxtaposition across faults exists on this side of the field. Pathways of high transmissibility leading to ANS-1, which cannot be identified from the sparse well control, might cause

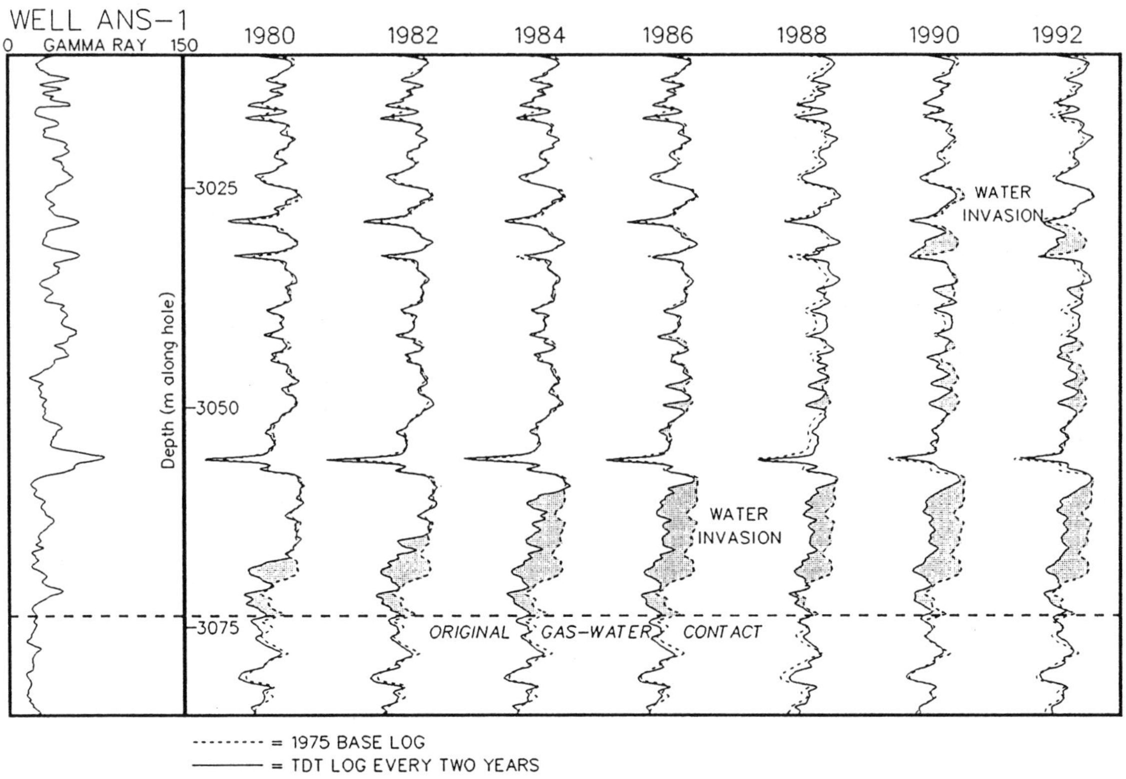

Fig. 11. Annerveen field, well ANS-1, time lapse Thermal Decay Time logs. All logs are compared to a 1975 run.

the gas saturation decrease in this well. The comparatively small rise in ZLN-1 can be attributed to the less favourable reservoir parameters (relatively low transmissibility) in this well.

Conclusion

The Annerveen field is at an advanced stage of production. At the time of writing over 75% of the GIIP has been produced. It is shown that geological evaluation and integration of data acquired during the lifetime of the field remains necessary to the point of field abandonment. In the case of Annerveen it can be concluded that semi-transmissive and/or sealing faults, e.g. the ANV-1 area at the eastern flank, as well as aquifer influx, e.g. the ANS-1 area at the western flank, could cause a loss of gas recovery towards the end of the field's lifetime.

Acknowledgements

The author is indebted to the Nederlandse Aardolie Maatschappij B.V., Shell Internationale Petroleum Maatschappij BV, Esso Nederland BV and Energie Beheer Nederland BV for granting permission to publish this paper. Particularly thanked are the colleagues within NAM, present and past, who have contributed to and reviewed this paper.

References

Gast, R.E. 1988 Rifting im Rotliegenden Niedersachsens – Die Geowissenschaften 6(4): 115–122

Glennie, K.W. 1972 Permian Rotliegendes of Northwest Europe interpreted in light of modern desert sedimentation studies – Am. Ass. Petroleum Geol. Bull. 56: 1048–1071

Glennie, K.W. 1983 Lower Permian Rotliegend desert sedimentation in the North Sea area. In: Brookfield, M.E. & T.S. Ahlbrandt (eds)

1983 Aeolian sands. Developments in sedimentology – Elsevier, Amsterdam: 521–541

Glennie, K.W. 1990 Lower Permian Rotliegend. In: Glennie, K.W. (ed.) Introduction to the petroleum geology of the North Sea – Blackwell, Oxford: 402 pp

Glennie, K.W., G.C. Mudd & P.J.C. Nagtegaal 1978 Depositional environment and diagenesis of Permian Rotliegendes sandstones in Leman Bank and Sole Pit areas of the UK southern North Sea – J. Geol. Soc. London 135: 25–34

Hancock, N.J. 1978 Possible causes of Rotliegend sandstone diagenesis in northern West Germany – J. Geol. Soc. London 135: 35–40

Hoetz, H.L.J.G. & D.G. Watters 1992 Seismic horizon attribute mapping for the Annerveen Gasfield, The Netherlands – First Break 10: 41–51

Lee, M., J.L. Aronson & S.M. Savin 1989 Timing and conditions of Permian Rotliegende Sandstone diagenesis, Southern North Sea: K/Ar and oxygen isotopic data – Am. Ass. Petroleum Geol. Bull. 73: 195–215

Nagtegaal, P.J.C. 1979 Relationship of facies and reservoir quality in Rotliegend desert sandstones, southern North Sea region – J. Petroleum Geol. 2: 145–158

NAM & RGD (Nederlandse Aardolie Maatschappij & Rijks Geologische Dienst) 1980 Stratigraphic Nomenclature of The Netherlands – Verh. Kon. Ned. Geol. Mijnb. Gen. 32: 1–77

Oele, J.A., A.C.P.J. Hol & J. Tiemens 1981 Some Rotliegend gas fields of the K and L blocks, Netherlands offshore (1968–1978) – a case history. In: Illing, L.V. & Hobson, G.D. (eds) Petroleum geology of the continental shelf of North West Europe – Heyden and Son, London: 289–300

Seemann, U. 1979 Diagenetically formed interstitial clay minerals as a factor in Rotliegend sandstone reservoir quality in the Dutch sector of the North Sea – J. Petroleum Geol. 1: 55–62

Stäuble, A.J. & G. Milius 1970 Geology of the Groningen gas field – Am. Ass. Petroleum Geol. Mem. 14: 359–369

Van Veen, F.R. 1975 Geology of the Leman Gas-Field. In: Woodland, A.W. (ed.) Petroleum and the continental shelf of Northwest Europe, Vol. 1, Geology – Elsevier Applied Science Publishers, Barking: 223–233

Van Wijhe, D.H. & M.J.M. Bless 1974 The Westphalian of the Netherlands with special reference to miospore assemblages – Geol. Mijnbouw 53: 295–328

Van Wijhe, D.H., M. Lutz & J.P.H. Kaasschieter 1980 The Rotliegend of The Netherlands and its gas accumulations – Geol. Mijnbouw 59: 3–24

Ziegler, P.A. 1990 Geological atlas of Western and Central Europe – Shell Internationale Petroleum Maatschappij, The Hague/Geol. Soc. Publ. House, Bath, 239 pp.

Rondeel et al. (eds), Geology of gas and oil under the Netherlands, 93–102, 1996.

Development of a tight gas reservoir by a multiple fracced horizontal well: Ameland-204, the Netherlands

S.V. Crouch[1], W.E.L. Baumgartner[1], E.J.M.J. Houlleberghs[1] & J.P. Walzebuck[1,2]
[1] *Nederlandse Aardolie Maatschappij BV (Business Unit Offshore), Grote Hout of Koningsweg 49, 1950 AA Velsen, the Netherlands;* [2] *Present address: BEB Erdgas und Erdöl GmbH, Riethorst 12, D-30659 Hannover, Germany*

Key words: bimodal productivity, Rotliegend, chlorite, kaolinite, well stimulation

Abstract

Poor productivity of a number of tight Permian gas reservoirs is a major constraint on field development in the Dutch offshore sector. In several fields economic development of Rotliegend reservoirs could not be achieved with conventional development techniques due to very low flow rates of less than 200 000 m^3/day. This case history covers the Ameland field in the Dutch North-Friesland concession area. The flanks of this layered reservoir can in general be classified as 'porous but tight'. The geological causes are poor sand facies development and pervasive clay mineral diagenesis, which affected the pore connectivity, resulting in significant untapped reserves. To improve this reservoir performance multiple-fracced horizontal well (MFHW) technology was applied. The project was approached in an integrated manner. The key problems addressed were: 1) verification of sufficient GIIP (gas initially in place); 2) modelling of reservoir architecture; and 3) analysis of productivity improvement by multiple-fraccing technology. The success of the project was confirmed when the MFHW well achieved economic gas production rates (transient) in the order of 2×10^6 m^3/day at 120 bar drawdown from this tight Rotliegend gas reservoir. The experience gained and the positive results indicate that multiple-fraccing of horizontal wells may be applicable to other tight gas reservoirs and could raise them from uneconomic prospects to potentially profitable developments.

Introduction

The Nederlandse Aardolie Maatschappij, in conjunction with partners Mobil Producing Netherlands Inc. and Energie Beheer Nederland BV, has recently drilled AME-204, the first, multiple propped, hydraulic fracture stimulated, horizontal well in the Shell Group of companies. This paper delineates the geological reasons why such a technically complex well was necessary, and how the well design was optimised.

The four Ameland fields are situated on the northern coast of the Netherlands in the North-Friesland concession (Fig. 1). The Ameland-Oost (AME) field is the third largest gas field in the Netherlands, with a volume of gas initially in place (GIIP) of 58.5×10^9 m^3. The gas reservoir is the Upper Slochteren Sandstone, which is a member of the Rotliegend Group and is of Permian age. The average isopach thickness is ca. 90–100 m and the reservoir depth varies between 3200–3500 m sub sea.

The structural configuration of the AME field comprises a series of NNW – SSE and WNW – ESE trending normal faults. These divide the field into six major southerly dipping fault-bounded blocks (Fig. 1). The depositional pattern has also been affected by these faults, indicating a degree of synsedimentary faulting.

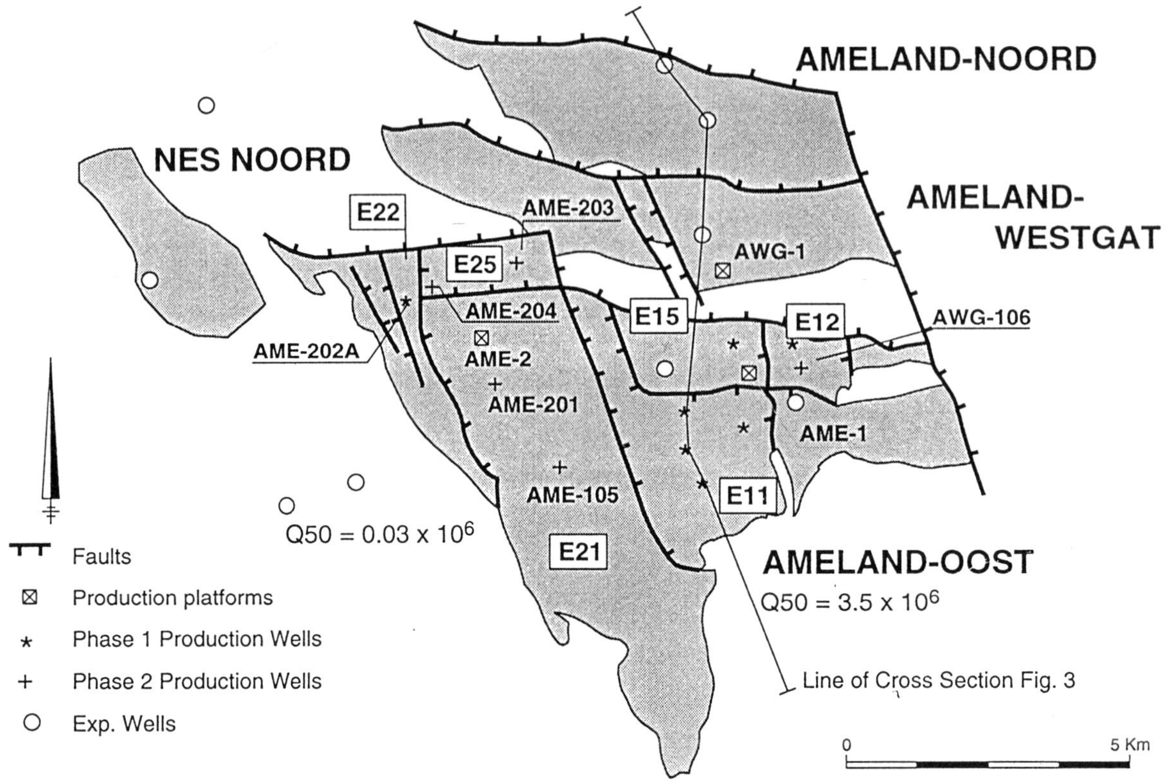

Fig. 1. Schematic map of the Ameland gas fields (grey) showing exploration wells (Exp.), phased development wells and major faults. Q50 = productivity under 50 bar drawdown in m^3/day.

Development history

The development of the Ameland field was carried out in two phases.

During the first phase, the central part of the field was developed (blocks E15 and E11, Fig.1). The wells had high productivities, in the range of $3.5–5.0 \times 10^6$ m^3/day at 50 bar drawdown. For the first five years of production the contractual capacity of the Ameland field could be maintained from these wells alone, and it was initially assumed that the connectivity in the field was adequate to ensure the effective drainage of the flank blocks towards the large pressure sink in the central area. However, after two years of production, pressure data indicated that the material balance GIIP was substantially lower than the volumetric GIIP. It was concluded that the reserves in the flank blocks were not being drained effectively. Combined with the progressive decline in reservoir pressure and well productivity in the central part of the field, this meant that a new strategy had to be developed for positioning future wells, which were required to maintain production capacity and to ensure proper drainage of the flank blocks.

Reservoir simulation showed that further infill drilling in the central blocks was not economically justifiable. Therefore, the second phase of development had to focus on the flank blocks, which the appraisal wells had shown to be less productive. The western flank of the field was developed by installing the AME-2 mini-satellite platform, tying-back of the appraisal well AME-201 and drilling the new well AME-203. However, when the first wells in this area showed well productivities as low as 0.03×10^6 m^3/day the emphasis of the planning phase changed.

Prior to further development it was necessary to understand the geological causes of the low productivity of the flank blocks compared to the central field, so that future well designs could be optimised.

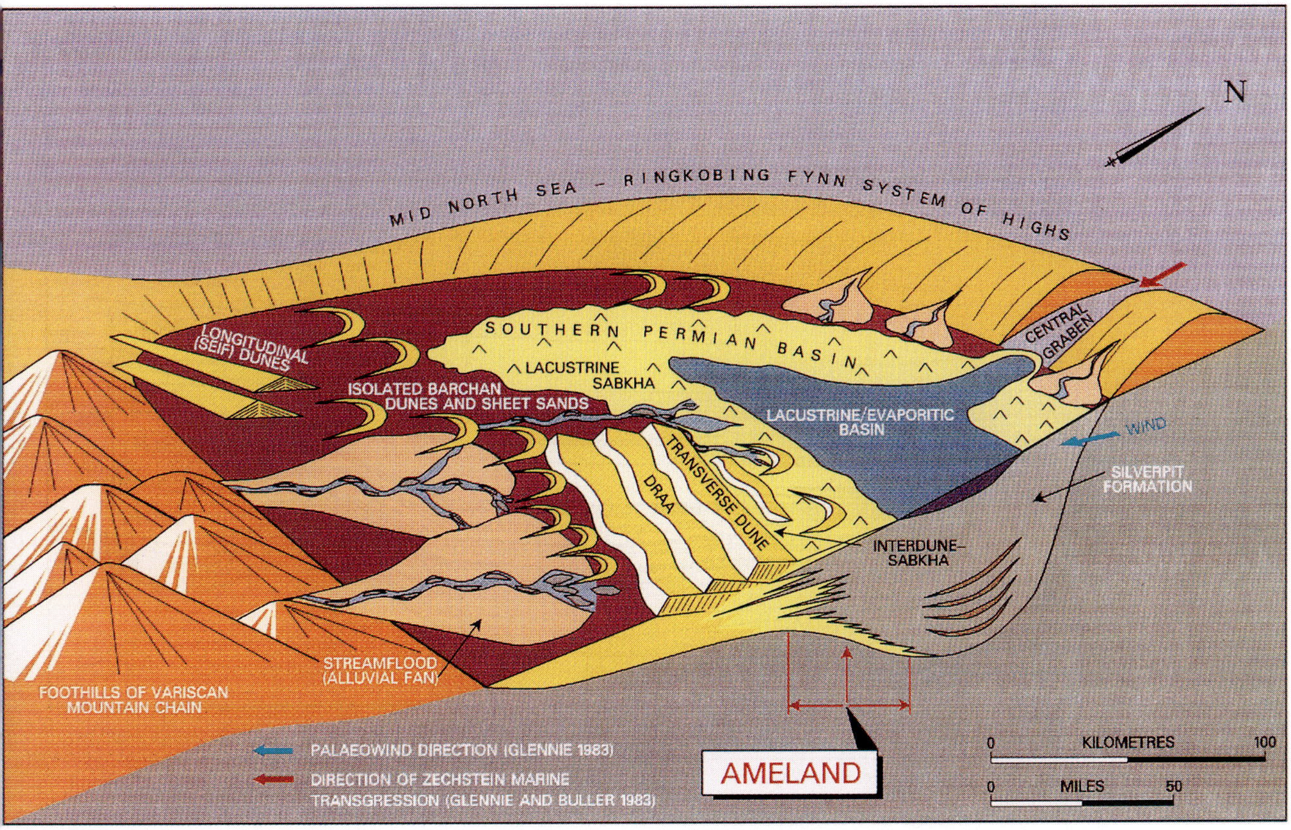

Fig. 2. Schematic diagram of the mid-European Southern Permian Basin showing the palaeogeographic setting of the Ameland reservoir.

Differences between the central blocks and the flanks of the field

The reservoir quality is determined by two main factors: depositional environment and diagenetic overprint.

Depositional environment

The Permian Rotliegend sandstones in the Ameland area have been deposited on the southern margin of the mid-European Southern Permian Basin (Fig. 2). The Upper Rotliegend Group consists of a locally conglomeratic sandy deposit along the basin margin, the Slochteren Sandstone Formation (this subdivides into two members, the Upper and Lower Slochteren Sandstone), and a shaly evaporitic formation in the basin centre, the Silverpit Claystone Formation (the Ten Boer Claystone and Ameland Claystone are members of this formation). These two formations grade into each other in an E – W trending belt that crosses the northern part of the onshore Netherlands (Glennie 1983, Glennie & Buller 1983, Glennie et al. 1978). The Ameland field is situated in the region where the deposits of the desert lake to the north and the more southerly depositional areas of wadi and aeolian sands interfinger (Fig. 2).

In the Rotliegend the best reservoirs normally consist of aeolian cross-bedded sandstones and homogeneous sands. However, in the Ameland area these represent only a small fraction of the lithofacies present, as sabkha deposits account for about 90% of the Upper Slochteren Sandstone sediments. Therefore, the best reservoir consists of wavy laminated sandstones deposited along the margins of the desert lake on dry aeolian sandflats, during periods of lower water table and a lesser degree of clay mineral adhesion. The deposits consist of a broad band of poorly bedded sands, silts and clays displaying many of the classical features of sabkha deposits, e.g. mudcracks, adhesion ripples and anhydrite nodules. This indicates a mainly

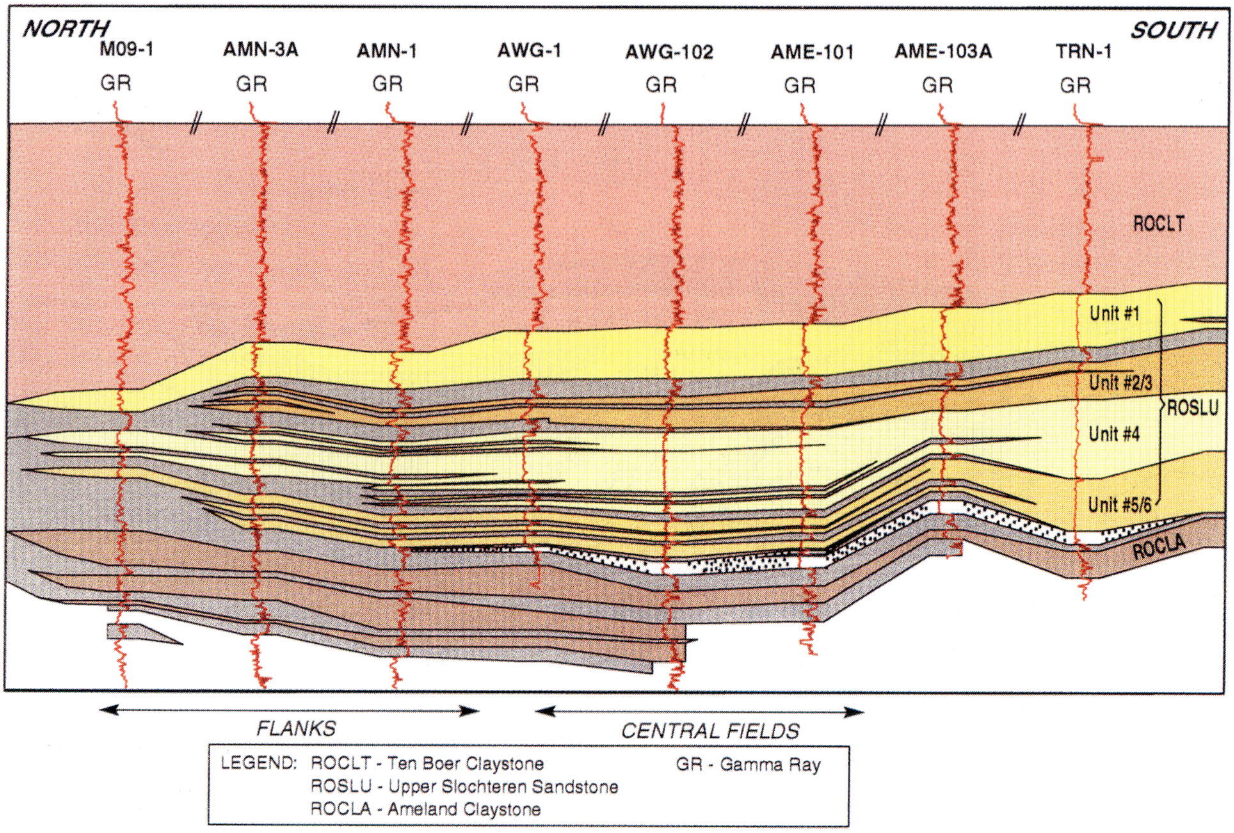

Fig. 3. Correlation panel through the Ameland field, showing the deterioration of the primary reservoir characteristics towards the flanks of the field. Grey shading indicates shale breaks.

aquatic influence subjected to limited aeolian deposition and subaerial desiccation in an arid climate.

The Upper Slochteren reservoir sequence has an isopach thickness of ca. 90–100 m and can be subdivided into six major units, correlatable field wide (Fig. 3). The reservoir units are separated by major dark-red claystones, deposited by playa lakes during periods of transgression across the sabkha surface. The shale breaks form effective vertical permeability barriers, and control differential depletion within the reservoir.

Regional mapping of the reservoir units shows a distinct trend towards the northeast with the sand units thinning basinward and a corresponding increase in the shale thickness. This conforms with the model of transition from sandflat environments in the south to playa lake deposits in the north. Isopach mapping of the major reservoir units shows that the shore-parallel trend is disrupted in places by local wadi systems, with a thicker sediment sequence and poorer reservoir properties. The presence of these wadi deposits in the wells AWG-106 and AME-203 (Fig. 1) appears to indicate that some of the NNW – SSE faults were active during deposition and controlled the drainage pattern.

Clay mineral diagenesis

Overprinting the deterioration of the primary reservoir characteristics towards the flanks of the field is a secondary parameter, diagenetic impairment. The sequence of authigenic clay mineral growth in the Ameland area is related to differences in the primary mineralogical rock composition, the influence of fresh water, the coalification of the Carboniferous and the burial history of the area. The diagenetic minerals with the greatest impact on productivity are chlorite and kaolinite. These two clay minerals are almost mutually exclusive and it is their relative distribution that controls the well productivity, as is shown in Fig. 4.

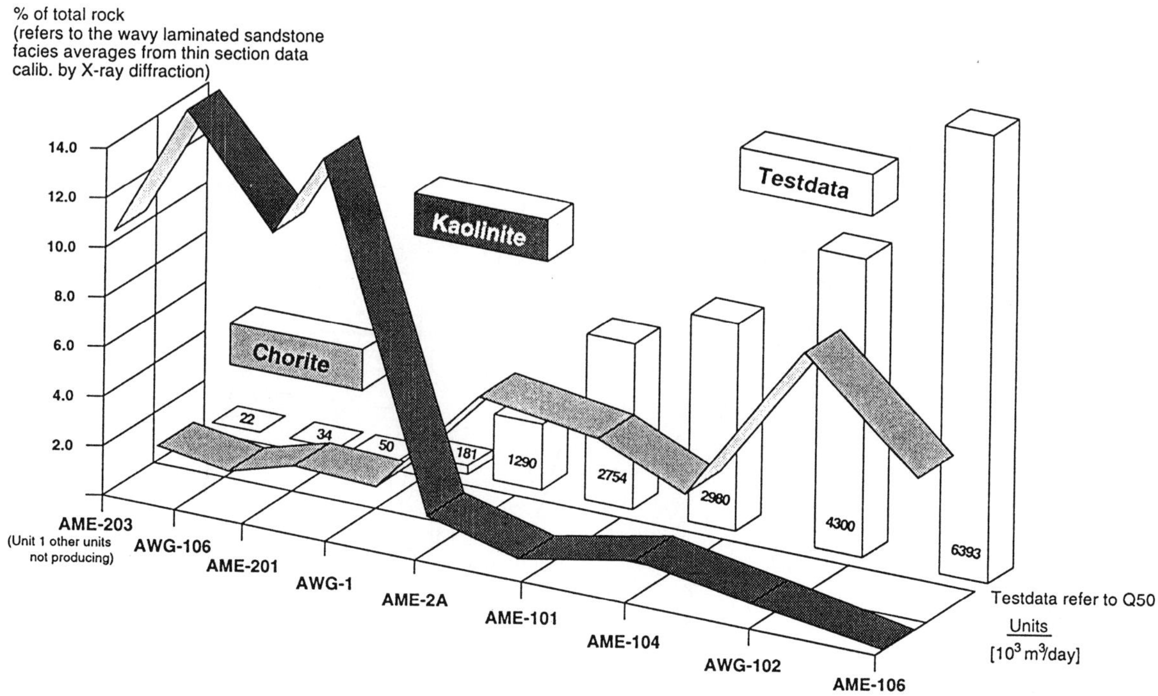

Fig. 4. Graph illustrating the mutually exclusive relationship between chlorite and kaolinite in the ROSLU reservoir and the control of these minerals on productivity.

From the diagenetic sequence it appears that early formed chlorite indirectly prevented subsequent kaolinite generation. Alternatively, two different physiochemical provinces are required to explain the mutually exclusive relationship.

Chlorite distribution
Chlorite distribution is limited to the central field, blocks E11 and E15 (Fig. 5), and its abundance is associated with an increased primary content of mafic minerals and rock fragments. In the Ameland region chlorite forms early pore-lining rims of crystals growing perpendicular to the grain surfaces (Fig. 6a). Its presence is associated with the replacement of lithic fragments (i.e. micas), sourced from the Lower Rotliegend clastics and volcanics which are well developed in the north-eastern part of the Netherlands.

Chlorite is recognised (Pittman et al. 1992, Ehrenberg 1993) as an effective mineral in preserving intergranular porosity. In the Ameland area this retention of the intergranular porosity has led to the preservation of permeabilities an order of magnitude higher than in the flanks of the field. The reason for this is that a well-developed and early formed clay grain coating forms an effective physical barrier. This barrier prevents quartz and, because the chlorite is Mg^{2+}-rich (Gaupp et al. 1993), it also prevents dolomite overgrowths from nucleating on detrital grains, and so preserves the primary intergranular porosity.

Kaolinite distribution
In comparison, the flanks of the field are associated with an almost total absence of chlorite and a much higher percentage of kaolinite (Fig. 5). Kaolinite occurs as authigenic pore fills in large secondary pores, thereby diminishing the intergranular porosity as it replaces framework grains (Fig. 6b). The dissolution of feldspars generates secondary porosity (Bjorlykke & Aagaard 1992) and therefore from log analysis the flanks of the field often appear to have the same or even better porosities than the central field. For an example the recently drilled well AME-203 was interpreted as having porosities in the range of 16–24%. However, productivity in this well was reduced to 0.04 $\times$ 10^6 m^3/day at 50 bar drawdown under steady-state conditions due to a reduction in the well permeability to 0.2 mD, caused by the presence of large amounts of kaolinite and the tendency of these fine particles

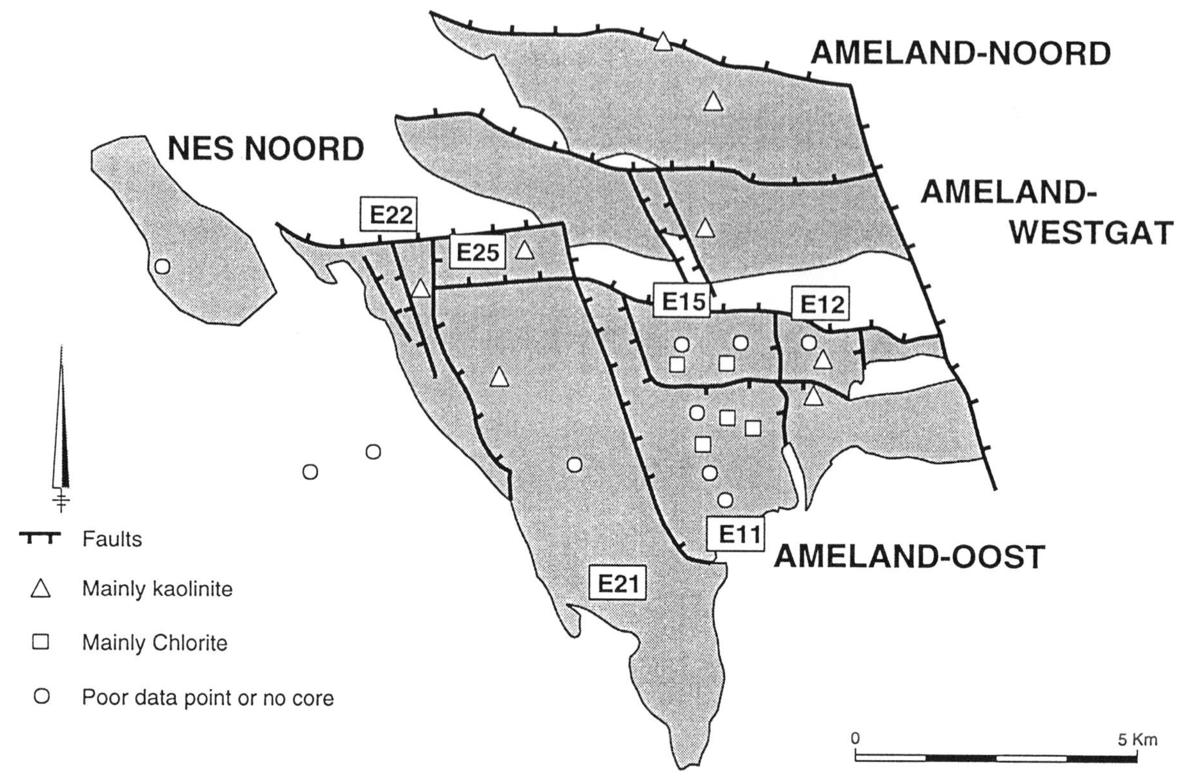

Fig. 5. Map of the diagenetic clay mineral distribution, showing the concentration of chlorite in the central field and kaolinite in the flanks of the field.

to detach and migrate under draw-down, thus further blocking already restricted pore-throats (Muecke 1979, Howard 1992).

Feldspar dissolution and replacement by kaolinite occurs because the water flushing through the system is acidic (Gaupp et al. 1993). The acidity can be increased by the addition of CO_2, and in the Ameland field this has two possible sources; the first resulting from the coalification of the Carboniferous; the second is from the Zechstein carbonates. Fluid analysis of hydrates from the Ameland field indicates some influence of fluids from the Zechstein carbonates.

In summary the poor productivities found in the flanks of the Ameland field reflect the:

1. Decrease of primary reservoir properties as a result of: i) a decrease in net/gross ratio, thinner beds associated with lower permeabilities, and ii) increasingly stratigraphically compartmentalised reservoir with thicker shale breaks creating permeability barriers.
2. Diagenetic clay mineral impairment associated with kaolinite.

Once the geological causes of the bimodal productivity distribution had been analysed, and in particular the reason for the poor productivity in the flank blocks, the challenge was how to combat it and successfully produce the untapped reserves.

Reasoning behind the multiple fracced horizontal well

As mentioned previously, the reservoir in the flanks of the field is porous, therefore, the problem is the gas productivity, not the lack of reserves. The block to be developed contained sufficient GIIP (7.7×10^9 m^3) to justify two drainage points, but a single horizontal well would be more cost effective. However, the Ameland East reservoir with its low dip, layercake structure and low vertical permeabilities is not a typical horizontal well candidate, as some layers within the reservoir would remain undrained with a standard

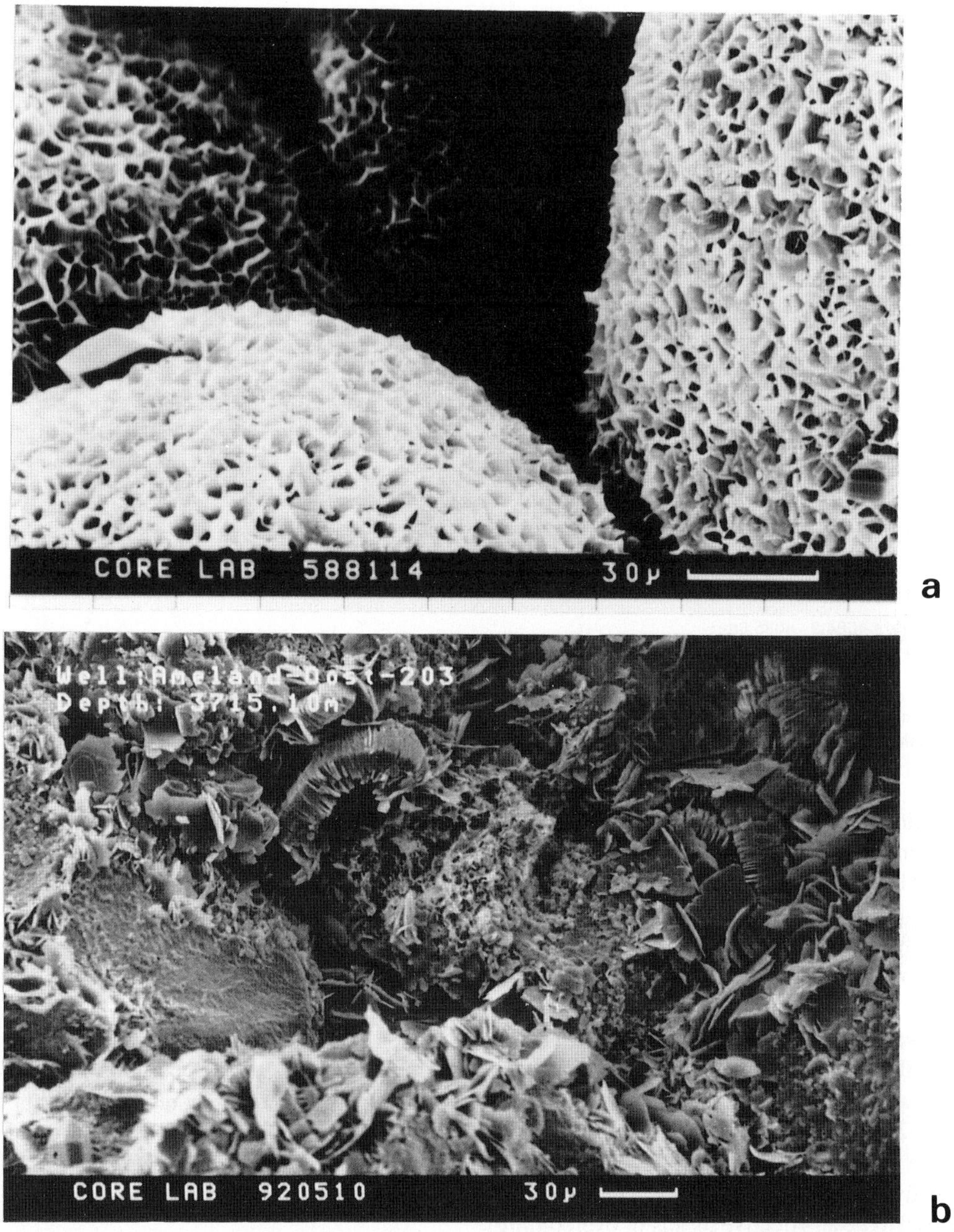

Fig. 6. SEM photographs showing porosity inhibition: (a) grain-coating chlorite development from a well in the central field, AME-104; (b) well developed kaolinite booklets from a well in the flanks of the field, AME-203.

horizontal well design. Therefore, an additional mechanism was required to connect all the reservoir layers; this was provided by the implementation of hydraulic-fracturing technology.

If the technologies of horizontal well drilling and multiple hydraulic fracture stimulation are combined then a horizontal well becomes a viable option, as it overcomes two of the major obstacles to productivity in this region, i.e. it provides the mechanism to connect reservoir compartments separated by vertical permeability barriers, and it increases the productivity.

The presence of a good seal, the Ameland Claystone (Fig. 7), between the planned well and the GWC (gas-water contact), was anticipated to provide ade-

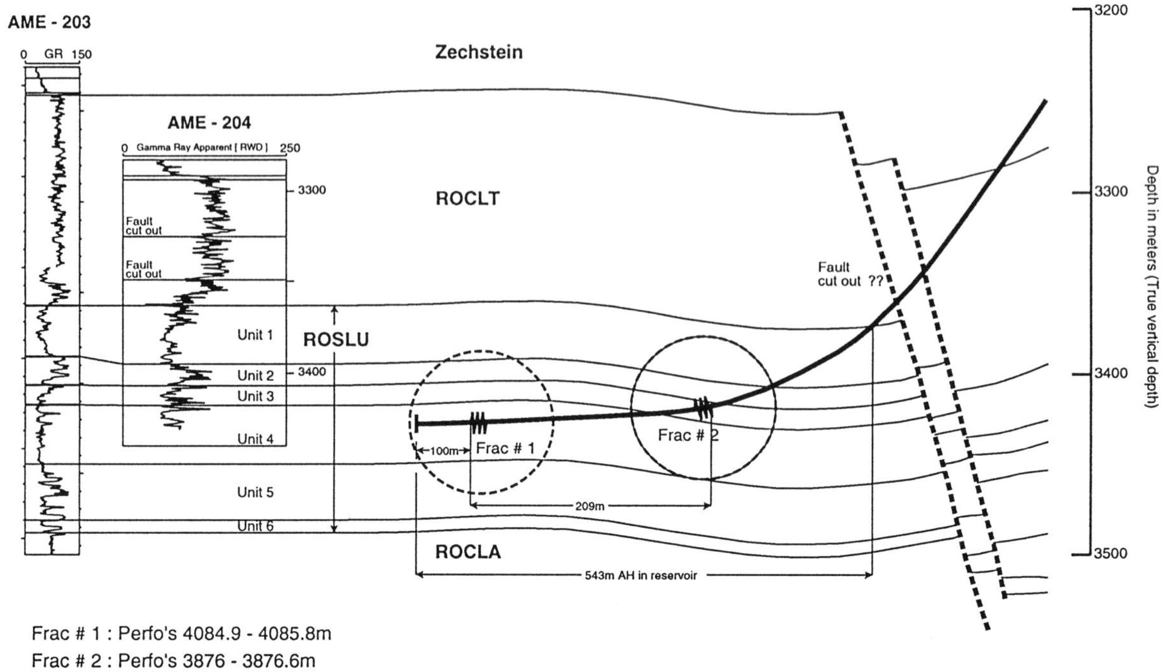

Fig. 7. Schematic diagram of the final well trajectory of AME-204, indicating fracture initiation points and showing the gamma-ray log correlation with the offset well AME-203. AH = Along hole depth.

quate protection against fracturing into the water leg, while good structural control over the area of interest was provided by the 3D seismic survey.

Productivity modelling

A detailed evaluation of well options was conducted, based on the geological model. Two options were screened: 1) a deviated well with a single hydraulic fracture, and 2) a horizontal well with two hydraulic fractures.

A single, hydraulically fractured, deviated well had already proved to be an effective means of overcoming poor productivity in a naturally impaired reservoir. The deviated well AME-201 with a single hydraulic fracture, increased its productivity from 0.05 (pre-fracture) to 0.75×10^6 m^3/day, while a horizontal well provides the opportunity to place two fracture points in the same block for a fraction of the cost.

In both options the fractures were assumed to propagate into all reservoir units, without penetrating the Ameland Claystone. Each case was screened for a high- and low-case permeability range. The high case was taken from the southern offset well AME-201 (0.9 mD) with the low-case values coming from the northern and closer offset well AME-203 (0.2 mD). A single well analytical model was used to establish productivities for all variations and combinations. This was then fed into a full field simulator and incremental profiles established for all cases, which were used as the foundations for an economic model.

In all cases a horizontal well with two hydraulic fractures emerged as the best option, showing the highest expected economic return. On this basis it was decided that a multiple fracced horizontal well was the optimum way of developing this block.

Well design

Two major principles were adopted. First to keep the well as simple as possible, and second to minimise deviation from the standard completion design used in the Ameland field.

The primary objective of the well was to achieve two fracture initiation points in the reservoir separated by a minimum of 300 m. The initiation points were in the middle of ROSLU unit 4 (the main reservoir unit), and in ROSLU unit 1, the unit separated by the thickest shale break (Fig. 3). This combination was evaluated to provide the greatest chance of connecting through all the permeability barriers and accessing the largest reserves. The well trajectory was planned to be as direct as possible, to avoid overpressured floating anhydrite/carbonate blocks within the Zechstein Salt caprock and to give maximum clearance from all mapped faults. A 500 m horizontal section was planned to allow sufficient separation between the fracture initiation points so that the fractures would not interfere with each other.

A Formation Microscanner (FMS) log run in the nearby AME-203 well showed indications of both drilling-induced and cemented natural fractures suggesting a NW – SE orientation for both sets. This would indicate a present-day minimum stress regime oriented NE – SW, suggesting any induced fractures would also propagate in a NW – SE orientation parallel to the maximum stress regime. Planning resulted in a well trajectory at an angle of less than 15 degrees to the assumed direction of fracture propagation. However, the actual fracture orientation was associated with large uncertainties. Therefore, provisions were made in the hydraulic fracture design, to allow for a bigger angle between the fracture and well, and to minimise the risk of the fracture turning away from the well path to conform with the local stress regime (Baumgartner et al. 1993). These are as follows:

1. A small perforated interval, of only 60 cm, to avoid the formation of multiple parasitic fractures. Sensitivity calculations on productivity loss due to such a small perforation interval highlighted the necessity of high fracture conductivity. For this reason a special, tip-screen-out fracture design was adopted. This maximises propped fracture width and therefore permeability.
2. The formation breakdown should take place at the highest possible rate to achieve a smooth turn in case the fracture would deviate away from the well path.
3. Pumping of sand squeezes to erode near-well-bore restrictions caused by the fracture.

Results

The drilling and stimulation of the well was carried out as planned, except that due to a poor cement bond over the upper reservoir unit (ROSLU 1) the second fracture initiation point had to be shifted down to a region of improved cement bond at the base of ROSLU 2. It is assumed that the fractures propagated as anticipated, although there is no mechanism to confirm this.

Table 1. Table comparing pre- and post-hydraulic fracture stimulation production performance.

Well name	Pre-fracture productivity Q50sss (10^6 m^3/day)	Post-fracture productivity Q50sss (10^6 m^3/day)
AME-201	0.05	0.750
AME-202A	0.042	Not available
AME-203	0.043	0.10
AME-204	Not available	1.2

Q50sss = productivity at 50 bar drawdown under semi-steady-state conditions.

The final Formation Evaluation While Drilling (FEWD) logs showed the reservoir characteristics to be in line with expectations, although the overall porosity was slightly higher and unit 1 was encountered slightly thicker than anticipated (Fig. 7).

After multiple hydraulic fracturing a maximum gas rate of 2.0 $\times$ 10^6 m^3/day was recorded during the cleanout period. Proppant-free production was achieved at a rate of 1.50 $\times$ 10^6 m^3/day at 50 bar drawdown; although still in transient flow this significantly exceeded the expectation of 0.9 $\times$ 10^6 m^3/day at 150 bar drawdown. This productivity was remarkable in view of the generally low productivity of the conventional fracture-stimulated vertical wells in the offset blocks (Table 1). Due to the low vertical to horizontal permeability (k_v/k_h) ratio and the short perforation interval, a meaningful pre- and post-frac performance comparison is not possible. However, a favourable result is concluded if post-frac performance is compared with nearby wells (Table 1).

Conclusions

1. A significant bimodal productivity variation is recognised in the Ameland field and is attributed to a combination of the depositional architecture and diagenetic effects. Two principal diagenetic clay minerals are recognised, chlorite and kaolinite. Early chlorite mineralisation preserved porosity and permeability, while later kaolinitization generated secondary porosity but reduced both permeability and productivity.

2. The low-dip, layercake nature of the Ameland reservoir with its poor vertical permeability and low cross-flow potential does not make it an ideal horizontal well candidate in the classical sense. However, a multiple-fractured horizontal well has proven to out-perform a conventional well.
3. The successful fracture stimulation of a deep, tight and layered sandstone gas reservoir has been demonstrated. Effective hydraulic fracture stimulation is possible even when the stress field is not accurately known, provided precautions are taken. Short perforation intervals were required to minimise the formation of multiple fractures; therefore fracture conductivity is vital because of the high velocities close to the well bore. For this reason a packed fracture design was necessary as the propped fracture width is the dominating factor.
4. Several other fields in the Dutch offshore are classified as porous but tight, i.e. these fields are associated with poor productivity and hence marginal economics. The technology employed in AME-204 is seen as an important step in enabling economic development of such fields.

Acknowledgements

This paper is published with the permission of the Nederlandse Aardolie Maatschappij BV, Shell Internationale Petroleum Maatschappij BV, and Esso Nederland BV. The authors gratefully acknowledge the cooperation of their partners in the Ameland field which are: Mobil Producing Netherlands Inc. and Energie Beheer Nederland BV.

References

Baumgartner, W.E.L, J. Shlyapobersky, I.S. Abou-Sayed & R.C. Jacquier 1993 Fracture stimulations of a horizontal well in a deep, tight gas reservoir; a case history from offshore the Netherlands – Society of Petroleum Engineers (SPE) Paper 26795

Bjorlykke, K. & P. Aagaard 1992 Clay minerals in North Sea sandstones. In: Houseknecht, D.W. & E.D. Pittman (eds) Origin, diagenesis and petrophysics of clay minerals in sandstones – Soc. Econ. Paleont. Mineral. Spec. Publ. 47: 65

Ehrenberg, S.N. 1993 Preservation of anomalously high porosity in deeply buried sandstones by grain-coating chlorite: examples from the Norwegian Continental shelf – Am. Ass. Petroleum Geol. Bull. 77: 1260–1286

Evans, P.F. 1989 Sedimentology and reservoir modelling of transitional aeolian sabkha sequences in Lower Permian (Rotliegend) gas reservoirs of the Southern North Sea – PhD Thesis, Keele University (unpublished)

Gaupp, R, A. Matter, J. Platt, K. Ramseyer & J. Walzebuck 1993 Diagenesis and fluid evolution of deeply buried Permian (Rotliegende) gas reservoirs, Northwest Germany – Am. Ass. Petroleum Geol. Bull. 77: 1111–1128

Glennie, K.W. 1983 Early Permian (Rotliegendes) paleowinds of the North Sea – Sediment. Geol. 34: 245–265

Glennie, K.W. & A.T. Buller 1983 The Permian Weissliegend of N.W. Europe: the partial deformation of aeolian dune sands caused by the Zechstein transgression – Sediment. Geol. 35: 43–81

Glennie, K.W, G.C. Mudd & P.J.C. Nagtegaal 1978 Depositional environment and diagenesis of Permian Rotliegendes sandstones in Leman Bank and Sole Pit areas of the UK Southern North Sea – J. Geol. Soc. London 135: 25–24

Howard, J.J. 1992 Influence of authigenic-clay minerals on permeability. In: Houseknecht, D.W. & E.D. Pittman (eds) Origin, diagenesis and petrophysics of clay minerals in sandstones – Soc. Econ. Paleont. Mineral. Spec. Publ. 47: 257

Muecke, T.W. 1979 Formation fines and factors controlling their movement in porous media – J. Petroleum Technology Feb. '79

Pittman, E.D, R.E. Larese & M.T. Heald 1992 Clay coats: occurrence and relevance to preservation of porosity in sandstones. In: Houseknecht, D.W. & E.D. Pittman (eds) Origin, diagenesis and petrophysics of clay minerals in sandstones – Soc. Econ. Paleont. Mineral. Spec. Publ. 47: 241

Rondeel et al. (eds), Geology of gas and oil under the Netherlands, 103–114, 1996.

CBIL logs: vital for evaluating disappointing well and reservoir performance, K15-FG field, central offshore Netherlands

Harm W. Frikken
Nederlandse Aardolie Maatschappij BV (Corporate Planning and Development), Postbus 28000, 9400 HH Assen, the Netherlands

Key words: borehole imaging, compartmentalization, fractures, Rotliegend, seismic attributes, stress anisotropy, wrenching

Abstract

The first two development wells of the K15-FG Rotliegend gas reservoir showed a large variation in productivity and sharply declining flow rates. Material balance data indicated only 30% of the field's expected Gas Initially In Place (GIIP) to be connected to the wells. This poor performance reflects stratigraphic and structural reservoir compartmentalization. A Circumferential Borehole Imaging Log (CBIL) was run in the second development well. The CBIL enabled identification of small-scale reservoir heterogeneity and thin, otherwise undetected highly productive aeolian layers. The log analysis also revealed the presence of a conjugate set of shear fractures, confirmed by cores. Furthermore, small-scale, sub-seismic resolution reverse faults were detected intersecting the well path, as well as anomalous structural dips around the well, indicating fault drag. The presence of near-wellbore reverse faults and sub-vertical shear fractures with cataclastic fill, restricting horizontal inflow, explains the disappointing production rates of the second well. Moreover, continuous borehole break-outs were observed from the log, indicating present-day stress anisotropy. The CBIL data in combination with lineations observed on seismic-attribute maps contribute to a concept of dextral wrenching across the field, with related reverse faulting and strike-slip faults or shear zones. Such a configuration of partly and completely sealing faults is a major cause of the poor reservoir connectivity as confirmed by pressure behaviour. The CBIL is an effective tool for evaluating both depositional and structural reservoir heterogeneity. The resulting interpretations created scope for optimization of recovery by resolving the causes of anomalous well behaviour and field architecture. Pressure analysis of a recently drilled third development well has once more confirmed the presence and sealing character of strike-slip faults within the field.

Introduction

The K15-FG field offshore the Netherlands (Fig. 1) consists of a tilted horst block, largely bounded by NW – SE and NE – SW striking normal faults. Gas is trapped in Permian, Upper Rotliegend Sandstones. The discovery well K15-FG-101 showed encouraging production rates on test and no signs of depletion, whereafter the field was developed with two wells from a mini-satellite platform. The Rotliegend consists of alluvial plain sediments, deposited in a gently northward sloping basin. The sediments consist of varying amounts of waterlaid deposits (fluvial, sheetflood and lacustrine) and windlaid, aeolian dune- and sheet-sand deposits (Glennie et al. 1978, Nagtegaal 1979, Oele et al. 1981).

The Rotliegend in the K15 area is sub-divided into five members of which the Upper Slochteren Sandstone (ROSLU, Fig. 2) is the main gas producing interval. The overlying, anhydritic and shaly Ten Boer (ROCLT) and the underlying Ameland (ROCLA) members (sabkha and lacustrine facies) are waste zones. Thick Zechstein salts provide both top and lateral seal (Fig. 2).

The Upper Slochteren Sandstone in the K15 area is strongly layered (Fig. 3) and three main rock-types,

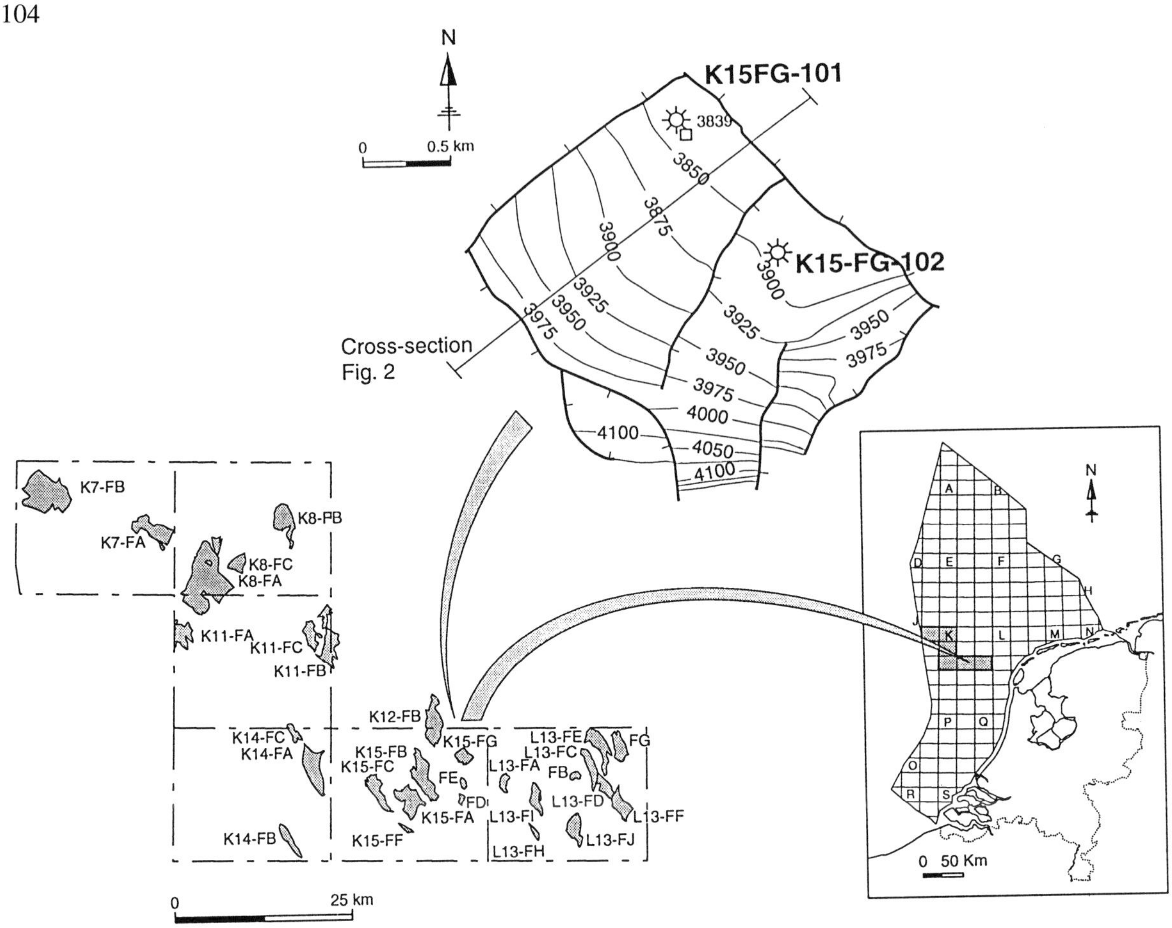

Fig. 1. Structural contour map at top Rotliegend, K15-FG gasfield, central offshore Netherlands. Contourdepths in meters.

characterizing the reservoir behaviour, are highlighted from top to bottom:

1. Relatively thin aeolian dune- and sheet-sands with good reservoir quality, which dominate gasflow into the wellbore (Fig. 3). These sands are largely defined by porosities in excess of 15%.
2. Thick packages of moderate to poor reservoir quality, predominantly waterlaid sands. Drainage of these sands is fully dependent on cross-flow, inside the reservoir, into depleted aeolian sands.
3. Fairly continuous shale layers, deposited in desert lakes and ponds. These shales reduce cross-flow potential, resulting in a series of stacked mini-reservoirs, with the overall drainage being dependent on connectivity to the thin aeolian sands.

The first development well K15-FG-101 showed severe capacity decline during the first year of production. Strong differential depletion of aeolian sands was noted (predominantly present in the ROSLU 2 sub-member, Fig. 3). Material balance data indicate limited connected volumes (Fig. 4; Frikken & Stark 1993). The second development well K15-FG-102 indicated a similar behaviour. However, this well had a productivity of only some 20% when compared to the first well (0.4 $\times$ 10^6 m^3/day and 2.0 $\times$ 10^6 m^3/day respectively at 50 bar drawdown), despite a larger number and better distribution of prolific sands especially in the ROSLU 1 sub-member (Fig. 3).

A CBIL (Circumferential Borehole Imaging Log, Atlas Wireline Services 1991) was run in oil-based mud in the second well. The CBIL was run to obtain a more detailed analysis of the reservoir architecture (stratigraphic heterogeneity, sedimentary structures, detection of possible fractures and stress-field anisotropy) for the planning of possible future infill wells and hydraulic fracturing design.

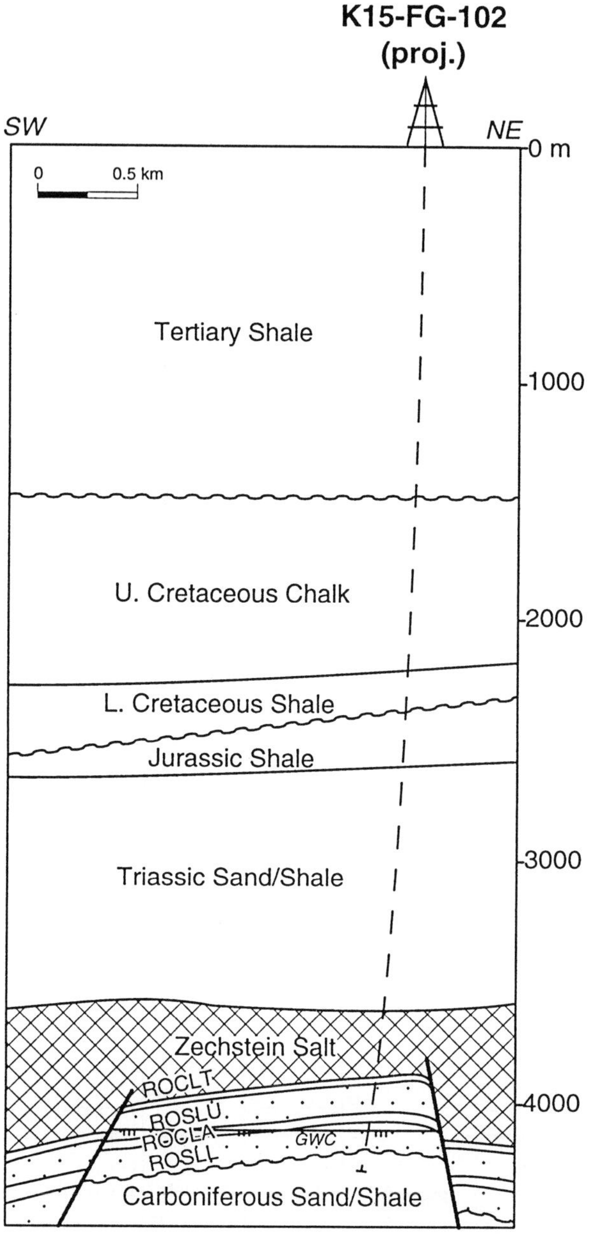

Fig. 2. Structural cross-section K15-FG area. For location see Fig. 1. GWC = gas-water contact.

CBIL data acquisition and analysis

The CBIL uses acoustic transducers, operating in a pulse-echo mode, to scan the entire circumference of the borehole wall. The transducers are hemispherically focussed to optimize the image resolution and rotate six times per second, acquiring 250 samples per revolution. The CBIL images are derived from reflections measured directly from the borehole wall.

Changes in amplitude of the obtained signal reflect variations in rock property, resulting in the identification of sedimentary structures and textural variations. Furthermore, the full borehole wall coverage of the CBIL log offers a particular advantage in the description and analysis of structural features in the wellbore.

The images are analysed on PC-based workstations. The displays represent cylindrical images, which are 'cut along a reference line (usually true north or the high or low side of the borehole) and unrolled' (Fig. 5). The true dip of planar geological features (corrected for borehole deviation and azimuth) is calculated either automatically or interactively. CBIL processing generates a large data set which needs careful interpretation and calibration with core data and 3D-seismic data.

Lithological analysis

The resolution of thin beds by the CBIL was excellent. Centimeter- to decimeter-thin shale intercalations, partly to fully cemented streaks and layers, alternating with porous sands, were detected from the images (Fig. 6) and these were confirmed by core data. Such thin beds and small-scale heterogeneities are beyond the resolution of conventional logs. Detailed knowledge of the internal heterogeneity, especially in the waterlaid sands, is of importance to better quantify the vertical permeability anisotropy in this rock-type.

Thin aeolian cross-bedded sands (characterized by porosities in excess of 15%) of only 25–50 centimeter thickness were identified from the images, showing dips towards the southwest, indicative of the palaeowind direction (Fig. 7). Again such thin beds are mostly beyond the resolution of porosity logs. Details about the distribution and character of such sands may enable reconstruction of the three-dimensional arrangement of such sands (Luthi & Banavar 1988). Outcrop studies provide further information about the dimensions of such deposits, helping to provide an improved calibration of well productivities.

Small-scale internal heterogeneity within these aeolian deposits appears to lie within the resolution of the CBIL. Alternations of more porous, coarser-grained slipface beds with finer-grained, partly cemented laminae have been recognized, as well as cemented bottom sets of small dunes at the interface with the

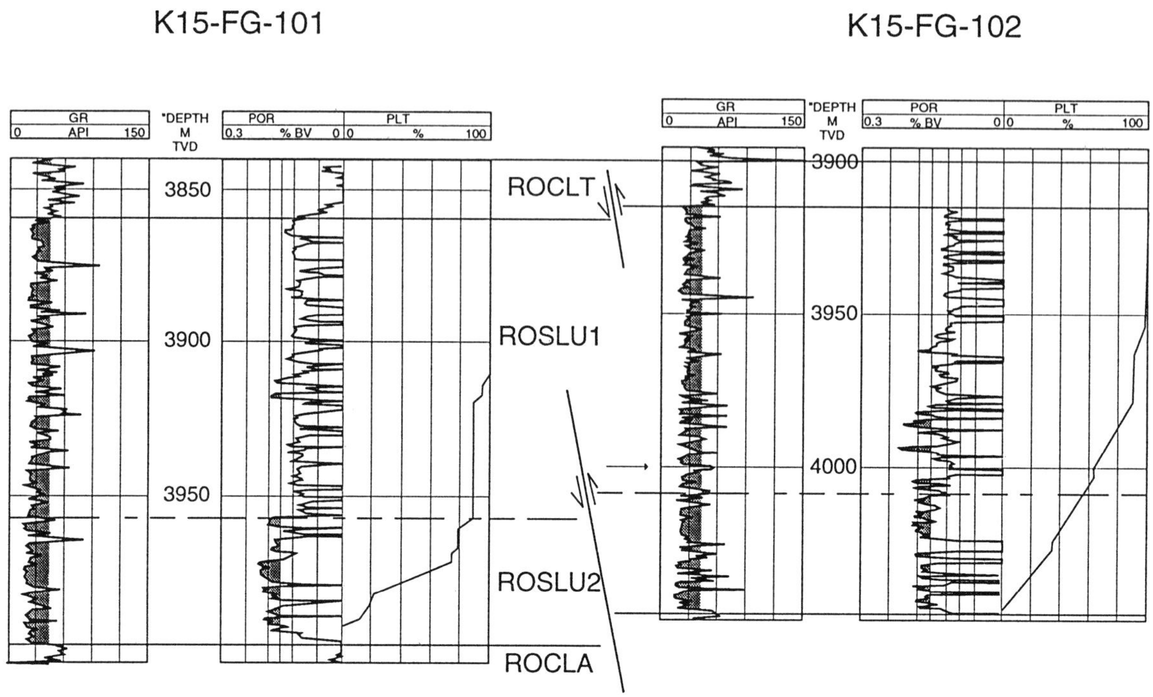

Fig. 3. Type log suites of Gamma Ray, Porosity and Production Logging Tool (PLT) logs of both K15-FG development wells, illustrating the strongly layered character and the dominant gas-flow from sands with porosities in excess of 15% (PLT signatures). Note the larger amount of prolific sands in the ROSLU 1 of well K15-FG-102 and the small fault cut-outs. TVD = true vertical depth.

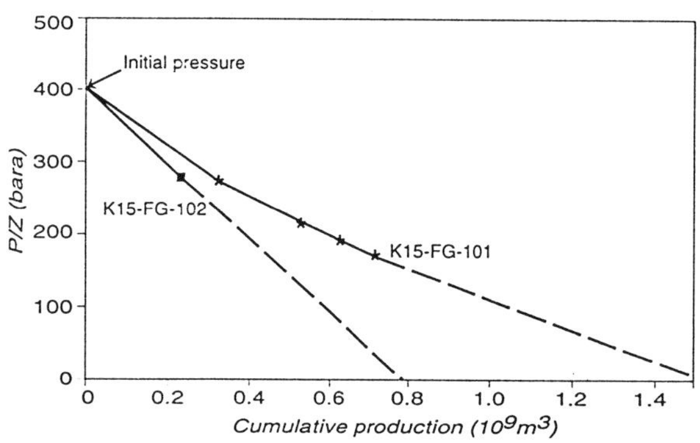

Fig. 4. Material balance data of the K15-FG reservoir, indicating a total connected gas volume of only 2.3×10^9 m^3, compared to an expected volume of 8×10^9 m^3.

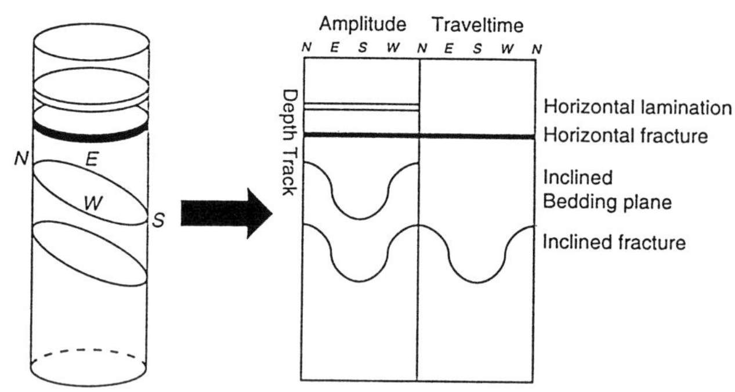

Fig. 5. Appearance of geological features on CBIL images as a function of dip angle, relative to the borehole axis.

interdune sabkha (Fig. 7). Considerable permeability anisotropy as a result of grainsize differences within cross-bedded sets has been recognized by previous authors (Van Veen 1975, Weber 1987), with better permeabilities parallel to the strike of the aeolian foresets.

Unfortunately, due to technical problems and poor hole condition across a cored section, details from the thicker sequence of aeolian sands in the ROSLU 2 submember (Fig. 3) could not be resolved by the CBIL. Recommended methods, like reducing the correlation interval and step distance in the analyses (Hoecker et al.

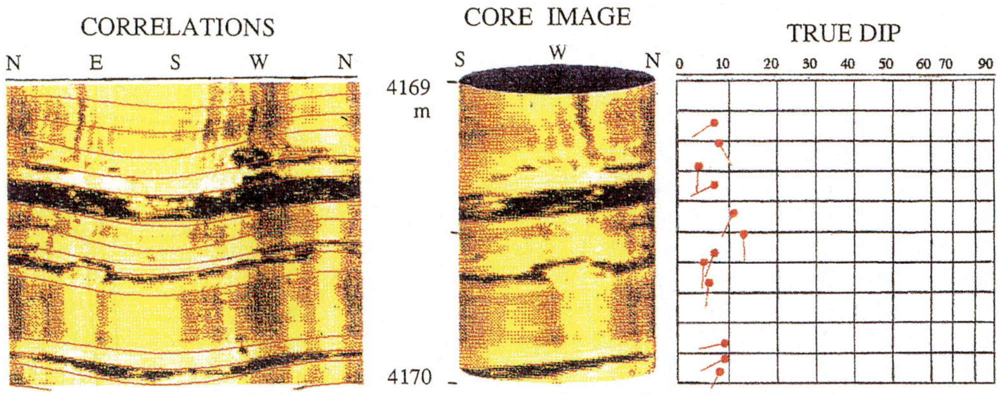

Fig. 6. CBIL images showing alternation of centimeter-thin shales (black), cemented patches (bright) and porous zones (yellow) in well K15-FG-102. The vertical, brown features are scrape-marks in the mud cake of the borehole wall, resulting from the drillstring (diminishing across the tight streaks). The tadpoles reflect true structural dips. The red lines on the correlation image mark the interpreted bedding surfaces.

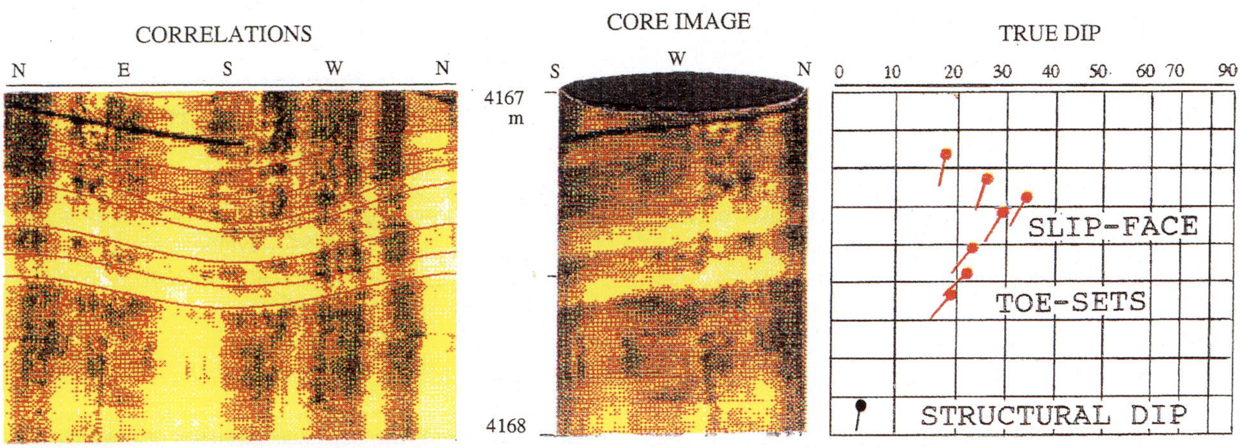

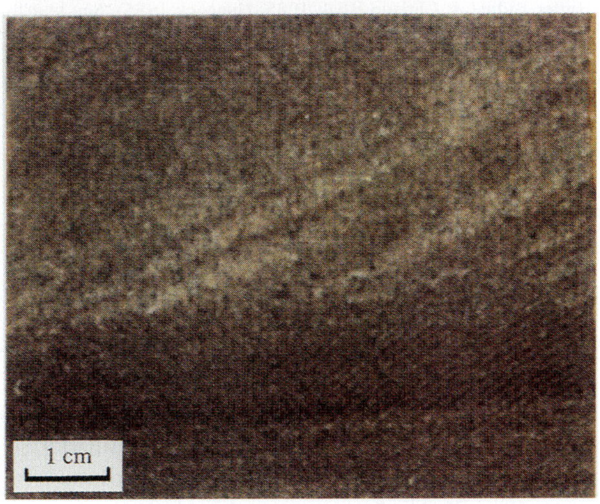

Fig. 7. CBIL images and core photograph for comparison, showing the presence of a small aeolian dune with foresets dipping southwest in well K15-FG-102. Note the alternation of cemented streaks (bright yellow) and porous laminae (orange-yellow). The red lines on the correlation image mark the interpreted bedding surfaces.

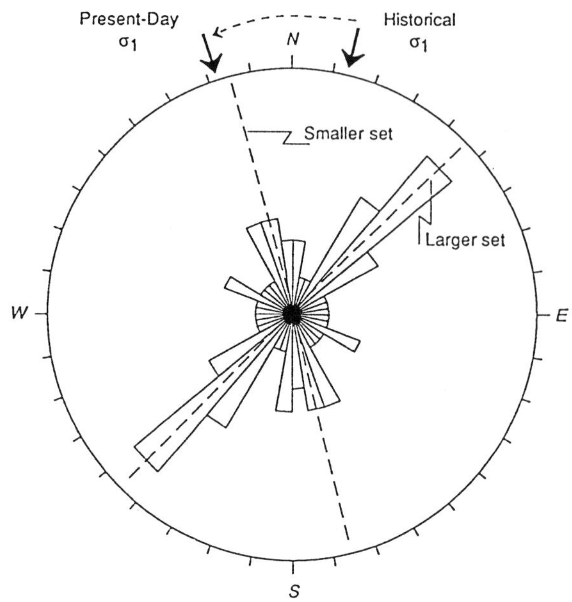

Fig. 8. CBIL fracture analysis showing a set of conjugate fractures in well K15-FG-102, resulting from historical stress anisotropy.

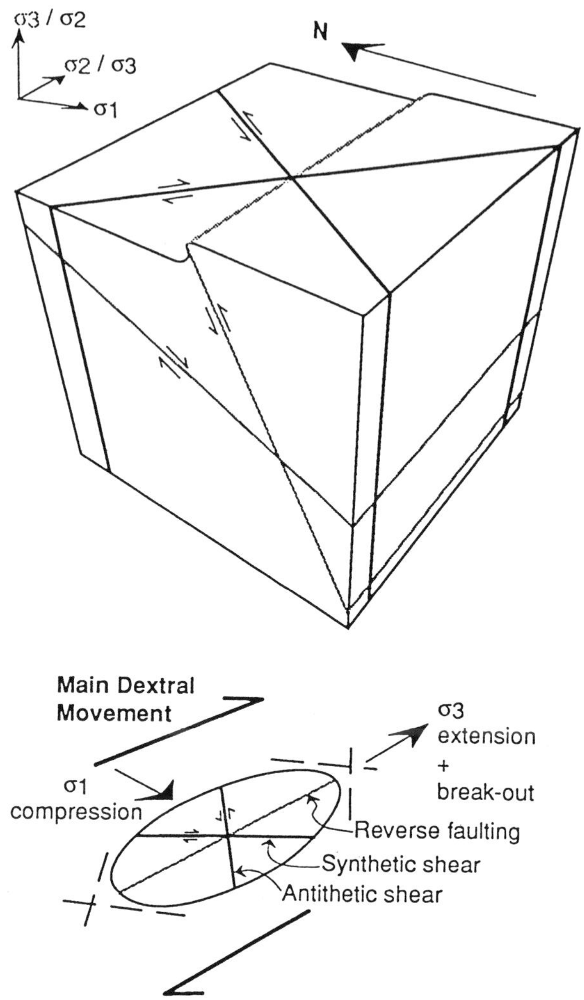

Fig. 9. Schematic representation of stress distributions related to dextral wrenching, resulting in the generation of strike-slip shear zones (faults and fractures), reverse faults and borehole break-outs (after Biddle & Christie-Blick 1985, Hancock 1985). See text for discussion of features.

1990) did not reveal details from this sand interval. Furthermore, the acoustic impedance contrasts between the foreset laminae in these more massive sands may be too small to be accurately detected.

Dip analysis

The depth contour map of the K15-FG field indicates structural dips in the area of well K15-FG-102 of around 5–10 ° to the southwest (Fig. 1). An initial CBIL bedding dip analysis showed too much scatter in dip-azimuth to accurately calibrate the mapped structural dips. This scatter was interpreted to be the result of the high resolution of the tool, detecting all kinds of scour, erosion and bounding surfaces. In order to ensure a better structural definition, a Gamma Ray log cut-off of 45 API was applied to highlight horizontally bedded shales. Thereby the mapped structural dip was largely confirmed by the shale layers showing dips of 5–10 degrees ranging from south to southwest (Fig. 6). Unexpectedly a number of anomalous northward dips were encountered, especially in the laminated and shaly ROCLT member overlying the Upper Slochteren Sandstone. These anomalous dips are interpreted to be the result of fault-drag and minor folding, related to reverse faulting (discussed in a following paragraph).

Fracture analysis

The image analysis showed the presence of a conjugate fracture set comprising: 1) a large set of NE – SW striking fractures (Fig. 8), dipping at angles of 50–80 degrees mainly to the southeast, and 2) a smaller set of subvertical (80 degrees) NNW – SSE striking fractures (Fig. 8). The frequency of these latter fractures is

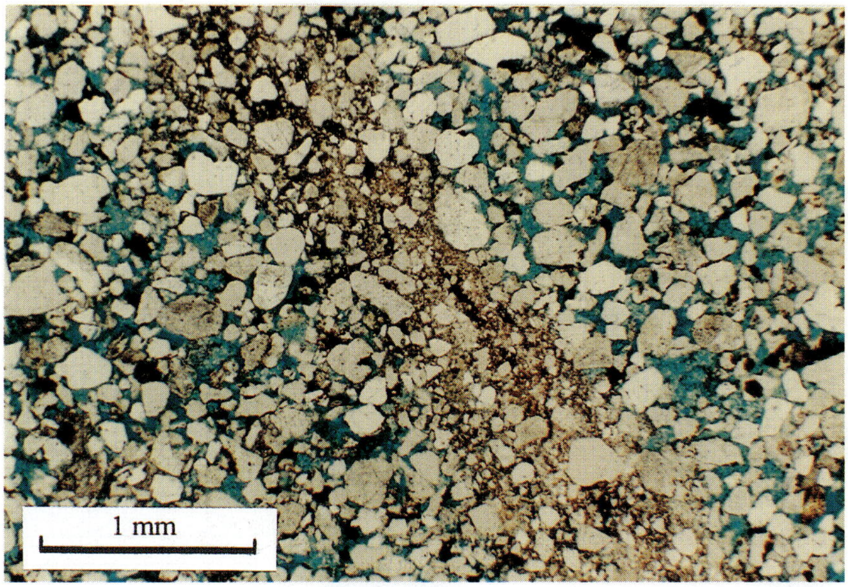

Fig. 10. Thin section photograph of a sub-vertical shear fracture in well K15-FG-102. The fracture fill consists of crushed framework grains and rock-flour (mylonite), creating an effective permeability barrier. Note the non-cemented matrix around the fracture.

probably underestimated due to the fact that they are oriented sub-parallel to the wellbore.

Any stress field consists of three orthogonal components (Hancock 1985): maximum principal stress (σ_1), intermediate principal stress (σ_2) and minimum principal stress (σ_3). Steep intersection of conjugate fractures indicates steep σ_2 and the conjugate set can be considered to result from the stress anisotropy between the subhorizontal σ_1 and σ_3, generating synthetic (dextral) and antithetic (sinistral) strike-slip faults and fractures (Fig. 9). The orientation of the conjugate fracture set from the CBIL analysis suggests a NNE – SSW trending maximum principal stress, referred to as historical σ_1 (Fig. 8).

The presence of these shear fractures has been confirmed by analysis of a spot core taken in well K15-FG-102. The fractures show predominant reverse displacement of bedding. No evidence was seen from thin sections of diagenetic alteration of the matrix surrounding the fractures (e.g. cement haloes), indicating that the fractures have not served as conduits for fluids. From thin sections it was recognized that the fractures reflect cataclastic deformation of the matrix and the fracture fill largely consists of extremely fine-grained rockflour (mylonite, Fig. 10). Therefore these fracture planes are typically effective permeability barriers.

Reverse faulting

Small fault cut-outs of a total of some 10m were identified from the logs of well K15-FG-102 (Fig. 3). The larger fault at the base of the ROSLU 1 in Fig. 3 was identified from the CBIL images to cut the wellbore (Fig. 11). The well test analysis of K15-FG-102 indicated the presence of a flow boundary at a distance of ca. 35 m away from the wellbore, furthermore indicative of the presence of faulting. The overall abundance of reverse shear displacements of the fractures in the core led to the interpretation of these faults as cutting the well in a reverse mode (Fig. 12). A deviated well penetrating a reverse fault from footwall to hanging wall would result in the omission of strata in the well track (Mulvany 1992).

The southward dipping events on the images in Fig. 11 are interpreted to represent synthetics to the main reverse fault plane (Fig. 9) and these indicate an east-west strike of the main fault. Such east-west trending reverse faults indicate north-south compression (i.e. σ_1

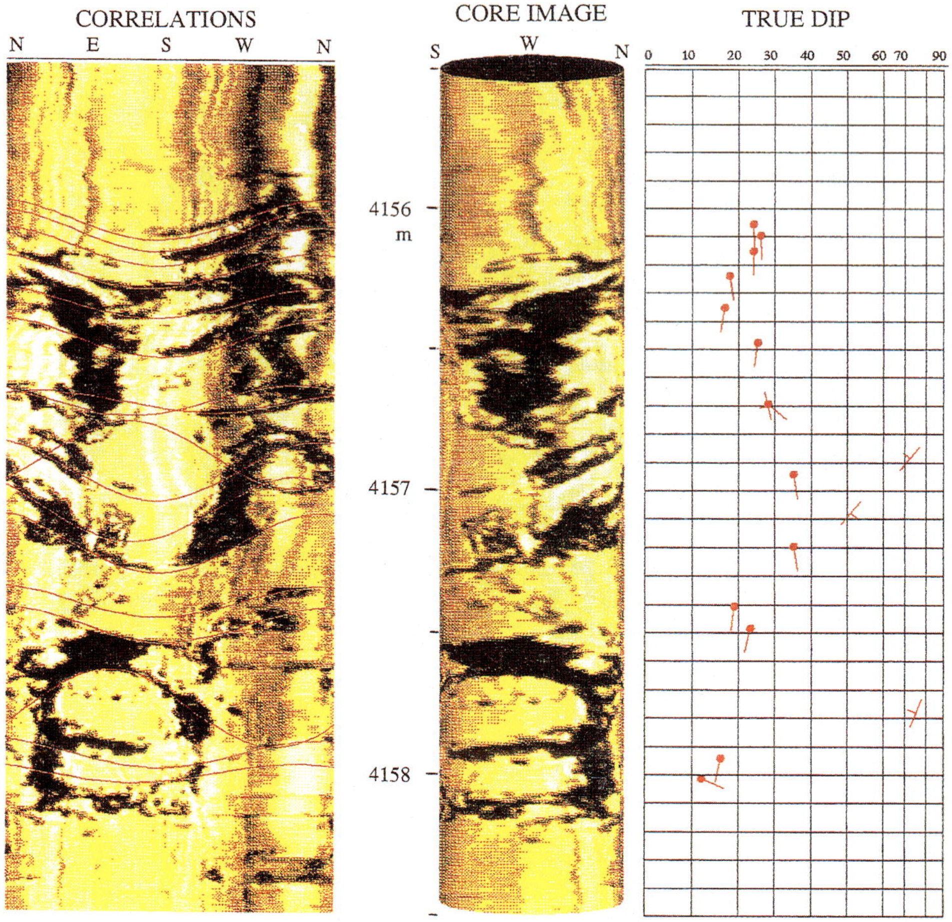

Fig. 11. CBIL images showing a brecciated zone (black patches), indicative of a fault cutting the wellbore of K15-FG-102. The features dipping ca. 30 degrees to the south (also marked by the red lines on the correlation image) are interpreted to represent shear planes which are synthetics to the main fault plane. They are represented by tadpoles. The dip symbols represent shear fractures which were generated prior to this faulting event.

oriented N – S). This suggests a slight anti-clockwise rotation of the maximum principal stress, when compared to the historical orientation responsible for the shear fractures (Fig. 8). It also indicates reversal of the two other stress components, the minimum principal stress (σ_3) being in a vertical position during reverse faulting.

Minor folding and fault-drag associated with the reverse faulting explain the anomalous northward structural dips (Fig. 12). The previously mentioned conjugate fracture set in well K15-FG-102 is interpreted not to be the result of folding, since folding and fault-drag in this case are only minor. Furthermore, our fractures show predominant reverse offsets, which is a less common feature in fracturing associated with folding (Stearns & Friedman 1972).

Borehole break-outs

Significant, uni-directional borehole break-outs were identified by the CBIL across a continuous interval of ca. 60 m (Fig. 13). Significant anisotropy between the horizontal stress components σ_1 and σ_3 may create rock-spalling and shear failure in the direction of σ_3, resulting in borehole break-outs (Mastin 1988), i.e. ellipsoidal widening of the borehole (Fig. 9). The borehole break-outs detected by the CBIL in well K15-FG-102 are oriented WSW – ENE (Fig. 14). This indicates present-day maximum principal stress oriented NNW – SSE, at a right angle to the break-outs. This suggests a renewed small anti-clockwise rotation of the maximum principal stress, when compared to the previous

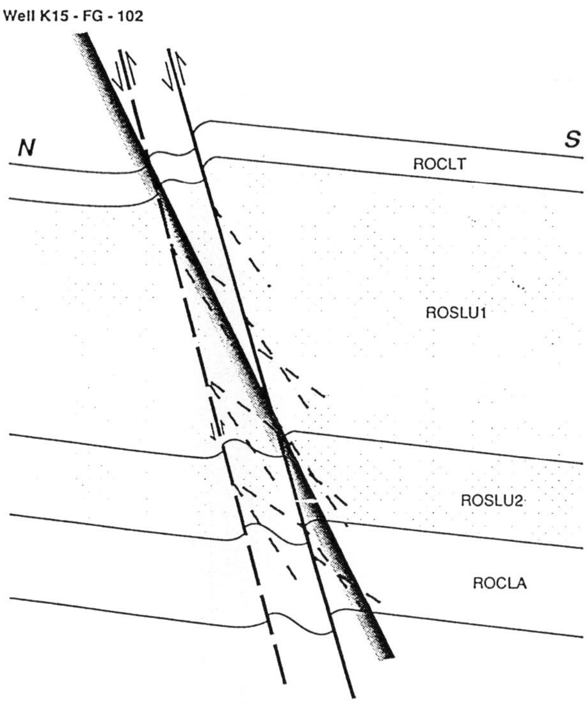

Fig. 12. Reconstruction of reverse faults cutting well K15-FG-102. The associated fault-drag and minor folding explain the anomalous northward structural dips. The fault planes and the subvertical shear fractures (dashed) severely hamper horizontal gas flow into the well-bore.

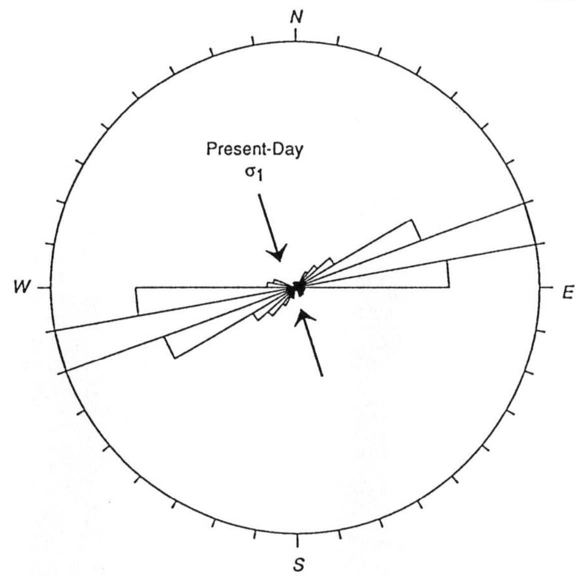

Fig. 14. Rose diagram of borehole break-outs trending ENE – WSW, ascribed to the present-day stress anisotropy.

orientations (NNE – SSW for the shear fractures, Fig. 8, and N – S for the reverse faults).

Borehole deviation may have a significant effect on break-out orientation (Mastin 1988). However, the deviation of well K15-FG-102 (ca. 20 degrees) still warrants the reliability of the break-out analysis and the conclusion of an overall, relatively minor stress field rotation of ca. 30 degrees. A stress field rotation of even up to 90 degrees has been identified from orientations of reverse faults in the neighbouring L13 block (Frikken & Stark 1993). The present-day stress field as analysed from well K15-FG-102 is in line with the regional stress regime across NW Europe (Klein & Barr 1986).

As a result of the stress field rotation, the present-day maximum principal stress became near-parallel to the historical, NNW – SSE striking, subvertical fracture set (Fig. 8). Consequently, this fracture set appears to have turned into minor dilational shear fractures as seen in cores (shear with subordinate dilation, Hancock 1985). However, no evidence for the presence of truly open fractures was found in cores or from the CBIL images. Nevertheless, evidence for the presence of open fractures, contributing significantly to productivities, has been recognized from a number of Rotliegend fields in the area (PLT data, cycle skips on sonic logs and losses while drilling, Frikken & Stark 1993).

Structural architecture

The overall structural style in the area is dominated by NW – SE trending normal faults with super-imposed smaller NE – SW trending faults. However, dextral wrenching as a result of basin inversion and opening of the Central North Sea rift system, has been recognized in the area (Ziegler 1978).

Dextral strike-slip, along NW – SE trending Variscan basement faults underlying the Rotliegend, is however not clearly evident and is largely camouflaged by the ubiquitous normal faulting. No clear evidence of wrenching is found in the overburden sediments in this area, probably largely due to the absorbing effect of the Zechstein salt packages overlying the Rotliegend (Fig. 2). As such, proper recognition of criteria for identification of wrenching (Harding 1990) is not possible. However in a number of fields in the area, the effects of super-imposed dextral wrenching and associated deformation features have been recognized, such as reverse faulting and strike-slip faults or shear zones (Frikken & Stark 1993).

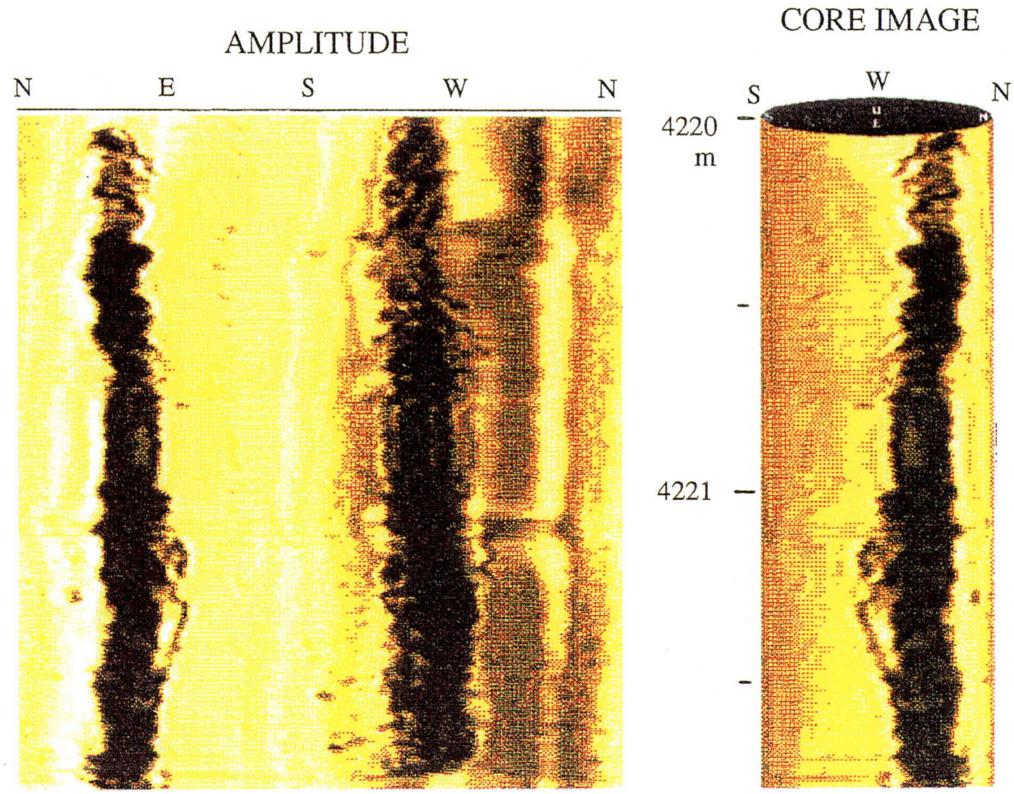

Fig. 13. CBIL images showing borehole break-outs in black. The vertical feature on the amplitude display (brown at NW) represents a groove in the mud cake of the wellbore, made by the drill-string at the downside of the hole.

The orientation of the stress field as indicated by the CBIL analysis of well K15-FG-102 is in accordance with the interpretation of dextral strike-slip along the NE boundary fault of the field. From a seismic Azidip map of the field an E – W trending, northward dipping lineation was recognized just south of well K15-FG-102, which is in line with the interpreted reverse fault and fault-drag zone (Fig. 15). A number of additional lineations on the Azidip map are in line with the strike of deformation features (synthetic and antithetic shear), associated with the stress anisotropy resulting from dextral wrenching (Biddle & Christie-Blick 1985). These lineations are interpreted to represent the presence of strike-slip faults or shear zones. Such faults are generally beyond the resolution of conventional seismic due to the very limited vertical throw.

Impact on reservoir behaviour

In combination with stratigraphic compartmentalization, the presence of strike-slip faults or shear zones is interpreted to be a major cause of the relatively poor reservoir connectivity in the area of the K15-FG wells. Such shear zones at least restrict repressurization of the wellbores, which may be time and pressure dependent. However, nearly completely sealing strike-slip faults have been recognized from well behaviour of the L13-FE field in the area (Fig. 1).

The NE – SW striking fault separating wells K15-FG-101 and -102 (Figs 1, 15) shows some normal displacement on seismic sections. However, this fault has most likely been reactivated by antithetic shearing, because only minor pressure communication across this fault was recorded from Repeat Formation Tester (RFT) measurements in well K15-FG-102 (10 bar depletion, compared to depletion of ca. 200 bars in well -101, Frikken & Stark 1993). The strike of the reverse faults to the south of well K15-FG-102 and the strike of the antithetic shear fractures and strike-slip faults are near-perpendicular to the strike of the aeolian foresets, restricting horizontal inflow. This explains the disappointing production rates of well K15-FG-102 when compared to K15-FG-101.

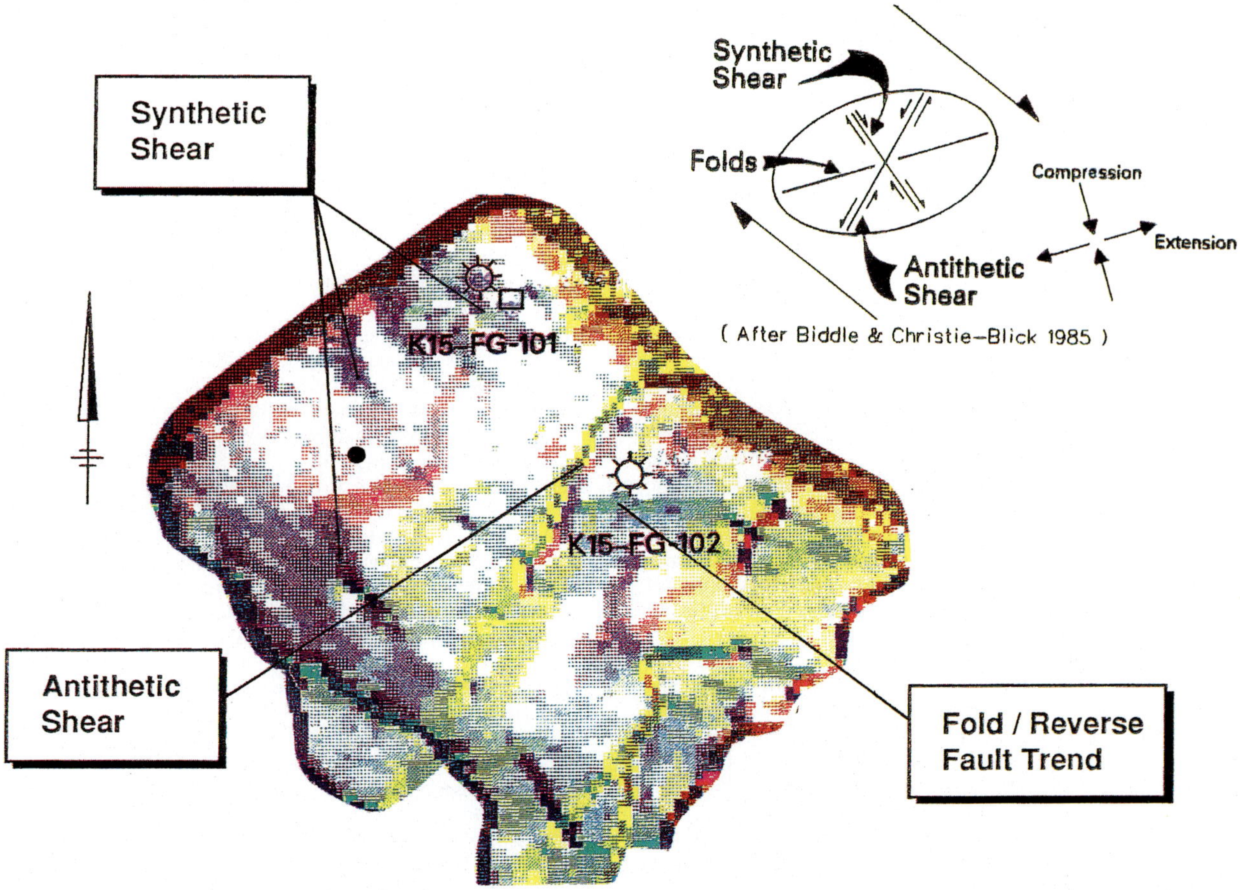

Fig. 15. 3D seismic Azidip map of top Rotliegend of the K15-FG field. Lineations are consistent with the deformation features, related to dextral wrenching. These lineations are interpreted to represent partly to completely sealing strike-slip faults or shear zones. The location of a recently drilled third development well is indicated by the black dot to the southwest of K15-FG-101 (see text for discussion of well results).

At the time of completing this paper, an additional development well K15-FG-103 had just been drilled to the southwest of well K15-FG-101 (Fig. 15). The well revealed a pressure depletion of only 20 bars (expected depletion 130 bars), which once more confirmed the presence and sealing character of strike-slip faults within the field.

Conclusions

The interpretation of CBIL log signatures, calibrated against core data and production behaviour, has assisted to define both stratigraphic and structural heterogeneity within the reservoir. The orientation of shear fractures and stress field anisotropy, coinciding with subtle expressions on seismic-attribute maps, indicates compartmentalization by strike-slip faults, which together with restricted vertical permeability, explains the anomalous reservoir behaviour.

This has provided guidance to the definition of further development activity and optimization of the overall recovery efficiency of the field. Consequently, an additional development well was recently drilled in the field and pressure data of this well once more confirmed the presence and sealing character of strike-slip faults within the field.

Acknowledgements

This paper is published with permission of the Nederlandse Aardolie Maatschappij BV (NAM), Shell Internationale Petroleum Maatschappij BV and Esso Nederland BV.

The author gratefully acknowledges the cooperation of the following companies, which together with NAM form partnerships in the K and L area: Clam Petroleum Company, Clyde Petroleum (Netherlands) BV, Energie Beheer Nederland BV and Oranje Nassau Energie BV. The author wishes to thank colleagues and especially H. Homann of Western Atlas for discussions and support.

References

Atlas Wireline Services 1991 The Circumferential Borehole Imaging Log (CBIL) – Western Atlas Int. Texas.

Biddle, K.T. & N. Christie-Blick 1985 Strike-slip deformation, basin formation and sedimentation – Soc. Econ. Paleont. Mineral. Spec. Publ. 37: 386 pp

Frikken, H.W. & J.B. Stark 1993 Character and performance of small Rotliegend gas reservoirs, Central Offshore Netherlands. In: North Sea oil and gas reservoirs III, Proceedings of the 3rd North Sea Oil and Gas Conference, Norwegian Institute of Technology, Trondheim, Norway, November 30 – December 2, 1992 – Kluwer, Dordrecht: 41–50

Glennie, K.W., G.C. Mudd & P.J.C. Nagtegaal 1978 Depositional environment and diagenesis of Permian Rotliegend Sandstone in the Leman Bank and Sole Pit areas of the U.K., Southern North Sea – J. Geol. Soc. London 135: 25–34

Hancock, P.L. 1985 Brittle microtectonics: Principle and practice – J. Struct. Geol. 7: 437–457

Harding, T.P. 1990 Identification of wrench faults using subsurface structural data: criteria and pitfalls – Am. Ass. Petroleum Geol. Bull. 74: 1590–1609

Hoecker, C., K.M. Eastwood, J.C. Herweijer & J.T. Adams 1990 Use of dipmeter data in clastic sedimentological studies – Am. Ass. Petroleum Geol. Bull. 74: 105–118

Klein, R.J. & M.V. Barr 1986 Regional state of stress in Western Europe. In: Proc. Int. Symposium on rock stress and rock stress measurements, Stockholm: 33–44

Luthi, S.M. & J.R. Banavar 1988 Application of borehole images to 3-dimensional geometric modelling of aeolian sandstone reservoirs, Permian Rotliegend, North Sea – Am. Ass. Petroleum Geol. Bull. 72: 1074–1089

Mastin, L. 1988 Effect of borehole deviation on breakout orientations – J. Geophys. Research 93: 9187–9195

Mulvany, P.S. 1992 A model for classifying and interpreting logs of boreholes that intersect faults in stratified rocks – Am. Ass. Petroleum Geol. Bull. 76: 895–903

Nagtegaal, P.J.C. 1979 Relationship of facies and reservoir quality in Rotliegend desert sandstones, Southern North Sea Region – J. Petroleum Geol. 2: 145–158

Oele, J.A., A.C.P.J. Hol & J. Tiemens 1981 Some Rotliegend gasfields in the K and L blocks Netherlands Offshore (1968–1978): A case history. In: Petroleum geology of the Continental shelf of N.W. Europe – Institute of Petroleum, London: 289–300

Stearns, D.W. & M. Friedman 1972 Reservoirs in fractured rock. In: Stratigraphic oil and gas fields: Classification, exploration methods and case histories – Am. Ass. Petroleum Geol. Mem. 16: 82–106

Van Veen, F.R. 1975 Geology of the Leman gas field. In: Petroleum and the continental shelf of N.W. Europe – Applied Science Publishers, London: 223–231

Weber, K.J. 1987 Computation of initial well productivities in aeolian sandstone on the basis of a geological model, Leman gasfield, U.K. In: Reservoir sedimentology – Soc. Econ. Paleont. Mineral. Spec. Publ. 40: 333–354

Ziegler, P.A. 1978 North-Western Europe: tectonics and basin development – Geol. Mijnbouw 57: 589–626

Rondeel et al. (eds), Geology of gas and oil under the Netherlands, 115–124, 1996.

Sub-horizontal drilling: remedy for underperforming Rotliegend gasfields, L13 block, central offshore Netherlands

Harm W. Frikken
Nederlandse Aardolie Maatschappij BV (Corporate Planning and Development), Postbus 28 000, 9400 HH Assen, the Netherlands

Key words: drilling mud, layered reservoir, small field, strike-slip faults, sub-horizontal well

Abstract

The first development well in the L13-FE gasfield showed a rapid production decline. Material balance data indicated less than 10% of the expected volumetric reserves to be connected. This poor connectivity is thought to be due to compartmentalization by sealing strike-slip faults, as indicated by faint lineations observed on seismic attribute maps. The presence of only a limited number of scattered, stratigraphically isolated, prolific layers within an overall rather tight and layered reservoir, resulted in a poor overall vertical permeability, which also contributed to the disappointing well performance. The vertical well was subsequently sidetracked sub-horizontally with the aim to connect a larger number of the scattered prolific sands, different fault-compartments and possible open fracture systems. Graded rocksalt drilling mud was used in order to minimize formation impairment. During drilling of the sub-horizontal section numerous problems were encountered due to mechanical failures and the heterogeneous, layered nature of the reservoir. A significant number of prolific, scattered sand layers were encountered and the presence of a small scissor-type fault was recognized from log correlations. Considering that only some 60% of the well could be completed due to mechanical problems, the well is now producing at acceptable rates. The attempt to challenge such a geologically complex, labyrinth-type reservoir by sub-horizontal drilling, has been cost-effective and successful. However, it is recommended that slimhole, sub-horizontal drilling in this type of layered reservoirs should be applied with caution.

Introduction

The L13-FE field was discovered in 1986 by exploration well L13–8 (Fig. 1). The field consists of an apparently undisturbed, NW – SE trending, tilted fault block, which is dip-closed to the southwest and fault-closed to the northeast and northwest. An expected volumetric Gas Initially In Place (GIIP) of ca. 4.6×10^9 m^3 is trapped in the Permian, Rotliegend, Upper Slochteren Sandstone reservoir.

Following the discovery in 1990 of the neighbouring L13-FG field, situated some 2 km to the east, simultaneous development of both fields was planned. The L13-FG field was developed by means of an extended-reach well (L13-FE-101) from a mini-satellite platform installed on the L13-FE field and two crestal wells were drilled to develop the L13-FE field. The first well L13-FE-102 was drilled ca. 250 m north of the abandoned L13–8 discovery well (Fig. 1). However, the well showed a disappointing test rate of 0.3×10^6 m^3/day (compared to 1.3×10^6 m^3/day of well L13–8). Subsequently, rapid production decline and marked pressure depletion occurred and material balance calculations indicated less than 10% of the expected reserves to be connected (Fig. 2). It is therefore worth contemplating whether development of the fields in this area would have taken place when the disappointing well L13-FE-102 would have been the discovery well of the field.

From production logs (PLTs) it was evident that the direct inflow into the wellbore was largely dominated by two highly permeable sand streaks (stratigraphically isolated between shales) in the topmost part of the reservoir (Fig. 3). As such the field showed all the

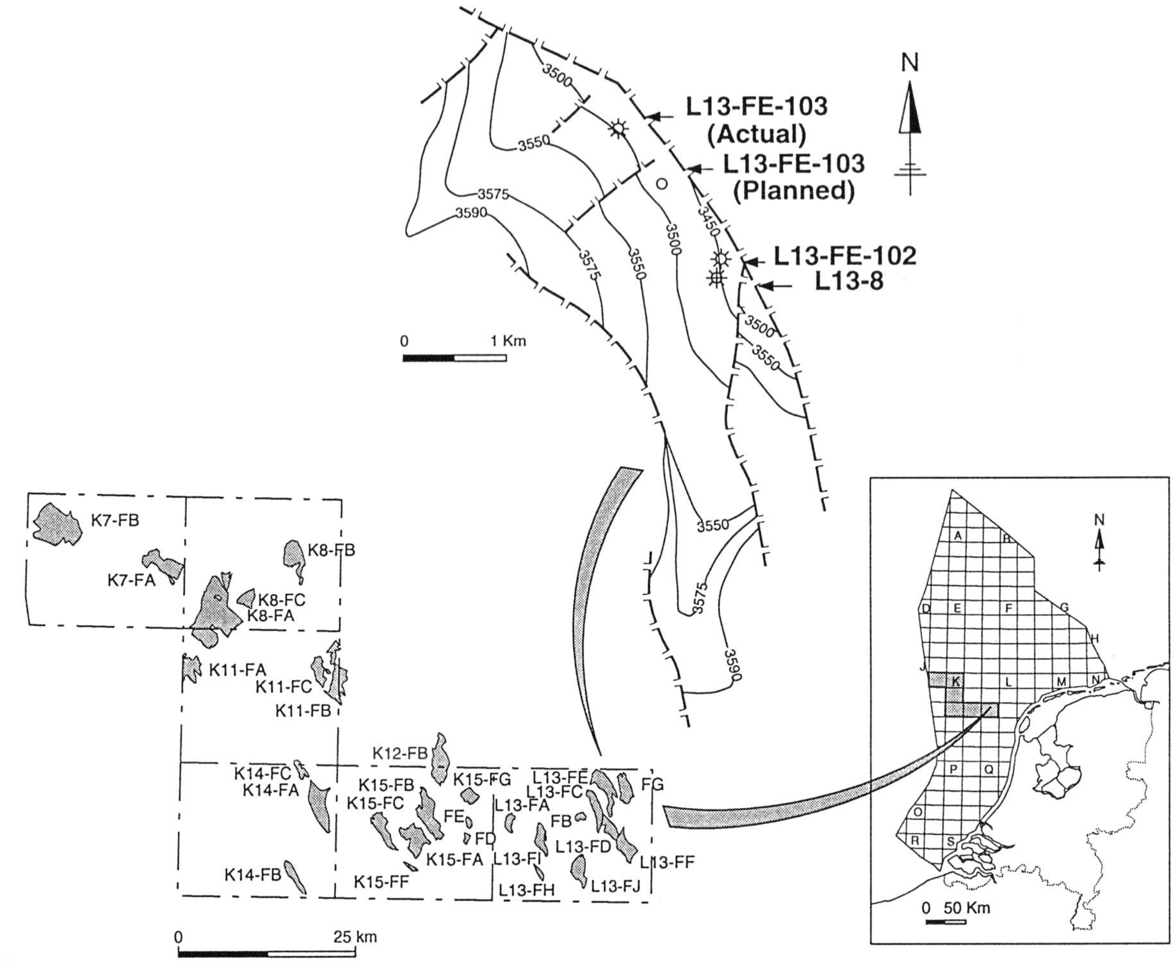

Fig. 1. Top Rotliegend structural map L13-FE gasfield, central offshore Netherlands, showing the abandoned discovery well L13–8 and development wells L13-FE-102 and L13-FE-103. Contour depths in metres.

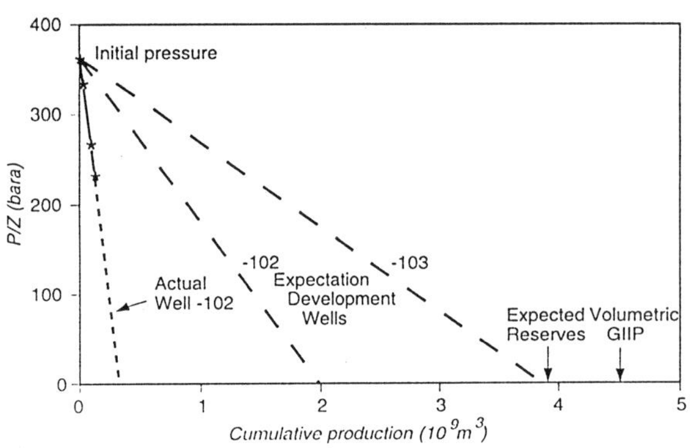

Fig. 2. Material balance data of the L13-FE field. Wells L13-FE-102 and -103 were expected to connect the field's reserves.

characteristics of suffering from the so-called 'small field behaviour', which was simultaneously observed in a number of recent offshore developments (Frikken & Stark 1993). The main causes for such behaviour are considered to be:

1. Significant areal variations in the amount of highly permeable, prolific layers. In the L13-FE area only minor amounts of such prolific layers occur, when compared to other areas (Fig. 4). The rather unpredictable, scattered nature of such layers reflects a 'labyrinth-type' reservoir (Weber & Van Geuns 1989);
2. Stratigraphic (vertical) compartmentalization of the prolific sands due to the presence of shale barriers, restricting cross-flow inside the reservoir from poor-quality sands into prolific sands; and

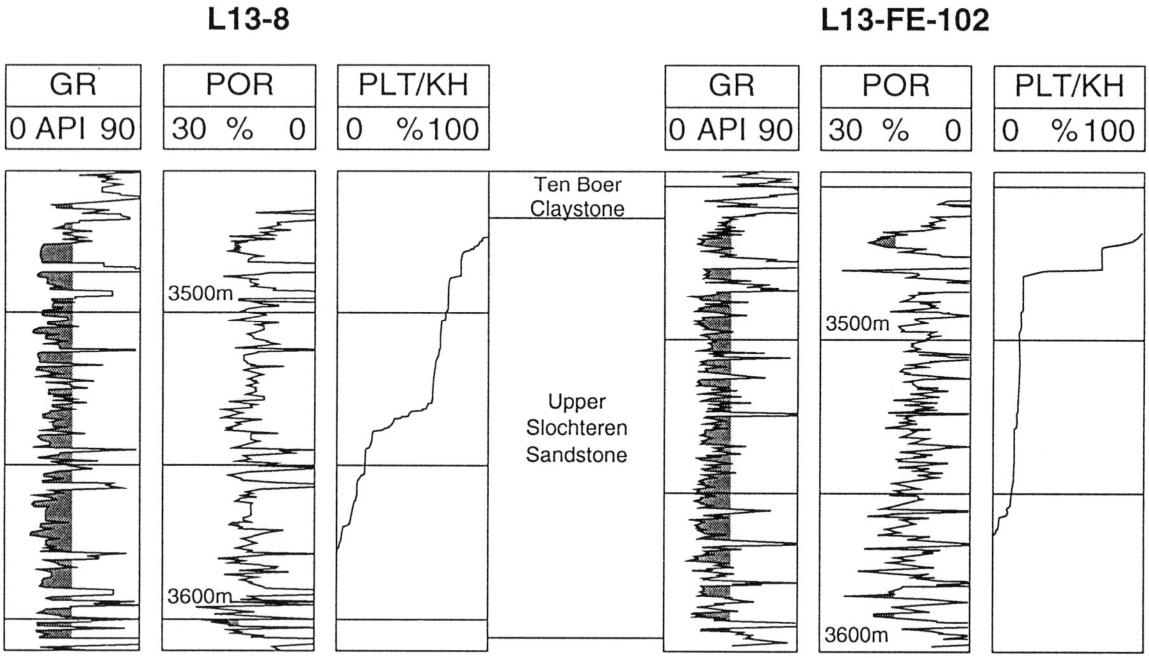

Fig. 3. Type log suite of Gamma Ray, Porosity, and Production Logging Tool (PLT) and calculated permeability thickness (KH), showing the difference in flow characteristics between L13-FE wells. Layers with porosities in excess of 15% dominate the flow into the well bore (PLT and KH signatures).

3. Compartmentalization (lateral) due to sub-seismic, largely sealing strike-slip faults or shear zones.

Shortly after production start-up of the well a seismic Azidip map became available, based on a recently acquired 3D seismic survey. This map showed N – S trending, en echelon lineations to the northwest of the well (Fig. 5). The lineations have been interpreted to represent largely sealing strike-slip faults or shear zones. These are probably the result of dextral wrenching (e.g. Biddle & Christie-Blick 1985, Hancock 1985), and have also been recognized in other fields in the area (Frikken & Stark 1993). Such faults are beyond the resolution of conventional seismic due to the limited vertical throw and their presence can only be identified by means of detailed seismic attribute mapping. Their sealing potential is analysed from pressure behaviour of wells. Taking into account the signature of the map and the production performance, it was obvious that the well had been drilled into an isolated fault compartment.

Planning

The second development well L13-FE-103 was planned ca. 1 km to the northwest of well -102. The location of this well was shifted some 600m further to the northwest, across a discontinuous fault (Fig. 1). This was done to reduce the likelihood of fault-interference at the proposed -103 location and to create an opportunity for either re-drilling or sidetracking the underperforming -102 well towards the northwest.

Well -103 encountered top Rotliegend ca. 45 m deeper than prognosis, due to uncertainties in the pick of an overburden unconformity and thickness variations of the overlying, wedging salt caprock. A Vertical Seismic Profile (VSP) survey obtained from this well (deviated along the proposed -102 sidetrack trajectory) furthermore confirmed the presence of fault disturbances.

Three options were considered for remedial action for the underperforming -102 well:

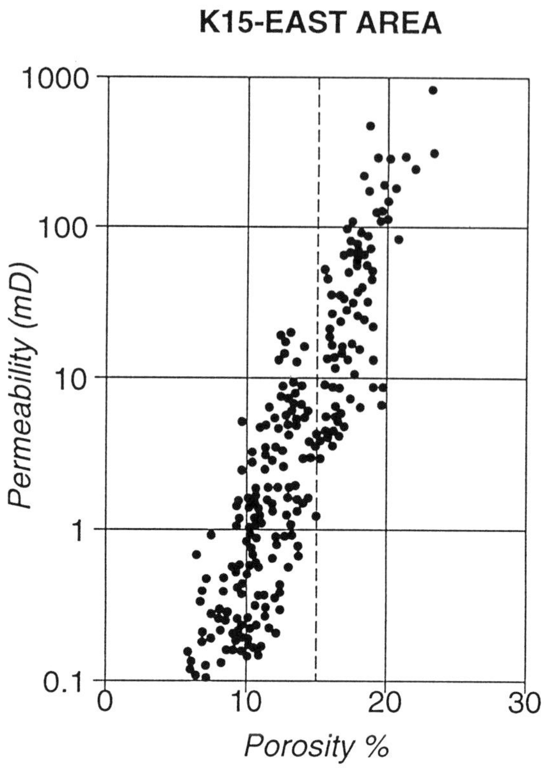

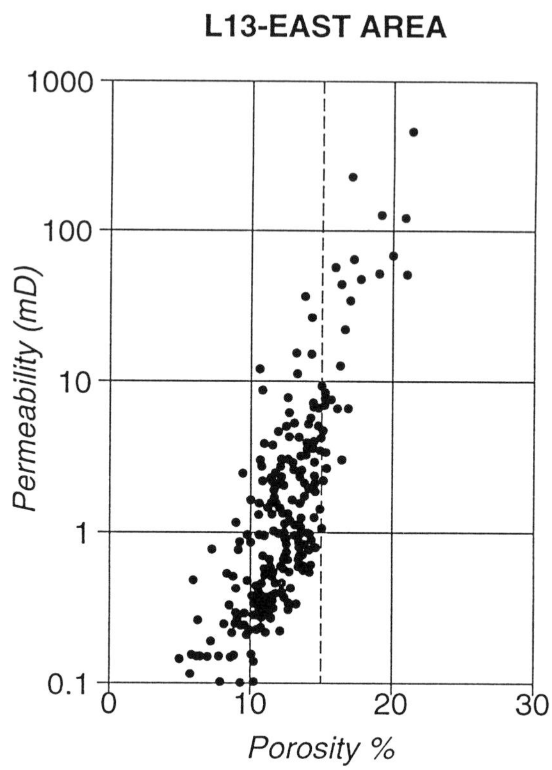

Fig. 4. Comparison of the variation in the amount of highly permeable layers (porosity > 15%) of the Upper Slochteren Sandstone in the K15 and L13 areas.

1. Stimulation by hydraulic fracturing. This option was ruled out due to uncertainties in fracture propagation distance and orientation (no stress-field data available) and due to technical restrictions of fracturing from a mini-satellite platform.
2. A new deviated well to the northwest of -102, together with maintaining the low and decreasing production rates of the existing well. Cost considerations and timing ruled out this option.
3. A sub-horizontal sidetrack of -102, towards the northwest along the almost horizontal crestal section of the field (Fig. 5). This option would have the advantage of: a) cost effectiveness (ca. 30% of the costs of a new well), b) early production, and c) the possibility of connecting a larger number of prolific sands, open fractures and fault-compartments, while still penetrating the full gas column. A completely horizontal well was not considered due to the rather layered and heterogeneous character of the reservoir and the limited penetration angle it would make with the almost horizontal reservoir section along the crest of the structure.

Design

Due to a major repair job in the $13\frac{3}{8}''$ casing of the original well L13-FE-102, the sidetrack was designed to kick-off relatively deeply, out of the $9\frac{5}{8}''$ casing (Fig. 6). The main build-up section was to be delayed until confirmation was obtained about the position of the uniformly thick Zechstein 2/1 Anhydrite/Dolomite section. This was done by means of Gamma Ray Measurement While Drilling (GR-MWD) log monitoring because of uncertainty in thickness of the overlying Zechstein 3 Anhydrite and the salt wedge (Fig. 6).

The $7''$ liner was planned to be set at the end of the build-up section, at the top of the Upper Slochteren Sandstone. Since no further MWD logging tools (besides GR-MWD) would be available for the $5\frac{7}{8}''$ subhorizontal hole section, pipe-conveyed logging of this section was selected. The logging was planned to include a Circumferential Borehole Imaging Log (CBIL) for fracture detection and analysis of stress-field anisotropy for possible design of hydraulic fracturing. It was decided to complete the hole with

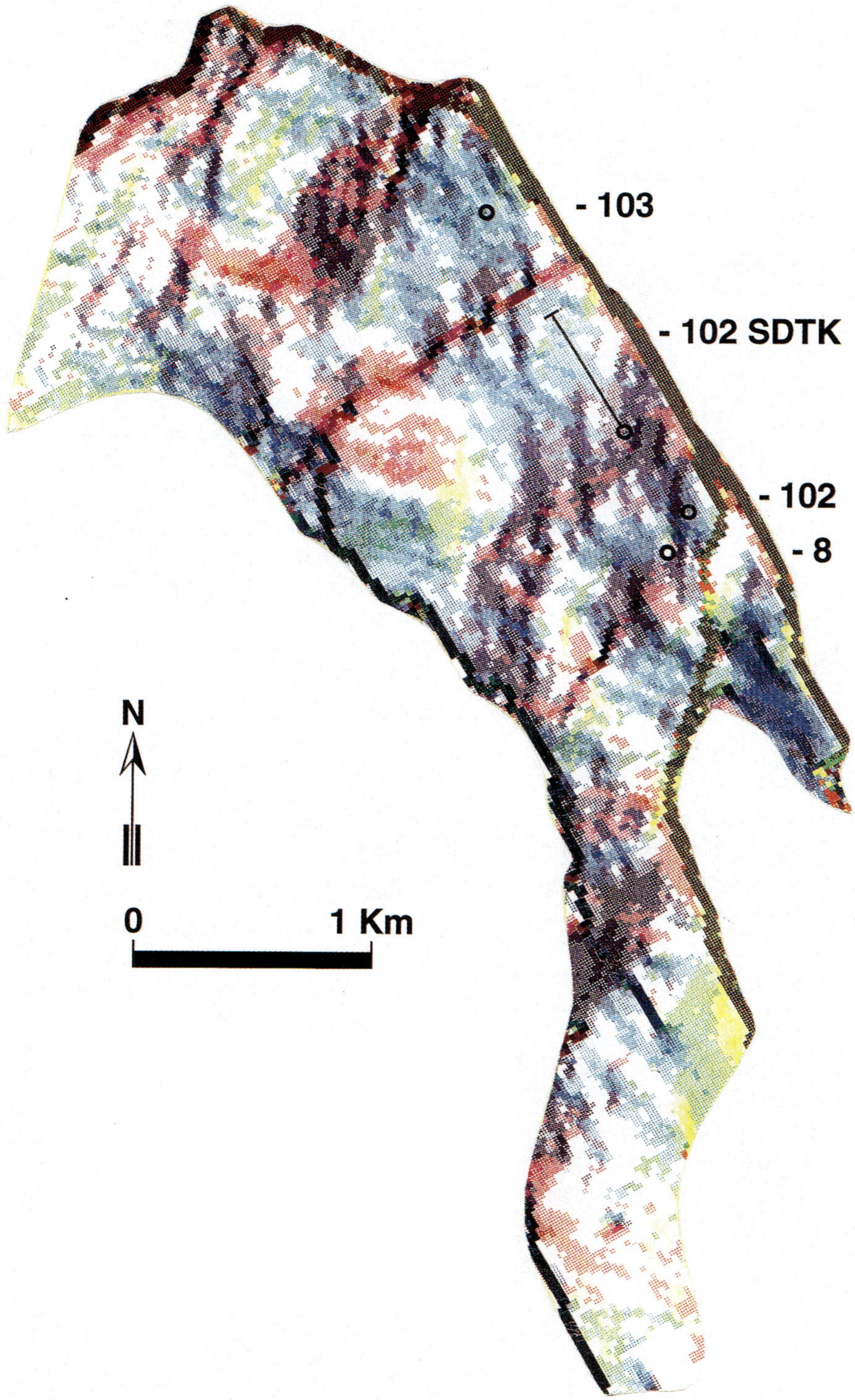

Fig. 5. Top Rotliegend 3D seismic Azidip map L13-FE field. The mainly N – S trending, enechelon lineations (purple) are interpreted to represent sealing strike-slip faults. The other lineations represent predominantly normal faults with distinct vertical offsets.

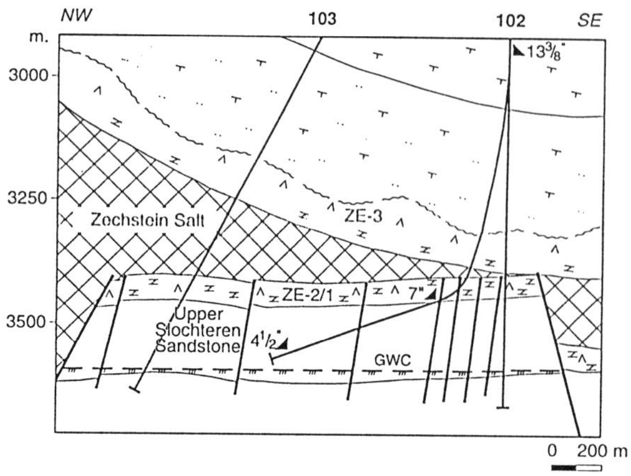

Fig. 6. Cross-section of the L13-FE field showing the planned sub-horizontal trajectory, the relatively complicated overburden and the presence of strike-slip faults (vertical exaggeration 2×). GWC = Gas-water contact.

an open hole slotted liner, in order to enable possible future retrieval of the liner for hydraulic fracturing.

Performance estimation

In order to design the optimum angle for the sub-horizontal well, sensitivities on the inflow performance were analysed with respect to permeability anisotropy, slant angle and skin (based on analyses by Kuchuk & Goode 1988):

1. The productivity improvement of a sub-horizontal well, when compared to a vertical well is predominantly governed by the vertical permeability (Fig. 7). Measurement of horizontal and vertical permeability from core-plugs did not show a significant anisotropy. It was recognized that effects of permeability anisotropy due to the rather extensive shale layers would be far more significant. It was evident that the layered and heterogeneous nature of the Upper Slochteren Sandstone would not be optimum for a horizontal (and even a sub-horizontal) well and in the worst case no productivity improvement could be expected when compared to a vertical well.
2. It was calculated that the slant angle could have a significant impact on inflow performance with a major productivity increase expected from slant angles in excess of ca. 75 degrees (Fig. 7).
3. Mechanical skin has a detrimental effect on (slanted) well performance (Renard & Dupuy 1991). In this respect the application of a slotted liner would carry an additional risk of poor productivity, because of the absence of perforations through the invaded zone.

Inverted Oil Emulsion Mud (IOEM) was at that time routinely applied in the offshore wells, mainly to reduce drilling problems. However, it became evident that oil-based muds could cause significant formation damage (McDonald & Buller 1992), and it was experienced that in a number of cases productivities of offshore wells were adversely affected due to use of oil-based muds (Frikken & Stark 1993; skin damage and possible effects on wettability). For this reason drilling fluid impairment tests on core samples were carried out in the laboratory with oil-based mud and graded rocksalt mud.

The test results indicated that graded rocksalt mud would maintain permeabilities, almost an order of magnitude higher, especially in the prolific, highly permeable sands, when compared to an oil-based mud (Fig. 8). For the less permeable waterlaid sands a near-reverse effect was indicated (Fig. 8). This was considered less relevant since these sands tend not to produce directly into the wellbore but are being depleted by cross-flow inside the reservoir.

Following these analyses it was decided to drill the sub-horizontal section at an angle of 80 degrees across ca. 600 m of the reservoir. Application of graded rocksalt mud was also decided, despite possibly poorer drilling lubrication properties when compared to oil-based mud. Stimulation by means of a coiled-tubing acid wash (Economides et al. 1991) was designed in order to further reduce formation impairment. From well-test pressure build-ups in the area, a ratio of ca. 100 between horizontal and vertical permeability (Kh/Kv, Fig. 7) and a permeability thickness product of 200 mD·m were calculated. The expected productivity at 50 bar drawdown, under semi-steady state conditions (Q50 sss) was estimated to be ca. 0.8×10^6 m^3/day (Fig. 7).

Execution

During sidetracking in the overburden, GR-MWD monitoring was applied for identification of the kick-off point for the main build-up section (top Zechstein 2/1 Anhydrite/Dolomite, Fig. 6). However, correlation by means of GR-MWD was not straightforward as no full logging suite is available during drilling. Together with poor drill-cuttings control as a consequence

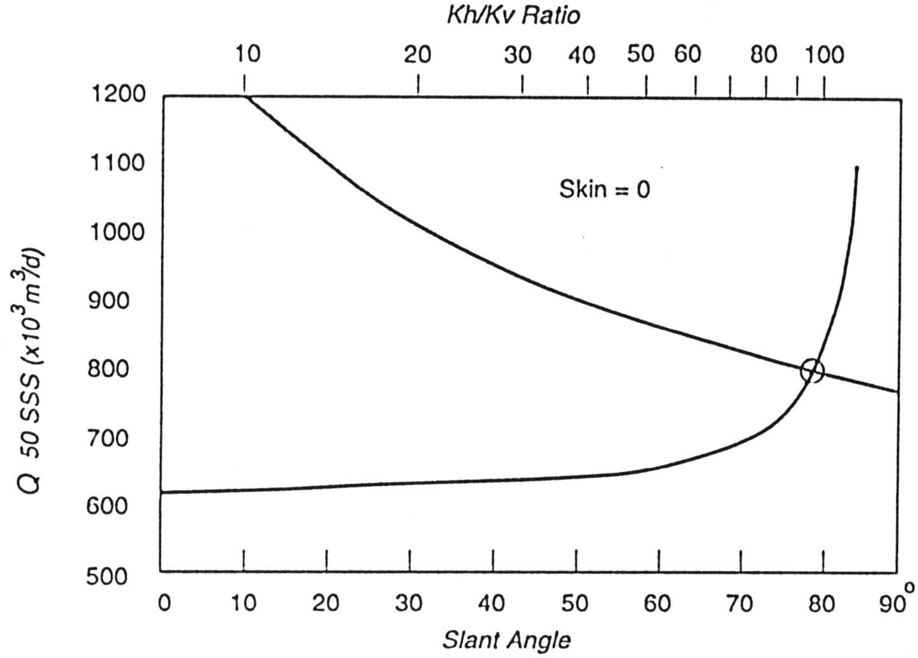

Fig. 7. Analysis of expected influence of slant angle (black line) and vertical permeability (grey line) on inflow performance for the sidetrack. See text for discussion of features.

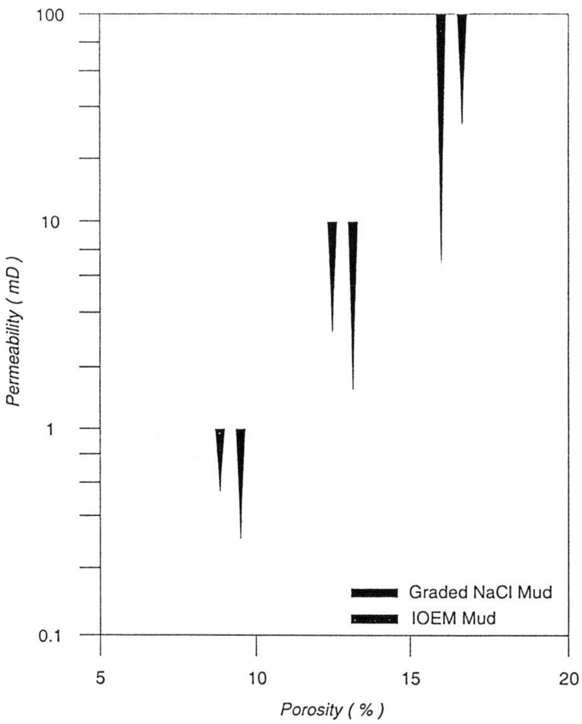

Fig. 8. Results of drilling mud impairment tests on core samples. See text for discussion of results.

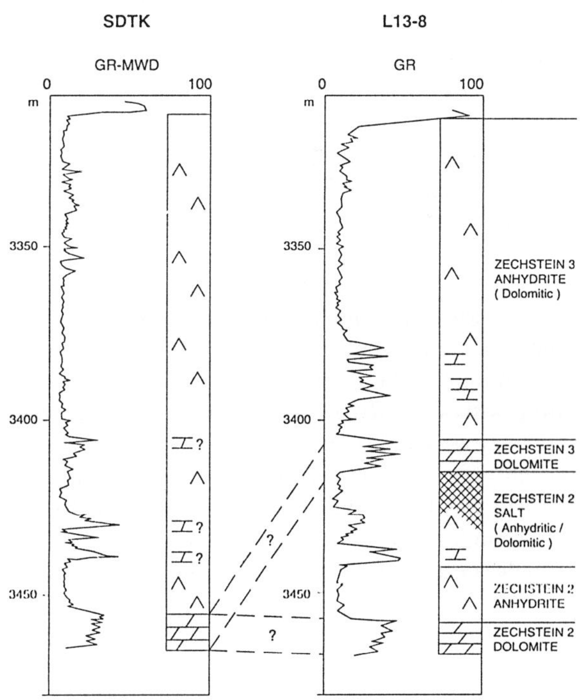

Fig. 9. Uncertainty in evaluating the Zechstein overburden stratigraphy of the sidetrack by means of the Gamma Ray-Measurement While Drilling (GR-MWD) technique.

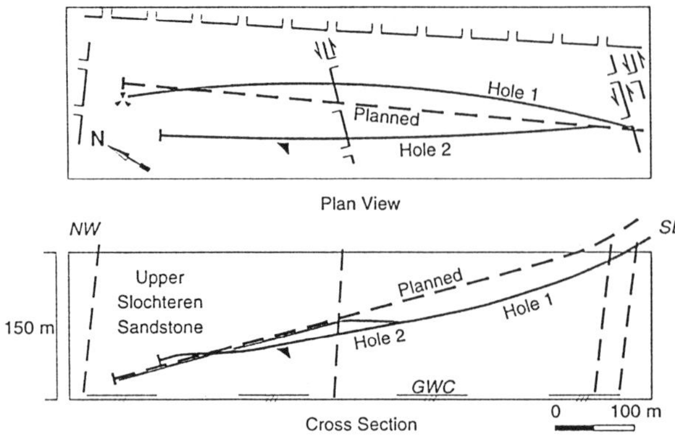

Fig. 10. Well trajectories in plan view and cross section of hole 1 and 2 of the sub-horizontal sidetrack showing considerable dog-legs (vertical exaggeration of cross section 2×).

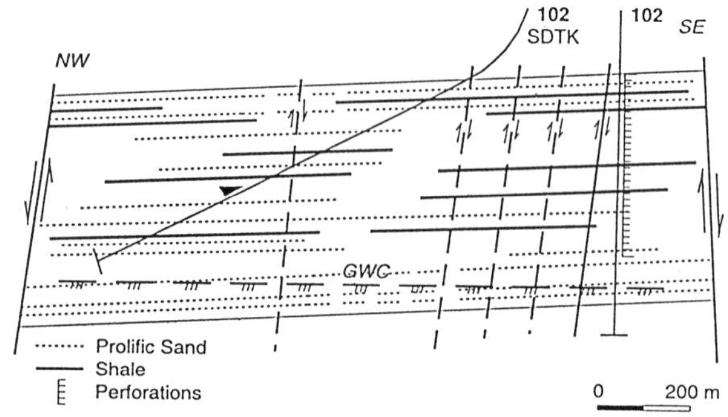

Fig. 11. Cross-section of the Upper Slochteren Sandstone showing the isolated position of prolific sands in the original well -102, the significant amount of prolific sands encountered by the sidetrack and the presence of a strike-slip fault in the middle of the trajectory. The black triangle indicates the slotted liner shoe (vertical exaggeration 3×).

of turbine drilling and an unexpected, poorly developed Zechstein 3 Dolomite, significant uncertainties evolved. Apparently the Zechstein 3 and 2 Dolomite markers were being encountered 50 m deep (Fig. 9), with the consequence of not being able to enter the reservoir at the required location and angle.

During the build-up phase the planned for build-up angle could not be achieved, resulting in angles exceeding 80 degrees across the reservoir section and consequently resulting in dog-legs. The drill-bit stood up several times in the harder shale layers, which resulted in problems of dropping of the angle (Fig. 10). The slim hole equipment (e.g. mudmotors and GR-MWD) proved unreliable under these drilling conditions, showing a considerable number of failures.

Furthermore, a laminar flow regime could not be maintained and turbulent flow occurred. This resulted in wash-outs in the more friable sand sections, coinciding with shale ledges. During pipe-conveyed logging these poor hole conditions resulted in the logging assembly parting and a density and neutron logging tool fish was left in the hole. During attempts to isolate the radioactive sources, a second fish was left in the hole, necessitating a re-drill of the sub-horizontal section (bottom hole target some 50 m away from the logging tool sources, Fig. 10). Further logging with pipe-conveyed tools (e.g. CBIL) was subsequently cancelled. Similar to the first hole, poor hole conditions of this re-drill section resulted (once more) in a fish in the hole (bottom-hole assembly) and the slotted liner could eventually only be run across ca. 60% of the open hole section (Fig. 10).

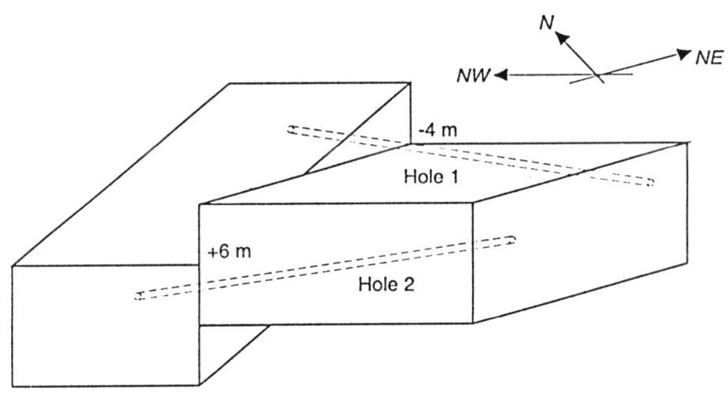

Fig. 12. The effect of a small scissor-type fault on apparent reservoir thickness: omission of strata in hole 1 (ca. 4 m) and repetition of strata in hole 2 (ca. 6 m).

Results

Minor (differential) pressure depletion of only a few bars was observed within the reservoir, possibly as a result of the production from the original -102 well and the 150–200 bar pressure drop in that well, indicating that the faults had become partly transmissive. Removal of the graded rocksalt filter cake and stimulation by means of a coiled-tubing wash with formic acid was successful. Evaluation of porosity logs obtained from the first hole indicated that the well had penetrated a significant number of prolific sand streaks (Fig. 11). Due to the absence of CBIL, Sonic and Production Logging Tool (PLT) data, no evidence could be obtained about the possible presence of open fractures,

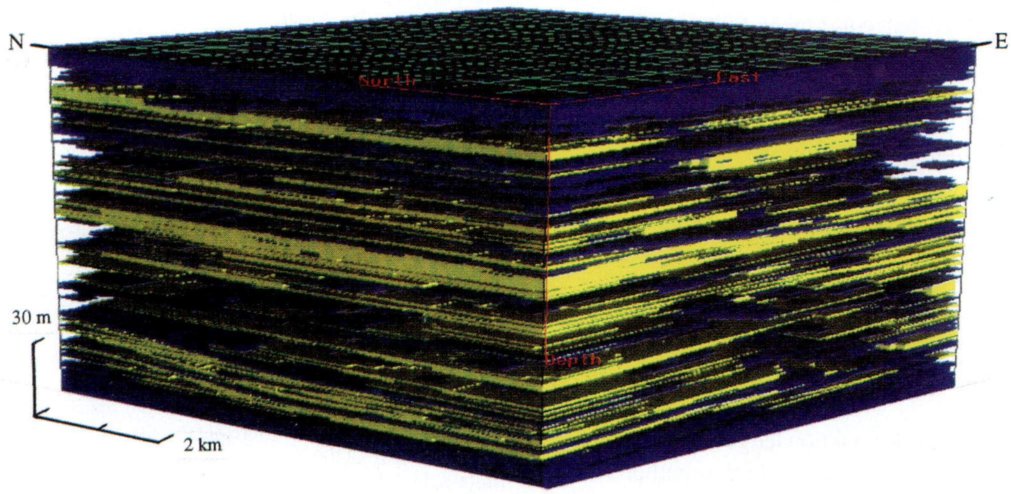

Fig. 13. Three-dimensional, structurally unconstrained reservoir-geological (MONARCH) model of the Upper Slochteren Sandstone of the entire L13 East area (11 × 10 km, vertical exaggeration 60×, view from SW). Prolific sands: yellow; shales: blue; poor quality sands: transparent. Note the prolific sands showing a decrease in quantity and a more scattered nature towards the east (L13-FE area).

which had previously been observed in surrounding fields (Frikken & Stark 1993).

From the GR log correlations a small fault was interpreted about half-way the well trajectory (Figs 10, 11). Omission of strata in hole 1 (ca. 4 m) and repetition of strata in hole 2 (ca. 6 m) point to a scissor-type fault (Fig. 12).

The well was brought into production and showed somewhat lower flow rates than predicted (ca. 0.5×10^6 versus 0.8×10^6 m^3/day at 50 bar drawdown), probably due to a lower than expected permeability × thickness product (120 versus 200 mD·m). This in turn is expected to be the result of obstruction of the open hole section due to the fish. No indications exist sofar from pressure behaviour of the well of any connectivity problems.

A three-dimensional reservoir-geological modelling study has recently been carried out, covering the entire L13 East area (Fig. 13). The results of the study were:

1. Stratigraphic compartmentalization in the L13-FE field is expected to be far more significant than fault compartmentalization;
2. A consistent match was obtained between material balance data from fields in the L13 East area and connectivity estimates, when N – S to NE – SW trending faults were considered to be sealing; and
3. Probabilistic modelling, excluding the results of the sub-horizontal well, confirmed a significant number of prolific sands to be expected in the area of the sidetrack.

The study furthermore predicted that wells L13-FE-102 Sidetrack and L13-FE-103 are expected to connect a volumetric GIIP of 1.5×10^9 m^3 and 1.4×10^9 m^3, respectively. The remaining 1.7×10^9 m^3 of gas is expected to be isolated in the southern part of the field, beyond the strike-slip faults or shear zones (Fig. 5). Material balance data over the next few years may validate the predicted connectivity and may prove the necessity (if economically attractive) of an additional, future drainage point in the southern part of the field.

Conclusions

The technically challenging, sub-horizontal sidetrack of well L13-FE-102 has been successful in obtaining cost-effective production from a geologically complex labyrinth-type reservoir.

Despite considerable technical problems the well has connected a significant number of prolific layers and appears to connect separate fault blocks. The application of graded rocksalt mud has been proven advantageous over the use of oil-based muds. The application of an openhole slotted liner allowed easy removal of the graded rocksalt mud-cake.

Technical and cost considerations necessitated a slimhole sidetrack, but equipment failures and trajectory instability occurred at drilling angles of 80 degrees. It is therefore recommended that slimhole, sub-horizontal drilling in this type of layered reservoirs should be applied with caution.

Acknowledgements

This paper is published with permission of the Nederlandse Aardolie Maatschappij BV (NAM), Shell Internationale Petroleum Maatschappij BV and Esso Nederland BV.

The author gratefully acknowledges the cooperation of the following companies, which together with NAM form partnerships in the K and L blocks: Clam Petroleum Company, Clyde Petroleum Netherlands BV, Energie Beheer Nederland BV and Oranje Nassau Energie BV.

The author wishes to thank all members of the multi-disciplinary area project team of the Business Unit Offshore for their efforts and colleagues of Shell Research for permission to use data from their modelling study.

References

Biddle, K.T. & N. Christie-Blick 1985 Strike-slip deformation, basin formation and sedimentation – Soc. Econ. Paleont. Mineral. Spec. Publ. 37: 386 pp

Economides, M.J., K.B. Naceur & R.C. Klem 1991 Matrix stimulation method for horizontal wells – J. Petroleum Techn. 43: 854–861

Frikken, H.W. & J.B. Stark 1993 Character and performance of small Rotliegend gas reservoirs, Central Offshore Netherlands. In: North Sea oil and gas reservoirs III, Proceedings of the 3rd North Sea Oil and Gas Conference, Norwegian Institute of Technology, Trondheim, Norway, November 30 – December 2, 1992 – Kluwer Academic Publishers, Dordrecht: 41–50

Hancock, P.L. 1985 Brittle microtectonics: principles and practice – J. Struct. Geol. 9: 437–457

Kuchuk, F.J. & P.A. Goode 1988 Pressure transient analysis and inflow performance for horizontal wells – Soc. Petroleum Eng. (SPE) Paper 18300

McDonald, J.A. & D.C. Buller 1992 The significance of formation damage caused by the adsorption of oil-based mud surfactant – J. Petroleum Sci. and Engin. 6: 357–365

Renard, G. & J.M. Dupuy 1991 Formation damage effects on horizontal-well flow efficiency – J. Petroleum Techn. 43: 786–789

Weber, K.J. & L.C. Van Geuns 1989 Framework for constructing clastic reservoir simulation models – Soc. Petroleum Eng. (SPE) Paper 19582

Rondeel et al. (eds), Geology of gas and oil under the Netherlands, 125–142, 1996.

Multidisciplinary exploration strategy in the northeast Netherlands Zechstein 2 Carbonate play, guided by 3D seismic

Jan M.M. van de Sande[1], Tom J.A. Reijers[1,2] & Neil Casson[1,3]
[1] *Nederlandse Aardolie Maatschappij (Business Unit Exploratie, XEX/1), Postbus 28000, 9400 HH Assen, the Netherlands;* [2] *Present address: Shell Petroleum Development Company of Nigeria (XGSW/2), P.M.B. 2418, Lagos, Nigeria;* [3] *Present address: Shell Internationale Petroleum Maatschappij (SEDXGS), Postbus 162, 2501 AN Den Haag, the Netherlands*

Key words: facies, fractures, hydrocarbon reservoir, porosity, quantitative prediction, thickness

Abstract

The Nederlandse Aardolie Maatschappij has actively pursued the exploration of the Zechstein 2 Carbonate play over the last 40 years. The effort concentrated largely on the SE Drenthe area. It has resulted in the discovery of 20 gas fields with cumulative reserves of some $57 \times 10^9 m^3$. The Zechstein 2 Carbonate Member is part of the Basal Zechstein Unit which developed mainly as an anhydrite platform at the southern fringe of the Southern Permian Basin. The facies within the member reflect platform, slope and basinal settings. The subsurface data collected in the exploration and development of the Zechstein fields have led to a detailed sedimentological model for the member. Prospects in the member are seismically defined as structural highs at Top Zechstein 2 Anhydrite level. However, as exploration moves towards the search for more subtle traps, 3D seismic becomes indispensable. Basal Zechstein isochore maps based on 3D seismic, display the topography of the Basal Zechstein platform. These maps are used to assess reservoir potential and gross reservoir thickness. Variations in reservoir parameters, especially porosity, of the Zechstein 2 Carbonate Member are reflected in the seismic response. Quantitative studies based on 3D seismic allow spatial prediction of specific reservoir parameters. Detailed fracture studies in gas fields have demonstrated a relationship between seismically detectable fault patterns and core-scale reservoir enhancing fracture patterns. The Collendoornerveen exploration well demonstrates the successful integration of the various disciplines.

Introduction

The Nederlandse Aardolie Maatschappij has actively pursued the exploration of the Zechstein 2 Carbonate play over the last 40 years. The first successful well, Coevorden-1, was drilled in 1948 and the most recent one, Collendoornerveen-1, was drilled in 1990. This exploration effort concentrated largely on the SE Drenthe area (Fig. 1). It has resulted in the discovery of 20 gas fields with cumulative reserves of some 57×10^9 m^3.

In the Zechstein 2 Carbonate play, abundant gas charge into the dip and fault-closed structures is sourced by the underlying coal measures of the Westphalian A and B. Seals are provided by the overlying Zechstein 2 Anhydrite Member and Zechstein 2 Halite Member. The main uncertainty in this play is the reservoir quality, especially the porosity and permeability of the carbonates.

The wealth of geological and geochemical data which has been collected in the exploration and development of the Zechstein fields has led to the establishment of detailed depositional and diagenetic models for the Zechstein 2 Carbonate Member (Clark 1986, Van der Baan 1990). More recently, interpretation of the extensive 3D seismic coverage and integration of quantitative prediction techniques as well as results of detailed fracture studies have added to the comprehension and provided powerful additional tools in the exploitation of the play.

The experience gained within NAM over the years of successful exploration and development of the Zech-

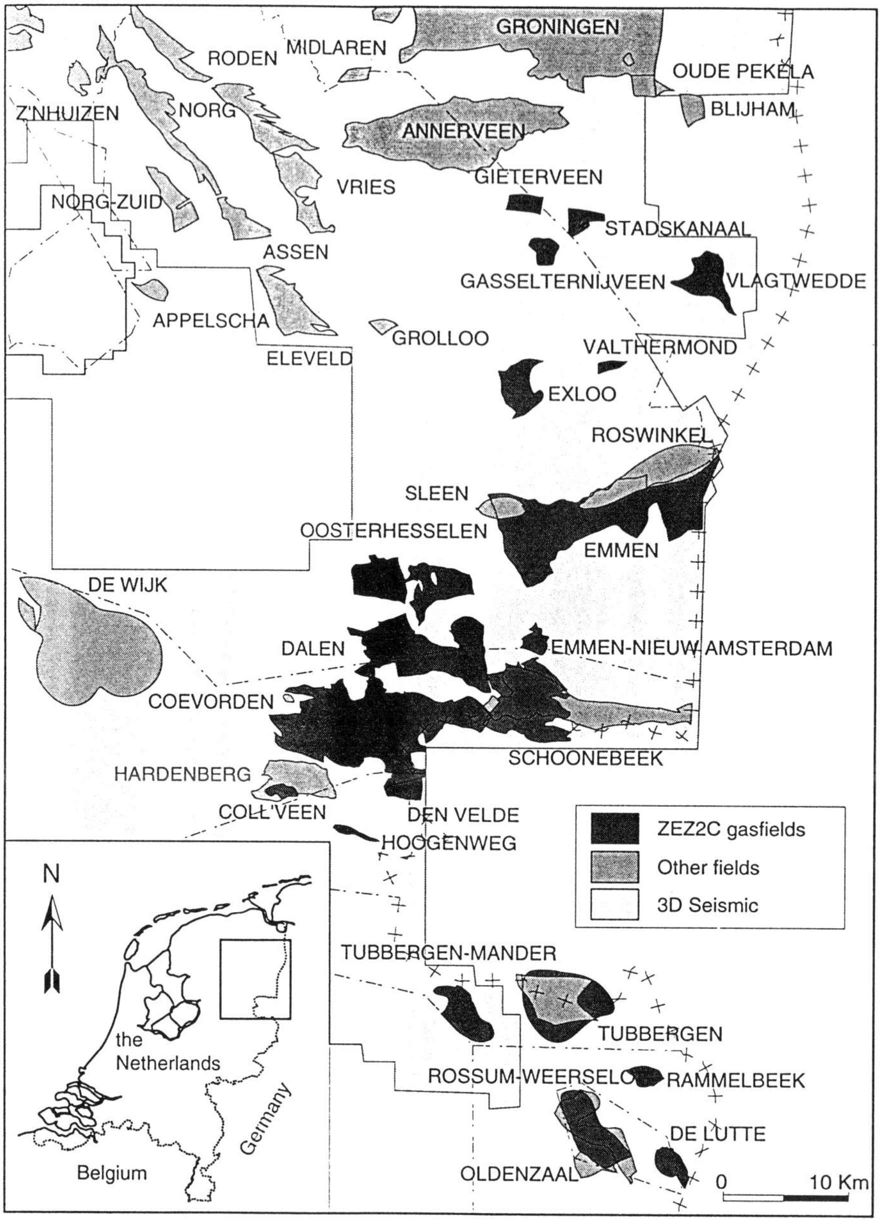

Fig. 1. Map of the NE Netherlands showing Zechstein 2 Carbonate Member (ZEZ2C) gas fields and other fields and the existing 3D seismic coverage.

stein 2 Carbonate Member yielded a comprehensive toolbox of predictive techniques that can be employed successfully in the continuing exploration programme in the Netherlands and elsewhere within the Zechstein hydrocarbon province.

The following sections describe i) the geological setting of the Zechstein 2 Carbonate Member, ii) the development of the Zechstein 1 Anhydrite platform, iii) the different facies of the Zechstein 2 Carbonate Member and their reservoir potential, iv) the interpretation techniques applied on 3D seismic data which provide information on reservoir parameters such as gross thickness and average porosity, and v) the application of these techniques in the Collendoornerveen case history.

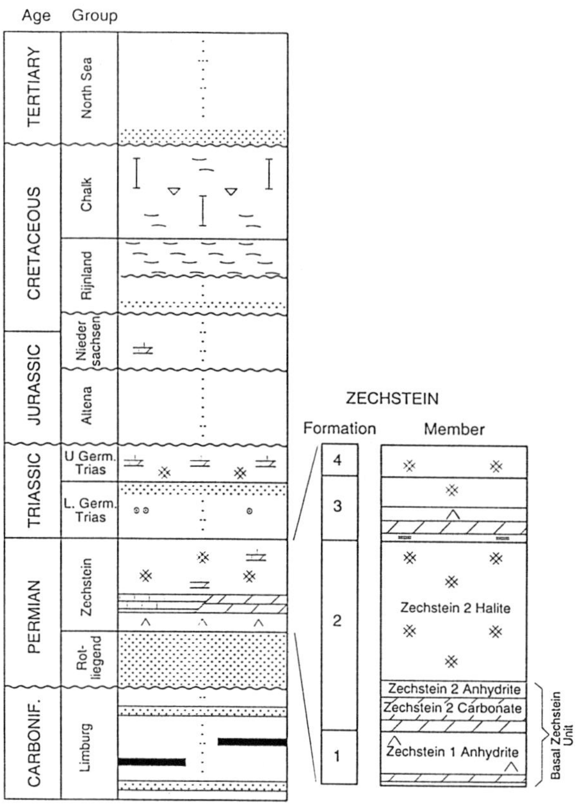

Fig. 2. Generalized stratigraphy of the NE Netherlands with details of the Zechstein Group. The Coppershale (approximately 1 m at base Zechstein) is not shown.

Geological setting

Within the Netherlands the Upper Permian Zechstein Group contains important hydrocarbon reservoirs, principally the Zechstein 2 and 3 Carbonate Members. In addition, the various salt members of this group are the seals to the vast majority of gas accumulations in Zechstein, Rotliegend and Limburg reservoirs, with the giant Groningen gas field as the most prominent example. Halokinesis in Zechstein salt causes most of the structuration for traps in the post-Zechstein interval and salt withdrawal provides the migration paths for gas into these traps. Although it is important to realize the significance of the total Zechstein Group, this paper will concentrate only on the lowermost part of the group, the Basal Zechstein Unit, which includes the important Zechstein 2 Carbonate reservoir (Fig. 2).

The Basal Zechstein Unit in the NE Netherlands includes the entire Zechstein 1 Formation and the lowermost part of the Zechstein 2 Formation (the Carbonate and Anhydrite Members, Fig. 2). It constitutes the non-mobile rock units formed prior to the deposition of the first thick, ductile Zechstein salt, the Zechstein 2 Halite Member. It overlies Carboniferous strata or the Lower Permian Rotliegend Group and was deposited during the Late Permian transgression which flooded the European Permian basins and led to the establishment of the major, intra-continental Zechstein sea. This inland sea extended some 1600 km from the United Kingdom in the west to the Baltic regions in the east (Ziegler 1990).

As can be seen on the palaeogeographic map for the Late Permian (Fig. 3), the NE Netherlands was situated on the southern fringe of the Southern Permian Basin in the zone of transition from basin centre to basin fringe with associated facies. Here, a sulphate and carbonate platform developed, represented by the Basal Zechstein Unit.

Platform carbonates are only a minor part of the Basal Zechstein Unit, the bulk being anhydrite. The general morphology of the setting in which the unit was formed is illustrated on the log correlation panel, showing the basin – slope – platform transition (Fig. 4).

Depositional model

Zechstein 1 Formation (ZEZ1)

The base of the Zechstein Group is marked by basin-wide occurrence of shales of the Kupferschiefer (Coppershale) Member which has everywhere approximately the same thickness (approximately 1 m). This was the first deposition following the transgression of the Zechstein sea which flooded the pre-existing low-relief, continental topography.

The Kupferschiefer is succeeded by a thin carbonate unit, the Zechstein 1 Carbonate Member, before increased salinity resulted in sulphate precipitation forming the Zechstein 1 Anhydrite Member. This anhydrite is the principal component of the Basal Zechstein Unit.

In the area of interest towards the southern fringe of the Southern Permian Basin, the Zechstein sea was relatively shallow (tens of metres). Due to the undulating nature of the pre-Zechstein topography in this area the water depth varied locally; however, it generally increased northwards towards the basin centre (Fig. 5B).

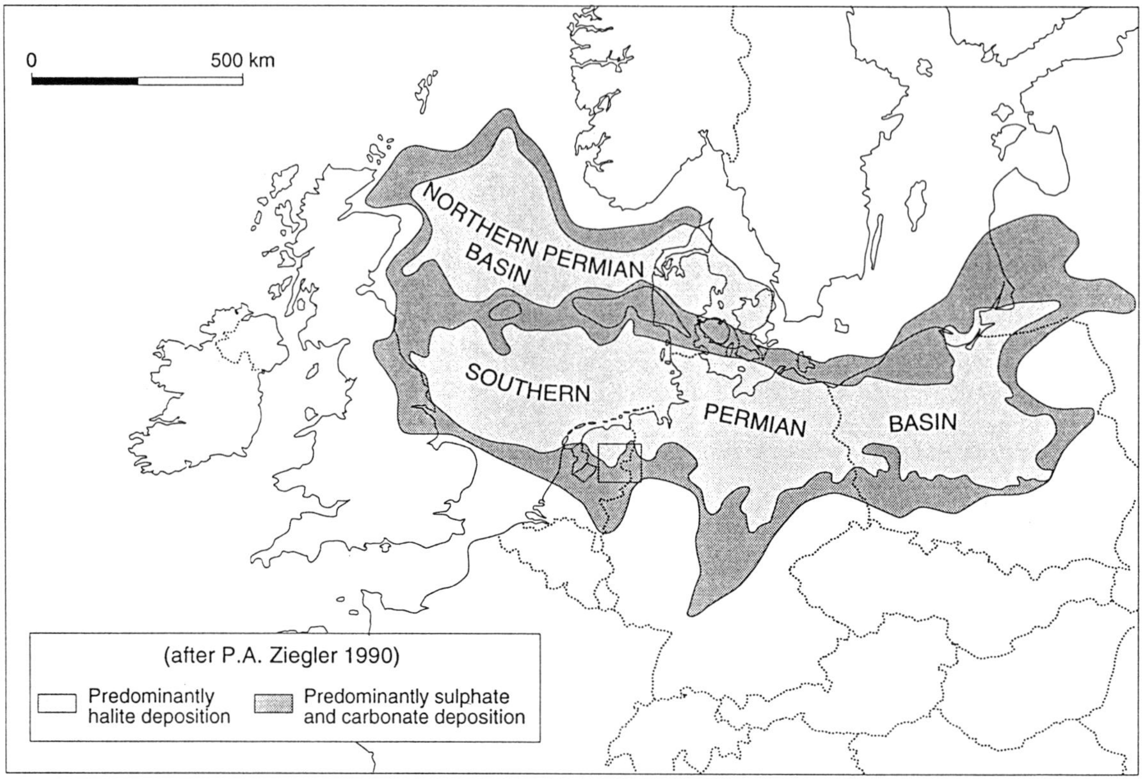

Fig. 3. Map showing the extent of the Zechstein evaporitic basin in Late Permian time.

In the hot, arid conditions that prevailed at the time, evaporation of the inland Zechstein sea was intense and triggered precipitation of gypsum close to the sea's surface (Van der Baan 1990; Fig. 6). The deposition rates of the gypsum greatly exceeded overall subsidence and sea-level rise, and in the shallow-water areas gypsum accumulation rapidly built up towards sea-level. However, sulphate-reducing bacteria proliferated around the chemocline, the transition from the oxic upper water layer to the anoxic deeper waters, which occurred in the Zechstein sea at some 15 m water depth. Gypsum crystals, precipitated in the shallow-water zone, sank into the chemocline and were effectively consumed by the sulphate-reducing bacteria. Thus the areas where the sea floor extended below the chemocline, received little or no gypsum deposition. Only at times when the chemocline was temporarily disturbed, probably as the result of storms, did gypsum deposition occur in the deeper water.

The marked lateral variation in rates of gypsum deposition between the shallow-water area and the starved basinal areas, resulted in a pronounced basin relief. The facies to the south, formed at or around mean sea-level within an extensive shallow-water and sabkha environment, continued to reflect this environment despite the ongoing overall subsidence. Over a distance of 2–3 km down-slope, this changed into basinal deposition with depositional rates virtually at a stand-still. During the period of gypsum deposition the described profile was progressively enhanced and the platform top stood some 250–300 m above the adjacent basinal areas towards the end of the development of the Zechstein 1 Anhydrite Member (Fig. 5C).

Zechstein 2 Carbonate Member (ZEZ2C)

A further rise in relative sea-level reduced the salinity, and sedimentation in the Zechstein basin changed from gypsum to carbonates (Fig. 5D). The resulting Zechstein 2 Carbonate Member forms the most important Zechstein reservoir in the NE Netherlands.

The facies of the Zechstein 2 Carbonate Member reflect a platform – slope – basin profile and owe their areal distribution to that profile, established by the

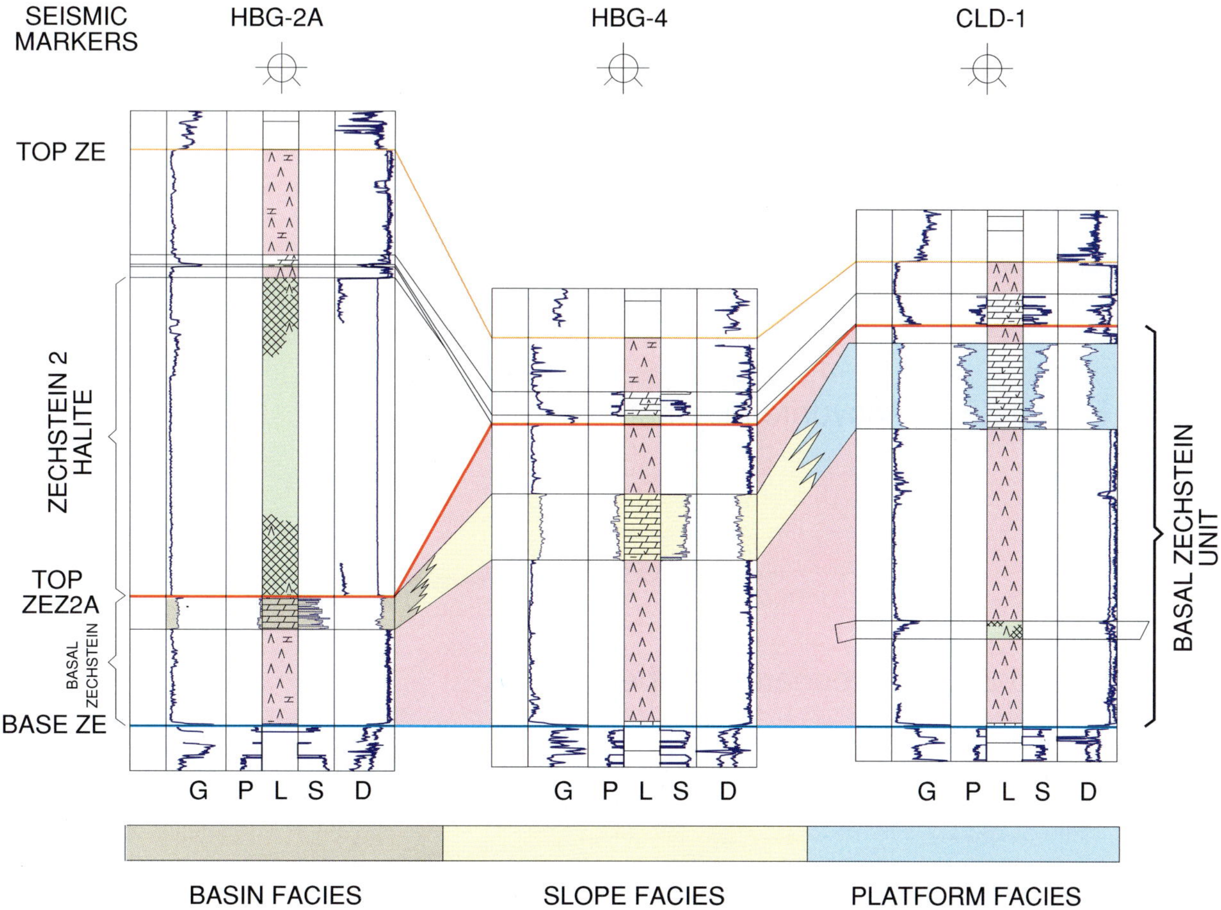

Fig. 4. Well log correlation panel of the Basal Zechstein Unit using gamma ray (G), porosity (P), lithology (L), gas saturation (S) and sonic and density (D) logs, showing basin, slope and platform carbonate facies of the Zechstein 2 Carbonate Member (wells: HBG-2A; HBG-4 (Hardenberg); CLD-1 (Collendoorn)). Coppershale at base Zechstein is not indicated.

underlying Zechstein 1 Formation, and to the position in the basin with respect to the prevailing northeasterly wind direction. Extensive descriptions and interpretations of the various facies can be found in the literature (e.g. Sanneman et al. 1978).

Zechstein 2 Anhydrite Member (ZEZ2A)

Carbonate deposition was replaced by gypsum precipitation, in response to increased salinities due to a relative sea-level fall. The influence of the chemocline and its associated sulphate-reducing bacteria again became apparent. The thickest accumulation of the Zechstein 2 anhydrite occurred in the shallow waters of the platform and upper slope settings, with the basinal area again starved of sediment (Figs 4, 5E).

Post-Basal Zechstein Unit

The development of the Basal Zechstein Unit ended with a further drop in relative sea-level. The increased salinity caused a change in the nature of evaporite precipitation from gypsum to halite. The sea-level fell below that of the platform surface and the Zechstein 2 halite precipitation was therefore largely restricted to the basin and slope areas where it filled the relief formed by the Basal Zechstein Unit. This resulted in a rather shallow, featureless basin (Fig. 5F). It is largely due to this flattening that the subsequent Zechstein 3 Carbonate Member does not show the same areal variation in facies as the Zechstein 2 Carbonate Member.

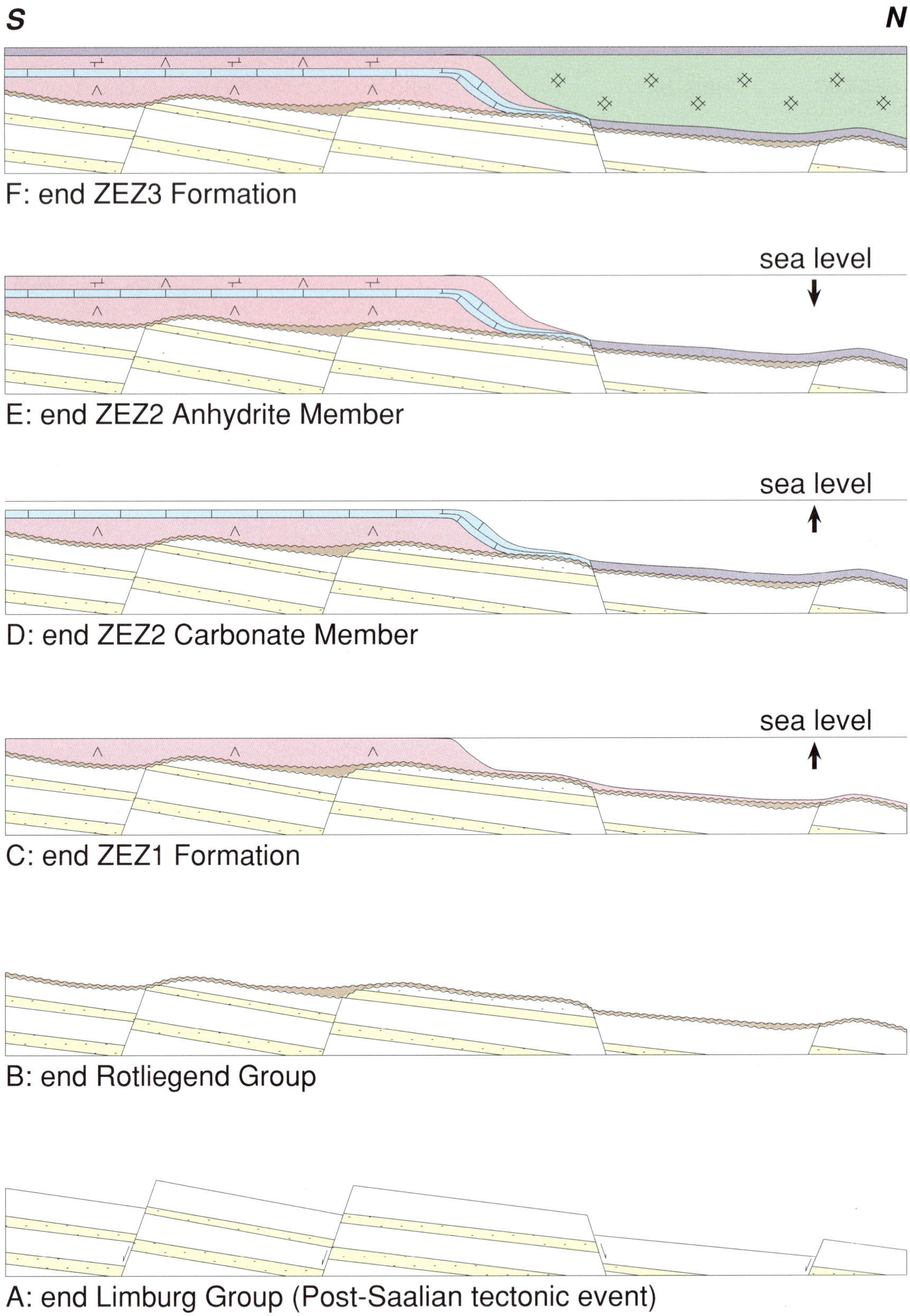

Fig. 5. Schematic sedimentation history of the Basal Zechstein Unit in the NE Netherlands. The vertical arrows indicate a relative sea level rise or drop.

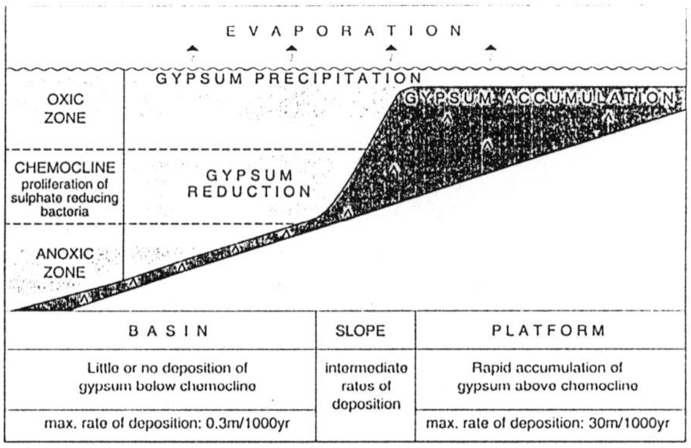

Fig. 6. Principles of the sulphate platform development of the Zechstein 1 Anhydrite Member in the NE Netherlands (after Van der Baan 1990).

Reservoir potential of the Zechstein 2 Carbonate Member

The facies of the Zechstein 2 Carbonate Member largely reflect water depth, salinity contrasts, sea level fluctuations and ensuing variations in carbonate production. Also the position relative to the prevailing northeasterly Late Permian wind direction was important. In the absence of any significant tides within the inland Zechstein sea, wind-generated wave action was the most important process determining environmental hydraulic energy contrasts between various areas.

High-energy ooidal grainstones develop along the windward face of the platform margin where wave action is strongest. Triggered by storms or long-shore currents, such deposits are constantly transported and winnowed over the platform. Ultimately they form wave-resistant bars. Comparatively little amounts of sediment are shed down the windward slope of the platform which is therefore steep, but due to the aggradational activity of accumulating wave-resistant bars, the carbonate rim of the platform is relatively thick. By contrast, in leeward settings, carbonate sediments are comparatively easily shed from the platform down the slope. This process results in rather thin platform facies and comparatively thick slope facies, and the platform slopes will have a more gentle dip than those on the windward side. The thicker reservoirs are thus formed on the slopes of the leeward sides of the platform and on the fringes of the windward sides.

The essential features of the facies model and its relevance to the reservoir potential of the Zechstein 2 Carbonate Member are illustrated in Figs 7 and 10. As with most carbonates, diagenesis largely controls the reservoir quality of the Zechstein 2 Carbonate Member. Each depositional facies (basin, slope, barrier, etc.) has a specific diagenetic overprint which results in carbonate units with enhanced or deteriorated reservoir quality, the Carbonate Fabric Units (Reijers & Bartok 1985; Fig. 7).

In the barriers on the windward sides of the platform, early cementation and repeated leaching of the lime grainstones, followed by dolomitization, led to good reservoir potential. On the open shelf on the leeward sides of the platform, dolomitization followed by dedolomitization of the lime wackestones resulted in a fair reservoir potential. Dolomitization of the lime packstones followed by leaching resulted in fair to good reservoir potential in the slope proximal to the platform. Good porosities can occasionally be found in slope slump deposits. In the basin and in the lagoon cementation has diminished the porosities, and reservoir potential is generally poor.

The link between depositional facies and diagenetic effects in specific Zechstein 2 carbonates provides a qualitative tool for reservoir prediction and adds to the general understanding of the overall geological model.

3D Seismic interpretation

This section describes the interpretation methodology applied to 3D seismic data in order to obtain information on the Zechstein 2 Carbonate reservoir.

Structural interpretation

Zechstein 2 Carbonate prospects are defined on seismic as structural highs at Top Zechstein 2 Anhydrite level. This level corresponds with a strong reflection at the base of the thick Zechstein 2 Halite Member over most of the region. However, at the edge of the Zechstein basin the limit of the thick Zechstein 2 Halite Member is seen to correspond with the thickening of the Basal Zechstein Unit as it passes from the basin into the platform area (Fig. 8). In the platform areas the Top Zechstein 2 Anhydrite reflection becomes less distinguishable and is less prominent than the overlying Top Zechstein 3 Anhydrite reflection. Consequently, prior to the advent of 3D seismic, most of the Zechstein 2 Carbonate discoveries were made in slope and basinal carbonates. The only platform discoveries were made in the lagoonal facies where Zechstein 1 Halite pil-

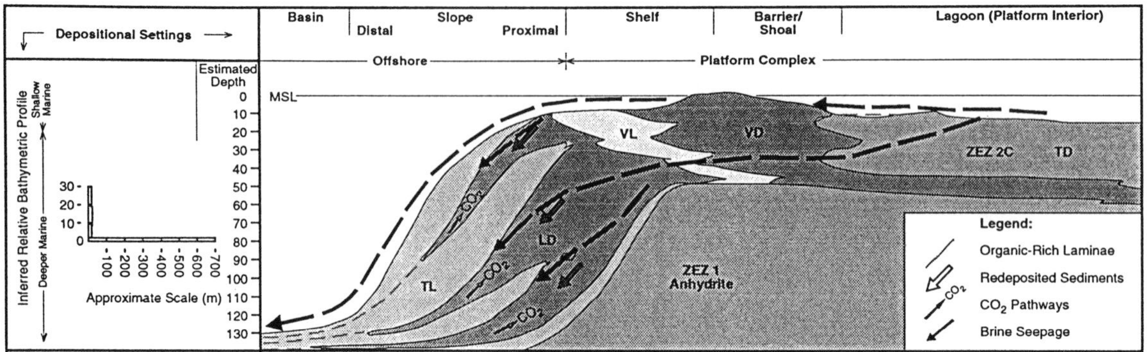

	Depositional Facies	Basinal	Slope: Distal	Slope: Proximal	Open Shelf	Barrier/Shoal	Restricted (Lagoonal) Shelf
Basic Depositional Parameters — Observed	Particles	Hemipelagics		Redeposited elements	Silt, mica		
Basic Depositional Parameters — Observed	Structures (bedding)	Laminae & wavy nod anhydrite beds	cm beds	Slumps, breccia, turbidites	Regularly bedded/ laminated organic-rich laminae	Massive, occasionally cross-bedded	Faint layering
Basic Depositional Parameters — Inferred	Depositional Texture	Mdst	Mdst	Allochthonous elements Mdst and Pkst/Wkst	Wkst (Bdst)	Pkst, Gst (Bdst?)	Pkst, Wkst (Bdst)
Basic Depositional Parameters — Inferred	Original Mineralogy	Organic-rich limestone and gypsum	Limestone (calcite) (+ gypsum)	Limestone (aragonite + calcite) (+ gypsum)		Limestone (aragonite)	Limestone and gypsum (aragonite + calcite)
Basic Diagenetic Parameters — Inferred	Sequence of Main Diagenetic Processes	-Dolomitisation -Anhydritisation -Dedolomitisation		-Dolomitisation -Anhydritisation -(Dedolomitisation) -Leaching -Anhydrite cementation		-Synsedimentary cementation -Repeated) leaching -Calcretisation -Dolomitisation -Anhydrite cementation	-Dolomitisation -Dolomite cementation
Basic Diagenetic Parameters — Inferred	Diagenetic Environments (Succession)	-Surface related -Burial		-Surface related -Burial		-Syndepositional -Surface related (vadose/phreatic)	-Syndepositional -Surface related Mg brines
Basic Diagenetic Parameters — Observed	Paragenesis of Diagenetic Products (Overprint)	-Gypsum ⟶ anhydrite -Dedolomite (dolomite + dolomite cement) ⟶ tight limestone -Anhydrite -Localised disconnected vugs		-Concretions -(leached) dolomite -Intercrystalline pores -Vuggy pores -Anhydrite cement	-Dolomite -Anhydrite -Intercrystalline pores -Dedolomite (vuggy) -Secondary vuggy pores -Anhydrite cement	-Beachrock -Drusy cement -Intraooid pores -Intraskeletal pores -Blocky cement -Blocky anhydrite cement	-Gypsum ⟶ (anhydrite) -Tight dolomite
Basic Diagenetic Parameters — Observed	Reservoir-Quality	Ø 0%		Exceptionally Ø 10%	Ø up to 10%	Ø 15%	Scarce disconnected interskeletal pores
	Carbonate Fabric Unit	Tight Secondary Limestone (TL)		Leached Dolomite (LD)	Vuggy Secondary Lst (VL)	Vuggy Dolomite (VD)	Tight Dolomite (TD)
	Prospectivity	Low		Medium		High	Low

Fig. 7. Sedimentological model for the Zechstein 2 Carbonate Member in the NE Netherlands with a description of the depositional and diagenetic aspects of the various carbonate facies (after Reijers & Bartok 1985). Mdst = mudstone; Pkst = packstone; Wkst = wackestone; Gst = grainstone; Bdst = boundstone; nod = nodular.

lows, interbedded within the Zechstein 1 Anhydrite, form simple anticlinal structures, easily identifiable on 2D seismic.

With all of the large, high-relief structures drilled, Zechstein 2 Carbonate exploration has increasingly focused on low-relief structures, particularly in the platform area where structural definition was previously difficult but where reservoir potential is high.

Interpretation of Basal Zechstein isochore maps

With 3D seismic covering the majority of the area in which the prospective Zechstein 2 Carbonate Member is present, it is now possible to accurately map both the Top Zechstein 2 Anhydrite and the Base Zechstein reflections together defining the Basal Zechstein Unit. On seismic the transition within this unit from basinal to platform setting is clearly visible as a divergence of the two reflections (Fig. 8). Over a distance of a few kilometres the two-way time thickness of the unit increases from some 20 m in the basinal areas, to around 100 m within the platform areas. When mapped over the entire area of the 3D surveys, the time difference between the top and base of the unit can easily be calculated and displayed. This is possibly the most powerful tool of the interpretational toolbox. The time isochore map (Fig. 9) illustrates clearly the details of the Basal Zechstein Unit, showing the intricate outline of the platform, the relative steepness of its slope, isochore anomalies at the base of the slope which might indicate the development of submarine fan deposits, and signs pointing to synsedimentary

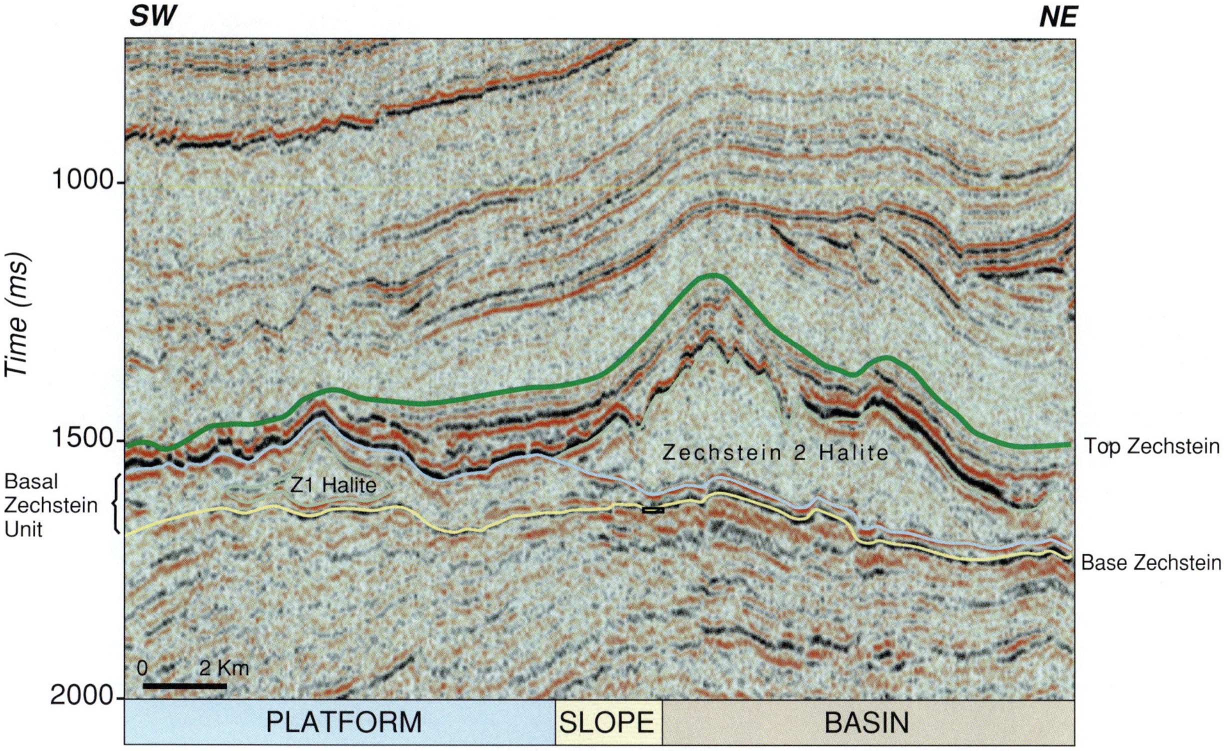

Fig. 8. Regional seismic line with interpretation of the Zechstein Group, showing the Basal Zechstein Unit in platform, slope and basinal setting. For location see Fig. 9.

faulting and the formation of Zechstein 1 Halite pillows within the lagoonal area. The slopes facing to the north and east are relatively steep compared to those facing to the west indicating that the prevailing winds came from the northeast.

Gross reservoir thickness prediction

Although 3D seismic enables the accurate mapping of the top and the base of the Basal Zechstein Unit it is usually not possible to map the reservoir body, the Zechstein 2 Carbonate Member, since there is usually little acoustic impedance contrast between this and adjacent members (Zechstein 2 Anhydrite Member above, and Zechstein 1 Anhydrite Member below).

In a simple model, based on the sedimentation model, the leeward side of a platform is prograding and thick carbonate sections are found on the slope (Fig. 10A). A relationship between the thickness of the carbonate member and that of the Basal Zechstein Unit can be established for a prograding platform edge (Fig. 10A). As the overall Basal Zechstein Unit thickens so does the Zechstein 2 Carbonate Member up to a maximum in the upper slope. Beyond this point the carbonate member thins, although the total Basal Zechstein Unit thickens. An equilibrium is reached on top of the platform where the two thicknesses remain approximately constant.

The windward side of the platform is likely to be aggradational, as argued above in the section on reservoir potential and as modelled in Fig. 10B. In this model the thickness of the carbonate member generally increases on the slope with increasing thickness of the overall Basal Zechstein Unit, and reaches its maximum on the platform fringe (Fig. 10B).

If these models are compared with well data from the NE Netherlands a clear relationship can be seen between the two variable thicknesses, depending on the type of platform margin present (Fig. 11). To effectively use the relationship between Zechstein 2 Carbonate reservoir thickness and total Basal Zechstein Unit thickness, the platform margin style (progradational or aggradational) has to be established. This is done on the basis of the isochore map in combination with the sedimentation model.

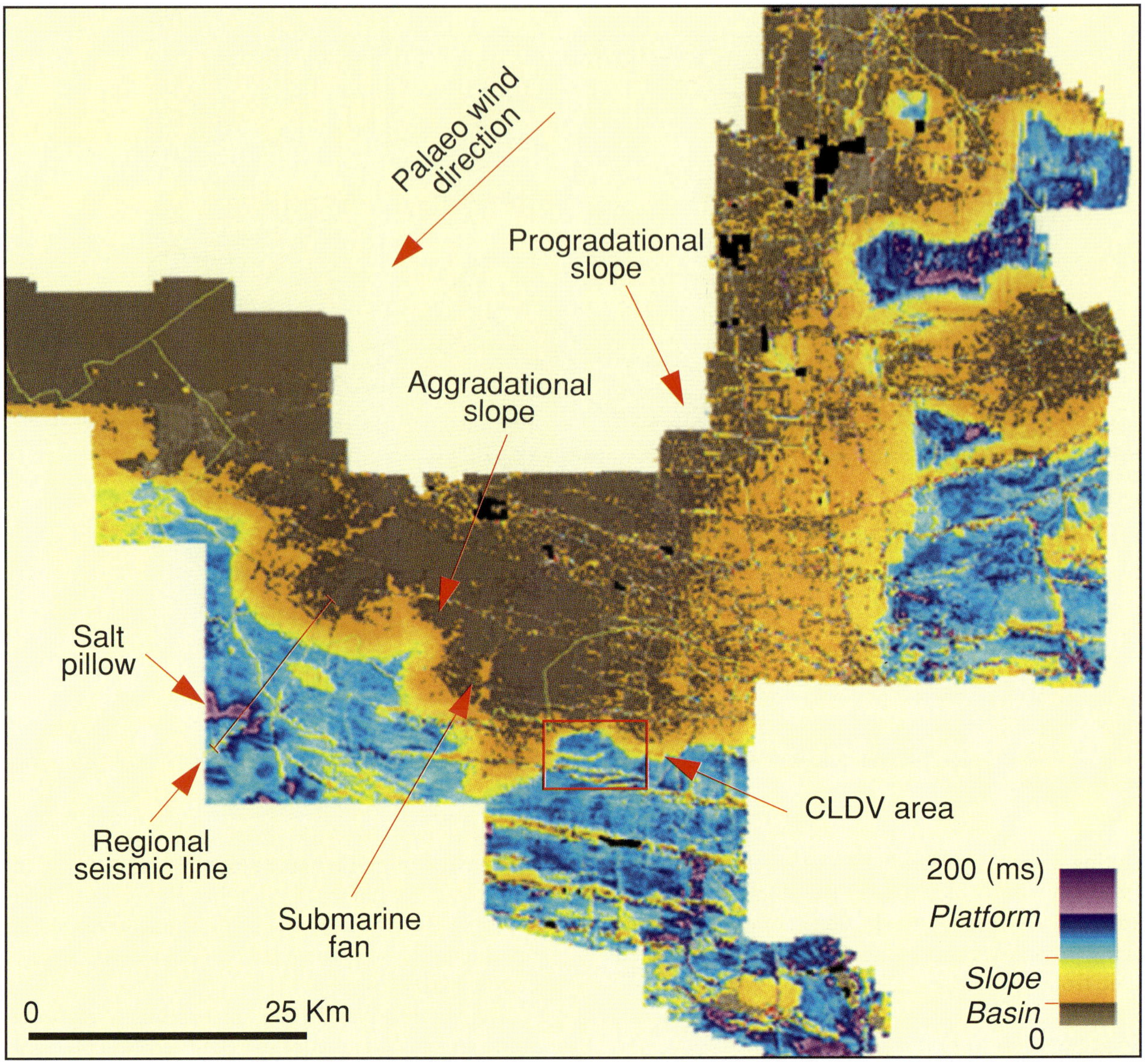

Fig. 9. Time isochore map of the Basal Zechstein Unit, derived from the 3D seismic, showing the areas with platform, slope and basinal depositional setting. The relatively steep slopes on the northeastern side of the platform and the more gentle slopes on the western side reveal the dominant northeasterly wind direction. Zechstein 1 salt pillows are visible as very high isochore values. Submarine fan deposits can be seen in the basin close to the platform edge. The regional seismic line (Fig. 8) and the Collendoornerveen (CLDV) area (Fig. 15) are indicated.

Reservoir quality prediction

Using the sedimentological model of Fig. 7 and the prevailing wind direction, the isochore data are interpreted in terms of prevailing facies. Based on this, a qualitative assessment of reservoir potential is possible in any particular area. This information is particularly helpful during prospect appraisal and ranking. A comparison of the actual average well porosities with the expected depositional facies, derived from the sedimentological model and the time isochore map, confirms this qualitative approach (Fig. 12).

Fracture prediction

The density of open fractures established in cored intervals correlates directly with reservoir permeability, allowing estimation of gas production rates. The platform edge zone can be delineated as a potentially highly fractured area (Fig. 13). Fracture density in the Zechstein 2 Carbonate Member increases from

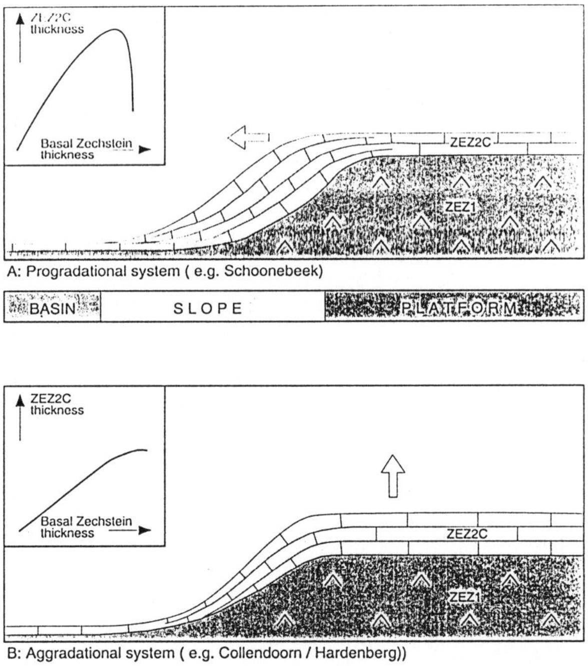

Fig. 10. Models of the progradational (A) and aggradational (B) platform margin styles. The thickness relationships between the Zechstein 2 Carbonate Member and the Basal Zechstein Unit are shown in the insets.

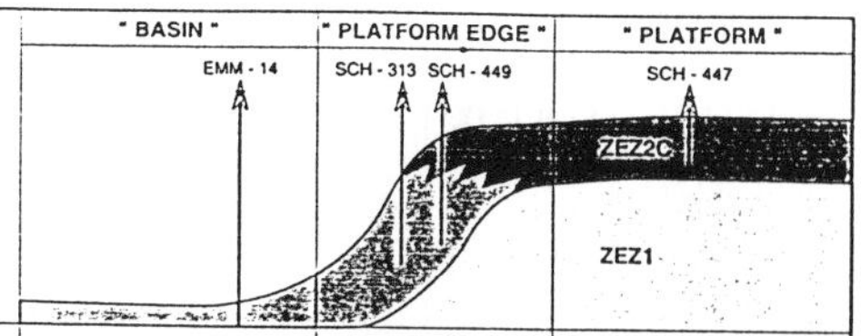

		" BASIN "	" PLATFORM EDGE "	" PLATFORM "
ZONE		EMM - 14	SCH - 313 SCH - 449	SCH - 447
OPEN FRACTURES	Type	- Mudstone related - Stylolite related extension fractures	- Tectonic related - Fractures with an extension and shear component	- Grainstone related - Extension fractures
	Origin	Loading (L)	Loading (L) with a tectonic (T) overprint	Loading (L)
	Size	Very small (cm)	Large (dm - m)	Small (cm - dm)
	Spacing L	dm - m	cm - dm	dm - m
	Spacing T	-	cm - dm	m
	Orientation L	± NS (sub) vertical	± NS (sub) vertical	± NS (sub) vertical
	Orientation T	-	± EW (sub) vertical	-
	Effect on permeability increase L	Relatively low	Relatively high	Relatively low
	Effect on permeability increase T	-	Relatively high	-

Fig. 13. Summary of the characteristics of open fractures, caused by loading (L) and tectonics (T) in the basin – distal slope, platform edge and platform lagoon zones. Observations are mainly based on cores from wells EMM-14 (Emmen) and SCH-313, SCH-447 and SCH-449 (Schoonebeek).

the basin, via the slope, towards the platform edge. Fractures are less densely developed on the platform interior than around the platform edge.

A series of detailed studies on fractures in cores in individual gas fields led to the realization that a genetic relationship exists between local (well) and sub-regional (field-scale) seismically undetectable fracture patterns on the one hand and fault patterns detectable on 3D seismic on the other hand. The orientation of fault-related fractures is associated with the (strike-) slip component on the nearest fault. The fracture patterns are influenced by nearby small faults. A local fault pattern detectable on 3D seismic therefore allows fracture predictions and ensuing reservoir permeability predictions in undrilled areas.

Reservoir porosity prediction

Although it is not possible to resolve the Zechstein 2 Carbonate Member on seismic, variations in reservoir parameters (porosity and thickness) are reflected in the seismic response, particularly in the platform areas where porosities are generally highest. In the basinal and slope areas, often with low carbonate porosities, there is little or no acoustic impedance contrast between the Zechstein 2 Carbonate Member, and the under and overlying Zechstein 1 and Zechstein 2 Anhydrite Members (Fig. 14A). However, when porosity in the Zechstein 2 Carbonate Member increases, as occurs on the platform, both density and velocity drop, giving rise to an acoustic impedance contrast that results in additional seismic reflections directly below the regional Top Zechstein 2 Anhydrite reflection (Fig. 14B).

This observation was initially made on 2D seismic where it was used as a qualitative guide to reservoir potential. Later, the first attempts were made to quantitatively predict reservoir quality, still using 2D seismic data (Maureau & Van Wijhe 1979). Obviously, these efforts faced the major limitation of being spatially restricted to the line of the seismic section on which the prediction was performed.

However, with 3D seismic and the progress made in quantitative prediction it is possible to make quantitative studies over an area. Attribute measurements were made on the regional Zechstein 2 Anhydrite reflection and on an additional reflection corresponding to the base of the Zechstein 2 Carbonate reservoir. Studies

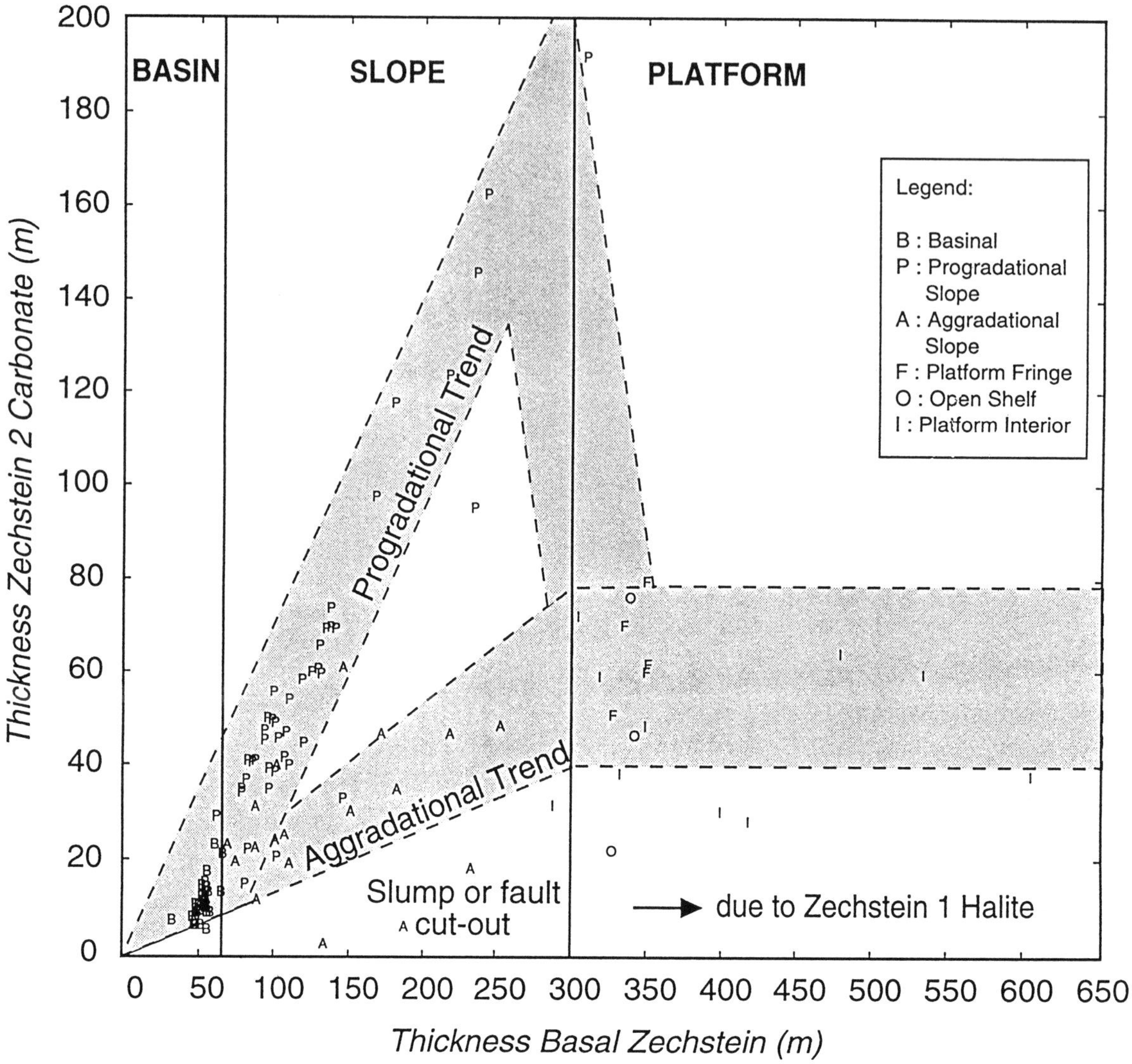

Fig. 11. Graph showing the thickness relationship between the Zechstein 2 Carbonate Member and the Basal Zechstein Unit found in wells in the NE Netherlands. The depositional setting of the Zechstein 2 Carbonate Member is determined from the time isochore map of the Basal Zechstein Unit (Fig. 9). The two platform margin styles can be discriminated. The Basal Zechstein unit becomes thicker than 350 m if Zechstein 1 Halite pillows are present.

have been carried out to quantify the variations in seismic response in terms of reservoir parameters.

In view of variations in thicknesses and of local variations in acoustic impedance values of the over and underlying layers of the Zechstein 2 Carbonate Member, it cannot be assumed that there is a simple relationship between reservoir porosity and reflection amplitude. It is therefore important to model the variation in the sedimentology over the study area in order to determine which seismic attributes (e.g. amplitude, loop area, loop width) are best suited to predict specific reservoir parameters (porosity, thickness, pore volume).

Collendoornerveen case history

The use of the 3D seismic as described above is demonstrated in this section for the assessment of the Collendoornerveen exploration prospect.

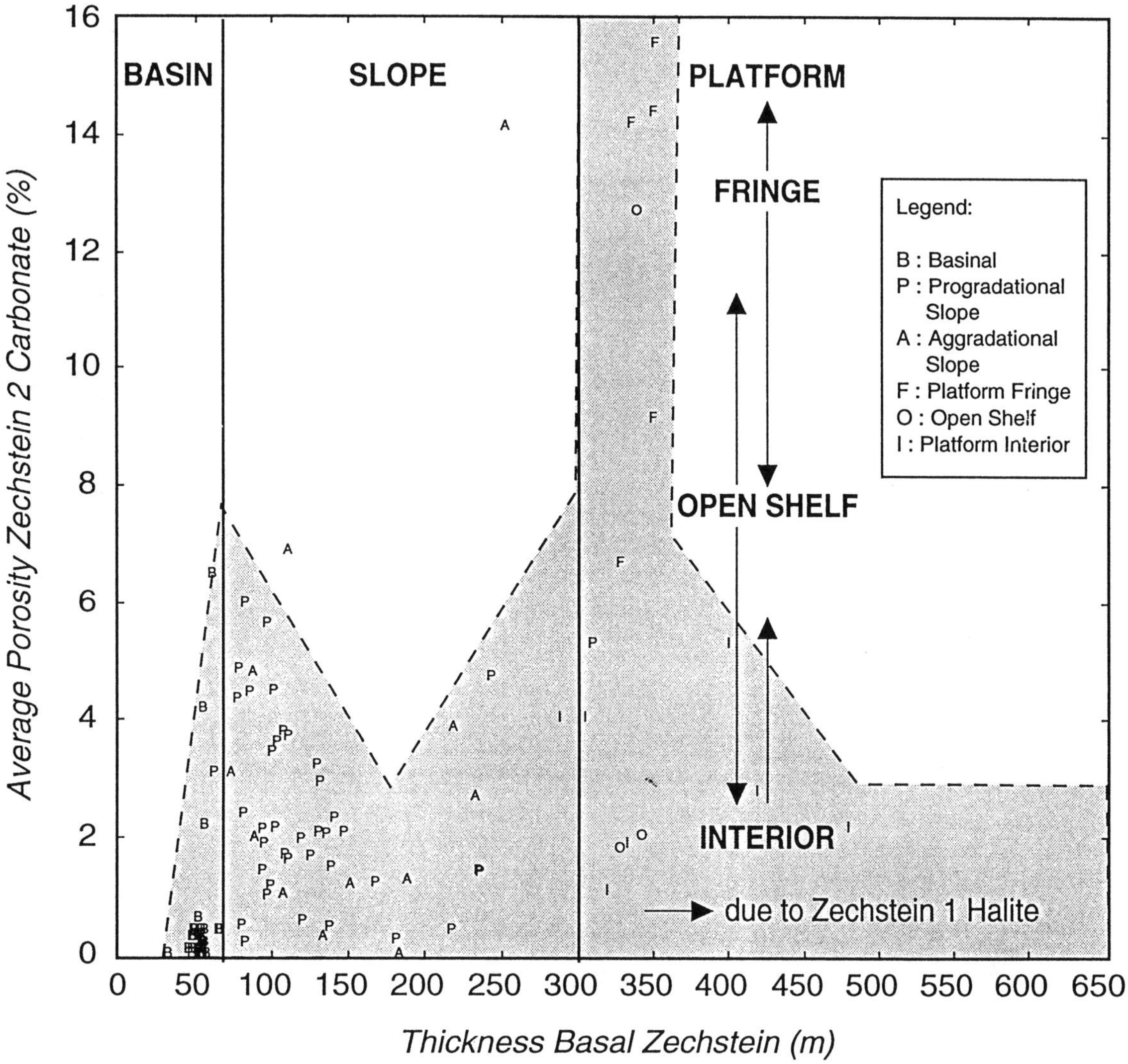

Fig. 12. Graph showing the relationship between the average porosity in the Zechstein 2 Carbonate Member in wells in the NE Netherlands and the member's depositional setting determined from the time isochore map of the Basal Zechstein Unit. The Basal Zechstein unit becomes thicker than 350 m if Zechstein 1 Halite pillows are present.

The availability of the Schoonebeek 3D seismic over the Collendoornerveen area enabled maturation of the Collendoornerveen prospect. A low-relief dip-closed structure was mapped at Top Zechstein 2 Anhydrite level (Fig. 15A). The Basal Zechstein time isochore map shows that the closure is situated on the fringe of the platform with a rather steep slope (Figs 9, 15B). It was concluded that the prospect is situated on the windward side of the platform where good reservoir potential and a relatively high fracture density can be expected. Using the aggradational platform edge model a gross reservoir thickness of some 70 m was predicted.

During the structural interpretation a strong reflection was noticed below the Top Zechstein 2 Anhydrite level (Fig. 16). This reflection was mapped and tied with the neighbouring wells Marslanden-1 (MRS-1) and Collendoorn-1 (CLD-1), where it was seen to correspond to the base of the Zechstein 2 Carbonate Member.

The highest amplitudes of this reflection lie in a narrow band immediately adjacent to the steep, northerly

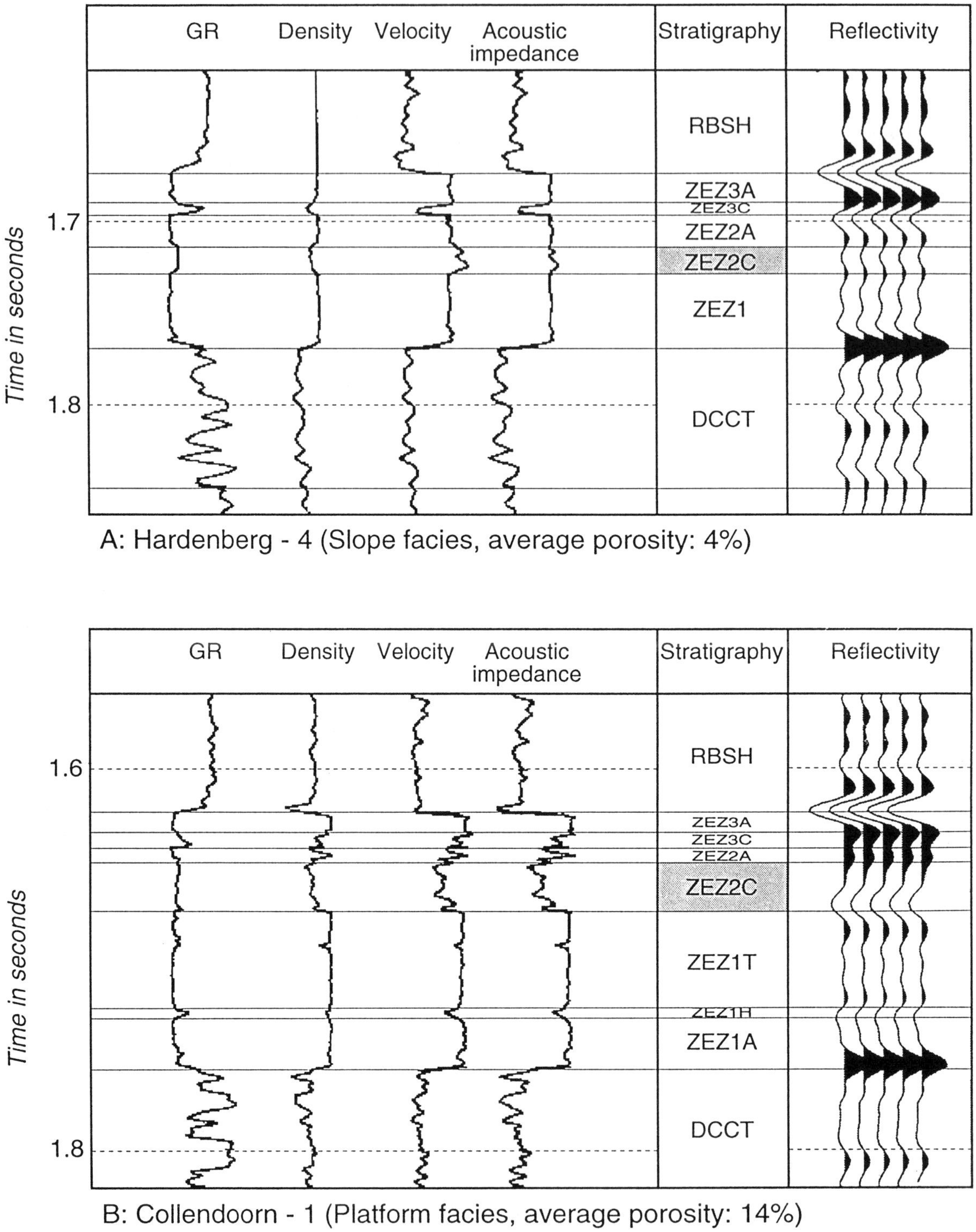

Fig. 14. Well logs and reflectivity synthetics of the Zechstein Group for slope (A) and platform setting (B). Density and velocity values are lower and the Base Zechstein 2 Carbonate reflection is stronger if the average porosity of the Zechstein 2 Carbonate Member is higher. Stratigraphic abbreviations according to Van Adrichem Boogaert & Kouwe (1993).

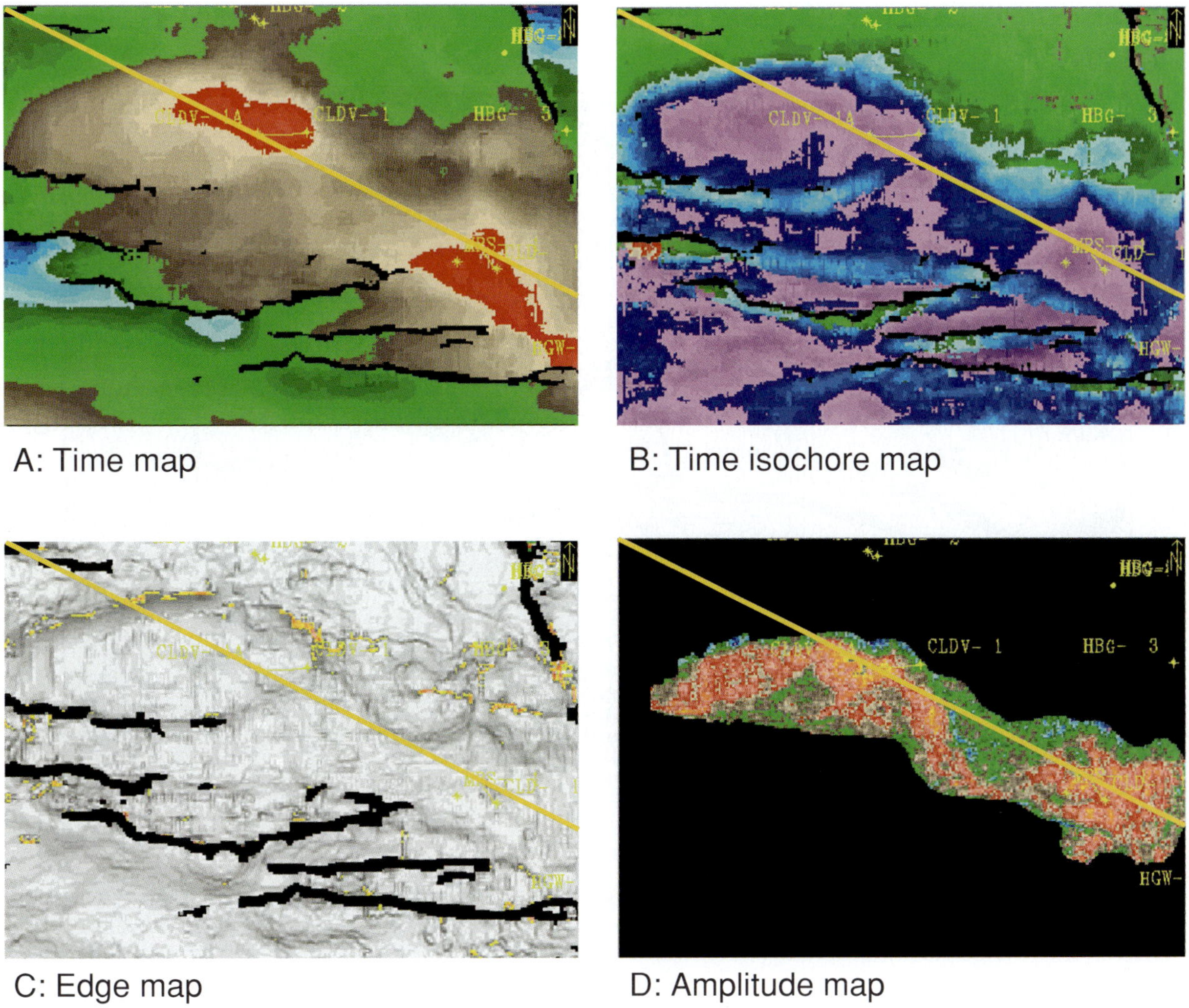

Fig. 15. Maps showing characteristics of the Collendoornerveen (CLDV) prospect determined from 3D seismic data. The area mapped measures 8 by 6 km. Its location is shown in Fig. 9. Indicated is the position of the seismic line presented in Fig. 16. (A) Time map of the Top Zechstein 2 Anhydrite reflection (red = shallow) showing the structural closure (assuming a laterally constant overburden). (B) Time isochore map of the Basal Zechstein Unit (purple = platform) showing the platform edge close to CLDV. (C) Edge map of the Top Zechstein 2 Anhydrite reflection showing the ESE – WNW strike of the minor faults. (D) Amplitude map of the Base Zechstein 2 Carbonate reflection (red = high) indicating the better porosities at the platform fringe.

facing slope (Fig 15D). The marked similarity of the reflection distribution and the established sedimentological model suggested that it was a direct indication of porous, high-energy platform-fringe deposits, with the reflection amplitude variations possibly being a function of reservoir porosity.

Detailed acoustic impedance modelling of the expected sedimentological variations over the prospect indicated that the best predictive seismic attribute is the loop area of the Base Zechstein 2 Carbonate reflection. A linear relationship was established which related the loop area to the average Zechstein 2 Carbonate porosity and hence a prediction of reservoir porosity was made over the extent of the mapped reflection (Fig. 17).

As a result the target location for the Collendoornerveen-1 exploration well (CLDV-1) was

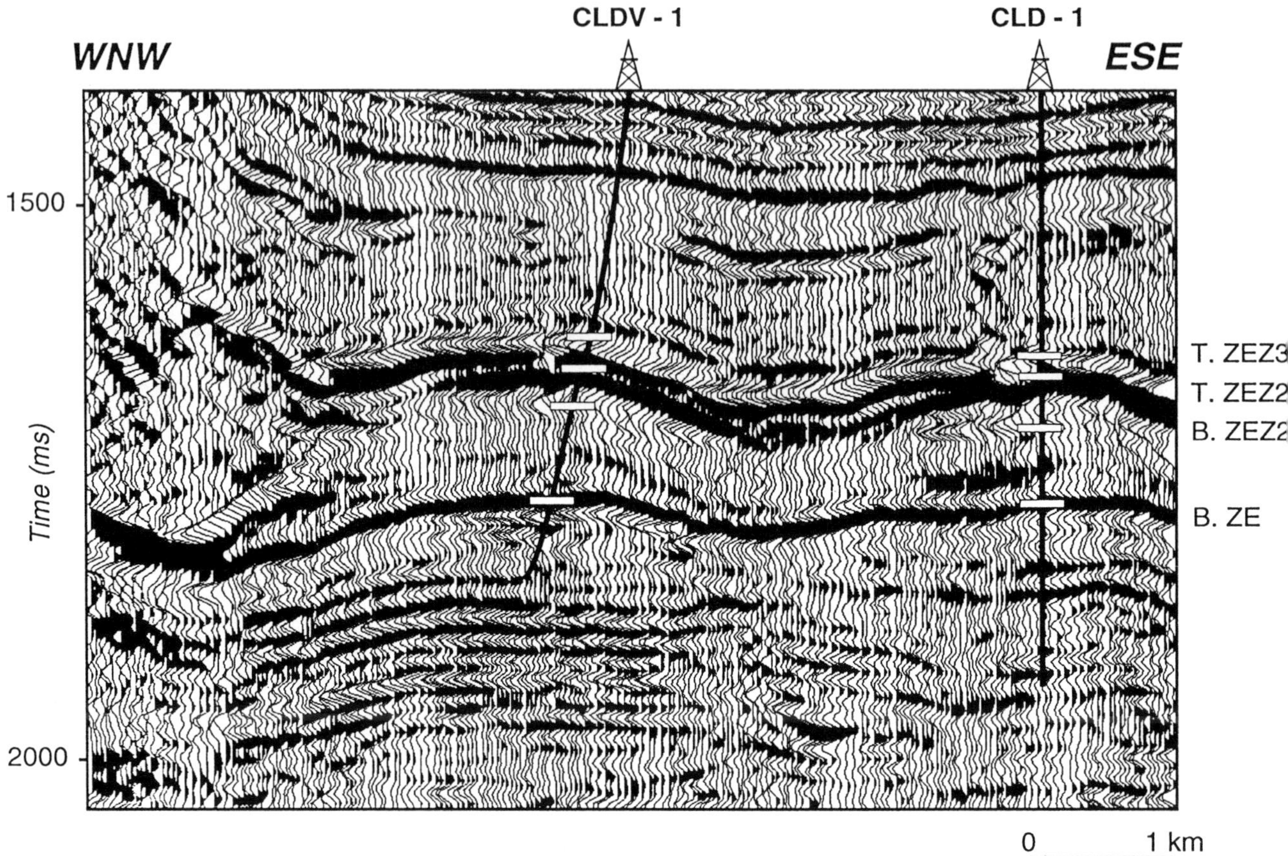

Fig. 16. Seismic line through the Collendoornerveen structure (Fig. 15) showing the development of the Base Zechstein 2 Carbonate (B.ZEZ2C) reflection. The seismic reflections mentioned in the margin have been indicated on the trajectories of the wells Collendoorn-1 (CLD-1) and Collendoornerveen-1 (CLDV-1).

moved slightly down-dip of the structural culmination in order to encounter a predicted local porosity maximum, estimated to be between 16–18%.

The exploration well CLDV-1 was successful. The Zechstein 2 Carbonate Member was found to be gas-bearing and consists of high-energy, platform-margin oolitic grainstones. The average porosity is 16% and the carbonate thickness is 63 m. The good porosity and the presence of open fractures resulted in a gas production rate of about 1×10^6 m^3/day.

Two types of fractures are observed in the cores of CLDV-1: cemented, vertical, N – S trending fractures which are related to the NW – SE Saalian extension, and open, vertical, ESE – WNW trending fractures related to the NNE – SSW Mid-Kimmerian extension. The open fractures show the same orientation as the strike of the local fault pattern (Fig. 15C).

Conclusions

A number of qualitative and quantitative interpretational tools have been established to aid the exploration and exploitation of the Zechstein 2 Carbonate play in the NE Netherlands. Elsewhere in the Zechstein hydrocarbon province these tools may not necessarily be used in the same specific form as presented here; however, the principles behind them can certainly be extrapolated to other areas. This toolbox is summarized in Fig. 18.

It is clear that the key to successful exploration of the Zechstein 2 Carbonate play is 3D seismic because:

1. with 3D seismic it is possible to define the subtle, low-relief structures that are the principal exploration targets in the now mature Zechstein hydrocarbon province;

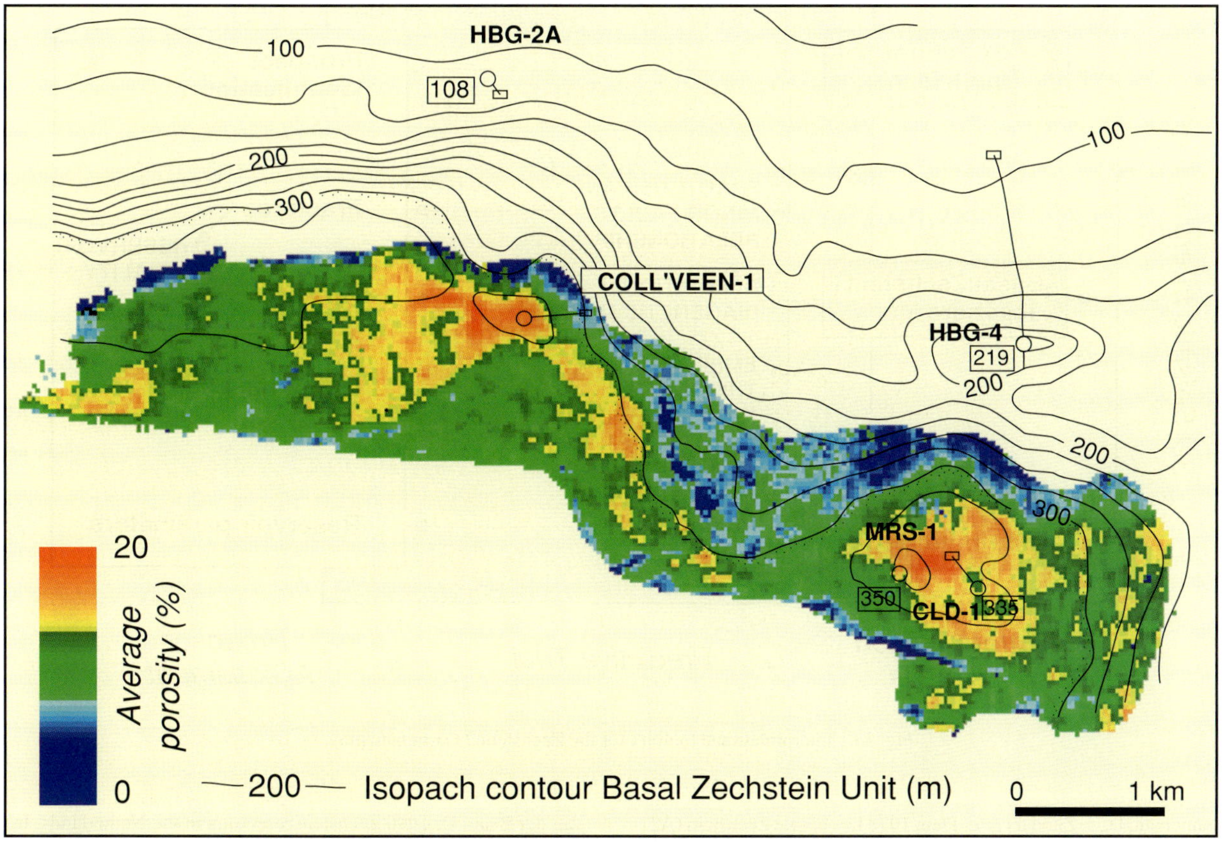

Fig. 17. Map showing the predicted average porosity of the Zechstein 2 Carbonate Member for the Collendoornerveen prospect. The thickness of the Basal Zechstein Unit is also indicated.

2. with 3D seismic it is possible to produce the detailed isochore maps that are needed, in combination with a sound sedimentological model, to trace facies distribution and to estimate gross reservoir thickness;
3. with 3D seismic it is possible to conduct quantitative prediction studies to map reservoir parameters over exploration prospects and existing fields; and
4. with 3D seismic, minor faults can be seen that have a causal relationship with productivity improving fractures.

Acknowledgements

The authors are indebted to the Nederlandse Aardolie Maatschappij, Shell Internationale Petroleum Maatschappij (SIPM) and Exxon Company International (ECI) for granting permission to publish this paper. Particularly thanked are the numerous colleagues, present and past, in NAM and the Koninklijke/Shell Exploratie en Produktie Laboratorium who have contributed to the findings presented in this paper. Students of the Vrije Universiteit, Amsterdam, have carried out detailed studies on fractures.

References

Clark, D.N. 1986 The distribution of porosity in Zechstein Carbonates. In: Brooks, J., J.C. Goff & B. van Hoorn (eds) Habitat of Palaeozoic gas in NW Europe. Geol. Soc. London Spec. Publ. 20: 121–143

Maureau, G.T.F.R. & D.H. van Wijhe 1979 The prediction of porosity in the Permian (Zechstein 2) Carbonate of Eastern Netherlands using seismic data – Geophysics 44: 1502–1517

Reijers, T.J.A. & P. Bartok 1985 Porosity characteristics and evolution in fractured Cretaceous carbonate reservoirs, La Paz field area, Maracaibo basin, Venezuela. In: Roehl, P. & Ph.W. Choquette (eds) Carbonate Petroleum Reservoirs – Springer, New York: 409–423

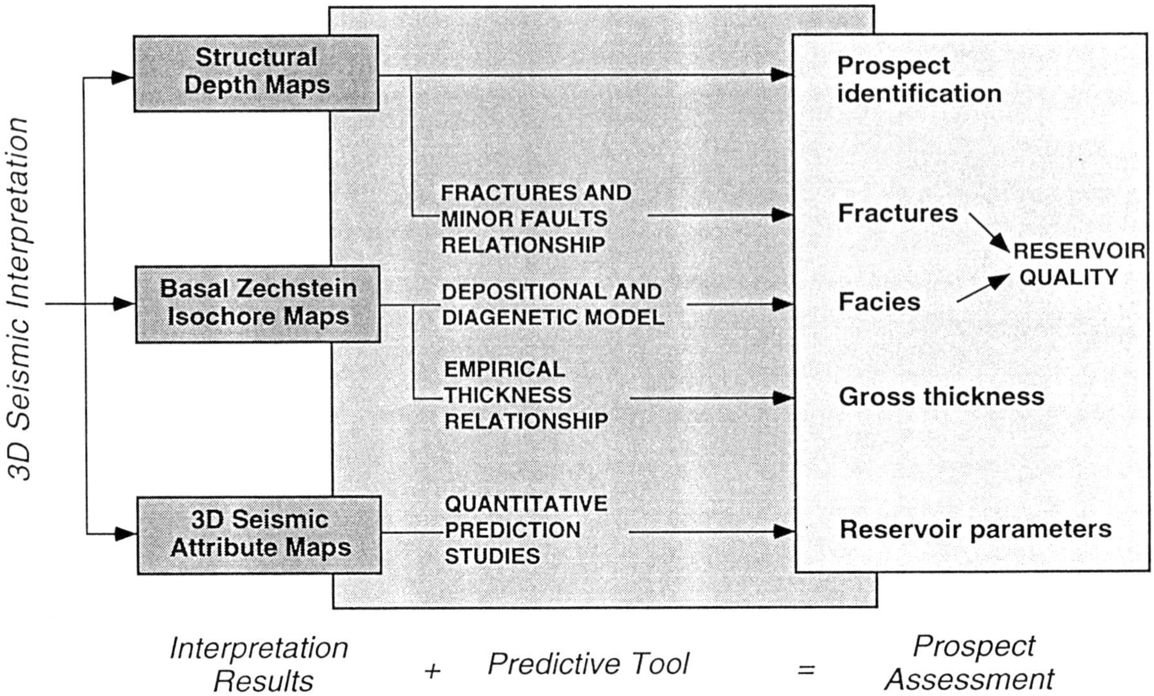

Fig. 18. Interpretational toolbox for the Zechstein 2 Carbonate play.

Sanneman, D., J. Zimdars & E. Plein 1978 Der basale Zechstein (A2 – T1) zwischen Weser und Ems – Zeit. Deut. Geol. Gesellschaft 121: 33–69

Van Adrichem Boogaert, H.A. & W.F.P. Kouwe (compilers) 1993 Stratigraphic nomenclature of the Netherlands, revision and update by RGD and NOGEPA – Meded. Rijks Geol. Dienst 50

Van der Baan, D. 1990 Zechstein reservoirs in the Netherlands. In: Brooks, J. (ed.) Classic petroleum provinces. Geol. Soc. London Spec. Publ. 50: 379–398

Ziegler, P.A. 1990 Geological Atlas of Western and Central Europe – Shell Internat. Petroleum Mij, Geol. Soc. Publ. House, Bath: 239 pp, 56 encls

Rondeel et al. (eds), Geology of gas and oil under the Netherlands, 143–158, 1996.

Salt tectonics in the southern North Sea, the Netherlands

G. Remmelts
Geological Survey of the Netherlands, Postbus 157, 2000 AD Haarlem, the Netherlands

Key words: basement faulting, petroleum geology, Zechstein

Abstract

In the Netherlands sector of the continental shelf salt structures are almost exclusively formed by Zechstein salt. Almost without exception these are related to basement faults, which have controlled the relative location of the salt structures, the triggering of salt movement, and the rate of salt movement. Differences in development of salt diapirs coincide with structural units. Dip-slip movement along basement faults enabled the upward flow of salt, because it weakened the overburden and created a differential stress field due to differential loading. Buoyancy related to density inversion alone seems insufficient to deform the overburden. Faulting is the trigger mechanism for the salt movement. The larger the throw on the fault, the further the salt structures will develop. A relationship has been proven between the timing of deformation phases and increased intensity of salt movement. It is shown that the intersection of fault systems in the basement is reflected in the geometry of the salt structures. In the study area, many hydrocarbon accumulations are related to salt structures. The hydrocarbons have been trapped in configurations, either structural or stratigraphic, that formed in response to the formation of the salt structures. Salt generally acts as a seal, but migration routes for hydrocarbons from below-salt source rocks into higher strata occur in areas of salt depletion. The study is mainly based on 2000 km of a regional 2D seismic survey (SNST83, Nopec/Geco-Prakla).

Introduction

The thick deposits of halite in the Late Permian Zechstein sequence of the southern North Sea have strongly influenced the structural development of the study area that occupies the northern half of the Netherlands sector of the continental shelf (north of 53°30′ N, Fig. 1). Salt movement, to a great extent related to basement faults, has had significant impact on the deformation of the Mesozoic sedimentary cover, and thus on the petroleum geology of the area. The area is an equivalent of the classic salt province of northern Germany as described by Trusheim (1957, 1960) and Sanneman (1963, 1968).

The majority of the gas discoveries in the Netherlands sector of the continental shelf were made in the Rotliegend fairway below the Zechstein Salt. However, the Rotliegend is shaled out in the northern part of the sector, and exploration for potential reservoirs has a target in the Mesozoic sequence overlying the Zechstein. Almost all hydrocarbon discoveries in this sequence are located in traps which have been formed in response to salt tectonics. For this reason knowledge of the development of the salt structures is essential for successful exploration.

The geology in the sector is dominated by a number of essentially Kimmerian (Jurassic) basins which are flanked by relatively stable platform areas (Fig. 1). Most faults occur in rather narrow zones separating the highs from the lows. Relatively minor faulting affected the platform areas where the post-Zechstein sequence consistently shows an incomplete section with a large hiatus between the Lower Triassic and Lower Cretaceous. In the basinal areas a more complete section is present, including Middle and locally Upper Triassic rocks as well as Lower to Upper Jurassic rocks. The latter can reach considerable thicknesses in rim synclines bordering salt structures.

The study is based on released seismic and well data. The framework for this research is provided by

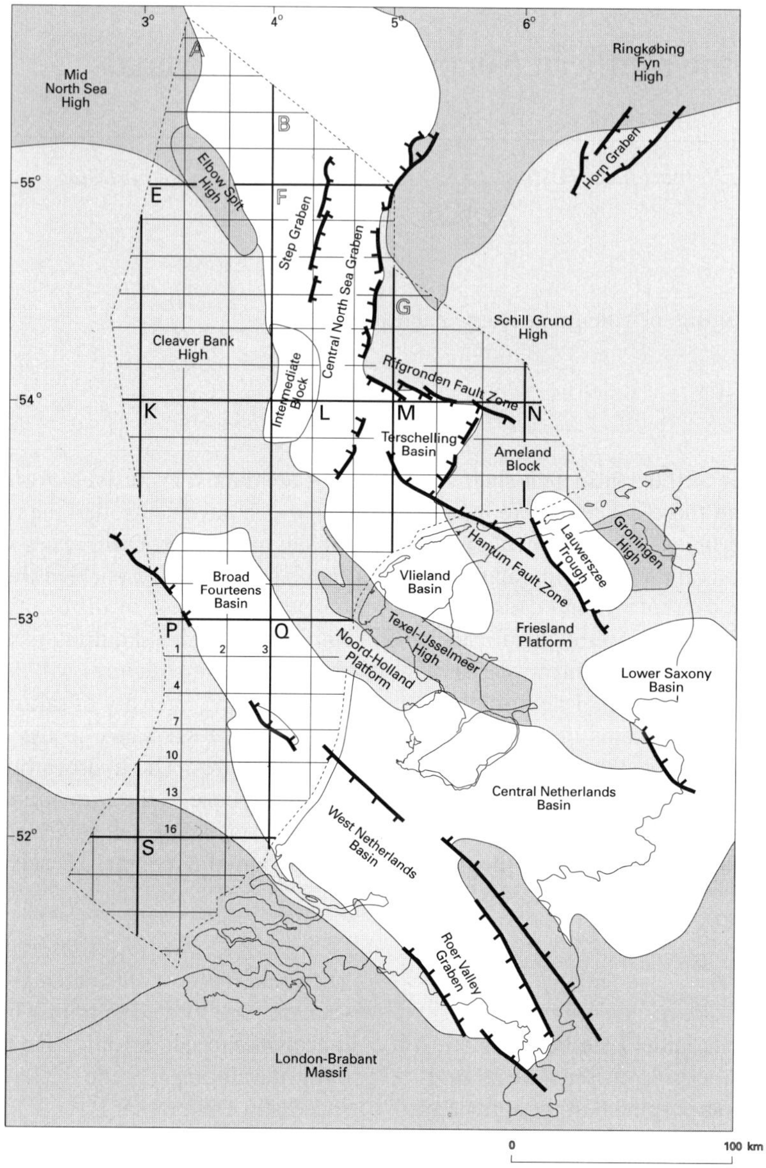

Fig. 1. Mesozoic structural units in the Netherlands on and offshore. Darkest shading indicates the larger massifs and highs. Subsided areas are white and intermediate areas are indicated with a light shading. The study area is located north of latitude 53°30′ N. From north to south a shift in dominant structural trend from N – S to NW – SE can be observed. Numbering system for blocks is shown in quadrant P.

some 2000 line kilometers of the regional SNST83 survey by Nopec/Geco-Prakla (Fig. 2), crossing a large number of salt structures. The study was carried out as part of the national research program on the feasibility of radioactive waste disposal in salt structures in the subsurface of the Netherlands (Remmelts et al. 1993).

Geological setting

The northern part of the Netherlands sector of the continental shelf is dominated by a Mesozoic rift basin, bordered by rather stable platforms (Fig. 1). The rift basin, the Central Graben, forms part of the Mesozoic North Sea rift system. The initial development

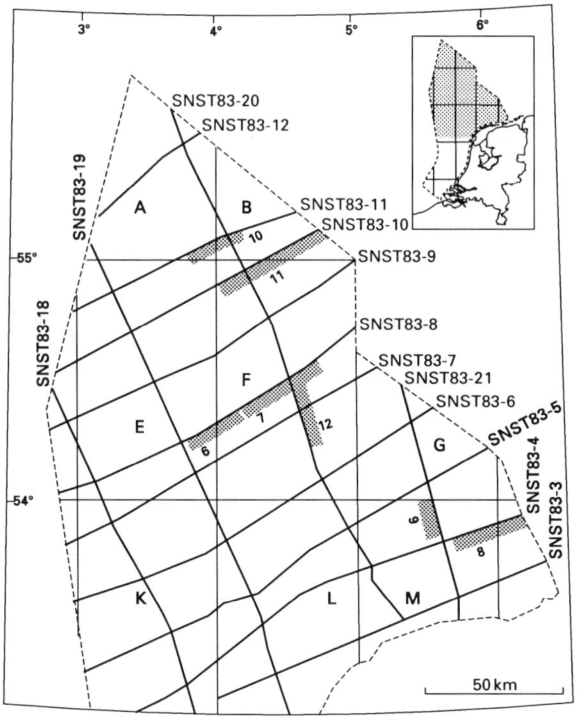

Fig. 2. Location map of the SNST83 seismic survey and of the seismic sections of Figs 6–12.

of this system is closely linked to the reorganization of plate boundaries and plate movements during the Late Permian to Early Triassic (Ziegler 1990). The graben is bounded in the north by a series of very large faults which show a diminishing vertical throw towards the south. The depth of the graben increases stepwise from the western flank (Intermediate Block and Step Graben) towards the central zone. In the east, this depth increase is more abrupt. The flanking platforms comprise the Elbow Spit High and the Cleaver Bank High in the west and the Schill Grund High in the east (Heybroek 1975, Van Wijhe 1987, Van Adrichem Boogaert & Kouwe 1993–1994). Towards the south, the boundaries of the Central Graben are poorly defined, and NW – SE trending structures start to dominate the structural grain. Here, the Terschelling Basin forms the southeastern extension of the Central Graben. The Terschelling Basin is bordered to the southeast by the Ameland Block.

The faults which delineate the Central Graben are more or less N – S trending in the north (Fig. 3). Towards the south they interfere with NNE – SSW and NW – SE oriented faults (Heybroek 1975). The NNE – SSW trend is a continuation of the Horn Graben and can be traced into the Off Holland Low south of the Central Graben, a Permo-Triassic basin which has been overprinted by the Broad Fourteens Basin (Heybroek 1975, Ziegler 1990). The NW – SE trend is related to the Broad Fourteens and West Netherlands Basins which have developed further to the south.

A schematic overview of the stratigraphic column in the study area is presented in Fig. 4. The Zechstein evaporites overlie a 'basement' consisting of Permian Upper Rotliegend anhydritic shales and continental to paralic Carboniferous deposits. The Zechstein Group comprises at least four depositional cycles. Most important for halokinesis are the second and third cycle with depositional thicknesses of rock salt varying between 400–800 and 200–400 m respectively.

Thickness changes in the Triassic sequence indicate that the first salt movements already took place during the Early Triassic (Dronkert et al. 1989, Fontaine et al. 1993) to Late Triassic (Heybroek 1975).

Lower Jurassic deposits are preserved only in the centre of the Central Graben. Erosion related to the Kimmerian tectonic phases, has removed all Jurassic sediments from the surrounding platforms. During the Mid-Kimmerian phase (Middle Jurassic), a gradual uplift of the Central North Sea took place. Subsequent erosion locally even reached the pre-Zechstein 'basement'. During the Late Kimmerian phase (Late Jurassic – Early Cretaceous), the Central Graben subsided further, and the platform areas were uplifted once again.

In Cretaceous times, the stress field gradually changed from extensional to compressional (Ziegler 1987, 1990). Differential subsidence, characteristic for Late Jurassic to Early Cretaceous times, was replaced by a more regional subsidence during Cretaceous and Tertiary times. However, as a result of the compressive stress field regional subsidence was interrupted by three inversion phases (Van Wijhe 1987): during the Late Cretaceous (Subhercynian phase, Santonian – Campanian), Early Tertiary (Laramide phase, Middle Paleocene) and Late Eocene – Early Oligocene (Pyrenean phase). The centre of the Central Graben was uplifted by reverse movement along pre-existing normal faults. Meanwhile, the platforms flanking the Graben continued to subside and NW – SE orientated dextral strike-slip movements, accompanied by accelerated salt movement, took place during the Pyrenean phase (Van Hoorn 1987, Oudmayer & De Jager 1993). During Neogene and Quaternary times the entire North Sea was a subsiding area.

Fig. 3. 'Basement' fault pattern at the Base Zechstein level.

Salt provinces

The salt structures in the northern half of the Netherlands sector of the continental shelf occur in two different settings, each of which is characterised by the morphology of the salt structures and their geometry and orientation. They are the platform areas and the basinal domains. These settings seem to be closely related to the structural grain of the area and the magnitude of differential movement of the 'basement'. The main structural elements and the main 'basement' faults are shown in Figs 1 and 2. More detailed fault patterns were taken from the seismic sections and literature (e.g. ECL 1983, Heybroek 1975, Oudmayer & De Jager 1993). The location, morphology and geometry of the salt structures are shown on the Top Zechstein depth map (Fig. 5, after ECL 1983).

On the platform areas, such as the Cleaver Bank High, the Schill Grund High and the Ameland Block, salt flow was minor and resulted in salt pillows only. In the basinal areas, the Central Graben, Step Graben, Intermediate Block and Terschelling Basin, strong salt movement occurred, and salt domes and salt walls were formed.

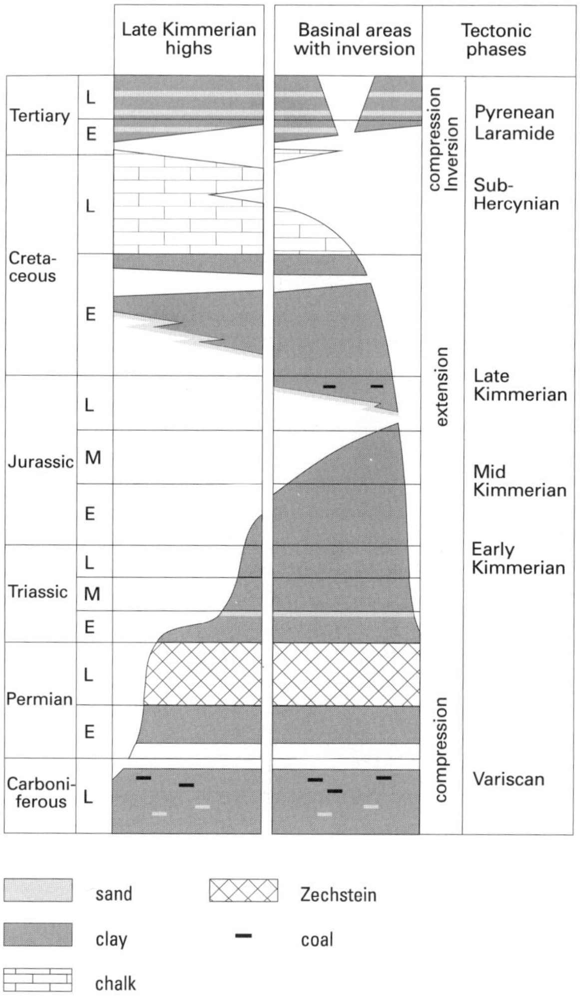

Fig. 4. Schematic stratigraphic column for the basinal and platform areas. Tectonic phases and stress regimes are listed in the right hand columns.

Salt structures on the platform areas

The Cleaver Bank High consists of a relatively stable platform with a structurally more deformed subarea in the northeast. Due to erosion related to the Kimmerian deformation phases, Lower Cretaceous directly overlies Lower Triassic and locally even rests on Zechstein (Fig. 4). The Zechstein sequence on the Elbow Spit High has been removed completely. In primary rim synclines, Middle Triassic may be present (northeastern subarea of the Cleaver Bank High, Fig. 6). The rest of the overburden consists of Lower Cretaceous (100–200 m), the Upper Cretaceous Chalk Group (300 m along the edge of the Elbow Spit High and approximately 700–900 m in the southeast) and the Cenozoic (1500 m in the south to 2000 m in the northeast). The highest salt swells are truncated by the Late Kimmerian unconformity, which allowed for the removal by solution of potentially large amounts of salt. On the Cleaver Bank High, the top of the Zechstein Group lies at a depth between 2000 m in the north and 2500 m in the south. The regional thickness of the Zechstein averages around 1000 m, though in the south it decreases to about 500 m. For a number of salt structures, the source layer became completely depleted during Tertiary times.

The Schill Grund High and Ameland Block, located east of the Central Graben, show a strong resemblance with the Cleaver Bank High. The sedimentary overburden consists of a comparable sequence (Fig. 4) with little or no Lower Triassic overlain by Lower Cretaceous (100–200 m), Upper Cretaceous Chalk (1000–1200 m) and Cenozoic deposits (1000–1500 m). The top of the Zechstein sequence on the Schill Grund High is generally formed by the Zechstein 3 Anhydrite (Main Anhydrite) or Carbonate (Platten Dolomite) Members. On the Ameland Block, erosion has not removed all of the Triassic sequence and consequently the underlying Zechstein 4 Formation is preserved. The depth of the top of the Zechstein Group is around 3000 m and gradually rises to 2500 m on the Ameland Block. The present-day thickness of the Zechstein generally varies between several hundreds of metres in the depletion areas to approximately 1250 m in the salt accumulations. In the Northwest German Basin, east of the Schill Grund High, the thickness of the Zechstein is 1000–1500 m of which 75% consists of the Zechstein 2 cycle (Jaritz 1987). The salt is almost absent in areas where severe erosion and/or dissolution has occurred (block G13). It measures more than 2000 m in the thickest accumulation (N4).

Only salt pillows have formed on the stable platform areas. Along the edges of the platforms, however, some diapiric structures have developed. In the northwest of the Cleaver Bank High (blocks E4, E5, E7, E8, E10 and E11, Fig. 5), the pillows are relatively small. They have not, or hardly, moved since the Pliocene. Their spacing is 5 to 7 km, with a general NW – SE orientation. Along the margin of the high, larger pillows have developed with a spacing of 10–15 km. The spacing of the salt pillows on the Schill Grund High and the Ameland Block is approximately 7.5 km (min. 6 km, max. 10 km) and their orientation is N – S and NNE – SSW. Variations in this spacing occur due to undulating salt ridges. Like on the Cleaver Bank High, the aver-

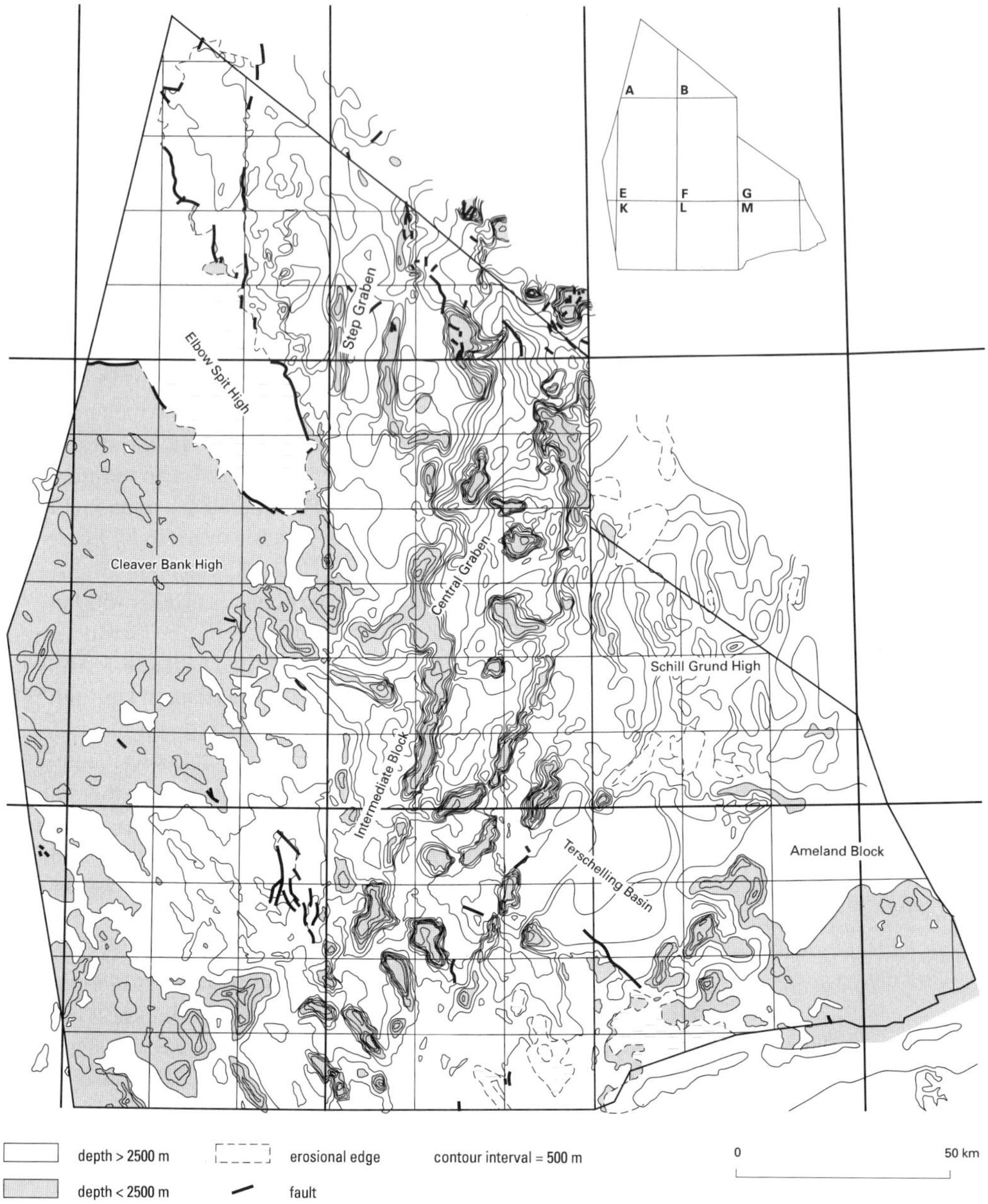

Fig. 5. Top Zechstein Group depth map (after ECL 1983), indicating the salt structures and the subareas. Note the long elongated salt walls in the north, and the smaller salt structures in the south where a NW – SE direction becomes dominant. Contour interval 500 m.

age spacing increases to 10 km towards the margins of the platform, which coincides with the spacing of the main faults in the 'basement'. The structural grain of the platform areas shows a close relationship with the salt structures. Where only minor salt movement took place, no significant 'basement' faults occur. However, larger salt structures exist along the fault-bounded margins of the high and in the northeast. The absence of large fault movements on the high hampered the development of differential loading and the relaxation of tectonic stresses took place elsewhere. The overburden of the salt deposits has not been weakened by faulting. In the absence of these triggering effects, the overburden was probably too strong to be breached by buoyancy forces only (Koyi et al. 1993, VUA 1992a,b, Tsai et al. 1987).

In the northeastern sector of the Cleaver Bank High, much stronger salt movement occurred than elsewhere

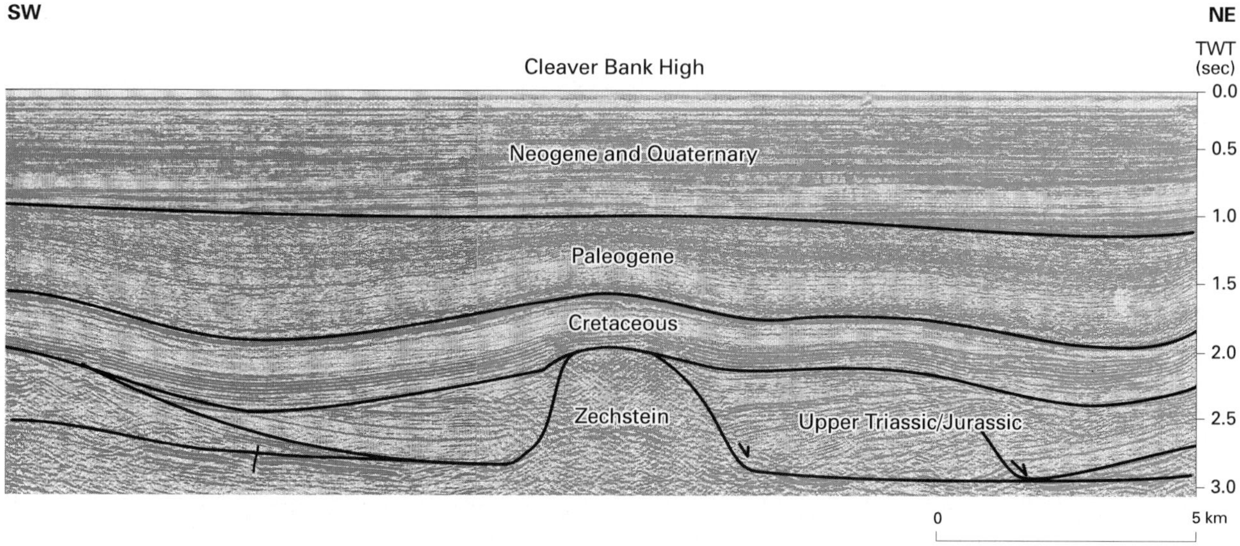

Fig. 6. Seismic section SNST83–8 crossing blocks E12 and F13, Cleaver Bank High. Early Cretaceous truncation down to the Triassic and sometimes Zechstein. In the northeastern rim syncline a thickening of the Upper Triassic indicates synsedimentary salt movement. For location see Fig. 2.

on this high. NW – SE trending salt walls coincide with 'basement' faults. This area shows interference of the NNE – SSW fault system related to the Central Graben and the NW – SE fault system related to the basins towards the south.

The first salt movements on the Cleaver Bank High must have occurred between the Middle Triassic and Early Cretaceous. A more precise estimate is hampered by the large hiatus in the sedimentary record. Timing is based on the angular unconformity at the base of the Cretaceous and the absence of internal thickness variations in the Lower Triassic in the rim synclines. From the Early Cretaceous onward, salt movement occurred in two phases: Late Cretaceous and Eocene – Oligocene. Generally salt movement ceased at the onset of the Miocene. The Base Miocene reflector is almost unaffected by further growth of salt structures. Only a few of the larger salt structures still show activity. Figure 6 shows a seismic line crossing blocks E12 and F13. The rim synclines on either side of the salt structure contain Triassic and possibly Lower Jurassic sediments. The thickening of the Upper Triassic section east of the salt structure is an indication for salt movement at the end of the Triassic. The last phase of salt movement is during the Eocene – Oligocene. For this specific structure salt movement ceased as a result of depletion of the source layer (Fig. 6).

The fault-bounded southern extension of the Elbow Spit High is shown in Fig. 7. It is a good example of salt flow from a down-thrown fault block upwards along the fault. Buoyancy forces drive this upward salt flow. These forces are normally insufficient to break and pierce the overburden, but in this case faulting had weakened the overburden. The timing of salt flow can here be deduced from the age of the sediments in the rim syncline at the eastern side of the salt wall (block F11, Step Graben). Based on well data (F11–03 and F08–02) and seismic character (for the Triassic sequence), a relatively constant thickness is assumed for the Lower Triassic. The Upper Triassic sequence is vaguely recognized on seismics to thin towards the salt structure, indicative of a first salt movement leading to the pillow stage. This stage continues throughout the Early and Middle Jurassic. A secondary rim syncline (thickening towards the salt structure) contains the Upper Jurassic Scruff Greensand Formation, indicating a breakthrough in Late Jurassic times.

Minor salt movement took place during the Early Cretaceous. Much stronger salt movement can be deduced for Late Cretaceous times. This acceleration of salt movement is thought to be related to the Subhercynian inversion phase. Many internal unconformities in the Upper Cretaceous Chalk sequence have recorded this event (e.g. blocks G13, G17, M3, M5, N1, see Fig. 8). Like in most of the study area, the Early Pale-

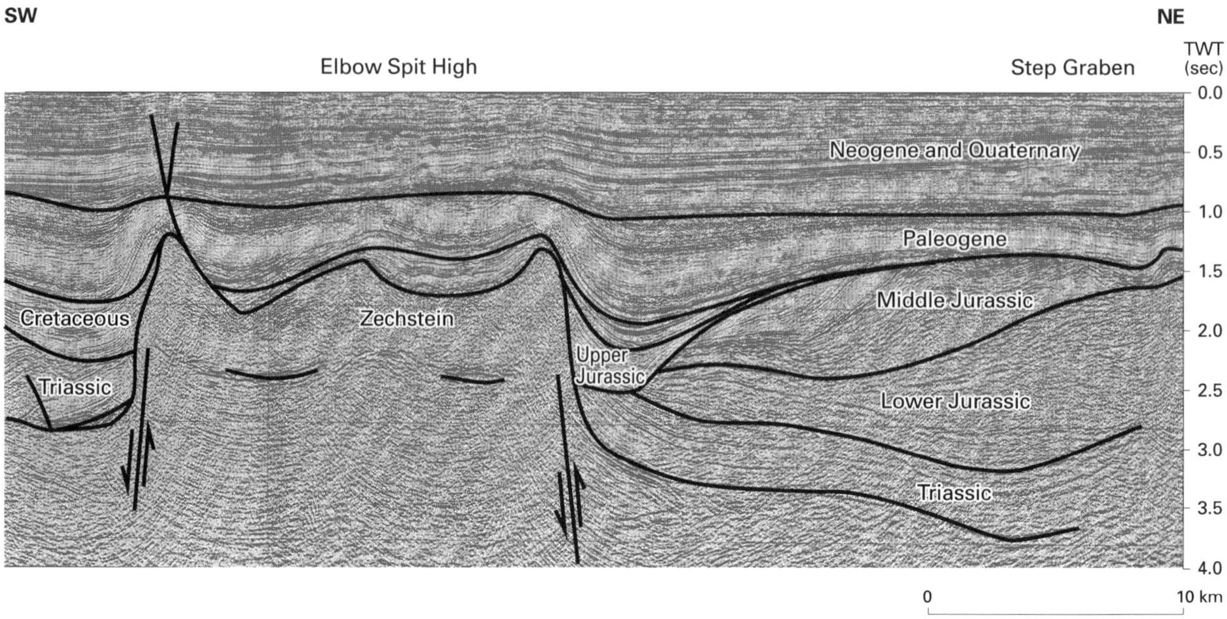

Fig. 7. Seismic section SNST83–8 over the fault-bounded southern extension of the Elbow Spit High. This is an example of salt flow from the downthrown block along the fault, towards the upthrown block. Note the Upper Jurassic fill of the secondary rim syncline indicating the timing of piercement of this structure. For location see Fig. 2.

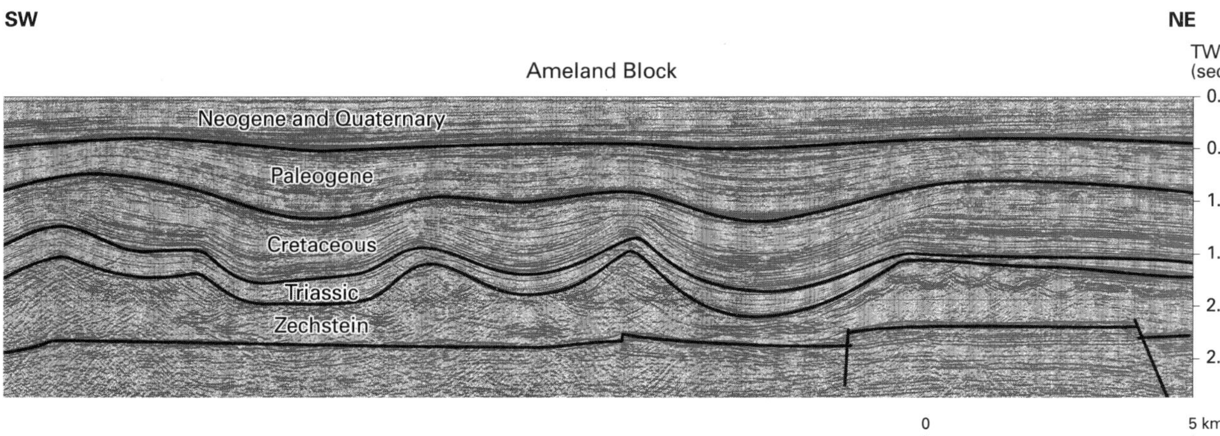

Fig. 8. Seismic section SNST83–21 over the Ameland Block. Late Cretaceous salt movement, related to the Subhercynian inversion phase, has resulted in numerous internal unconformities in the Upper Cretaceous Chalk. For location see Fig. 2.

ogene hardly witnesses salt flow, but the Late Paleogene again shows enhanced salt movement. In general, halokinesis has ceased by the end of the Paleogene. In a few salt structures related to recently active faults (e.g. Fig. 9), salt flow has still been active recently. Undisturbed layers can only be found in the first 150 m below seabed.

Salt structures in basinal areas

The Central Graben is bounded by faults with very large offsets, in general many hundreds of metres at the base Zechstein. This configuration causes differential loading of the salt on either side of these faults thus generating an excellent situation for salt movement.

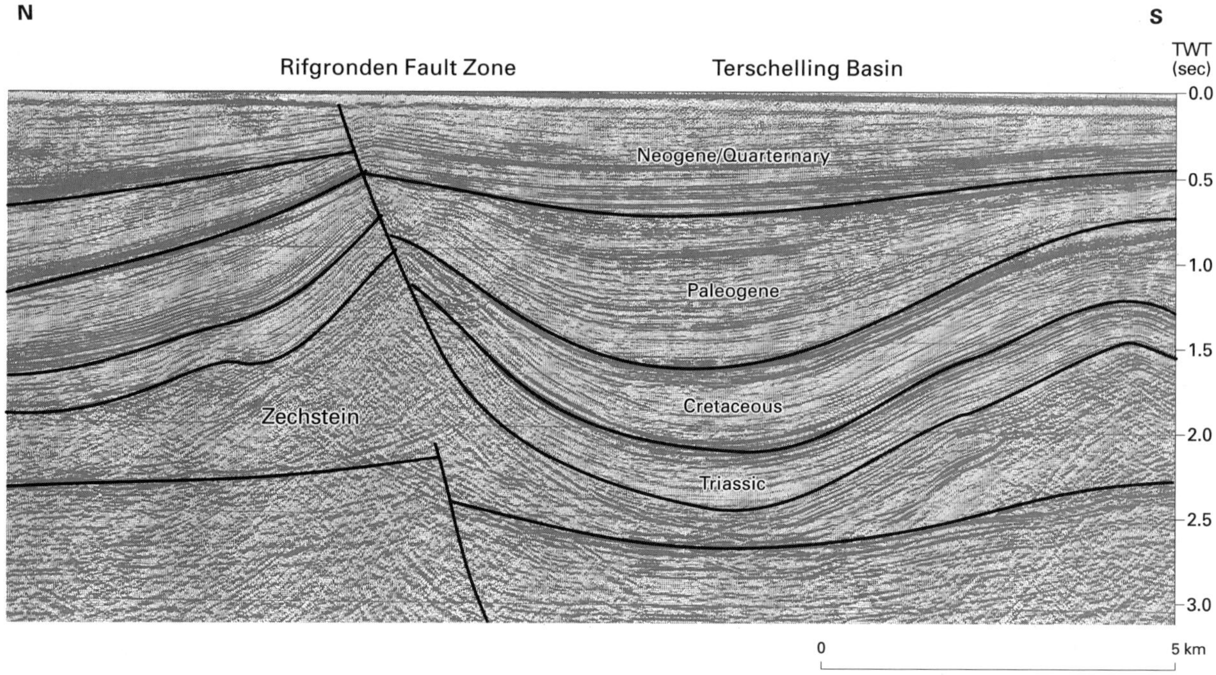

Fig. 9. Seismic section SNST83–21 crossing the Rifgronden Fault Zone which is accompanied by a salt structure. Listric faulting in the overburden can be traced up to approx. 150 m below the sea bed. Note that the salt accumulation is above the upthrown block. For location see Fig. 2.

The stratigraphic sequence in the graben is much more complete than on the bordering platform areas (Fig. 4). Thick sequences of Triassic are overlain by Lower Jurassic claystones and shales. A hiatus corresponding with the Mid Kimmerian phase separates the Lower Jurassic from the Middle and Upper Jurassic. This hiatus, related to the Late Kimmerian phase, is much less evident than on the adjacent platforms. It is best observed close to the salt structures. In the area of maximum Late Cretaceous inversion (a N – S axis in the centre of the Central Graben), erosion locally removed the complete Cretaceous sequence. In addition, both the total thickness of the sedimentary sequence and the thicknesses of the individual stratigraphic groups vary strongly as a result of salt movement.

Due to the extreme volume of salt that has moved from its original position into the salt walls, and given the accompanying erosion and/or dissolution, it is not possible to calculate the original thickness of the Zechstein deposits in the Cental Graben. Regional, relatively undisturbed, thicknesses on the neighboring platforms give an indication of the original salt thickness, which is estimated at more than 1000 m. Heybroek (1975) uses a value of 1200 m in his paleoreconstructions. The recent thickness in the central part of the graben is commonly less than 500 m in depletion areas. One has to keep in mind that these estimations are based on seismic data which, at the depth of these depleted occurrences, have lost a significant part of their quality. Moreover, due to the lack of economic interest, the Zechstein Group has never fully been drilled in the area. The thickness (or height) of the salt in the salt structures is generally more than 2500 m. The largest height reaches more than 5000 m in block F17, in a salt wall related to the western boundary fault between the Central Graben and the Intermediate Block (Fig. 5).

North – south trending diapiric salt walls have developed in the north of the Central Graben and in the Step Graben (Fig. 5). Towards the south this orientation shifts to a NNE – SSW direction. The salt structures near the southern end of the Central Graben and in the Terschelling Basin as well as those in the very north (B-quadrant and German waters) have a NW – SE trend. This conforms to the structural grain of the 'basement'. A number of circular salt domes has developed in the Central Graben (F14, F9). Smaller, more or less ellipse-shaped, isolated domes occur both in the south and in the extreme north. Specifically in these areas, the dominant N – S structural trend of the Central Graben interferes with NW – SE fault trends. The alternating movements along these two fault systems seem to be the reason for the development of solitary salt domes instead of salt walls along the 'single fault trend'. Intersection of faults results in a local maximum salt flow at that point. The salt structures tend to line up along both fault trends, but the elongation of each structure follows the direction of the dominant trend: in the south of the Central Graben mainly NW – SE and to a lesser extent NNE – SSW, in the north more N – S oriented. Since the stratigraphic sequence in the Central Graben is more complete than on the surrounding highs, it is possible to give a more accurate indication of the timing of the salt flow.

Seismic lines show subtle thickness changes in the Lower Triassic, indicating salt movement as early as that. No indications for salt movement in Early Jurassic times have been observed, but this may very well be due to the poor seismic quality at the Lower Jurassic level. During the Late Jurassic, most of the salt structures reached their diapiric phase. The inversion phases also caused a period of enhanced growth in the graben. The youngest phase of salt flow occurred during the Late Paleogene. In Neogene times, only some gravitational faulting occurred along the salt structures, for which the salt is assumed to have acted as a passive detachment level (e.g. SNST83–6). Indications for Miocene salt flow have only locally been observed above large structures (e.g. in block L7).

In the south of the Central Graben, in the blocks L-5, 6 and 7, salt pillows probably formed in Late Jurassic times, whereas piercement of salt domes and walls occurred during the Early Cretaceous (SNST83–5). Salt structures in this part of the Central Graben formed slightly later which corresponds to the observation that the structural development of the graben is prograding from the north towards the south (Ziegler 1990). The structure of the graben is also less pronounced in the south resulting in less strongly developed salt structures. In addition, the interference of two fault trends appears to have influenced the development of the structures in this southern domain.

In a transitional position between the rift shoulder and the Central Graben is the so called Step Graben (Van Wijhe 1987, Van Adrichem Boogaert & Kouwe 1993–1994). For the southern part we have used the informal name 'Intermediate Block'. The faulted steps are accompanied by strong salt movements which resulted in long salt walls (Fig. 5).

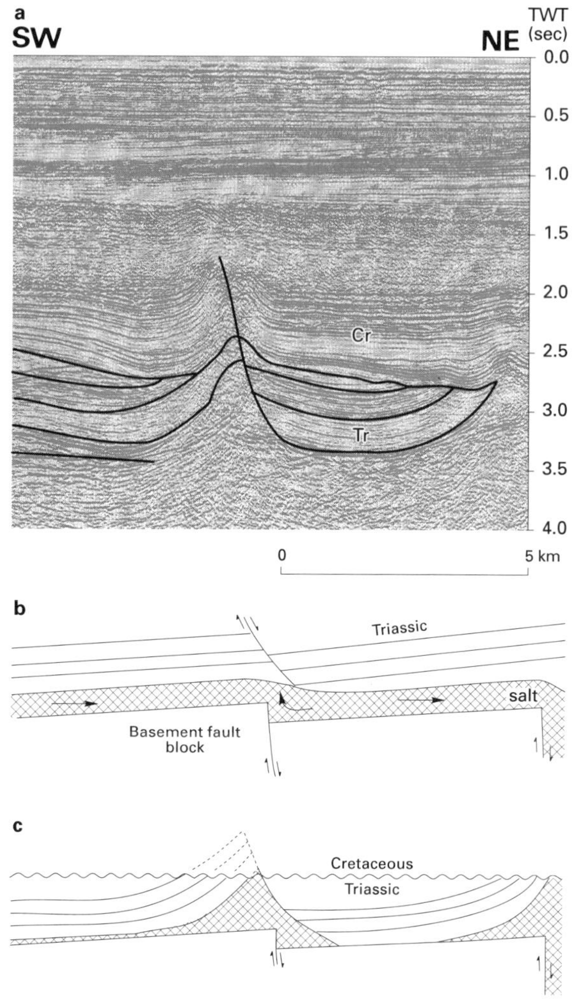

Fig. 10. Seismic section over the Step Graben and its structural explanation. For location see Fig. 2. Cr for Cretaceous, Tr for Triassic. (a) Seismic section SNST83–11. (b) Block faulting and accompanying rotation generates salt movement (see arrows, mainly along the bedding plane, partly up the fault plane). (c) Geometry of the overburden after salt movement and subsequent truncation by erosion.

The stratigraphy is characterized by a thick sequence of Triassic and thin or no Lower Jurassic. Cretaceous deposits are mostly preserved, as little or no inversion took place, and are around 1000 m thick. The Cenozoic cover has a general thickness of about 2000 m.

The two N – S salt walls in the east of the Step Graben (blocks B10, B13, B16 and F1) are most prominent. Towards the Elbow Spit High (in the A-quadrant) the salt is getting thinner, and only a few isolated salt structures formed. The salt walls are very clearly linked to large fault displacements in the 'basement'. However, no salt wall is present above the fault zone which separates the Elbow Spit High from the Step Graben. Chaotic structures in the sediment directly overlying the fault indicate that salt movement had occurred. Severe erosion at the beginning of the Cretaceous probably has removed a significant amount of salt, which must have caused the salt wall to collapse. West of the boundary fault, another N – S fault transects the Step Graben. This fault has a recent vertical throw of 150–200 m, and is accompanied by a salt wall. Figure 10a-c shows a section through this part of the Step Graben, where tilting of the overburden is the cumulative effect of salt flow and tectonics. Comparison of the geometries on either side of the salt structure shows that the angle between the Triassic and the base of the Cretaceous is much larger in the west than in the east. This may be explained by tilting of the basement blocks resulting in differential loading and subsequent updip directed salt flow. Depletion of the salt on the western side of the fault block has resulted in subsidence of the overlying sediments which saved them from Early Cretaceous erosion. Accumulation of the salt on the eastern side steepened the overburden which was subsequently eroded. Apart from this salt flow along the top of 'basement', salt flow also occurred along the fault plane from the downthrown block upwards. Piercement (diapiric stage) took place in Cretaceous times.

The boundary fault separating the Step Graben from the Central Graben shows some very large salt structures (blocks B14, B17) but in contrast to the above mentioned salt walls, they are discontinuously aligned along the fault. This may be related to the interference of different fault trends. The salt wall above the boundary fault shows an enormous rim syncline at the side of the Central Graben (Fig. 11). The large local thickness of Upper Jurassic sediments is an indication for salt flow towards this structure. There is however no well data which confirms the age of the rim syncline fill. In agreement with the theory of Trusheim (1960), the subsequent secondary rim synclines gradually move towards the dome. The diapiric stage seems to have been reached in Cretaceous times. The structure shows high growth rates during the Eocene – Oligocene. From Neogene times onward, no rim synclines can be observed on the seismic sections. The slight updoming of very young sediments indicates recent movement of the salt.

The main salt structures on the Intermediate Block are related to the boundary fault with the Central Graben. The structures here are slightly smaller than those along the boundary between Step Graben and Central Graben, which is thought to be the result of more restricted differential loading on opposite sides of smaller 'basement' faults than at the Step Graben – Central Graben boundary. In the Intermediate Block, due to erosion, there is no witness of the first phase of salt movement. The diapiric stage was reached in Late Jurassic times, with a period of major salt flow, which continued during the Early Cretaceous.

The Terschelling Basin is a fault-bounded structural element, situated at the southeastern end of the Central Graben (Fig. 5). The Rifgronden Fault in the north (Fig. 9), the Hantum Fault to the south, and finally an unnamed NNE – SSW fault towards the east, delimit the basin. It shows a gradual deepening towards the Central Graben from which it is separated by a normal fault.

Besides faulting, halokinesis played an important role in controlling deposition. Contrary to the adjacent Central Graben, uplift occurred prior to the end of the Middle Jurassic, and the older Jurassic sequence has been removed by erosion. Consequently, Upper Jurassic deposits, which are thinner than in the Central Graben, generally rest on the Triassic. Subsidence and faulting continued during the Early Cretaceous, and deposits from that period are thicker than in the Central Graben (Van Adrichem Boogaert & Kouwe 1993–1994). In large parts of the basin advanced salt movement has led to almost complete withdrawal of Zechstein salt. It is assumed that many salt structures started to develop before Mid Jurassic times and were truncated during the Kimmerian phases of uplift and erosion in Mid Jurassic and Late Jurassic – Early Cretaceous times. Chaotic structures in the Mesozoic sediments are indications for this early salt movement and subsequent dissolution. Most salt structures are located along the fault-bounded edges of the Terschelling Basin.

As can be deduced from the truncation of the Triassic below the Cretaceous, salt flow occurred above 'basement' faults prior to the Cretaceous. Seismic line SNST83–21 (Fig. 9) shows the asymmetric depletion of the salt. Most material from the salt wall seems to have been derived from the down-faulted block, as a result of differential loading. Like most of the salt structures in the basin, the wall probably started to develop before Cretaceous times but its main movement has been since the Late Cretaceous, especially in

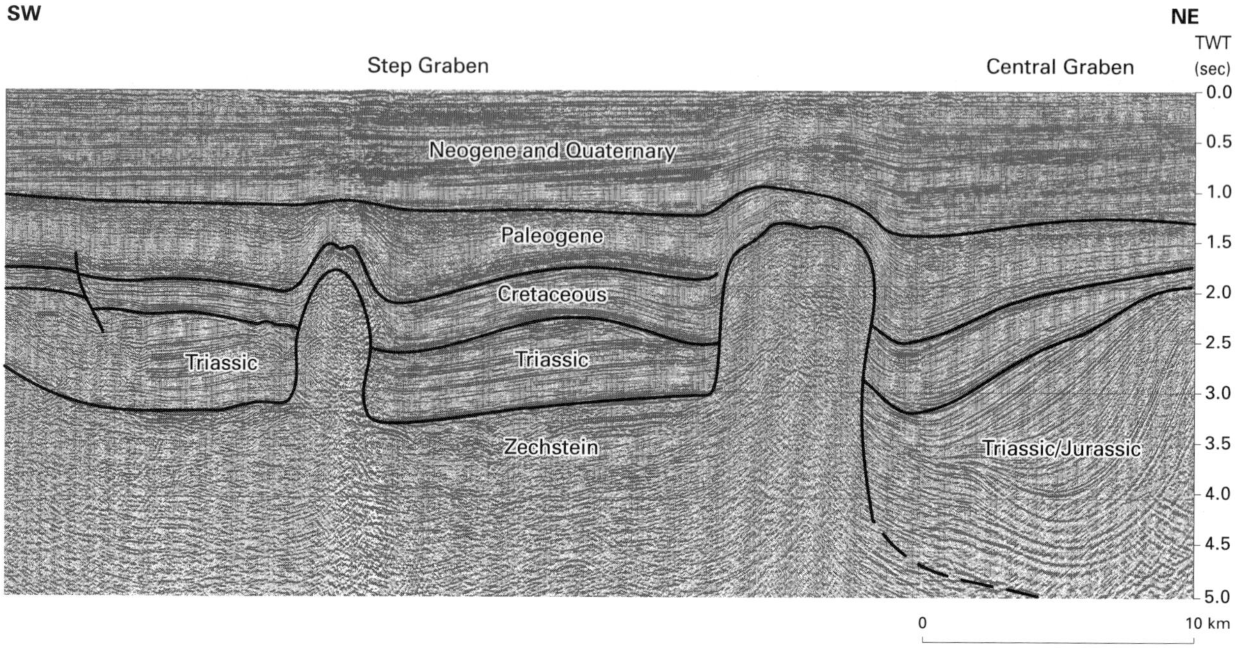

Fig. 11. Seismic section SNST83–10 over the Step Graben and the western border fault of the Central Graben. An extensive salt wall directly overlies the 'basement' fault. Note the progradation of the secondary rim synclines toward the salt structure on the northeastern side. For location see Fig. 2.

the Eocene – Oligocene, when the 'basement' faults were rejuvenated.

Petroleum geology

There is a significant potential for gas and oil in the northern half of the Dutch sector of the continental shelf. Finds in this area are mainly in post-Zechstein reservoirs and are all in structural traps related to salt movement. Not only is the salt a perfect seal, but salt movement has resulted in numerous structural traps. Knowledge of the history of salt movement and the parameters involved is a necessity to evaluate these salt-structure-related prospects. Gas trapped in the pre-Zechstein sequence (mainly Upper Carboniferous) is located in tilted fault blocks along basement faults underneath the Zechstein salt. Successful wells tapping these occurrences are mainly located on the platform areas. The Rotliegend, the main reservoir for hydrocarbons in the Dutch offshore located in the K and L quadrants, has lost its reservoir potential in the northern area due to the shaling out of the Rotliegend sandstones. However, reservoirs can be expected in the Triassic Main Buntsandstein (Volpriehausen and Detfurth Sandstone Formations), as well as in the Mid and Late Jurassic Lower and Upper Graben Formation, Puzzle Hole Formation or Scruff Greensand Formation and in the Early Cretaceous Vlieland Sandstone Formation (Van Adrichem Boogaert & Kouwe 1993–1994). The Late Cretaceous Chalk Group has not yet shown commercial success, although it is a prospective reservoir level in the Danish waters just northeast of the study area. A more recent play are the Tertiary reservoir sands.

The Carboniferous has been the source for most of the gas in the area. Originally the Zechstein salt was a barrier between the source and Mesozoic reservoirs but migration routes are thought to have been created in areas of complete salt withdrawal. This is deduced from gas finds in the Mesozoic sequence for which no other source is available but the Carboniferous. For the Tertiary gas fields the Namurian is suggested to be the source rock in the absence of the Westphalian (Duyverman et al. 1991). Oil has been generated from the Lower Jurassic Posidonia Shale.

The importance of strong salt movement as a factor in trap formation can be illustrated with a number of examples. The seismic section of Fig. 12 over a salt structure in block F14 illustrates various potential

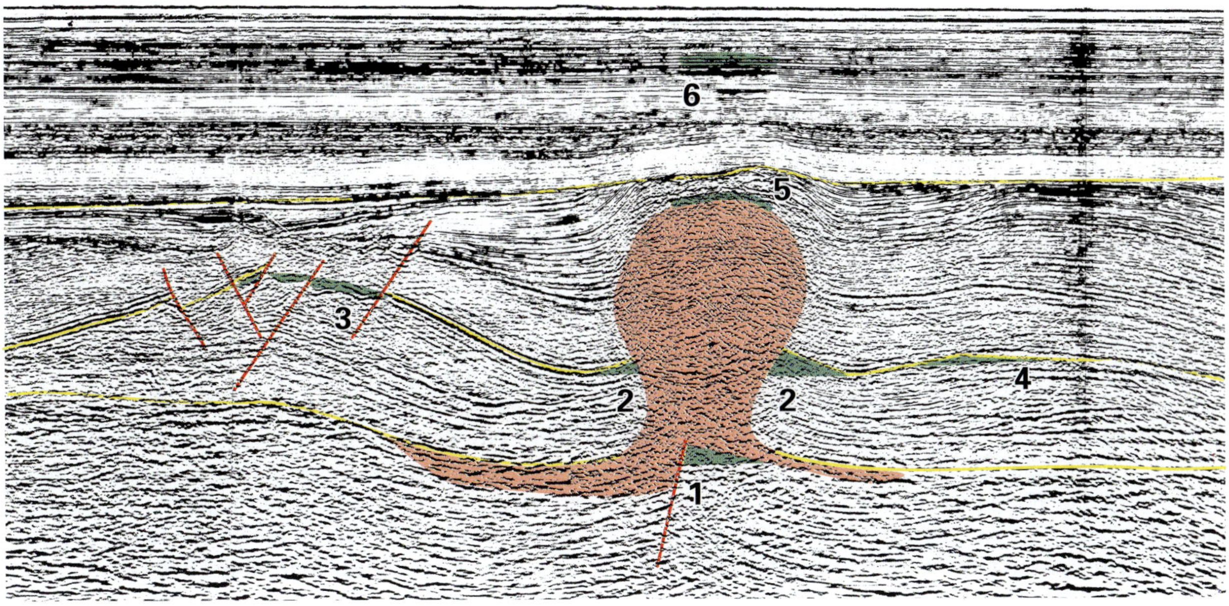

Fig. 12. Seismic section across a salt structure in the Central Graben, showing various possible structural traps related to salt movement. 1. Footwall block of sub-salt basement fault (visibility introduced as example). 2. Triassic sandstones, dip closure. 3. Dip closure over turtle back structure, compartmentalized. 4. Dip closure over turtle back structure. 5. (4-way) dip closure over the crest of the salt structure, may be strongly fractured. 6. Tertiary sandstone reservoir dip closure related to the salt movement. Yellow markers enhance the structuration. For location see Fig. 2.

Table 1. Oil and gas fields.

Block	Reservoir	Structure	Oil-gas	Operator	Status	Field
A12[3]	Tertiary	4-way dip	gas	NAM	–	A12-FA
A18[3]	Tertiary	4-way dip	gas	NAM	–	A18-FA
B13[2]	Tertiary	4-way dip	gas	NAM	–	B13–3
B16[2]	Tertiary	4-way dip	gas	NAM	–	B16–1
B17[2]	Tertiary	4-way dip	gas	NAM	–	B17–5
F3[1]	Jurassic	turtle back	gas	NAM	–	F3-FA
F3[1]	Jurassic	turtle back	oil & gas	NAM	+	F3-FB
F14[2]	Jurassic?	dip	oil	Statoil	–	F14–5
F15[3]	Triassic	turtle back	gas	Elf	+	F15-a
F17[1]	Jurassic	turtle back	oil	NAM	–	F17-FAB
F18[1]	Jurassic	turtle back	oil	NAM	–	F18-FA
L1[2]	Rotl/Carb?	fault block	gas	NAM	–	L1–4A
L1[2]	Jurassic	dip	oil	NAM	–	L1–3
L2[1,3]	Triassic	turtle/dip	gas	NAM	+	L2-FA
L2[1,3]	Triassic	dip	gas	NAM	–	L2-FB
L4[1]	Rotliegend	fault block	gas	Elf	+	L4-A/B
L4[1]	Rotliegend	fault block	gas	Elf	–	L4–3
L5[1]	Jurassic	turtle back	oil	NAM	–	L5
L5[3]	Triassic	turtle back	gas	NAM	+	L5-FA

[1] released well data; [2] Ministry of Economic Affairs (1984–1993) and Annual Reports of the State Inspectorate of Mines (1984–1993); [3] literature, see text. + = producing.

traps. Note that the section is merely chosen for its geometry, to illustrate the configuration of such traps, but that no hydrocarbon occurrences have been proven here. The structural configuration in some of the oil and gas fields as deduced from released data is listed in Table 1.

The footwall fault block of the 'basement' below the salt structure is a potential structural trap with a dip closure away from the fault, and a lateral and top seal formed by the Zechstein evaporites. The main period of trap formation is Late Jurassic during the Late Kimmerian tectonic phase. Many of these traps especially in the inversion zone of the Central Graben, have been modified during inversion in Late Cretaceous and Early Tertiary times. Prospective reservoir levels are Upper Westphalian sandstones and the lower part of the Rotliegend Formation in the south of the study area. The reservoirs are sourced by Carboniferous gas. This type of trap is most prospective in the southern part of the Central Graben where the Rotliegend has not completely shaled out, and at the margins of the Central Graben and platform areas where the Carboniferous is not too deeply buried to be overcooked. The top seal is Zechstein salt. This trap type (Fig. 12, no. 1) has successfully been drilled in L4 and probably in L1 (Table 1).

Above the Zechstein salt there are essentially three structural trap configurations. Firstly, abutment of reservoir rocks against a salt diapir (Fig. 12, no. 2). This mainly involves Triassic Volpriehausen and Detfurth Sandstone intervals and in some cases Jurassic reservoirs (Table 1). The trap has a dip closure away from the salt piercement and a lateral seal against the salt; top seal has to be provided by overlying shales. The Triassic sequence is tilted during the pillow phase of the salt structure. During the first stage of the diapiric phase (Late Jurassic mostly) the Triassic sequence is pierced and abuts on the salt structure. Further depletion of the salt layer may result in collapse of the trap. An example of this type of trap is the L2-FB field, a (non-producing) gas field in the Triassic. Dissolved salt may precipitate in reservoirs. This so-called salt plugging occurred in the Triassic Volpriehausen and Detfurth Formations in several wells in the study area (Dronkert & Remmelts, this volume, Fontaine et al. 1993). Salt plugging has been observed up to several hundreds of metres away from the salt structure in block L2 (Dronkert & Remmelts, this volume).

Secondly, there are the four-way dip closures in turtle back anticlines (Fig. 12, nos. 3, 4), many examples of which are found in the Central Graben area: fields F15-a (Fontaine et al. 1993), F3-FA, F3-FB, F17, L2-FA, L5-FA (Houllebergĥs et al. 1993). These traps form by collapse of the edges of the primary rim syncline fill into the depleted secondary rim synclines (Trusheim 1960). Extension of reservoirs in turtle back anticlines may result in a break up into a number of compartments, requiring more production wells in order to drain each compartment. An example of such a reservoir is the F18 field; the setting of such a trap is illustrated in Fig. 12, no. 3.

Thirdly, four-way dip closures above salt structures. Prospective reservoirs may be present directly on top of diapiric structures. They are the Upper Jurassic and Lower Cretaceous sands, and the Chalk, or the Tertiary with subtle highs above salt domes (Fig. 12, nos. 5, 6). Although no commercial hydrocarbon finds have been reported from the Upper Cretaceous Chalk in the Netherlands sector of the continental shelf; it is an important play in the Danish sector. The Chalk is normally tight (permeability < 1 mD at 2000 m, Hancock & Scholle 1975) but above salt domes it is locally fractured by the upward movement of salt, thus creating secondary permeabilities which may be up to 1000 times higher than usual (Megson 1991, Jørgensen & Andersen 1991, Foster & Rattey 1993). Five Tertiary reservoirs have been discovered in the A- and B-quadrants (Table 1). Trap structuration may be related to the youngest, Paleogene, diapiric movements (Fig. 12, no. 6). Gas is trapped in 4-way dip closures over the updoming salt structures. Growth of the salt structure is related to compressional stresses during Late Cretaceous and Tertiary times. Enlargement of the 4-way dip closure may have occurred due to later differential compaction. Traps above salt structures may also be delineated by extensional faulting of collapse grabens or by listric normal faulting along one of the flanks of the dome, both of which often occur in this setting. Miocene and Pleistocene sands are favourable targets for gas exploration. The presence of gas in these sands causes phase reversals and pull-downs on seismic (Duyverman et al. 1991). No exact information on the source rock is available. Duyverman et al. (1991) suggest Namurian coals, since Westphalian coals are absent from the area. None of these Tertiary fields is on production yet, but the recently constructed pipeline to block F3 brings the infra-structure within reach.

Concluding remarks

In the Dutch sector of the North Sea, salt structures follow the structural grain of the region, and are directly related to 'basement faults' (see also Jenyon 1986, Jaritz 1987, Koyi 1991). The height of salt structures is proportional to the throw of the underlying 'basement fault'. On the relatively stable platforms, relatively small salt pillows are found, while along the margins larger salt pillows occur above larger faults. In the basinal areas, elongated salt walls have developed above long uninterrupted fault zones with large vertical offsets. Local extension above salt structures, and possibly gravitational instabilities, created grabens or listric faults in the overburden.

Periods of enhanced salt flow correspond with tectonic phases. In the basinal areas the first salt movement took place in Triassic times. Dating of the first salt movements on the platforms is hampered due to a large hiatus in the sedimentary record. Some indications exist for Early Triassic (Schill Grund High) and Late Triassic movements (NE corner of the Cleaver Bank High). The strongest salt flow occurred during the Late Kimmerian rifting (Late Jurassic to Early Cretaceous) and the Subhercynian, Laramide and Pyrenean inversion phases (Late Cretaceous and Eocene – Oligocene). Locally, some minor salt movement took place during the Neogene. Salt movement in the southern part of the Central Graben has started slightly later than in the north. This is in agreement with the southward progradation of regional deformation, which further illustrates the tectonic control on halokinesis. The fact that salt flow increased during tectonic phases suggests a relationship between intraplate stresses and salt movement, and/or a relationship between differential loading (induced by fault movement) and salt movement. It was not possible to distinguish the influence of either two factors from the seismic data. Accelerated rates of salt doming, coinciding with Late Cretaceous and Eocene – Oligocene inversion, are not clearly reflected in significant vertical throws along the 'basement' faults. This could imply a direct role of the intraplate stresses on the flow of salt during compressive deformation. During Late Kimmerian rifting (Late Jurassic to Early Cretaceous), most of the salt structures in the basinal areas reached a diapiric stage. Strong erosion and dissolution (subrosion) of Zechstein salt occurred during this stage. Many salt pillows in the Terschelling Basin were truncated during this phase.

Salt flow induced by buoyancy forces only, in a tectonically undisturbed area, will generally be inhibited by the strength of the overburden. 'Basement' faulting with large differential movement will deform and weaken the overburden. In combination with increased buoyancy forces generated by differential loading, it will enable salt movement parallel to 'basement' faults (Fig. 13). The position of areas of salt withdrawal may shift laterally either due to alternating normal or reversed movement along the underlying 'basement' fault, or due to interference of two or more fault systems.

Almost all hydrocarbon finds in this part of the North Sea are related to salt movement. Salt and salt movement may positively affect trap formation, creation of migration routes, secondary permeability, sealing capacity and topography-induced facies changes.

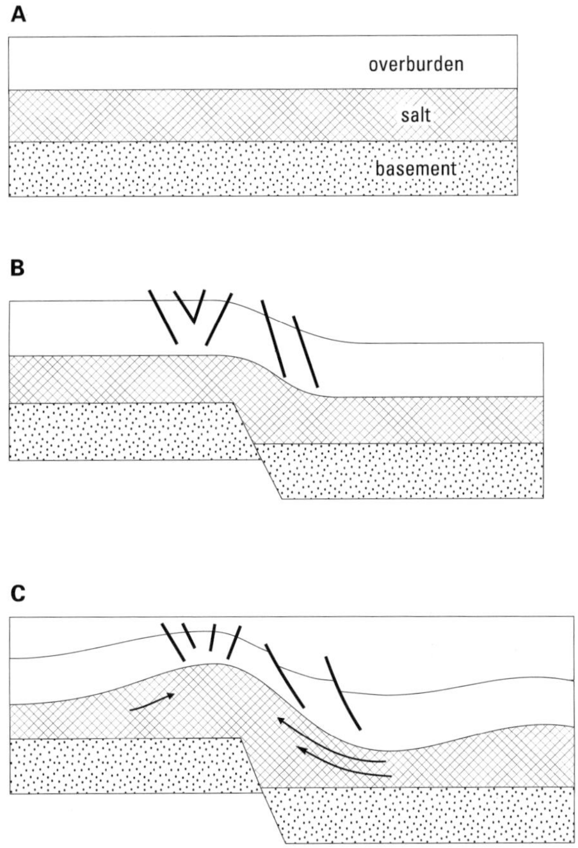

Fig. 13. Model for basement-fault-related salt movement in the southern North Sea. (A) A density inversion is created when a sufficiently thick sedimentary overburden is overlying a salt sequence. The buoyancy forces in the salt will generally not be large enough to overcome the strength of the overburden. (B) Extensional basement faulting will be accommodated by the salt layer, but the overburden will be weakened due to stretching and faulting. (C) Differential loading combined with tectonic stresses will enable the salt to move.

Negative effects are also to be expected in salt plugging, reservoir compartmentalisation, leakage along reactivated faults, topography-induced facies changes and poor seismic imaging of the structures.

Acknowledgements

This work is based on research executed within the program on radioactive waste disposal on land (OPLA, Phase 1A). Nopec/Geco-Prakla is thanked for the permission to use their SNST83 survey. My colleagues Douwe van Rees, Dick van Doorn and Mark Geluk are thanked for the constructive discussions. Comments on the draft version by Dick Batjes and Jan de Jager certainly improved the manuscript. Figures were drafted by Han Bruinenberg, Els Berendse and Merijn Kerlen.

References

Dronkert, H., S.D. Nio, W. Kouwe, N. Van der Poel & Y. Baumfalk 1989 Buntsandstein of the Netherlands Offshore. Non-exclusive Study Intergeos BV, Leiderdorp, 3 vols.: 450 pp

Dronkert, H. & G. Remmelts 1995 Influence of salt structures on reservoir rocks in block L-2, Dutch continental shelf. In: Rondeel, H.E., D.A.J. Batjes & W.H. Nieuwenhuijs (eds) Geology of gas and oil under the Netherlands – this volume

Duyverman, H.J., K. Geil, O. Michelsen & K. Sørensen 1991 Tertiary geology and prospectivity of the Netherlands' northern offshore – Paper presented at the Eur. Ass. Petroleum Geosc. Conference at Florence. No E021.

ECL (Exploration Consultants Ltd.) 1983 Offshore Netherlands, petroleum exploration appraisal. vol. 1. Regional exploration review – Exploration Consultants Ltd., Henley on Thames, 110 pp

Fontaine, J.M., G. Guastella, P. Jouault & P. de la Vega 1993 F15-A: a Triassic gas field on the eastern limit of the Dutch Central Graben. In: Parker, J.R. (ed.) Petroleum Geology of Northwest Europe: Proc. 4th Conference London 1992 – Geol. Soc., London: 583–593

Foster, P.T. & P.R. Rattey 1993 The evolution of a fractured chalk reservoir: Machar oilfield, UK North Sea. In: Parker, J.R. (ed.) Petroleum geology of Northwest Europe: Proc. 4th Conference London 1992 – Geol. Soc., London: 1445–1452

Hancock, J.M. & P.A Scholle 1975 Chalk of the North Sea. In: Woodland, A.W. (ed.) Petroleum and the continental shelf of North West Europe, vol. 1 Geology – Applied Science Publishers, London: 413–427

Heybroek, P. 1975 On the structure of the Dutch part of the Central North Sea Graben. In: Woodland A.W. (ed.) Petroleum and the continental shelf of North West Europe, vol. 1 Geology – Applied Science Publishers, London: 339–351

Houlleberghs, E., P. Haltmeier & J. Van De Sande 1993 Use of Prediction Techniques in an Integrated Study of the L5 Field, Offshore Netherlands – Am. Ass. Petroleum Geol. Bull. 77: 1631 (Abstr.)

Jaritz, W. 1987 The origin and development of salt structures in northwest Germany. In: Lerche, I., & J.J. O'Brien (eds) Dynamical geology of salt and related structures – Academic Press, Orlando: 480–493

Jenyon, M.K. 1986 Salt tectonics – Elsevier Applied Science Publ., London/New York: 191 pp

Jørgensen, L.N. & P.M. Andersen 1991 Integrated study of the Kraka Field – Soc. Petroleum Eng. Paper 23082

Koyi, H. 1991 Gravity overturns, extension and basement fault activation – J. Petroleum Geol. 14: 117–142

Koyi, H., M.K. Jenyon & K. Petersen 1993 The effect of basement faulting on diapirism – J. Petroleum Geol. 16: 285–312

Megson, J.B. 1991 The North Sea Chalk-play: Examples from the Danish Central Graben. In: Hardman, R.F.P. (ed) Exploration Britain: Geological insights for the next decade – Geol. Soc., London, Spec. Publ. 67: 247–281

Ministry of Economic Affairs 1984–1993 Olie en Gas in Nederland opsporing en winning/Oil and gas in the Netherlands exploration and production – Ministry of Economic Affairs, The Hague (published yearly)

Oudmayer, B.C. & J. De Jager 1993 Fault reactivation and oblique-slip in the Southern North Sea. In: Parker, J.R. (ed.) Petroleum geology of Northwest Europe: Proc. 4th Conference London 1992 – Geol. Soc., London: 1281–1292

Remmelts, G., E. Muyzert, D.J. van Rees, M.C. Geluk, C.C. de Ruyter & A.F.B. Wildenborg 1993 Evaluation of salt bodies and their overburden in the Netherlands for the disposal of radioactive waste, b. Salt movement – Rijks Geologische Dienst, Haarlem, report 30.012B/ERB: 87 pp

Sanneman, D. 1963 Über Salzstock-Familien in NW-Deutschland – Erdöl. Z. 79: 499–596

Sanneman, D. 1968 Saltz-stock Families in Northwestern Germany. In: Diapirism and diapirs – Am. Ass. Petroleum Geol. Mem. 8: 261–270

Trusheim, F. 1957 Über Halokinese und ihre Bedeutung für die structurelle Entwickelung Norddeutschlands – Z. dt.geol. Ges. 109: 111–151

Trusheim, F. 1960 Mechanism of salt migration – Am. Ass. Petroleum Geol. Bull. 44: 1519–1540

Tsai, F.C., J.E. O'Rourke & W. Silva 1987 Basement rock faulting as a primary mechanism for initiating major salt deformation features. In: Proc. 28th Symposium on Rock Mechanics, Tucson, USA: 621–631

Van Adrichem Boogaert, H.A. & W.F.P. Kouwe (compilers) 1993–1994 Stratigraphic nomenclature of the Netherlands, revision and update by RGD & NOGEPA – Med. Rijks Geol. Dienst 50 (loose-leaf publication)

Van Hoorn, B. 1987 Structural evolution, timing and tectonic style of the Sole Pit inversion – Tectonophysics 137: 239–284

Van Wijhe, D.H. 1987 Structural evolution of inverted basins in the Dutch offshore – Tectonophysics 137: 171–219

VUA (Vrije Universiteit, Amsterdam), Sectie Structurele Geologie/Tektoniek 1992a Intraplaattektoniek, bekkenmodellering en zoutdiapirisme. Implementatietechnieken. Tussenrapportage OPLA project februari 1992, OPLA92–56

VUA (Vrije Universiteit, Amsterdam), Sectie Structurele Geologie/Tektoniek 1992b Intraplaattektoniek, bekkenmodellering en zoutdiapirisme. Inhoudelijke deelstudie: Modellering Deel 1 en 2. Tussenrapportage OPLA project augustus 1992, OPLA92–76

Ziegler, P.A. 1987 Late-Cretaceous intra-plate compressional deformations in the Alpine foreland – a geodynamic model – Tectonophysics 137: 389–420

Ziegler, P.A. 1990 Geological Atlas of Western and Central Europe – Shell Int. Petr. Mij. The Hague: 239 pp

Rondeel et al. (eds), Geology of gas and oil under the Netherlands, 159–166, 1996.

Influence of salt structures on reservoir rocks in Block L2, Dutch continental shelf

Hans Dronkert[1] & Gijs Remmelts[2]
[1] *Faculty of Mining & Petroleum Engineering, Delft University of Technology, Postbus 5028, 2600 GA Delft, the Netherlands;* [2] *Geological Survey of the Netherlands, Postbus 157, 2000 AD Haarlem, the Netherlands*

Key words: Netherlands continental shelf, petrophysics, sandstone reservoirs, diagenesis, halite cement

Abstract

In the subsurface of the Netherlands continental shelf, thick layers of Zechstein salt have developed into salt domes and ridges that occasionally pierce through the overlying formations. To measure the influence of the salt on the laterally adjacent Mesozoic sandstone reservoir rocks, a 'cementation model' was developed. The target area, Block L2, was chosen for the presence of salt domes, reservoir rocks and five representative wells. All available well information (wire-line log, test and core data) has been used to detect the presence of salt in the Lower Triassic and Upper Jurassic reservoir intervals. This was done mainly by combining gamma ray, sonic, resistivity and density data from the well logs into a computer model. The cementation model produces displays on which the presence of halite cement is clearly indicated. Only one well, L2–2, located within a few hundred metres of a salt dome, showed salt plugging. In the main reservoir rock, the Volpriehausen Sandstone (Lower Triassic), four other wells, located at more than 1.5 km from a salt dome or ridge, did not show signs of halite cementation in this reservoir. Therefore, the influence of salt domes on the surrounding reservoir rocks is believed to be restricted to less than 1.5 km at 1–2 km depth. The salt plugging of the Detfurth Sandstone (Lower Triassic) can be attributed to early seepage from the evaporitic Röt Formation. Because the Detfurth Sandstone in the L2–2 well is not salt-cemented, seepage of Röt brines into the Volpriehausen Sandstone reservoirs can be excluded.

Introduction

As part of a much larger study for waste disposal on land in the Netherlands (OPLA research programme) a study was undertaken to investigate the state of cementation of reservoir sands around salt domes. Unfortunately, the limited accessibility of onshore subsurface data is a serious obstacle for these studies. The Dutch mining law for the Netherlands onshore which dates from Napoleonic times in the early eighteenhundreds, does not provide the obligation for concession holders to release information from the subsurface. For the Dutch offshore, however, a new mining law dating from the sixties, guarantees that most information on the subsurface is released after ten years. Since the early eighties, political agreements between the Dutch government, the northern provinces of the Netherlands, and the environmentalists have forbidden all research relating to salt structures on land because this might initiate developments leading to disposal of radioactive waste in some of these structures. Therefore, public investigation of the effect of salt structures on surrounding reservoir rocks proved to be impossible for the Dutch onshore area. In order to partly overcome these restrictions a region was selected in the Dutch offshore with an analogous geological setting. The main objectives of the study were:

a) to define the influence of the salt structure on the neighbouring aquifers,
b) to determine the lateral extension of this effect,
c) to outline the consequences for the northern Netherlands.

The investigation has been directed towards the decrease of permeability due to salt plugging which may occur when dissolved salt precipitates in the pores of permeable rocks in the vicinity of a salt body. Sever-

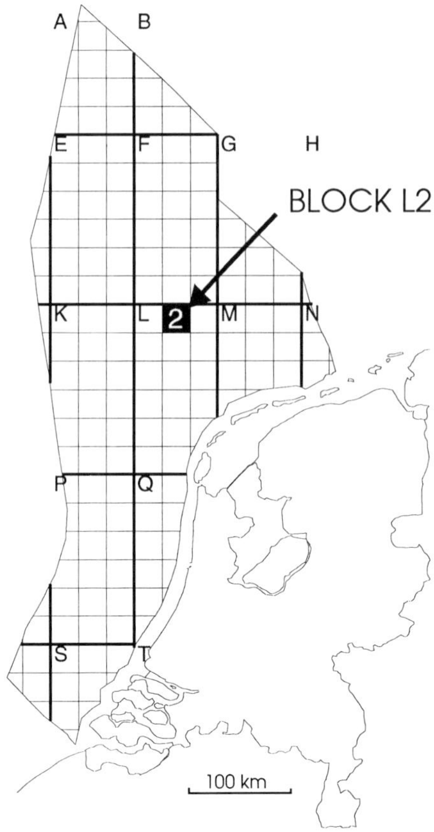

Fig. 1. Location of Block L2 in the Netherlands sector of the continental shelf.

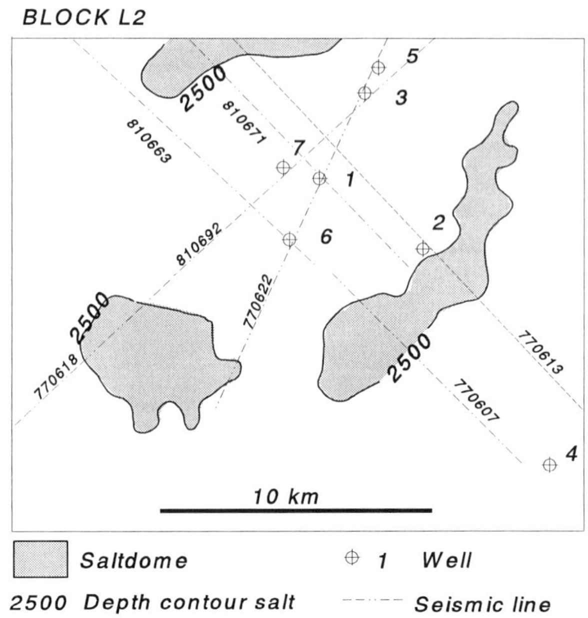

Fig. 2. Map at 2500 m depth of Block L2 showing locations of Zechstein salt structures, wells and seismic lines (after ECL 1983).

al important Triassic and Jurassic aquifers which occur near the flanks of salt structures in the subsurface of the Netherlands were studied. All released seismic and well data of the Netherlands sector of the continental shelf were at our disposal. An area was selected which shows a geological setting comparable to that of the northern Netherlands. Block L2 was chosen since it encompasses a number of Mesozoic aquifers (or reservoir rocks), several salt structures and an acceptable data density. It contains seven released wells and sufficient seismic data (Figs 1, 2).

Although a host of papers has been published on all facets of salt and salt structures, little is known on the hydrogeological parameters of rocks surrounding salt structures. Much confidential research, however, has been carried out by oil companies on salt cementation of oil and gas reservoirs. Only indirect information is available from petroleum geological studies (Dronkert 1990).

In a wider sense much has been written on the salt structures of northern Europe (Trusheim 1957, 1960, Hentchel & Kleinitz 1976, Jenyon 1986, Dronkert 1985, 1987, 1990, Dronkert et al. 1989, 1990, Jaritz 1987). Both the geometry and composition as well as the internal configuration of salt structures, such as salt pillars and cushions, have been described. Petrophysical and geophysical parameters of salt deposits have been extensively studied (petrophysics: Dronkert 1990, Fertl 1979, 1987, Hilchie 1982, Jensen 1983, Lee et al. 1989, Serra 1986a,b; seismics: Lohmann 1979). Much of this research has taken place in Germany and the USA because the salt mines in these countries facilitated detailed investigations. Only little is published on the Dutch salt structures (Coelewij et al. 1978, Schmoll & Schuermann 1980, Haile & Blunden 1984, RGD 1984, 1988, 1991, 1993, Van Doorn et al. 1985, Geluk 1986, Dronkert et al. 1989). No salt mines are present in the Netherlands. Salt exploitation is by solution mining and consequently only seismic surveys, well log data and cores are available.

To enable the selection of an analogue area in the Netherlands offshore, a location map of all released well data was produced for each reservoir unit, indicating the thickness and the depth interval. These maps were combined with the depth map of the top of the Zechstein to show the relative position of the wells with respect to the salt structures. The possibilities narrowed down to blocks F18 and L2, the latter of which was most favourable. Block L2 contains three salt structures of a shape and depth comparable to those in the onshore area. The top of the Zechstein salt is at a regional depth of 4 to 5 km, and the salt structures have their tops between 1200 and 2000 m. Two reservoir intervals occur in the block: the Triassic Main Buntsandstein Formation, containing the Volpriehausen and Detfurth Sandstone Members, and the Late Jurassic Delfland and Scruff Greensand formations. The Early Cretaceous Vlieland Sandstone is absent in L2. The seven released wells from Block L2 were drilled between 1968 and 1982.

Method of investigation

To demonstrate the cementation of salt in reservoir rocks, several log measurements can be used (Serra 1986a,b). The gamma-ray was used for correlation purposes and to determine sand to shale ratios. The acoustic or sonic logs (DT) were used to estimate the porosity, and the formation density logs (FDC, RHOB) to measure rock density. The compensated neutron log (CNL) had not been run in all wells and could therefore not be used for porosity estimates. Caliper (borehole diameter) and micro-resistivity logs were used to check on borehole irregularities. Deep resistivity and induction logs were used in the calculation of water saturation. To calculate water saturation, porosity and halite content, the log evaluation program ES-Log© from Energy Systems (USA) was implemented on the VAX© system of the Geological Survey of the Netherlands (RGD). In this program, the single-porosity model of Raymer/Hunt was chosen (Raymer et al. 1980) because the reservoir rocks consist mainly of clastics with a low content of carbonate. Parameters used are listed in Table 5. For the calculation of the water saturation the Indonesian formula was applied, best suited for siliciclastic reservoirs. It is a standard procedure at the RGD. To calculate the halite content, a computer program was written which is referred to as the 'cementation model' throughout this paper.

Since the porosities of the analysed reservoir rocks are generally below 20%, it is not possible to determine the presence of halite directly from the borehole measurements. This is done indirectly with the cementation model using the fact that every lithology has a specific density and a specific transit travel time for acoustic waves (Fig. 3, Table 1). The Triassic reservoir rocks consist of shale and sandstone and have a density from 2.5 to 2.7 g/cm^3. Rocks of the Delfland

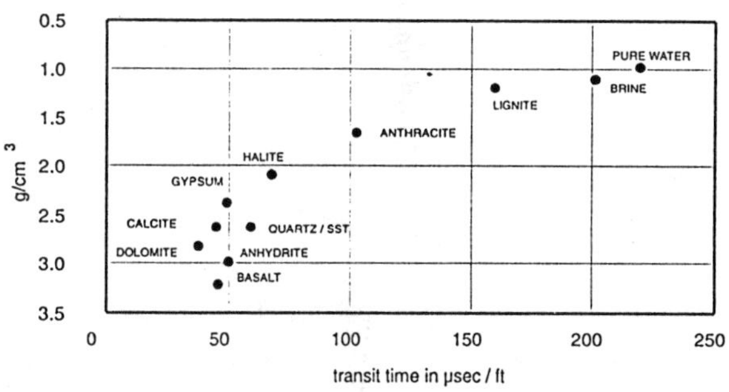

Fig. 3. Cross plot of acoustic transit time versus density of minerals, rocktypes and water (after Serra 1986a).

Table 1. Log parameters of common reservoir rocks and fluids (after Serra, 1986b).

Rock/fluid	Density (kg/dm^3)	Gamma ray (API)	Velocity (m/μsec)
Anhydrite	2.96	0	119
Halite	2.17	0	127
Sylvite	1.98	500	139
Dolomite	2.87	0	115
Calcite	2.71	0	110
Shale	2.65	50–140	134
Quartz	2.65	0	139
Water	1.00	0	33
Brine	1.14	0	30

Subgroup and the Scruff Greensand Formation contain sand, clay and some coal; the density varies from 2.0 to 2.5 g/cm^3. Halite is characterized by a relatively low density: just over 2 g/cm^3. The densities of coal and lignite vary between 1.5 and 1.1 g/cm^3.

The cementation model is based on the hypothesis that the theoretical density of a rock follows from its specific porosity, its lithologic composition and texture. The difference between the measured density of a rock interval and its theoretical value, will be an indication for the mineralogical composition of the cement. This comparison is made by cross-plotting the bulk densities against the interval transit time. If the measured weight is higher than the theoretical value, the cement has a relatively high density such as anhydrite. If the measured weight is lower, it indicates a light weight cement such as halite. Another way to highlight the presence of halite is to combine in a graph the theoretical density log with the actual density log

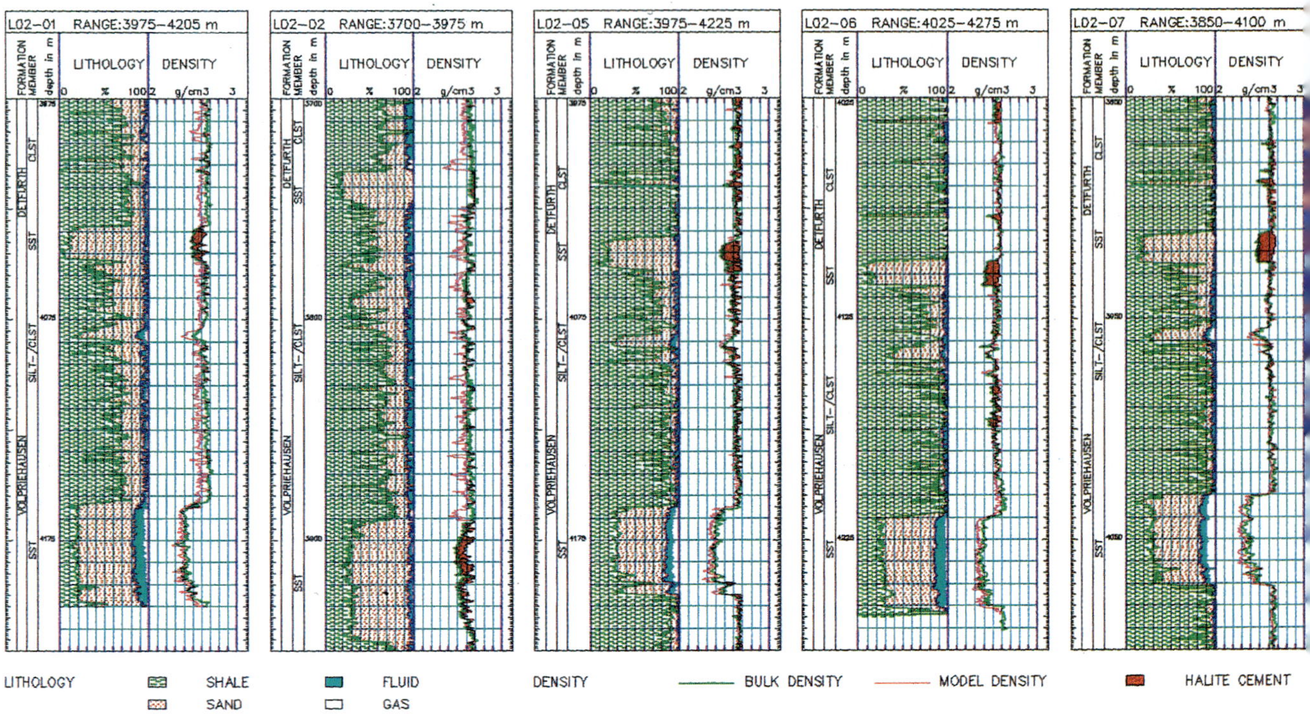

Fig. 4. Log diagrams based on the 'cementation model' for the reservoir rocks of the Volpriehausen and Detfurth Sandstone Members of the Triassic Buntsandstein Formation in wells L2–1, L2–2, L2–5, L2–6 and L2–7. The columns on the left of each diagram indicate stratigraphic position, gross lithology and depth below sea level in metres. The 'lithology' column gives the approximate shale (green), sand (red), water (blue) and gas/oil (no colour) composition. The fifth column presents the logs of the measured bulk density (= green line) and the calculated theoretical or model density (= purple line) of the rock. The difference between these values is an indication for the cement composition. The cement is interpreted to consist of halite (red) when the calculated density is higher than the measured density.

for the specific aquifer (Fig. 4). Input data for these graphs are the porosity of the rock to calculate the volume of solid rock, its density and the volumetric fractions of the mineralogical components and fluids in the rock. The theoretical density curve is based on the composition of the stratigraphic interval as depicted in the second column (Fig. 4). In the presence of halite cement, the theoretical density curve shows higher values than the measured density curve. In that case, the space between the two density logs has been shaded (in red in Fig. 4). If heavier cement minerals are present the measured density is higher than the theoretical density, and the enclosed area has not been shaded (Fig. 4).

A number of assumptions has been made for the cementation model. The density of the formation water in the Triassic reservoir has been taken as 1.1 g/cm^3, which is the value for nearly halite-saturated brine, generally present at several kilometres depth. When the rock consists of over 85% of clay minerals, quartz and feldspars, the density has been set at 2.65 g/cm^3. Microscopic analysis of the Main Buntsandstein has shown that the remaining 15% or less, consist mainly of dolomite (density 2.87 g/cm^3), anhydrite (2.96 g/cm^3) and traces of barite and haematite (> 4 g/cm^3). Therefore, the density of the Triassic reservoir rocks was set from 2.65 to 2.7 g/cm^3 (Dronkert et al. 1989). Since the Delfland and Scruff Greensand consist of younger, less compacted rocks, a different density was estimated. The clays in this interval are lighter, less heavy minerals occur and organic material is abundant. Standard values for the density were taken at 2.5 g/cm^3 for the rocks and 1.0 g/cm^3 for the pore fluid.

Results

The most important Mesozoic reservoirs in the Netherlands are the Volpriehausen and the Detfurth Sandstone Members of the Lower Germanic Triassic Group, Main

Table 2. Reservoir summation of the Volpriehausen Sandstone Member.

Well	Log depth interval (m)	Distance to salt structure (km)	Net thickness (m)	Porosity (%)	Shale (%)	Sw (%)
L2–1	4157–4205	5	47	14.6	19.5	65
L2–2	3891–3946	0.1–0.2	39	4.8	29.7	66
L2–5	4160–4200	1.5–2	38	14.4	30.6	43
L2–6	4229–4259	5	27	13.5	28.3	76
L2–7	4030–4071	3–3.5	40	15.7	29.8	35

Table 3. Reservoir summation of the Detfurth Sandstone Member.

Well	Log depth interval (m)	Distance to salt stucture (km)	Net thickness (m)	Porosity (%)	Shale (%)	Sw (%)
L2–1	4032–4047	5	12	4.5	10.7	51
L2–2	3737–3751	0.1–0.2	1.5	4.1	39.5	54
L2–5	4038–4053	1.5–2.0	3.8	4.5	21.0	37
L2–6	4098–4120	5	1.4	3.9	28.7	46
L2–7	3911–3929	3.0–3.5	0.5	3.0	12.6	37

Buntsandstein Formation, and the sandstones of the Delfland and Scruff Greensand (Upper Jurassic and Lower Cretaceous). For the stratigraphy see Herngreen & Wong (1989) and PGK (1993, this volume). The cementation model has been run for all these reservoir units.

Volpriehausen Sandstone Member

The wells used show similar reservoir characteristics for the Volpriehausen Sandstone (Table 2). Only well L2–2 shows a remarkable porosity reduction to 4.8%, indicating severe cementation. Water saturations vary strongly but are relatively low, due to the presence of gas in most wells.

Detfurth Sandstone Member

The Detfurth Sandstone Member apparently combines a low porosity with a relatively low water saturation (Table 3). Cross plots of density against interval transit-time indicate salt plugging for well L2–1, and partial plugging of the interval in wells L2–6 and L2–7 (Fig. 4).

Interpretation of Volpriehausen and Detfurth Sandstone Members

From the calculations in the cementation model it appears that the Volpriehausen Sandstone Member in well L2–2 is cemented with halite, while the other wells are cemented with anhydrite or dolomite (Fig. 4). No indications for primary evaporite deposition such as anhydrite nodules, dolomite beds or halite crystal molds have been found in the Volpriehausen Sandstone and in the lower part of the Siltstone Member (Dronkert et al. 1989). Seepage of brine from the overlying Röt salts is unlikely since the intermediate claystones and siltstones form a barrier sealing off the Volpriehausen Sandstone gas reservoirs. Moreover, the Detfurth Sandstone Member is not halite-cemented in well L2–2, but the Volpriehausen Sandstone Member is. The salt structures in Block L2 started to rise in the later part of the Early Triassic (Dronkert et al. 1989). During the Late Jurassic, they pierced the Triassic formations. From that time the Main Buntsandstein reservoirs have been in close contact with the Zechstein salt. In view of the mainly fluvial environment of deposi-

Table 4. Reservoir summation of the Delfland and Scruff Greensand formations.

Well	Log depth interval (m)	Distance to salt structure (km)	Net thickness (m)	Porosity (%)	Shale (%)	Sw (%)
L2–1	1932–1958	5	20	20.4	30.4	64
L2–2	2100–2500	0.1–0.2	157	14.4	32.2	75
L2–3	1734–2007	2.0–2.5	100	18.5	34.6	73
L2–4	2000–2475	3.0–4.0	165	11.7	32.8	76
L2–5	1576–1968	1.5–2.0	196	18.8	31.6	80
L2–6	2140–2188	5	35	14.9	40.4	77
L2–7	2015–2053	3.0–3.5	6	16.2	44.0	34

Table 5. Programming parameters used with ES-Log© for the Buntsandstein and Delfland/Scruff reservoir rocks.

Depth interval increment	0.1523 m
Temperature	per well, Hilchie algorithm (ES-Log©)
Sonic clay value	Buntsst. 70 μsec/ft, Delfl./Scruff 120 μsec/ft
Sonic sandstone (Raymer-Hunt)	55.6 μsec/ft, 'c' = 0.7
Gamma ray sand/shale line	20/120 API
Resistivity pore fluid	0.1 ohm, Rclay = 0.5 ohm, Wiley-Rose = 79
Indonesian Formula	a = 1, m = n = 2
Summation 'cut-offs'	3% < phi < 30%
	Sw < 90%
	Vshl < 50%

tion of the Volpriehausen Sandstone Member during a relatively wet climatic period (Dronkert et al. 1989), and in view of the timing of the salt movements and the distance of the reservoir to the salt bodies, the halite in the Volpriehausen Sandstone Member is most likely of secondary origin and may have been derived from the Zechstein salt.

Density readings in the Detfurth Sandstone Member are, in general, lower than in the Volpriehausen Member. Core analyses have shown that the sediments at the top of the Volpriehausen Clay-Shale interval, and especially in the Detfurth Claystone and the Hardegsen and Solling Formations, show evidence for synsedimentary evaporitic minerals such as anhydrite nodules and calcareous concretions (Dronkert et al. 1989). A tendency towards more saline circumstances occurring from the deposition of the Volpriehausen Sandstone Member to that of the Detfurth and onwards, is deduced from these observations. This trend eventually culminated in the Röt salt, stratigraphically not far above the Detfurth Sandstone Member. Contrary to the case for the Volpriehausen Sandstone Member, this suggests early, but secondary, salt precipitation in the Detfurth Sandstone Member. Moreover, dissolved halite from the Röt could more easily percolate into the Detfurth than into the Volpriehausen Sandstone. The salt in the Detfurth Sandstone probably cannot be attributed to the Zechstein salt structures. Compared to the Rotliegend, faults are relatively rare in the Main Buntsandstein and hardly ever seen in the cores, which confirms the good sealing capacities of the silt- and claystones within the Main Buntsandstein.

The results of well L2–7 show deviating values for the porosity. Calculated from the borehole measurements, these are twice as high as measured in core analyses. Hence, the cementation model gives the impression that, apart from the Detfurth Sandstone Member, the interval has been cemented mainly by heavier minerals such as anhydrite and dolomite. Core analysis, on the other hand, shows 3–5% of halite, and a porosity of 12%. However, these small quantities of halite could also have precipitated due to pressure release and dehydration after storage.

Delfland Subgroup and Scruff Greensand Formation

The Delfland and Scruff Greensand formations appear to have a high porosity (> 15%; Table 4). Since the water saturations are high (> 70%), one can assume that no hydrocarbons are present. The clay content of the formations is also high. A clay cut-off value of 50% was used to distinguish reservoir and non-reservoir intervals. The non-reservoir was included in the calculations.

The cementation model yielded unsatisfactory values for the interval of the Delfland and the Scruff Greensand formations. The calculated density values are too high, which is ascribed to the fact that the model cannot cope with the presence of light organic materials such as lignite. Due to the absence of sufficient appropriate well-logs, cores and core analyses, the cementation model could not be calibrated properly.

Significance for the northern Netherlands

On the basis of the large geological similarity between the case-study area in Block L2 and the northern Netherlands onshore, many of the results of this study can be extrapolated to the latter area. However, the present-day hydrological settings of both regions clearly differ from each other. It cannot be totally excluded that these differences are also reflected in the hydrogeological characteristics of the Mesozoic rocks studied.

Hydrological data for model simulations have to take into account the reduced porosity in reservoir rocks directly bordering a salt structure. This is especially the case for deeper reservoirs that have a nearly halite-saturated pore brine. The present research suggests that the maximum reach of salt cementation induced by a salt structure, is less than 1.5 km. The phenomenon of salt plugging has often been reported by the oil and gas exploration industry (Dronkert et al. 1989). In order to get a good grip on this problem in the northern Netherlands, a much larger number of well-log measurements and core analyses of widely distributed wells is needed than was available for this study.

Conclusions

1. A 'cementation model' has been developed which enables the identification of halite as cement in reservoir rocks. The model has well-log measurements as input data and is calibrated with core analyses. The model has been successfully applied to two Lower Triassic sandstone reservoirs: the Volpriehausen Sandstone and the Detfurth Sandstone members. The model did not function with the data from a third group of reservoirs, the Delfland – Scruff sequence, which contains a significant quantity of lignite, apart from sand and clay.
2. From the analysis of the Volpriehausen Sandstone in the L2 block, the presence of an appreciable amount of salt cement, up to a distance of a few hundred metres from salt structures is demonstrated. The salt plugging, a major diagenetic feature, resulted in a reduction of pore space. The maximum reach of the observed salt plugging is less than 1.5 km.

   The Detfurth Sandstone Member also shows salt plugging. This can be attributed to an early enrichment (Triassic) of possibly primary origin or by a later influx from the overlying Röt salt deposits. The results from the study area may be carefully extrapolated to the northern part of the Netherlands, because of its similarity in stratigraphic and structural framework and corresponding geological history.
3. Because gas generation and migration (post-Jurassic) in the area postdate the formation of salt structures (since Late Triassic), the gas now present in most of the reservoirs could not have prevented the reservoirs from becoming salt plugged.

Recommendations

1. To efficiently study the effects of salt cementation on the pore space in a reservoir rock, use should be made of measurements produced by the most modern logging tools available, e.g. spectral logs for gamma ray and density measurements. With the latter technique, an analysis of the mineralogy of the reservoir is possible.
2. To refine the results of this study, 3D seismics together with the cores and logs of several bore holes on and near salt structures in the northern part of the Netherlands should be incorporated.

3. Offshore and onshore wells from areas underlain by salt structures should be analysed in larger numbers to establish a widely applicable model for the migration of, and the cementation from saline pore brines in reservoir rocks.

Acknowledgements

This investigation was carried out by the Geological Survey of the Netherlands (OPLA programme) as part of a wider feasability study of storing low-grade nuclear waste in salt structures. A.F.B. Wildenborg was the project coordinator and G. Remmelts was the project leader of this study. Mrs N. Parker-Witmans digitized many of the logs used. J.J.F. Rademaekers and A. Meinster installed the data sets on the RGD VAX© for use with Energy Systems ES-Log©. J.N. Breunese was the petrophysics supervisor. H. Dronkert programmed the cementation model, performed the data processing and wrote this paper. The suggestions of the reviewers W.K. Campbell, J.A. Mulock Houwer, D.A.J. Batjes and H.E. Rondeel significantly improved the manuscript.

References

Coelewij, P.A.J., C.M.W. Haug & H. Van Kuyk 1978 Magnesium-salt exploration in the northeastern Netherlands – Geol. Mijnbouw 57: 487–502

Dronkert, H. 1985 Evaporite models and the sedimentology of Messinian and Recent evaporites – Thesis, Univ. Amsterdam – GUA Papers, Ser. 1/24: 283 pp, 199 figs

Dronkert, H. 1987 Diagenesis of Triassic evaporites in northern Switzerland – Ecologae geol. Helv. 80: 397–413

Dronkert, H. 1990 The effects of brine migration and salt cementation on hydrocarbon reservoirs – Eur. Oil & Gas Conf. Palermo, 9–12 Oct. 1990 (Abstr.)

Dronkert, H., S.D. Nio, W. Kouwe, N. Van Der Poel & Y. Baumfalk 1989 Exploration and production potential of the Buntsandstein – Non-exclusive Study Intergeos BV, 3 Vols.: 450 pp

Dronkert, H., H.-R. Bläsi & A. Matter 1990 Facies and origin of Triassic evaporites from the NAGRA boreholes, northern Switzerland – Geologische Berichte Landeshydrologie und -geologie 12: 119 pp, 20 encls

ECL (Exploration Consultants Ltd.) 1983 Petroleum exploration appraisal 1983. Offshore Netherlands, revision and update 1985. Depth maps, Encls 72–83

Fertl, W.H. 1979 Gamma Ray spectral data assists in complex formation evaluation – The Log Analyst 20 (5): 3–37

Fertl, W.H. 1987 Log-derived evaluation of shaly clastic reservoirs – J. Petroleum Technol., Feb. 87: 187–194

Geluk, M.C. 1986 Vormen en afmetingen van inhomogeniteiten in zoutvoorkomens. – Internal Confidential Report – Rijks Geol. Dienst Project BP 10570 (OPLA REO-4): 101 pp

Haile, P.M. & H.A. Blunden 1984 Zechstein magnesium rich evaporite deposits of Northern Netherlands and their volumetric analysis by Global – SAID Symp. on Form. Eval. (Paris), Trans. paper 37: 5 pp

Hentchel, J.H. & W.F. Kleinitz 1976 Aufbau der Salzgesteine des Salzstockes Etzel, abgeleitet aus Kernuntersuchungen und Logininterpretationen – Kali und Steinsalz 7: 28–39

Herngreen, G.F.W. & Th.E. Wong 1989 Revision of the 'Late Jurassic' stratigraphy of the Dutch Central North Sea Graben – Geol. Mijnbouw 68: 73–105

Hilchie, D.W. 1982 Advanced well log interpretation – Publ. D. W. Hilchie Inc., Colorado: 135 pp

Jaritz, W. 1987 The origin and development of salt structures in north-west Germany. In: Lerche, I. & J.J. O'Brien (eds) Dynamical geology of salt and related structures – Academic Press, Orlando: 480–493

Jensen, P.K. 1983 Calculations on the thermal conditions around a salt diapir – Geoph. Prosp. 31: 481–489

Jenyon, M.K. 1986 Salt Tectonics – Elsevier Appl. Science Publ., London.: 191 pp

Lee, M., J.L. Aronson & S.M. Savin 1989 Timing and conditions of Permian Rotliegende sandstone diagenesis, southern North Sea: K/Ar and oxygen isotopic data – Am. Ass. Petroleum Geol. Bull. 73: 195–215

Lohmann, H.H. 1979 Seismic recognition of salt diapirs – Am. Ass. Petroleum Geol. Bull. 63: 2097–2102

PGK (Petroleum Geological Circle) 1993 Synopsis: Petroleum geology of the Netherlands – 1993 In: Rondeel, H.E., D.A.J. Batjes & W.H. Nieuwenhuijs (eds) Geology of gas and oil under the Netherlands – this volume

Raymer, L.L., E.R. Hunt & J.S. Gardner 1980 An improved sonic transit time-to-porosity transform – SPWLA 21st Ann. Log. Symp. Trans., Paper P: 13 pp

RGD (Rijks Geologische Dienst) 1984 Inventarisatie van slecht doorlatende laagpakketten in de ondergrond van het Nederlandse vasteland – Rapport nr. OP6009, Haarlem: 131 pp

RGD (Rijks Geologische Dienst) 1988 Geologische inventarisatie en ontstaansgeschiedenis van zoutvoorkomens in Noord- en Oost-Nederland – Rapport nr. 10568, Haarlem: 272 pp

RGD (Rijks Geologische Dienst) 1991 Geologische atlas van de diepe ondergrong van Nederland. Toelichting bij kaartblad II Ameland-Leeuwarden – RGD, Haarlem: 87 pp

RGD (Rijks Geologische Dienst) 1993 Evaluatie van Nederlandse zoutvoorkomens en hun nevengesteente voor de berging van radioactief afval; overzicht van de resultaten. Programma van onderzoek OPLA, Fase 1A – Rapport nr. 30012/ER, Haarlem: 115 pp

Schmoll, J. & W. Schuermann 1980 Geophysik bei der Planung eines Kavernenfeldes – Fifth symposium on Salt I: 365–371

Serra, O. 1986a Fundamentals of well-log interpretation 1. the acquisition of logging data – Developments in Petroleum Science 15A – Elsevier, Amsterdam: 423 pp

Serra, O. 1986b Fundamentals of well-log interpretation 2. the interpretation of logging data – Developments in Petroleum Science 15B, Elsevier, Amsterdam: 684 pp

Trusheim, F. 1957 Ueber Halokinese und ihre Bedeutung fuer die strukturelle Entwicklung Norddeutschlands – Z. dtsch. geol. Ges. 109: 111–151

Trusheim, F. 1960 Mechanism of salt migration in northern Germany – Am. Ass. Petroleum Geol. Bull. 44: 1519–1540

Van Doorn, T.H.M., C.I. Leyzers Vis, N. Salomons, W. Van Dalfsen, H. Speelman & W. Zijl 1985 Aardwarmtewinning en grootschalige warmteopslag in Tertiaire en Kwartaire afzettingen – Rapport nr. 85 KAR 02 EX, RGD, Haarlem: 108 pp

Rondeel et al. (eds), Geology of gas and oil under the Netherlands, 167–178, 1996.

The environments of deposition of the Triassic Main Buntsandstein Formation in the P and Q quadrants, offshore the Netherlands

R. Ames[1] & P.F. Farfan[2]
[1] *Amoco Eurasia Petroleum Co, Moscow, Russian Federation;* [2] *Amoco (UK) Exploration Company, Amoco House, West Gate, London W5 IXL, England*

Key words: core interpretation, log interpretation

Abstract

Interpretations of core and log data within the P and Q quadrants were used to quantify the relative change in the distribution of environments of deposition within each of the five members of the Main Buntsandstein Formation. This Triassic formation was deposited in a semi-arid continental basin. Sediments reached the P and Q quadrants in braided stream complexes from a hinterland to the southeast. The braided stream complexes were flanked by aeolian, interdune, flood-plain and crevasse-splay environments and terminated in playa lakes. The distribution and systematic arrangement of environments of deposition migrated subtly across the study area. This is attributed to slight fluctuations in climatic conditions (rainfall and windstrength or direction).

Introduction and regional geology

The P and Q quadrants cover about 12 000 km^2 of the Dutch southern North Sea (Fig. 1). Oil is produced mainly from Lower Cretaceous reservoirs. Gas is produced from the Permian, Zechstein Platten Dolomite and the Triassic, Main Buntsandstein Formations. Until recently, Main Buntsandstein production was restricted to Block P6, but in the past few years the productive Buntsandstein fairway has been extended southward, as far as Block P18. In the Triassic, the study area was on the low-relief northern flank of the London-Brabant Massif, which was part of a major topographic feature that separated the intermontane Mid-European Basin from the Tethys Basin (Fig. 1).

Stratigraphy

The Main Buntsandstein Formation in the study area, consists of intercalated sandstones, claystones and siltstones, which are subdivided into five members: the Volpriehausen Sandstone, the Volpriehausen Clay-siltstone, the Detfurth Sandstone, the Detfurth Claystone, and the Hardegsen Sandstone (Fig. 2). These rest conformably on the Lower Buntsandstein Formation.

The original distribution and thickness of the Main Buntsandstein Formation was modified by two phases of erosion, as follows:

a) In north-eastern parts of the study area, the Early Triassic, Main Buntsandstein Formation was partially eroded and unconformably covered by the Late Triassic, Röt Formation (Fig. 2);
b) The Kimmerian, Late Jurassic and Early Cretaceous uplifts partially or totally denuded some areas and fault blocks of Triassic sediments.

Consequently, a complete Main Buntsandstein section is preserved only in the central and western parts of the Broad Fourteens and West Netherland Basins (Fig. 3).

An east-west line, which broadly corresponds to the southern limit of underlying Zechstein evaporites, called the 'zero salt line' in this study, subdivides the Main Buntsandstein into two stratigraphic provinces (Fig. 3). North of this line, in the basin centre, there are two persistent clay-siltstone beds which allow the Main Buntsandstein to be subdivided into five discrete members: the Volpriehausen Sandstone, the Volpriehausen Clay-siltstone, the Detfurth Sandstone, the Detfurth

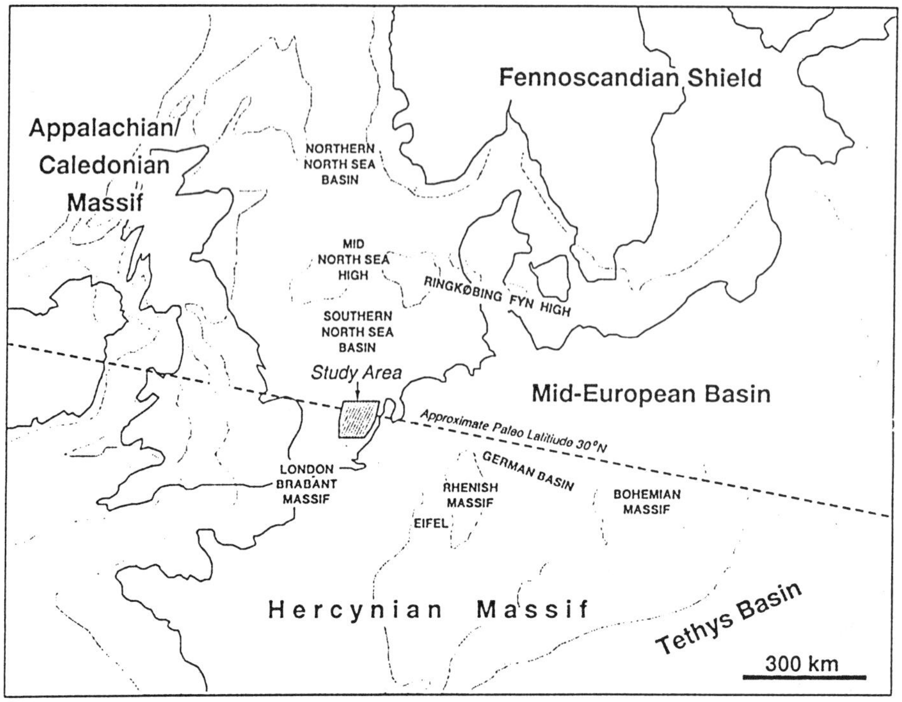

Fig. 1. Paleogeographic map of the Early Triassic of Europe (after Mader 1985b). The broken line over the study area represents the approximate 30° N palaeolatitude.

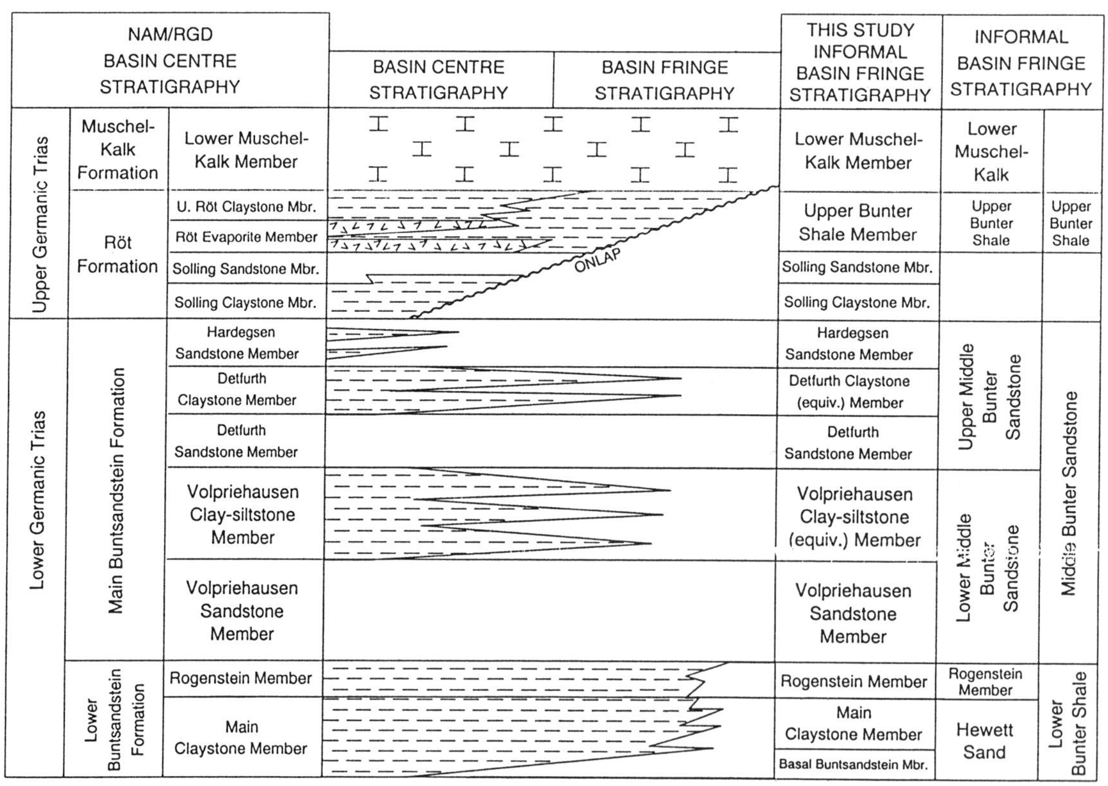

Fig. 2. Stratigraphic chart for the Lower and Upper Germanic Trias Group, P and Q blocks offshore Netherlands.

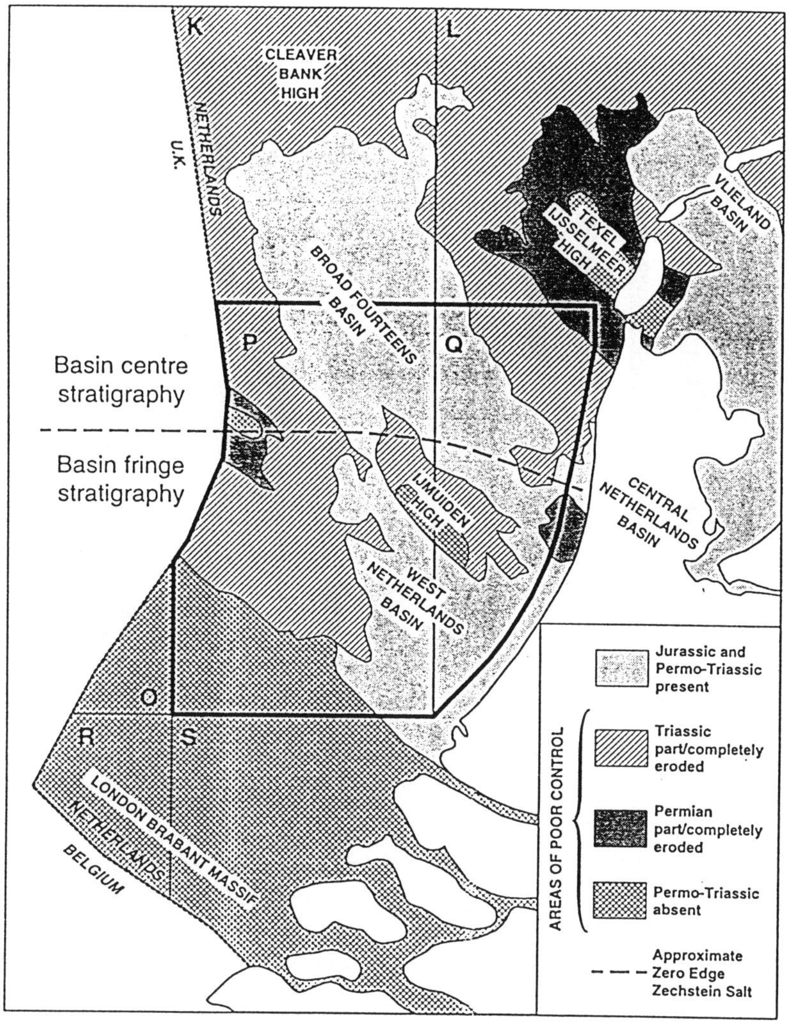

Fig. 3. Structural elements of the southern Dutch North Sea and distribution of pre-Cretaceous sediments.

Claystone and the Hardegsen Sandstone (Fig. 2). South of the 'zero salt line', in the basin fringe, siltstone and claystone beds grade into sandstones and lithostratigraphic subdivision of the Main Buntsandstein is more difficult. However, subtle regionally persistent log characteristics have been correlated from the basin centre into the basin fringe province. This permitted an informal subdivision of the previously undifferentiated Bunter Sandstone Formation into the Volpriehausen Sandstone, Volpriehausen Clay-siltstone equivalent, Detfurth Sandstone, Detfurth Claystone equivalent, and Hardegsen Sandstone Members, all of which are lateral equivalents of the members recognised in the basin centre.

Previous studies

The 'Stratigraphic Nomenclature of the Netherlands' (NAM & RGD 1980) provided the framework for evaluating and defining the stratigraphy of the Main Buntsandstein Formation. However, at the time the study was being completed, new well control in the basin fringe was changing our understanding. Consequently, an informal modification of the existing stratigraphic nomenclature was developed. A formal, revised stratigraphy is currently being published (Van Adrichem Boogaert & Kouwe, in press).

There is little published work on the Main Buntsandstein in the P and Q quadrants. However,

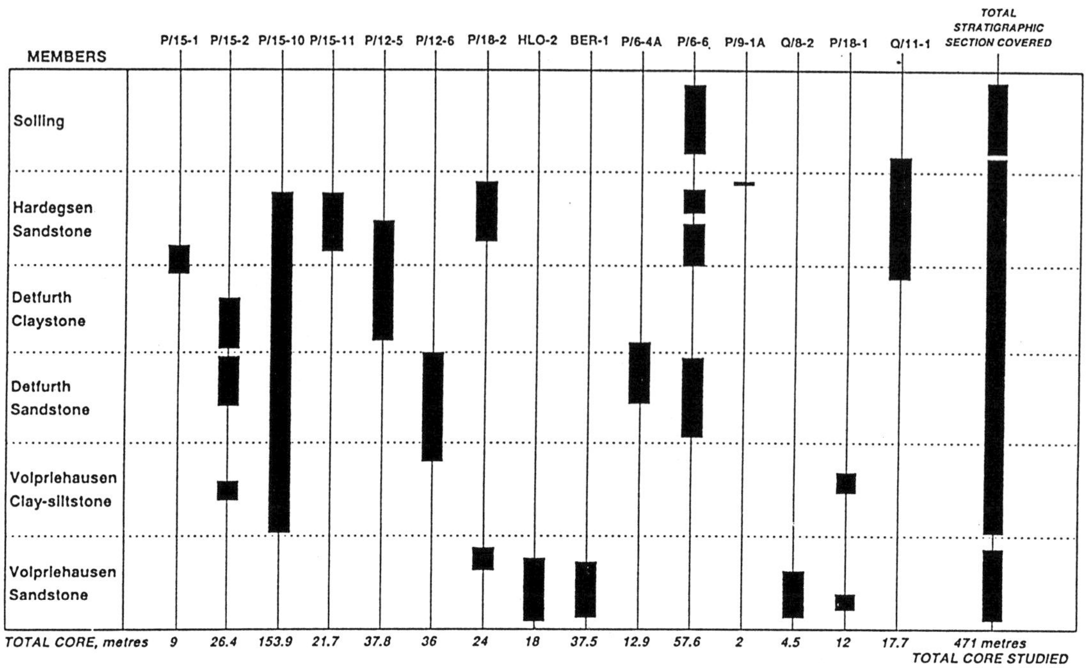

Fig. 4. Stratigraphic distribution of cores in the Main Buntsandstein Formation used in this study. For location see Fig. 5.

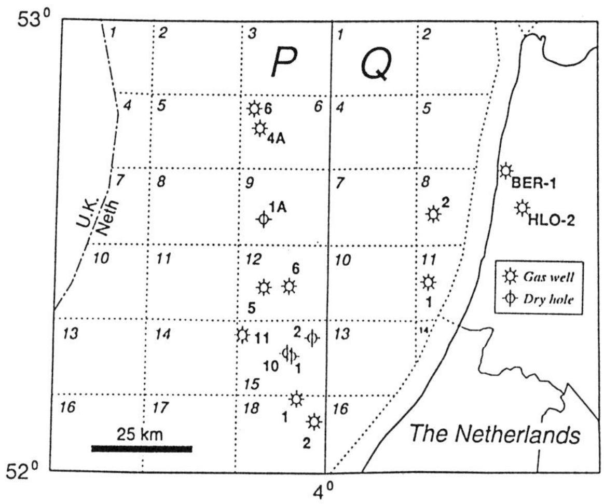

Fig. 5. Distribution of wells with core in the Main Buntsandstein Formation (Fig. 4) used in this study.

the Buntsandstein is well studied and documented in the German Basin of northwest Europe (Gall et al. 1977, Mader 1981, 1982, 1983a,b, 1985a,b). In the Eifel region it is usually described as braided-fluvial and aeolian sandstones or clays derived from high ground further south and east. During the deposition of the early Middle Buntsandstein (different nomenclature), the provenance regions were subjected to high and frequent rainfall. Later, with increasing aridity, stream discharge declined and aeolian dunes evolved.

The Buntsandstein Formation in the Eifel region is assumed to be correlative and comparable with the Main Buntsandstein Formation in the southern parts of the P and Q quadrants. Based on our interpretations, both areas were affected by similar changes in climatic conditions; fluvial sedimentation of the lower Main Buntsandstein (Volpriehausen) was superseded by predominantly aeolian deposition of the middle Main Buntsandstein (Detfurth and Hardegsen Sandstones).

Data base and methods of analysis

A total of 471 m of core from the Main Buntsandstein Formation and the overlying Solling Member of the Röt Formation in 15 wells were analysed in this study (Fig. 4). Available core coverage was adequate to ensure that:

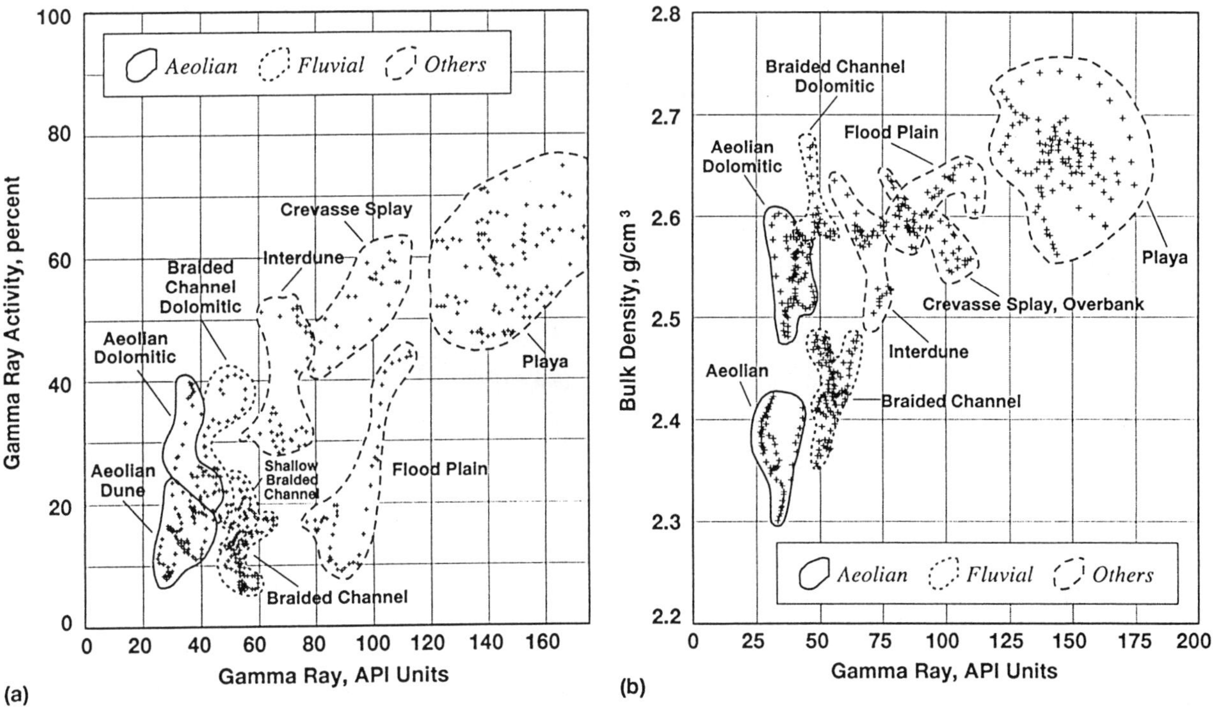

Fig. 6. Discrimination matrices: gamma ray versus gamma ray activity, and gamma ray versus bulk density.

CORE DEFINED: ENVIRONMENTS OF DEPOSITION	HORIZON PREDICTED: Aeolian	Aeolian Dolomitic	Braided Channel	Braided Channel Dolomitic	Shallow Braided Channel	Interdune	Flood Plain	Crevasse Splay	Playa Lake	Total %
Aeolian	*99*	0	1	0	0	0	0	0	0	100
Aeolian Dolomitic	0	*89*	2	8	1	0	0	0	0	100
Braided Channel	0	3	*79*	0	18	0	0	0	0	100
Braided Channel Dolomitic	0	11	0	*89*	0	0	0	0	0	100
Shallow Braided Channel	0	2	42	0	*56*	0	0	0	0	100
Interdune	0	0	0	0	0	*99*	0	1	0	100
Flood Plain	0	0	0	0	0	0	*100*	0	0	100
Crevasse Splay	0	0	0	0	0	1	0	*99*	0	100
Playa Lake	0	0	0	0	0	0	0	0	*100*	100

Aeolian *Fluvial* *Others*

Fig. 7. Matrix of relative frequency with which the environments predicted by the HORIZON log interpretation module coincided with those interpreted from core.

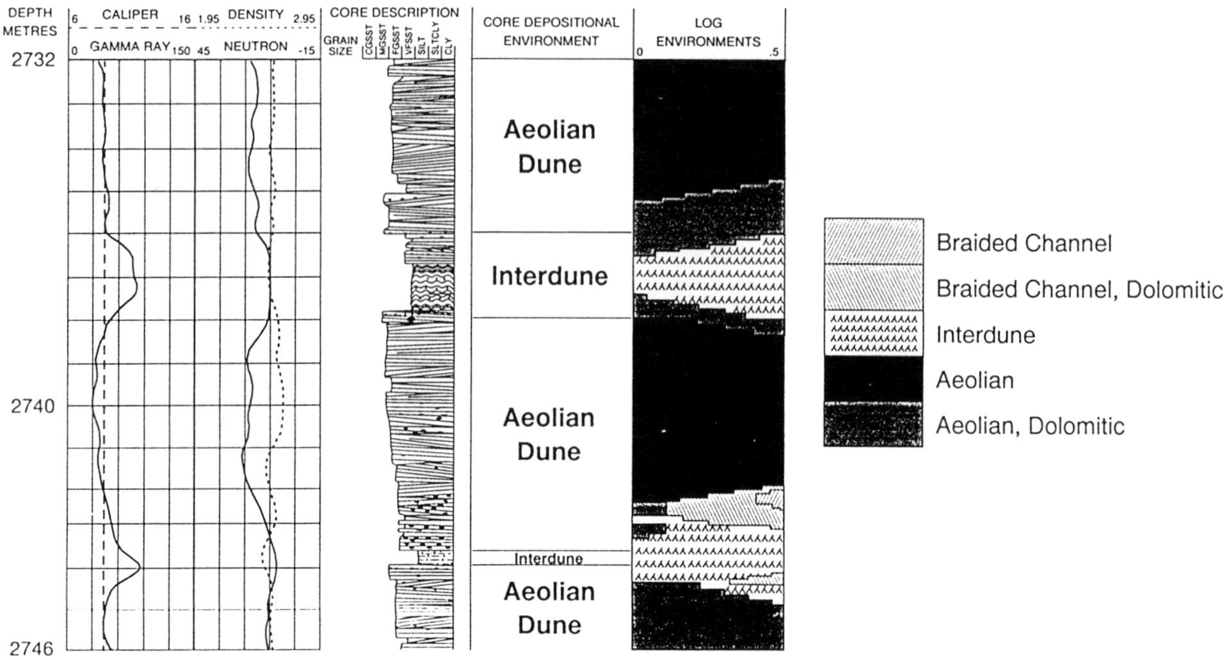

Fig. 9. 'Lithplot' showing a comparison of the core-interpreted and log-interpreted environments in the P/15–10 well.

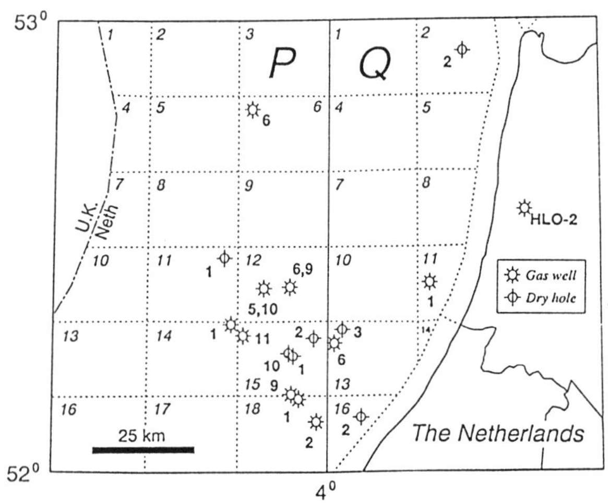

Fig. 8. Distribution of wells analysed using the HORIZON log interpretation module.

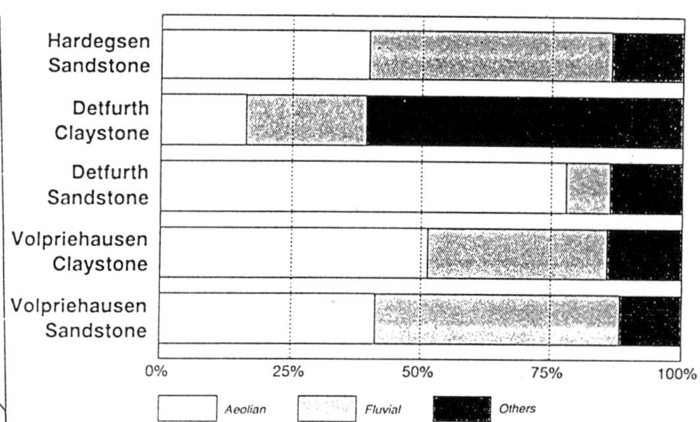

Fig. 10. Relative abundance of each of the categories of environment of deposition within the members of the P/15–10 well.

a) all facies types were represented;

b) the intervals provided coverage of almost the entire formation (Fig. 4); and

c) there was a good geographic spread in the data base (Fig. 5).

To achieve a consistent data set, all the cores were logged and interpreted by Reservoirs Inc. The data collected included analyses of lithology, mineralogy, texture, sedimentary structure, grain-size, porosity, permeability and the interpreted environments of deposition. A total of seven environments of deposition

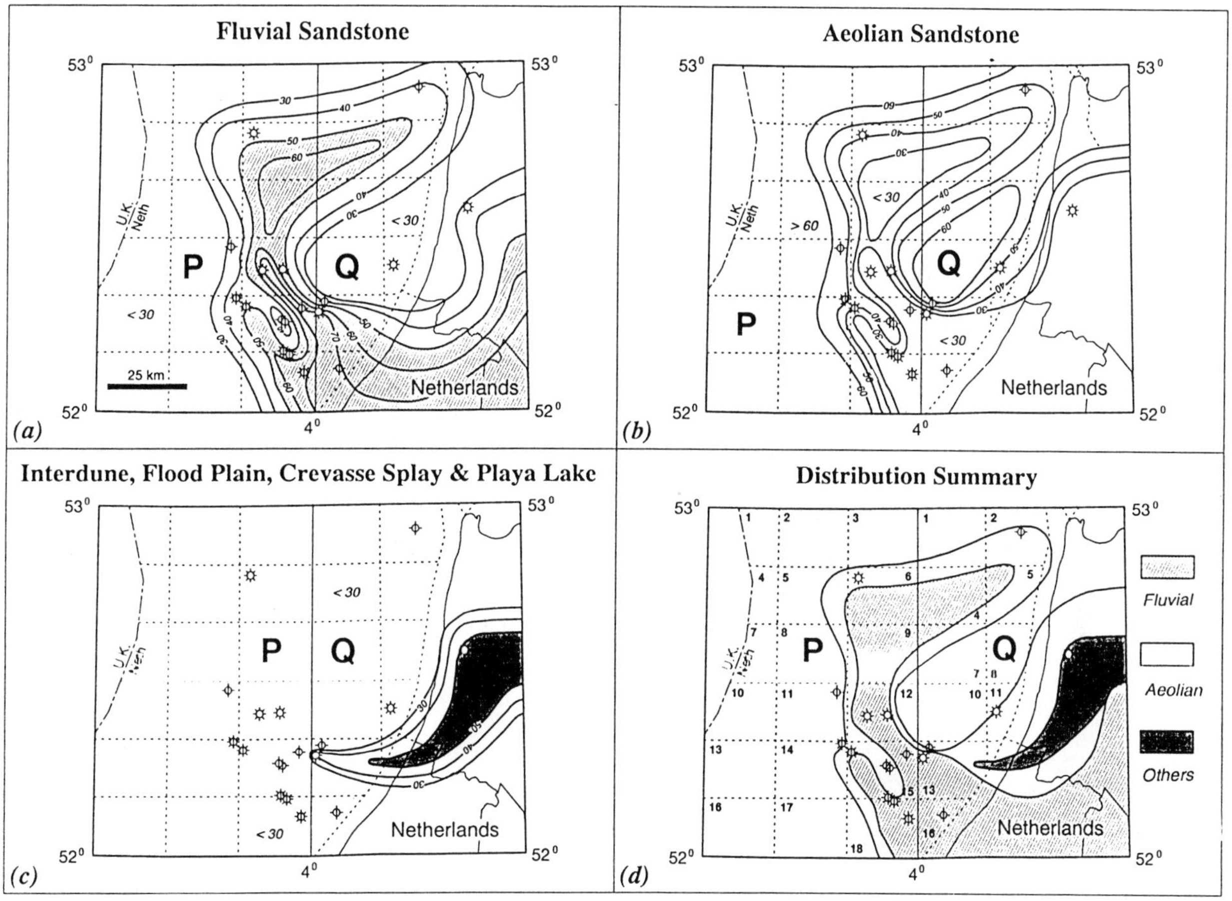

Fig. 11. Distribution of environments of deposition – Volpriehausen Sandstone Member. The distribution map (d) indicates those facies that occupy more than 50% of the rock unit.

were identified: aeolian, interdune, braided-channel, shallow-braided channel, flood-plain, crevasse-splay and playa.

Aeolian sandstones in the Main Buntsandstein Formation are usually in stacked sets which rarely exceed 5 m in thickness, and dune sets are either rare and low-relief, or absent. Consequently, the aeolian sands are interpreted as predominantly sheet-sands. Fluvial sandstones are organised in thick packages of repeated shallow (1–2 m) channels and are most frequently interpreted as braided-stream deposits. Clay clasts are commonly found at the bases of both aeolian and braided-stream deposits, either as deflation or channel lags. Interdune, flood-plain and crevasse-splay deposits are generally inter-bedded fine-grained sandstones, laminated silts and red or green shales, sometimes capped by well-developed palaeosoils. Playa-lake sediments are predominantly shales. Dolomitic cements are present to variable extents in all the environments of deposition.

The Main Buntsandstein Formation is only partially cored in some of the wells and consequently provided an insufficient database for regional mapping purposes. The HORIZON interpretation module from the Western Atlas 'Well Data System' (WDS) was selected to expand the data base by calibrating the environments of deposition interpreted from core, through logged intervals that were not cored. Most log analysis programs apply pre-established formulas to determine petrophysical properties from log responses. HORIZON takes a different approach. Descriptive features such as facies and environment of deposition are determined by performing non-parametric statistical estimations, in this case, utilising logs over the cored intervals in which the relevant rock properties (e.g. environments) are reliably known (Tatzlaff et al. 1989). By a pro-

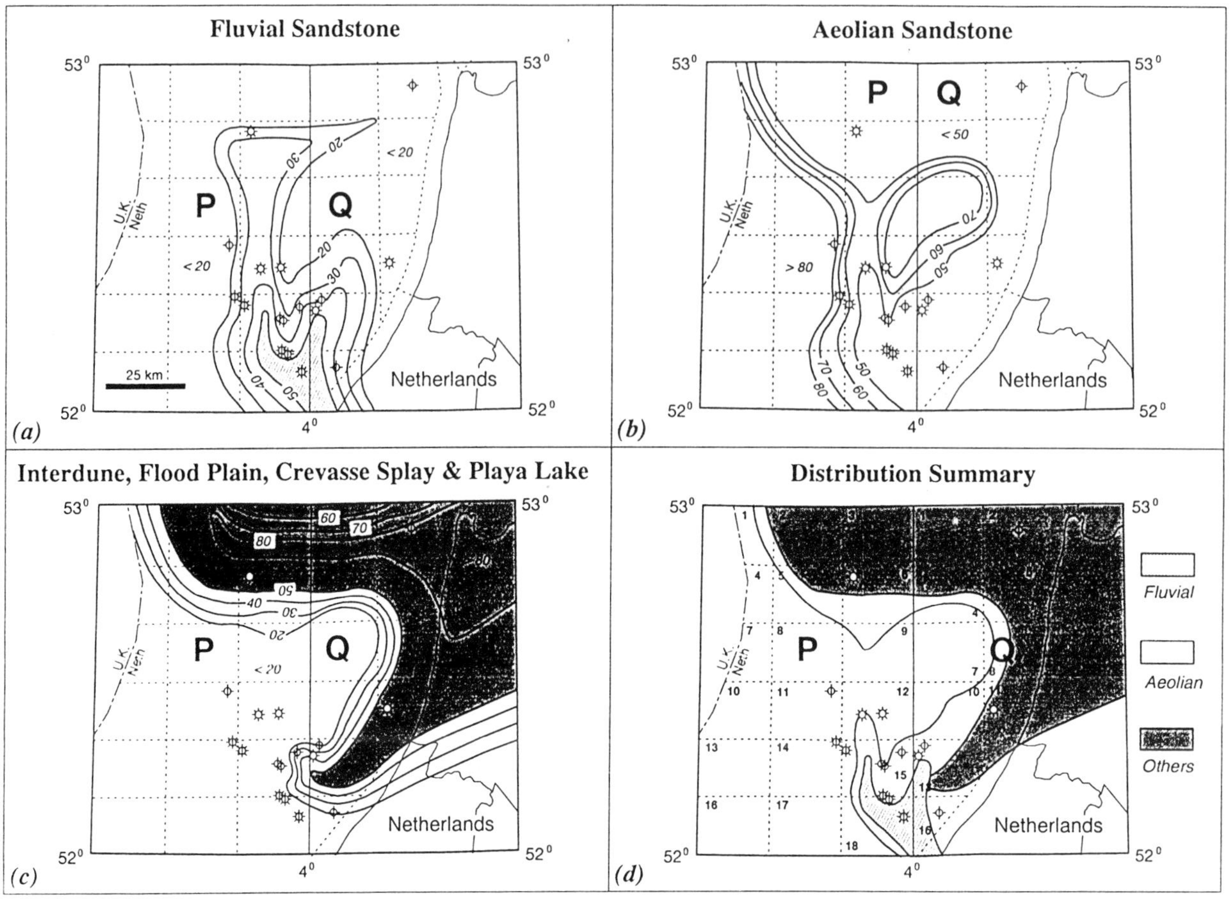

Fig. 12. Distribution of environments of deposition – Volpriehausen Clay-siltstone Member. The distribution map (d) indicates those facies that occupy more than 50% of the rock unit.

cess of elimination, the combination of log data which most consistently discriminated between environments and/or facies were the gamma ray, bulk density and a derivative of the gamma ray curve called the gamma ray activity. Gamma ray activity is a synthetic curve derived from the gamma ray curve, generated by the shapes program of the HORIZON module.

A cross-plot of gamma ray activity (%) versus gamma ray units illustrates the relationship between core-interpreted environments and their log facies (Fig. 6a). As expected, sandstone-rich, aeolian and fluvial deposits have lower gamma ray values and are less gamma ray active than the mud-prone and interbedded, interdune, crevasse-splay, flood-plain and playa sediments. Discrimination between the environments in the muddier facies is good, each cross-plotting as a relatively discrete population, with only slight overlap between crevasse-splay and interdune deposits, whereas discrimination between the sandstones is less certain. A cross-plot of bulk density (g/cm^3) versus gamma ray shows a clear distinction between the high-density dolomitic or muddier facies and the lower-density, low API, non-dolomitic sandstones (Fig. 6b). There is a clear distinction between the non-dolomitic, aeolian and braided-channel sandstones in this cross-plot. Some uncertainty remains in the distinction between the dolomitic braided-channels and the aeolian sediments. However, the dolomitic braided-channel sandstones are in general, denser and more radioactive than the dolomitic aeolian sandstones. This uncertainity is tabulated in Fig. 7, which is a matrix that cross-plots the relative frequency with which the environments interpreted from core, were consistent with the environments predicted by the discriminators and the HORIZON module. The consistency between the interpreted core and HORIZON

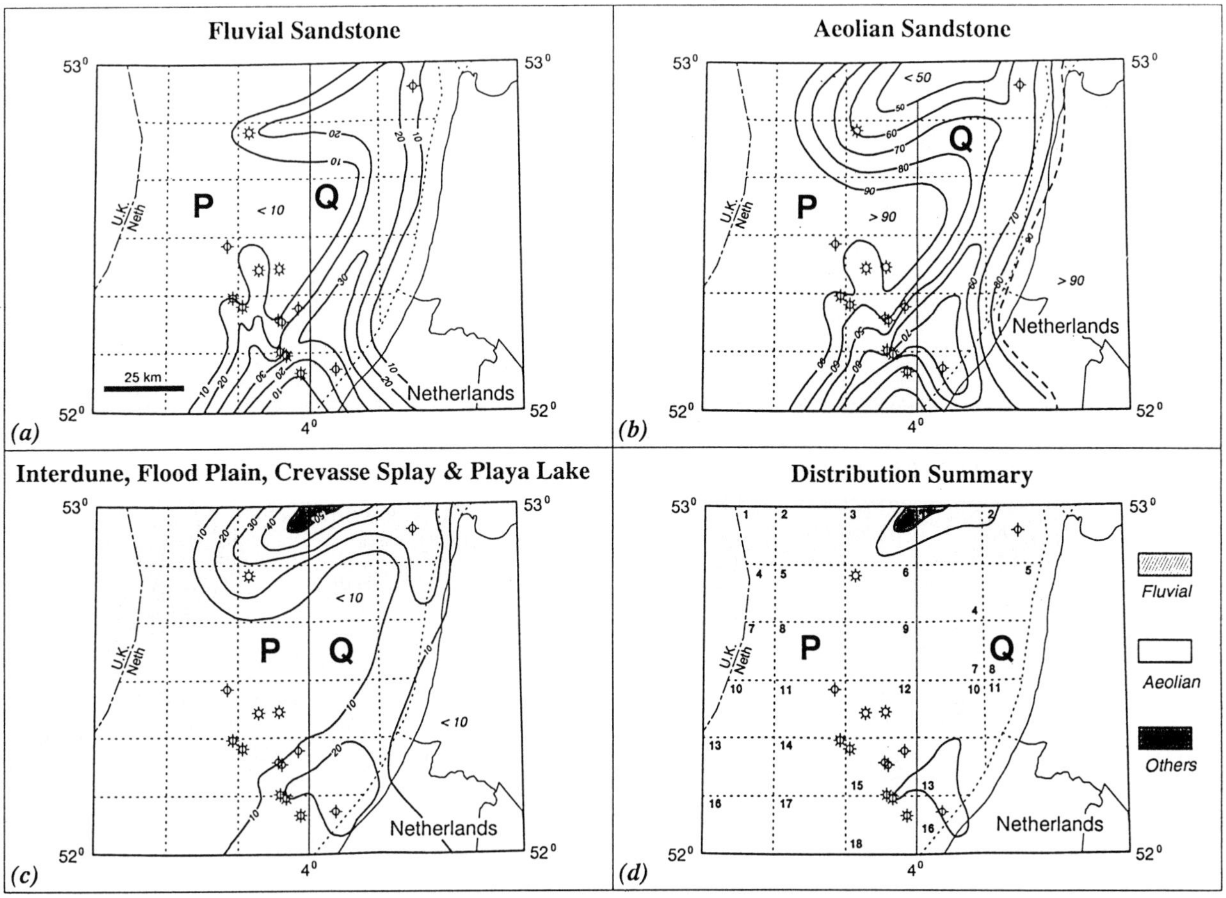

Fig. 13. Distribution of environments of deposition – Detfurth Sandstone Member. The distribution map (d) indicates those facies that occupy more than 50% of the rock unit.

environments in non-dolomitic aeolian sandstones is high (99%), while the aeolian dolomitic sandstones and fluvial sandstones were misinterpreted more frequently. Furthermore, the discriminators were poor at distinguishing between braided-channels and shallow-braided-channels (42 versus 56%), making this subdivision invalid for mapping purposes. There is good consistency between core and HORIZON interpretations in muddier facies, with only insignificant confusion between interdune and crevasse-splay sediments. Once a satisfactory discrimination matrix was developed, HORIZON was used to analyse the log data and to interpret environments of deposition across the entire Main Buntsandstein Formation in 18 wells (Fig. 8).

The reasons for the good correlation between the core and the HORIZON-defined environments are difficult to quantify. The log responses indicate a subtle difference between aeolian sandstones and fluvial sandstones which contain more radiogenic minerals (e.g. clay and/or mica). These criteria are different from those used to distinguish between the environments of deposition in the cores (e.g. sedimentary structures, and textures). The coincidence in the correlation between the log and core-defined environments is attributable to aeolian winnowing processes, which have removed the radiogenic constituents from the original fluvial deposits and produced a less radiogenic, aeolian sandstone deposit. The results of the log analysis were plotted in a log format ('Lithplot'), with each environment or facies identified by a symbol (Fig. 9). Nine rock types were discriminated using this method. To simplify the results for mapping purposes, these nine rock types were grouped into three categories: aeolian (including 'aeolian-dolomitic'), fluvial

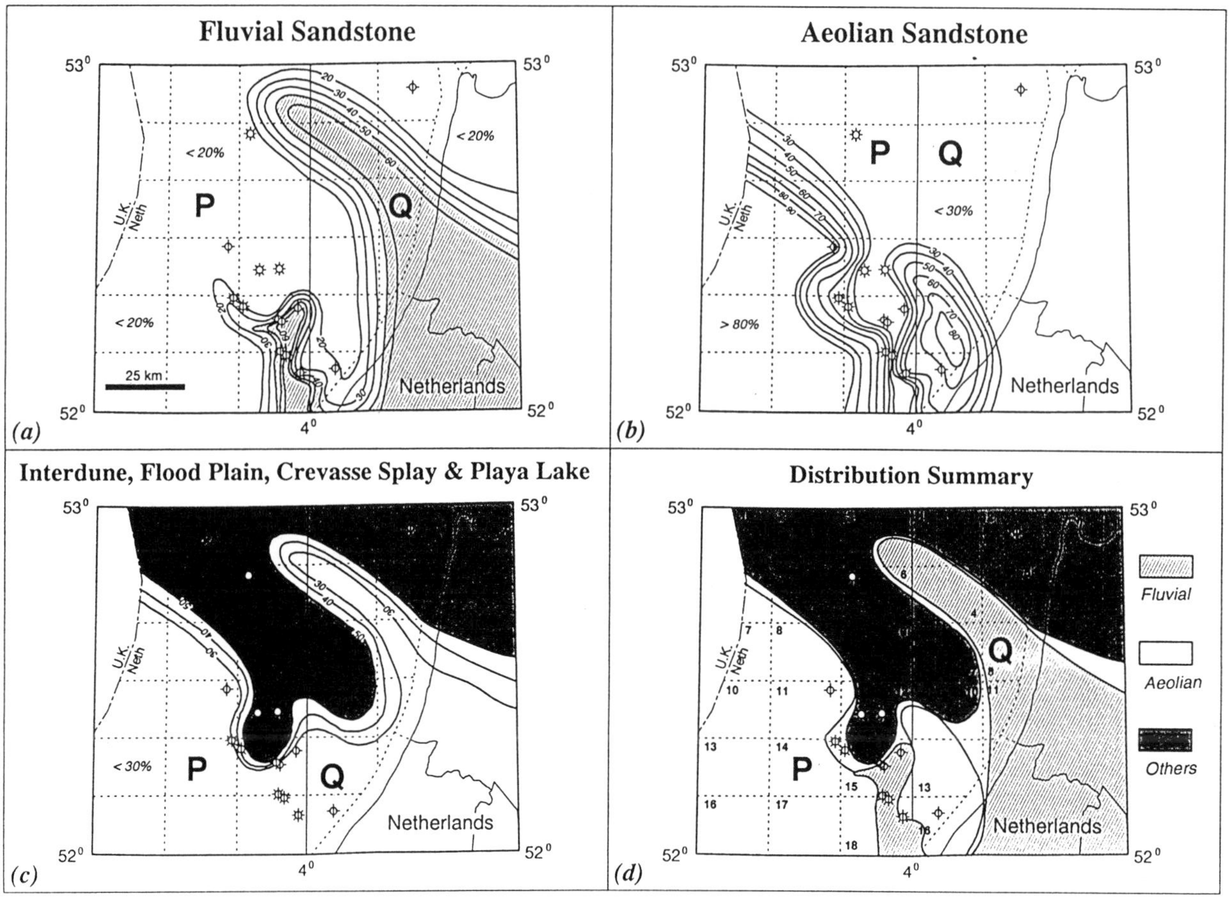

Fig. 14. Distribution of environments of deposition – Detfurth Claystone Member. The distribution map (d) indicates those facies that occupy more than 50% of the rock unit.

(incorporating 'braided-channels', 'shallow-braided-channels' and 'dolomitic-braided-channels'), and other (including 'interdune', 'flood-plain', 'crevasse-splay' and 'playa' environments). The relative abundance of each category was calculated as a percentage of the gross interval thickness for each of the five Main Buntsandstein Members in all wells (e.g. Fig. 10), and contoured between the 18-well data base.

Results

The relative abundance of the three categories of environment of deposition for each member of the Main Buntsandstein Formation is illustrated in Figs 11–15. A distribution summary map illustrates where a particular environment is most abundant (i.e. $> 50\%$), and unshaded patches represent areas where no environment is dominant.

The suite of maps shows a reasonable and consistent arrangement of environments through time. Generally, playa-deposits were preserved in the central, northern parts of the study area and were fringed by a swath of aeolian sediments, transected by narrow tracts of fluvially dominated sediments. Through time, the width and location of each environmental tract changed slightly, probably in response to changes in aridity or wind velocity. The aeolian tract reached its maximum extent during deposition of the Detfurth Sandstone Member which restricted playa-deposits to the northernmost part of the study area (Fig. 13). Playa deposits were most extensive during deposition of the Volpriehausen Clay-siltstone and Detfurth Claystone Members (Figs 12, 14). The Hardegsen Member shows only a narrow tract of fluvial environments, flanked by extensive aeolian deposits (Fig. 15).

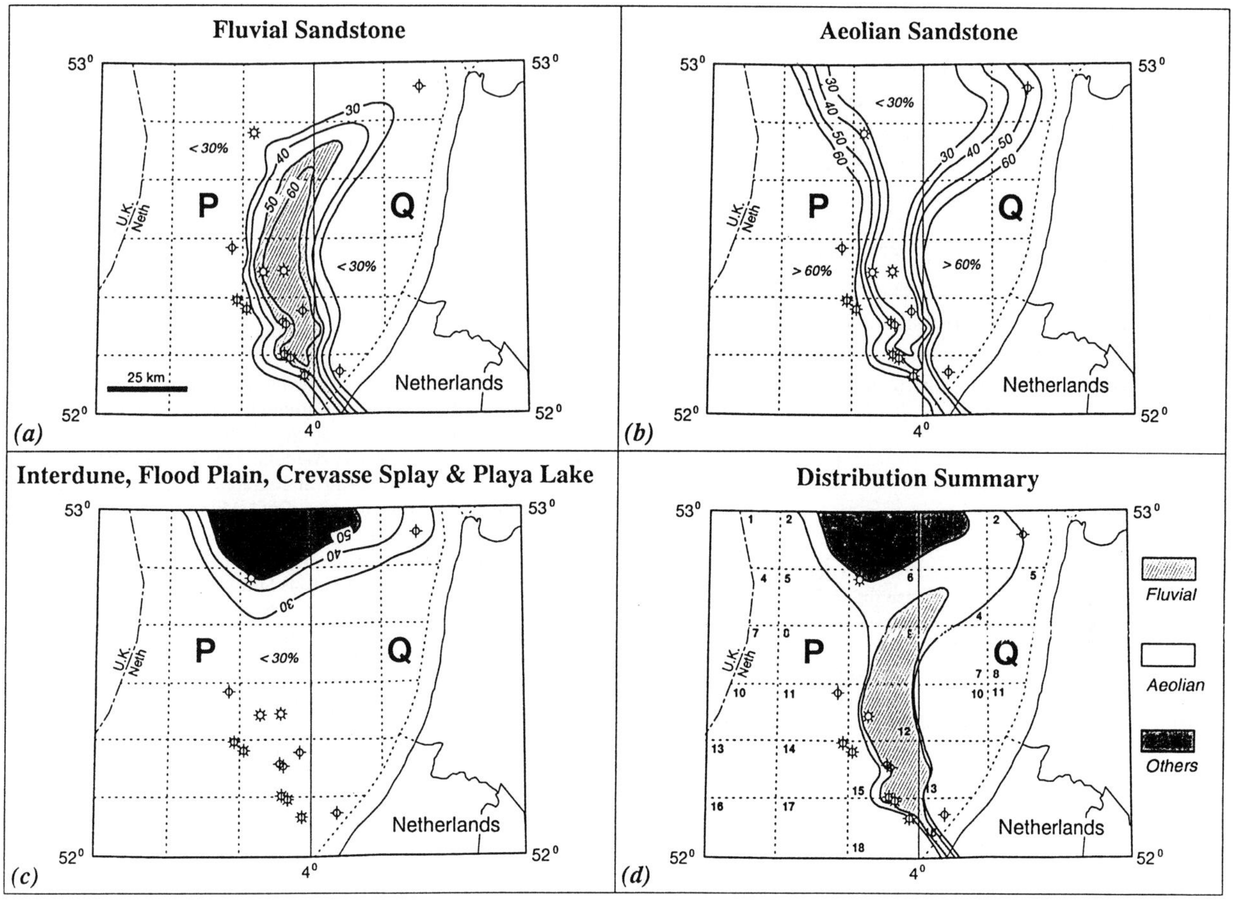

Fig. 15. Distribution of environments of deposition – Hardegsen Sandstone Member. The distribution map (d) indicates those facies that occupy more than 50% of the rock unit.

The contours on the maps are generally confined to areas of good well control but shading has been extended to fill the study area. The control at younger stratigraphic levels decreases because some members of the Main Bundsandstein are locally absent, eroded at the unconformity at the base of the Upper Germanic Trias Group.

Conclusions

Minor variations in the log facies (gamma ray activity, density and total radio activity) of the Main Buntsandstein have been correlated with environments of deposition interpreted from cores and used to interpret sections where no core exists. These analyses produce quantified results suitable for generating maps which have been used to predict facies where no well control exists.

The major lithological variations in the Main Buntsandstein Formation, north of the 'zero salt line', were caused by expansion and contraction of a mud-dominated playa-lake. Further south, in the 'basin fringe', the differences in the log facies are more subtle and were caused by changes in the mineralogy in the predominantly sandstone beds. These changes are, at least in part, caused by fluctuations between environments of deposition which coincided with advances and retreats of the playa lake.

Acknowledgements

The authors thank Mobil Producing Netherlands Inc, Dyas B.V., Veba Oil Nederland BV, Oranje Nassau Energie Participatie BV, Oranje Nassau Energie Participatie Mij, Petron Exploratie BV, Clyde Petroleum (North Sea) Ltd, DSM Energie BV, Energie Beheer Nederland BV, and Amoco Netherlands Petroleum Co. for their permission to publish this study. We are indebted to R. Miller of Reservoirs Inc and D. Beaty of Western Atlas International for their contributions to this project. We also thank J. Williams, J. Klasen and the editors for their helpful suggestions with the text.

References

Gall, J.C., M. Durand & E. Muller 1977 Le Trias de part et d'autre du Rhin. Corrélations entre les marges et le centre du bassin germanique – Bulletin du BRGM (deuxième série) Section IV, 3: 193–204

Mader, D. 1981 Genesis of the Buntsandstein (Lower Triassic) in the Western Eifel (Germany) – Sediment. Geol. 29: 1–30

Mader, D. 1982 Aeolian sands in continental red beds of the Middle Buntsandstein (Lower Triassic) at the western margin of the German Basin – Sediment. Geol. 31: 191–230

Mader, D. 1983a Aeolian sands terminating an evolution of fluvial depositional environment in Middle Buntsandstein (Lower Triassic) of the Eifel, Federal Republic Germany, In: Brookfields, M.E. & T. S. Ahlbrandt (eds) Eolian sediments and processes. Developments in Sedimentology 38 – Elsevier, Amsterdam: 583–612

Mader, D. 1983b Evolution of fluvial sedimentation in the Buntsandstein (Lower Triassic) of the Eifel (Germany) – Sediment. Geol. 37: 1–84

Mader, D. 1985a Fluvial conglomerates in continental red beds of the Buntsandstein (Lower Triassic) in the Eifel (F.R.G.) and their palaeoenviromental, palaeogeographical and palaeotectonic significance – Sediment. Geol. 44: 1–64

Mader, D. 1985b Aspects of fluvial sedimentation in the Lower Triassic Buntsandstein of Europe. In: Mader, D. (ed.) Lecture notes in earth sciences 4 – Springer-Verlag, Berlin: 1–12

NAM & RGD (Nederlandse Aardolie Maatschappij BV & Rijks Geologische Dienst) 1980 Stratigraphic nomenclature of the Netherlands – Verh. Kon. Ned. Geol. Mijnb. Gen. 32: 5–77

Tatzlaff, D.M., E. Rodriguez & R.L. Anderson 1989 Estimating facies and petrophysical parameters from integrated well data. In: Transactions of the Log Analysis Software Evaluation & Review (LASER) Symposium, London, Dec. 13–15, 1989 – Soc. Professional Well Log Analysts, London.

Van Adrichem Boogaert, H.A. & W.F.P. Kouwe, in press. Stratigraphic nomenclature of the Netherlands, revision and update by RGD and NOGEPA – Med. Rijks Geol. Dienst 50, Section E

Rondeel et al. (eds), Geology of gas and oil under the Netherlands, 179–189, 1996.

Inversion of reservoir quality by early diagenesis: an example from the Triassic Buntsandstein, offshore the Netherlands

K. Purvis[1] & J.A. Okkerman[2]
Shell Research BV, Postbus 60, 2280 AB Rijswijk, the Netherlands; [1] *Present address: Petroleum Development Oman, P.O. Box 81, Muscat, Sultanate of Oman;* [2] *Present address: Shell Petroleum Development Company, Port Harcourt, Nigeria*

Key words: anhydrite, dolomite, halite, isotopes, sandstone, porosity reduction

Abstract

Sandstones of the Triassic Main Buntsandstein form a major gas reservoir in the Netherlands offshore. The sequence is dominated by siliciclastics deposited in a semi-arid continental setting, and includes dune, interdune, sheetsand and fluvial sandstones. Reduction in reservoir quality is caused primarily by dolomite, halite and anhydrite cementation, with minor authigenic illite and chlorite. Integration of petrographic and isotopic data has allowed the origins and relative timing of the different cements to be constrained. The carbon and oxygen isotopic composition of dolomite ($\delta^{13}C = -3$ to +2.9‰ PDB, $\delta^{18}O = -3.7$ to -9.3‰ PDB) combined with strontium isotopic data (0.7091 to 0.7109 $^{86}Sr/^{87}Sr$) suggests that it precipitated from meteoric groundwater. Halite and anhydrite formed from a mixture of meteoric water and saline fluids expelled from underlying evaporites and claystones. Sulphur isotopic data (δ^{34} S = +4.2 to +12.1‰ CDT) support this interpretation for the origin of the anhydrite. Precipitation of the major authigenic minerals occurred during early diagenesis, prior to burial depths of 500 m. Cementation and groundwater flow preferentially followed the zones of highest permeability and caused an inversion of reservoir quality. Sandstones with the highest depositional porosity and permeability (i.e. dune sandstones) are the most cemented, and have poorer reservoir quality than the fluvial and interdune sandstones which originally had lower depositional porosity and permeability. Formation of authigenic illite and chlorite occurred during burial and has significantly reduced permeability further.

Introduction

The Lower Triassic Buntsandstein gas play is a major exploration objective in the Netherlands with proven gas reserves of about 65×10^9 m^3 in the West Netherlands Basin alone (Vreeken & Kong 1993), and a number of Triassic fields are currently coming on stream in the Dutch offshore (Fontaine et al. 1993). As the Main Buntsandstein is very uniform in thickness with individual members that can easily be recognised (Roos & Smits 1982), and Buntsandstein structures are relatively easily defined by 3-D seismic, one of the most important exploration risks is reservoir quality. Consequently, the present study was undertaken in order to identify the main geological factors controlling the distribution of porosity and permeability in the Main Buntsandstein reservoirs.

Geological background

The study area is situated in the south-eastern North Sea, northern Dutch sector, and covers parts of blocks F17, F18, L2, L5, L6 and L9 (Fig. 1). The area forms part of the southern North Sea Basin and lies at the southern end of the Central Graben system. The axis of the basin is aligned SW – NE and the structure of the basin is influenced to a large extent by the halokinesis of the thick Zechstein salt sequence.

The Lower Trias in the southern North Sea Basin consists of two units; the Lower Buntsand-

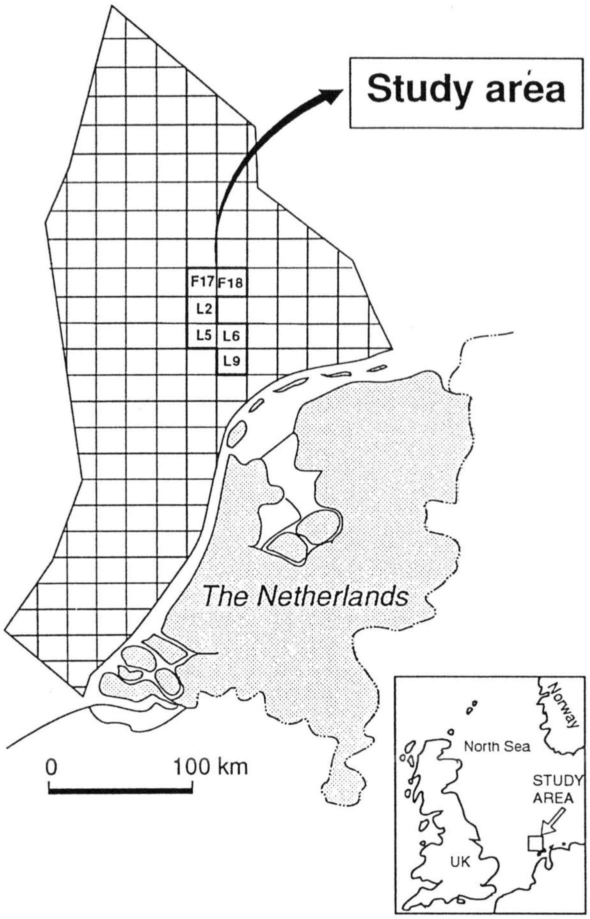

Fig. 1. Map showing location of study area, southern North Sea.

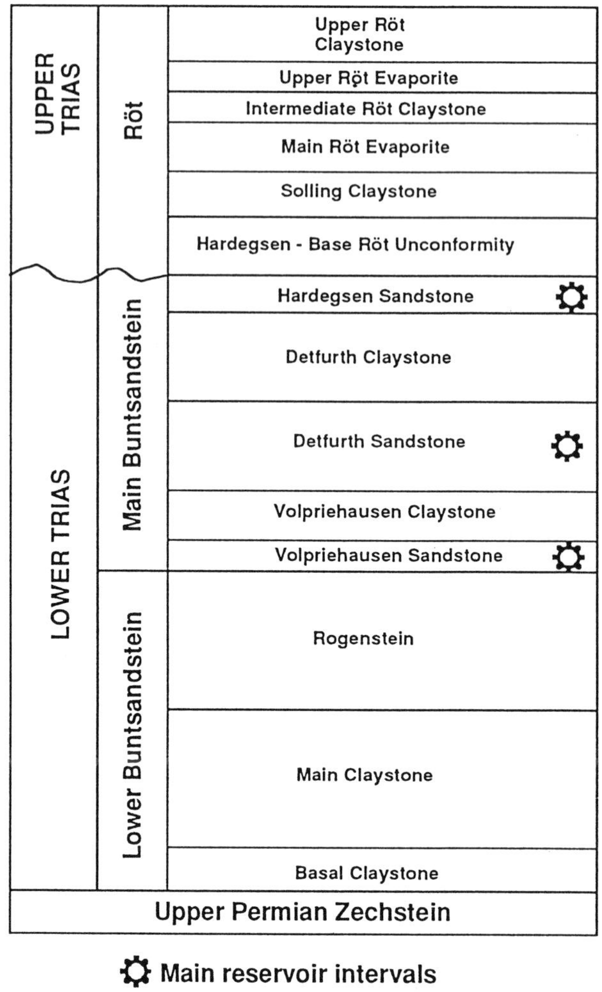

Fig. 2. Lithostratigraphic subdivision of the Germanic Trias in the Netherlands offshore. Modified from NAM & RGD (1980).

stein Formation and the Main Buntsandstein Formation (Fig. 2). The standard lithostratigraphic subdivision of these formations used in this paper follows the subdivision proposed by NAM & RGD (1980).

The Lower Buntsandstein Formation consists mainly of red, anhydritic silty claystones. Three members are distinguished; the Basal Buntsandstein Member which directly overlies the deposits of the last Zechstein cycle, the Main Claystone Member, and the Rogenstein Member. Only the Rogenstein Member, which consists of red-brown and green anhydritic silty claystones, has been recovered in cores from the wells included in this study.

The Main Buntsandstein Formation is characterised by the presence of well-correlatable sandstone units separated by non-reservoir intervals consisting of shaly sediments (Roos & Smits 1982). Three reservoir intervals are recognised, from the base to the top, the Volpriehausen Sandstone Member, the Detfurth Sandstone Member and the Hardegsen Sandstone Member (Fig. 2).

The rocks present in the Lower and Main Buntsandstein Formations of the study area were deposited in an arid to semi-arid, continental environment. Sedimentation was climatically controlled, with the Armorican Massif, Massif Central and London-Brabant Massif to the south and south-west, and the Rhenish Massif to the east being the main source areas (Ziegler 1978, 1982). Whilst mid-Triassic climates were probably the most arid in the Earth's history (Frakes 1979), there is evidence that Pangean climates were monsoonal (Kutzbach & Gallimore 1989). This suggests that climatic conditions during deposition of the Buntsand-

Table 1. Major lithofacies in Triassic Main Buntsandstein reservoir intervals.

Lithofacies	Main features	Origin
Cross-bedded sandstone	Planar cross-bedded, well to very well sorted, medium- to fine-grained sandstones, with dips of up to 30°. Grains are well rounded and frosted	Small (2–3 m) aeolian dunes
Irregular horizontally laminated sandstone	Fine-grained, poorly to well sorted sandstones, showing poorly defined horizontal and irregular, undulatory lamination. Thin clay drapes are locally developed	Sandsheets formed by a combination of aeolian and fluvial sheetflood processes
Pin-striped sandstone	Well sorted and well stratified fine-grained sandstones with low angle cross-bedding. Erosive bases are overlain by asymptotic laminations, passing up into low angle cross-bedding	Fluvial deposits of unconfined sheet flow and/or minor channelised flow
Cross-laminated sandstone	Trough cross-laminated, fine-grained sandstones with erosive bases. Typically overlain by irregular horizontally laminated sandstones	Deposits of confined, channelised flow. Possibly related to fluvial bars
Non-stratified sandstone	Fine-grained, structureless, moderately to well sorted sandstones	Fluvial sandstones, homogeneous as a result of absence of grain size variation, or post-depositional homogenisation
Mudstone	Laminated, red-brown claystone with minor admixed or interlayered siltstone and fine-grained sandstone. Structures include siltstone dykes, asymmetrical ripples and local mudstone conglomerates	Shallow lake or ponds with frequent fluxes of fluvial sands and silts

stein were seasonal, with markedly arid and wetter periods.

The main lithofacies present in the sandier intervals of the Buntsandstein represent the deposits of a desert outwash plain characterised by sheetflood deposition of clastic material with minor aeolian and fluvial reworking. The overall setting is envisaged as a relatively flat plain with shallow depressions which were probably flooded with sheetflood-runoff during infrequent rainstorms similar to that proposed by Clemmensen (1979). Lakes may have been ephemeral during dry climatic periods and perennial during wetter climatic periods. The major lithofacies present in the Buntsandstein of the study area, and their probable mode of formation, are shown in Table 1.

Samples and analytical techniques

Cores were sampled from ten wells located in blocks F17, F18, L2, L5, L6 and L9 in the Dutch sector of the southern North Sea (Fig. 1). Porosity, permeability and grain density were measured by conventional core-plug analysis.

Resin-impregnated thin sections were examined using transmitted light microscopy and their modal compositions evaluated by point-counting 300 grains. Polished thin sections were also examined using cathodoluminesence (CL) using a Technosyn MK II cathodoluminoscope. Samples were studied by scanning electron microscopy (SEM) in secondary electron imaging mode, with mineral identification aided using Energy Dispersive X-Ray Analysis (EDX).

X-ray diffraction analysis (XRD) of powdered whole-rock samples and different size fractions of the clay grade was undertaken on samples that were air-dried, glycolated and heated to 375 °C for 30 min. Clay-mineral identification followed the method of Starkey et al. (1984).

Carbon, oxygen and strontium isotope analyses were carried out on fifteen samples of separated dolomite cements, and sulphur isotope analysis was undertaken on eleven samples of anhydrite (Tables 2, 3). Isotopic results are reported in the relevant notations; carbon and oxygen in PDB (Peedee Belemnite standard), oxygen in SMOW (Standard Mean Ocean Water standard), and sulphur in CDT (Canyon Diablo Troilite standard).

Table 2. Carbon, oxygen and $^{87}Sr/^{86}Sr$ isotopic composition of authigenic dolomite.

Well	Depth (m)	Sample	Formation	$\delta^{13}C$ (PDB)	$\delta^{18}O$ (PDB)	$^{87}Sr/^{86}Sr$
L2–2	3732.6	1	Detfurth	+0.71	-6.12	0.709096
	3786.5	5	Detfurth	-3.02	-7.45	
	3892.5	8	Volpriehausen	-2.88	-7.58	0.710327
	3892.0	9	Volpriehausen	-0.57	-5.88	
	3943.0	16	Volpriehausen	+2.97	-6.52	0.709581
L2–5	4179.6	104	Volpriehausen	+1.93	-6.56	0.710924
L2–6	4109.0	201	Detfurth	+1.11	-8.57	
	4124.6	204	Detfurth	-1.97	-5.82	0.710506
	4224.6	207	Volpriehausen	-0.87	-9.37	
	4236.6	210	Volpriehausen	+0.66	-8.17	0.710840
L9–6	3624.1	701	Volpriehausen	+2.05	-4.44	0.710100
	3628.95	709	Volpriehausen	+1.54	-4.50	0.711142
	3628.91	710	Volpriehausen	+2.08	-3.74	0.710788
	3643.05	729	Rogenstein	+1.11	-4.66	0.710317
	3654.05	732	Rogenstein			0.710013

Petrography and diagenesis

Detrital mineralogy

The sandstones are predominantly lithic-arkosic and arkosic arenites according to the classification scheme of Nagtegaal (1978), and are dominated by monocrystalline quartz grains, with subordinate polycrystalline quartz. Feldspar is common to abundant with orthoclase, albite and sanidine dominant, whilst plagioclase is rare and microcline and antiperthite are sparsely present.

Rock fragments are common and include quartz-mica schist, metamorphic quartz, granitic rock fragments and rare sandstone and shale fragments. Mica is rare and consists only of muscovite.

Detrital hematite and carbonate ooids are relatively rare in most samples, although sandy oolitic grainstones are occasionally found in the Detfurth Sandstone Member. The ooids are typically undeformed by compaction, 150–300 μm in diameter, and composed of dolomite and/or calcite.

Authigenic mineralogy

The authigenic mineralogy of the Main Buntsandstein sandstones is outlined below, and the interpreted paragenetic sequence is shown in Fig. 3. The isotopic compositions of the authigenic dolomite and anhydrite samples analysed are shown in Tables 2 and 3, respectively.

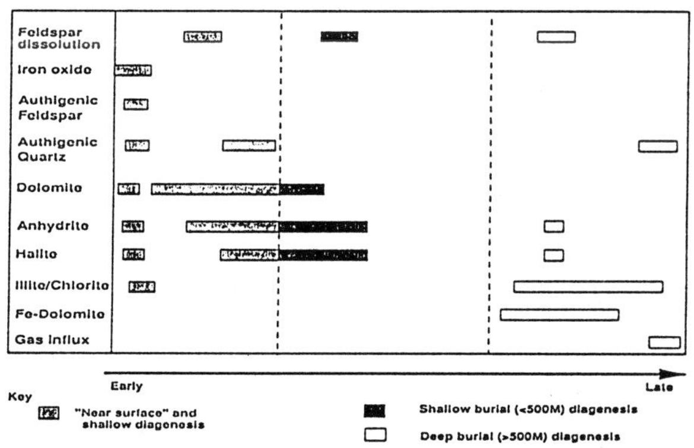

Fig. 3. General paragenetic sequence for sandstones of the Main Buntsandstein.

Halite: Halite occurs as a pore-filling and poikilotopic cement (Fig. 4a), ranging in abundance from 0 to 24.5%. Halite often encloses dolomite and quartz overgrowths and is clear to grey in colour. Halite cementation appears to have been texturally selective, generally occurring in the lithofacies with the best primary porosity and permeability characteristics. Consequently, it occludes most of the porosity in samples from dune sandstones, giving this lithofacies a high minus-cement porosity (Table 4).

Table 3. Sulphur isotopic composition of authigenic anhydrite.

Well	Depth (m)	Sample	Formation	δ^{34}S (CDT)
L2–2	3732.6	1	Detfurth	+12.1
	3892.0	9	Volpriehausen	+10.4
	3943.0	16	Volpriehausen	+11.6
L2–5	4177.94	103	Volpriehausen	+10.7
L2–6	4124.6	204	Detfurth	+10.6
	4224.6	207	Volpriehausen	+5.6
	4232.6	209	Volpriehausen	+4.2
L9–6	3625.3	703	Volpriehausen	+5.7
	3641.8	727	Volpriehausen	+11.8
	3643.05	729	Rogenstein	+11.8
	3654.05	732	Rogenstein	+10.9

Dolomite: Dolomite is the most pervasive diagenetic mineral occurring throughout the Buntsandstein. It occurs as rhombic crystals and as a pore-filling cement in amounts ranging from 0 to 30%. It is often enclosed by both anhydrite and halite (Fig. 4a,b). The distribution of the dolomite is partly fabric-selective, occurring predominantly in finer grained laminae (Fig. 4c). Under CL both rhombic and pore-filling dolomite luminesce a bright orange-yellow colour . Outer margins are sometimes non-luminescent, and typically ferroan when analysed by EDX.
Calcite: Calcite is relatively rare in the Buntsandstein, occurring only in samples of the Upper Detfurth, in amounts ranging from 13.7 to 18.6% as a pore-filling cement and as calcitic ooids.
Anhydrite: Anhydrite occurs as a pore-filling, often poikilotopic, cement (Fig. 4d), in amounts ranging from 0 to 34.7%. Like halite, anhydrite appears to have preferentially cemented those lithofacies with the best primary porosity and permeability characteristics, resulting in the occlusion of porosity in the dune sandstones. This is shown by the high minus-cement porosities for this lithofacies (Table 4). Anhydrite often encloses quartz overgrowths (Fig. 4e) and both encloses and partly replaces ooids. Whilst pore-filling anhydrite occurs as a major cement in the Buntsandstein sandstones, anhydrite nodules also occur in the claystones of the Rogenstein and in some fine-grained sandstones in the Volpriehausen Sandstone.
Iron oxide: Iron-oxide and clay coatings occur in the samples studied as red/brown to red coloured grain coating cements that give most of the cores their distinctive colour. The clay coatings are absent at grain point contacts and thicken on some grain surfaces.
K-feldspar: Detrital K-feldspars from the Volpriehausen Sandstone locally possess extensive overgrowths (0–4%). Detrital feldspars also exhibit evidence of dissolution, and are often enclosed by anhydrite (Fig. 4f).
Quartz: Authigenic quartz occurs in amounts ranging from 0 to 18%. It is present as small euhedral overgrowths on detrital grains (Fig. 4e) and as outgrowths (*sensu* McBride 1989) in primary pore space (Fig. 5a).
Illite: Illite occurs as detrital or grain-tangential clay, and in an authigenic platy, boxwork and fibrous form. Grain-tangential illite occurs as grain-coatings that are absent at grain contacts and show meniscus bridges. Authigenic illite occurs as a pore-lining clay with several distinct morphotypes, similar to the hierarchy proposed by Macchi (1987), ranging from grain-rimming clay with irregular upturned plates to plates and laths with fibrous projections that are pore-bridging (Fig. 5b). The fibrous illite is found intergrown with quartz outgrowths (Fig. 5a) and ferroan dolomite rhombs.
Chlorite: Authigenic chlorite predominantly occurs with a boxwork (or honeycomb) texture (Fig. 5c), and is ferroan when analysed by EDX on the SEM. Chlorite was also observed as an alteration product of muscovite in the mica flakes present in silt and fine-grained sandstone samples.

Origin of the major porosity-occluding cements

Dolomite

Dolomite occurs as an early cement, often being enclosed by anhydrite and halite, and is the dominant cement in the finer-grained laminae of most sandstones.

The bright yellow-orange luminescence of both forms of dolomite when seen under CL suggests that manganese is the activator (Nickel 1978). The similarity of the luminescence in the Rogenstein, Volpriehausen and Detfurth of all wells in the study area suggests that they formed under similar conditions of Eh and pH. The rarity of non-luminescent ferroan dolomite suggests that most of the dolomite formed under relatively oxidizing conditions and suggests an early origin.

The oxygen isotopic composition of the authigenic dolomite is light (^{18}O-depleted) compared to a marine

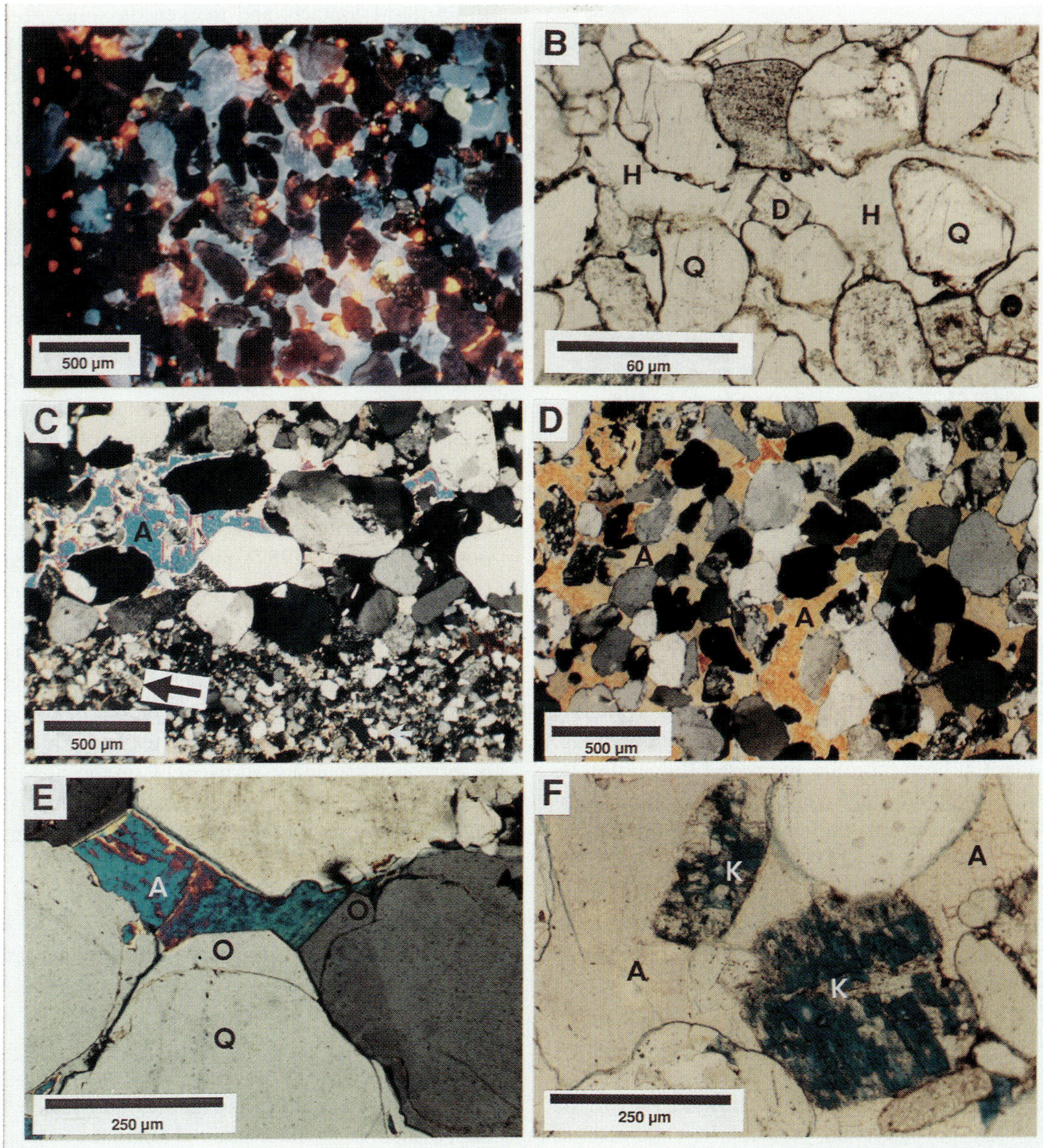

Fig. 4. Thin-section photomicrographs showing the main authigenic cements present in sandstones from the Main Buntsandstein. (A) Cathodoluminescence photomicrograph showing light blue halite enclosing orange-yellow dolomite. Sample 707, depth 3628.5 m, Well L9–6. (B) Dolomite rhomb (D) enclosed by pore-filling halite (H). Q = detrital quartz. Plane polarised light. Sample 727, depth 3641.8 m, Well L9–6. (C) Fabric-selective cementation. Anhydrite (A) in coarse-grained lamina and dolomite (arrowed) in fine-grained lamina. Crossed nicols. Sample 712, depth 3631.65 m, Well L9–6. (D) Extensive anhydrite (A) infilling primary porosity. Crossed nicols. Sample 714, depth 3632.45 m, Well L9–6. (E) Anhydrite (A) enclosing quartz overgrowths (O). Q = detrital quartz. Crossed nicols. Sample 304, depth 3917.66 m, Well L6–1. (F) Anhydrite (A) enclosing partially dissolved detrital K-feldspars (K). Plane polarised light. Sample 713, depth 3631.92 m, Well L9–6.

signature, indicating that the pore waters from which it precipitated were probably meteoric. Comparison of the Buntsandstein stable-isotope data with that reported by Taylor (1983), shows that the dolomite compositions fall in the field of 'evaporitic continental dolomite', albeit with a slightly lighter ^{18}O signature. The relationship between the measured $\delta^{18}O$ of the dolomite, porewater $\delta^{18}O$ and temperature was cal-

Table 4. Cement distribution and reservoir quality of the major lithofacies present in Triassic Main Buntsandstein reservoir intervals.

Core plug data			
Lithofacies	Porosity (%)	Permeability (mD)	Grain density (g/cm^3)
Dune sandstone	11.0	11.7	2.58
Outwash sandstone	14.9	46.4	2.69
Pin-striped sandstone	8.3	2.6	2.59
Cross-bedded fluvial sandstone	11.3	8.1	2.62
Massive fluvial sandstone	13.1	33.7	2.63

Petrographic data						
Lithofacies	Porosity (%)	Minus-cement porosity (%)	Halite (%)	Dolomite (%)	Anhydrite (%)	Total cement (%)
Dune sandstone	4.9	30.7	13.9	3.6	8.3	25.8
Outwash sandstone	10.7	26.9	3.3	9.6	3.3	16.2
Pin-striped sandstone	10.1	33.2	17.0	3.5	2.6	23.1
Cross-bedded fluvial sandstone	6.5	27.1	8.5	9.7	2.4	20.6
Massive fluvial sandstone	12.0	24.6	7.0	5.3	0.3	12.6

culated using the fractionation equation of Dutton & Land (1988), and is shown in Fig. 6. Assuming the dolomite formed from evaporitic meteoric water ($\delta^{18}O = -5‰$ SMOW; Harwood & Coleman 1983), precipitation would have occurred at temperatures of 30 to 45 °C, i.e. near-surface temperatures. The carbon isotopic composition of the dolomites is relatively heavy (^{13}C-enriched), suggesting a lack of organic matter in the pore waters from which the dolomites precipitated. This is in accord with the semi-arid to arid continental setting of deposition of the Buntsandstein.

Strontium-isotope analyses preclude a marine origin for the dolomite as the average $^{86}Sr/^{87}Sr$ composition (> 0.709) is greater than the maximum $^{86}Sr/^{87}Sr$ value for Triassic seawater (0.7078; Burke et al. 1982). The relatively high radiogenic strontium isotopic composition of the dolomites is more indicative of a meteoric water origin, with ^{87}Sr supplied to the pore waters by the dissolution of unstable lithic fragments and feldspars, which are typically enriched in ^{87}Sr (Sullivan et al. 1994).

Whilst the bulk of the dolomite formed under near-surface conditions, it is possible that minor amounts of non-luminescent, ferroan dolomite formed later during burial diagenesis, as conditions became reducing.

Anhydrite

Samples that are anhydrite-cemented typically have very high minus-cement porosities (up to 48.3%), and a floating grain texture which suggests that anhydrite cementation occurred under near-surface conditions, prior to any compaction within the sediments.

The majority of the anhydrite samples analysed have an isotopic composition that is lighter than coexisting Triassic sea water (+16 to +19 $\delta^{34}S‰$ CDT, Claypool et al. 1980, Nielsen 1978), and is similar to Zechstein anhydrite (+10 to +12 $\delta^{34}S‰$ CDT, Claypool et al. 1980, Kramm & Wedepohl 1991; Table 3). This suggests that most of the anhydrite observed in the Buntsandstein was recycled from the underlying Zechstein. It is unlikely that the anhydrite was sourced from the overlying Röt evaporites as these have a distinctly heavier sulphur isotopic composition (+18 to +30 $\delta^{34}S‰$ CDT, see data in Taylor 1983). The light isotope values (+4.2 to +5.6 $\delta^{34}S‰$ CDT) suggest that light sulphate, derived from the oxidation of sulphides exposed in shales or igneous rocks in the source area, was involved in the precipitation of some of the anhydrite (Taylor 1983).

The inflow of Zechstein-derived surface or ground water is an attractive scenario which would account

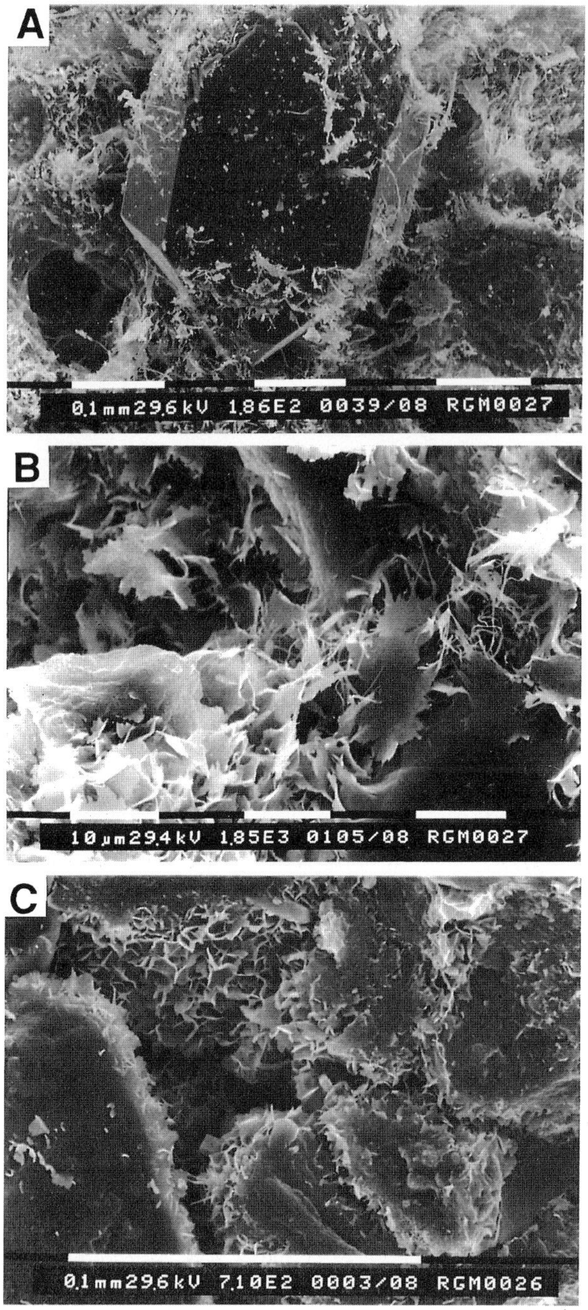

Fig. 5. SEM photomicrographs of sandstones from the Main Buntsandstein. (A) Prismatic quartz outgrowth partially enclosing authigenic illite. Sample 207, depth 4224.6 m, Well L2–6. (B) Pore-bridging, fibrous illite. Sample 505, depth 3621.3 m, Well F17–2. (C) Authigenic, grain-rimming boxwork chlorite. Sample 701, depth 3624.1 m, Well L9–6.

for the origin of the anhydrite present in the Buntsandstein. Although gypcretes are a common feature in arid and semi-arid environments (Watson 1983), they

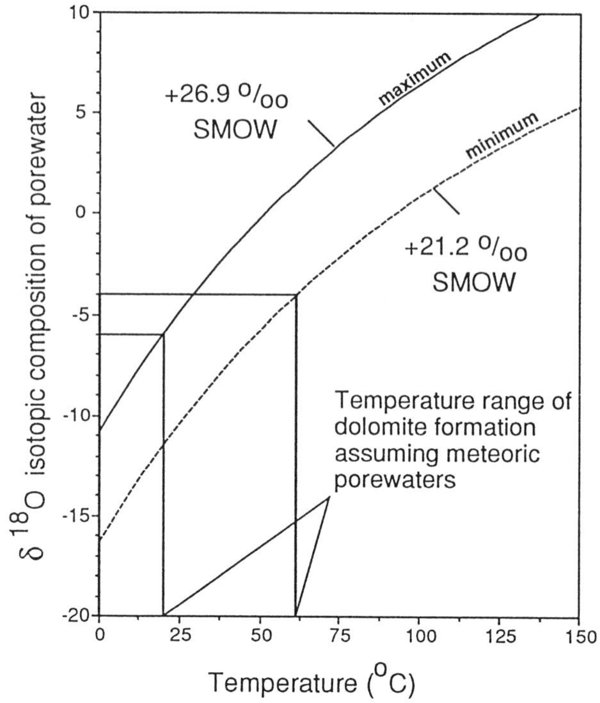

Fig. 6. $\delta^{18}O$ of porewater versus temperature for authigenic dolomite from the Main Buntsandstein sandstones. The curves show the maximum and minimum $\delta^{18}O$ values of dolomite using the equation: $10^3 \ln \alpha$ dolomite $- H_2O = 2.78(10^6)T^{-2} + 0.11$ (T= degrees K; Dutton & Land 1988). Also shown are the likely temperatures of formation of the dolomite assuming precipitation from continental meteoric water.

have not been observed in the cores analysed in this study. As it is likely that a considerable fraction of these cements would be dissolved during burial, early diagenetic gypsum cements remain a potential source of anhydrite cement during later diagenesis.

Another possible origin of the early anhydrite cements is by precipitation from the brines derived from the underlying Zechstein during shallow burial. The transformation of primary Zechstein gypsum to anhydrite during burial would result in the formation of overpressured saline brines, which may have entered the Buntsandstein as a result of hydrofracturing through the underlying claystones. This scenario is supported by the observation of hydrofractured and anhydrite-cemented rocks in cores of the Rogenstein. In the Buntsandstein, anhydrite may have precipitated consequent to a combination of pressure and temperature decrease. Further movement of the pore fluids would have been controlled by pressure gradients and compactional forces, and the porosity and permeability of the sandstones. Consequently, the anhydrite-bearing pore waters entered the originally higher-permeability facies and precipitated anhydrite.

Halite

Halite cementation in Triassic sandstones has been noticed as a potential problem by Laier & Nielsen (1989) in the Buntsandstein-equivalent sandstones of Tønder, southwest Denmark, and by Bifani (1986) in similar sandstones of the Esmond gas complex, UK southern North Sea. The origin and mechanism of halite precipitation in the Buntsandstein is poorly understood, in spite of the fact that halite cementation is a major porosity-reducing process.

The high minus-cement porosities observed in the halite-cemented sandstones (up to 33.9%) indicate an early origin for the halite. The source of the halite is poorly constrained owing to the lack of any isotopic or geochemical techniques that can be used to constrain its origin. The most likely source, given the petrographic and stratigraphic data, is the underlying salt-rich Zechstein sequence. Whilst it is unlikely that the overlying Röt sequence was the source for the anhydrite observed in the Buntsandstein (see discussion above), the possibility that this was the source for the halite cannot be ruled out. Assuming that the Zechstein was the major source for the halite, how was this cement transported into the Buntsandstein reservoir sequence and how did it precipitate?

Whilst Laier & Nielsen (1989) propose that halite cementation occurred in the Triassic of the Tønder structure as a result of hyperfiltration, this mechanism does not appear to be valid in the study area. There is no apparent correlation between halite cement and shale layers in the Buntsandstein comparable to that observed by Laier & Nielsen (1989), and the model of hyperfiltration does not account for the fabric-selective nature of the halite in the Buntsandstein.

The most feasible scenario that accounts for the distribution of the halite present in the Buntsandstein is:

- Halite was introduced in surface and ground water during deposition of the Buntsandstein.
- Halite was also introduced into the Buntsandstein from compactional waters derived from the underlying mudstones and from overpressured brines from the Zechstein during shallow burial (< 500 m). Mass balance calculations suggest that the saline pore water expelled from the Rogenstein and Main Claystone would have been insufficient to precipitate the entire volume of halite observed in the Buntsandstein.
- Flow of these brines was controlled by pressure gradients, compactional flow and the porosity and permeability of the sandstones. It is considered that the flow of brines and ground water through the Buntsandstein would have been concentrated along those beds with the highest initial porosity and permeability. Consequently these beds contain the most halite cement, an observation also made by Bifani (1986).

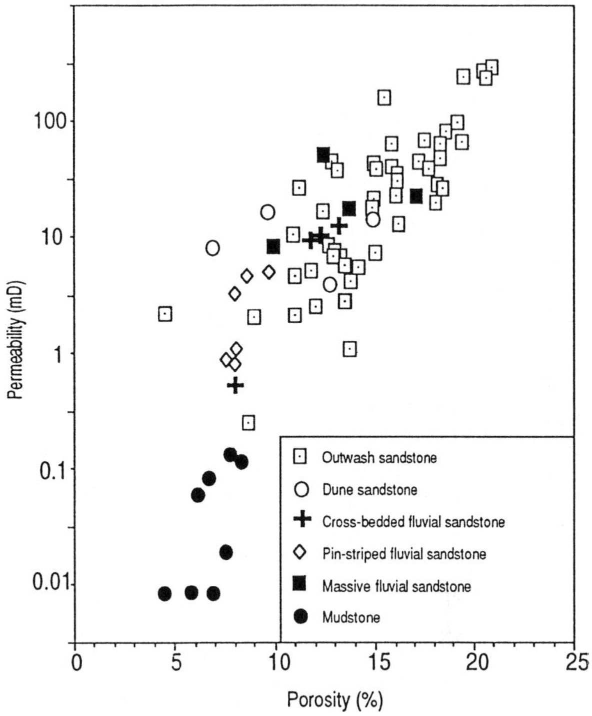

Fig. 7. Porosity – permeability cross plot showing the relationship between lithofacies and reservoir quality. Note that the best reservoir quality exists in the outwash sandstones, which originally had lower depositional porosities than the dune and fluvial sandstones.

Effect of diagenesis on porosity and permeability

The reservoir quality of the Volpriehausen Sandstone in one of the wells studied is shown in the porosity-permeability cross-plot in Fig. 7. This clearly shows that the originally best reservoir sandstones (dune sandstones) have a significantly reduced reservoir quality,

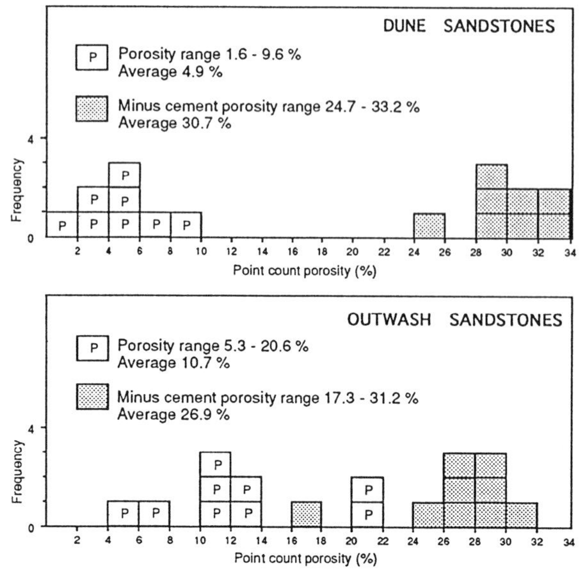

Fig. 8. Comparison of point-counted minus-cement porosity and porosity observed in dune sandstones and outwash sandstones from the Volpriehausen Sandstone. Note increased porosity reduction in dune sandstones compared to outwash sandstones.

which is explained by the increased volume of anhydrite and halite in these samples (Table 4).

Intergranular porosity has been severely reduced by pore-filling cements and partly by mechanical diagenesis (compaction). The intergranular porosity is observed to be almost completely obliterated in dune and fluvial sandstones by halite and anhydrite (Table 4). In the fine and very fine-grained sandstones, the main reduction of intergranular porosity has been caused by dolomite and clay cementation.

Minor intragranular porosity developed mainly in feldspar grains and in some ooids as a result of dissolution by undersaturated meteoric water. This process was effectively stopped by the precipitation of the pore-filling cements. There is no evidence of significant dissolution of the intergranular cements, and consequently only minor secondary porosity was developed.

Permeability in the reservoir lithologies in the Main Buntsandstein is reduced by the porosity-occluding halite and anhydrite cements and by the grain coating, pore-filling and pore- bridging authigenic clay minerals.

The halite and anhydrite cements have effectively destroyed the permeability of the lithofacies with the best initial reservoir properties (Table 4, Fig. 8). The infiltration of grain-coating detrital clay and the authigenesis of grain-coating, pore-filling and pore-bridging authigenic clays has reduced the permeability of the sandstones through the increased tortuosity of the flow path.

Conclusions

1. The sandstones of the Main Buntsandstein reservoirs were deposited in a semi-arid continental setting and consist predominantly of mineralogically immature lithic arkosic and arkosic arenites.
2. The reservoir properties in the Main Buntsandstein reservoir intervals are determined mainly by the presence of three pore-filling cements: dolomite, anhydrite and halite. Other authigenic minerals, except authigenic illite, chlorite and quartz, are only present in trace amounts.
3. Carbon, oxygen and strontium isotopic compositions of the authigenic dolomite cements indicate early cementation by meteoric water at relatively low temperatures ($< 50\,^{\circ}C$).
4. Sulphur isotope values obtained from anhydrite cements suggest that the sulphate is sourced predominantly from the underlying Zechstein deposits with a minor contribution from the oxidation of sulphides exposed in shales or igneous rocks in the source area.
5. Cementation by anhydrite and halite has effectively destroyed the reservoir properties of the sandstones with the best depositional porosity and permeability characteristics, causing an inversion of primary reservoir properties.

Acknowledgements

This research was undertaken for the Nederlandse Aardolie Maatschappij BV (Exploration), whose co-operation is gratefully acknowledged. Geochemical analyses were undertaken at the following institutes; PRIS, Reading University, UK (XRD), Stable Isotope Group, University of Miami, USA (carbon and oxygen isotope analyses), Department of Geology and Geochemistry, Florida University, USA (strontium isotopic analyses) and SURRC, East Kilbride, UK (sulphur isotopic analyses). The authors would like to thank W. Blendinger (Shell Research BV) for commenting on an earlier version of this paper, and journal referees

H. Dronkert and A. van de Weerd. This paper is published with the permission of Shell Research BV and Nederlandse Aardolie Maatschappij BV.

References

Bifani, R. 1986 Esmond gas complex. In: Brooks, J., J.C. Goff & B. van Hoorn (eds), Habitat of Palaeozoic gas in N.W. Europe – Geol. Soc. Spec. Publ. 23: 209–221

Burke, W.H., R.E. Denison, E.A. Hetherington, R.B. Koepnick, H.F. Nelson & J.B. Oto 1982 Variation of seawater $^{87}Sr/^{86}Sr$ throughout Phanerozoic time – Geology 10: 516–519

Claypool, G.E., W.T. Holser, I.R. Kaplan, H. Sakai & I. Zak 1980 The age curves of sulphur and oxygen isotopes in marine sulfate and their mutual interpretation – Chem. Geol. 28: 199–260

Clemmensen, L. 1979 Triassic lacustrine redbeds and paleoclimate: The 'Buntsandstein' of Helgoland and the Malmros Klint Member of East Greenland – Geol. Rundschau 68: 748–774

Dutton, S.P. & L.S. Land 1988 Cementation and burial history of a low-permeability quartzarenite, Lower Cretaceous Travis Peak Formation, East Texas – Bull. Geol. Soc. Amer. 100: 1271–1282

Fontaine, J.M., G. Guastella, P. Jouault & P. de la Vega 1993 F-15A: a Triassic gas field on the eastern limit of the Dutch Central Graben. In: Parker, J.R. (ed) Petroleum Geology of Northwest Europe: Proc. 4th Conference – Geol. Soc. Lond.: 583–594

Frakes, L.A. 1979 Climates throughout geologic time – Elsevier, New York: 310 pp

Harwood, G.M. & M.L. Coleman 1983 Isotopic evidence for UK Upper Permian mineralisation by bacterial reduction of evaporites – Nature 301: 597–599

Kramm, U. & K.H. Wedepohl 1991 The isotopic composition of strontium and sulphur in seawater of Late Permian (Zechstein) age – Chemical Geology 90: 253–262

Kutzbach, J.E. & R.G. Gallimore 1989 Pangean climates: Megamonsoons of the Megacontinent – J. Geoph. Res. 94: 3341–3357

Laier, T. & B.L. Nielsen 1989 Cementing halite in Triassic Bunter Sandstone (Tønder, southwest Denmark) as a result of hyperfiltration of brines – Chemical Geology 76: 353–363

Macchi, L. 1987 A review of sandstone illite cements and aspects of their significance to hydrocarbon exploration and development – J. Geol. 22: 333–345

McBride, E.F. 1989 Quartz cement in sandstones: a review – Earth Sci. Rev. 26: 69–112

Nagtegaal, P.J.C. 1978 Sandstone-framework instability as a function of burial diagenesis – J. Geol. Soc. Lond. 135: 101–105

NAM & RGD (Nederlandse Aardolie Mij. & Rijks Geologische Dienst) 1980 Stratigraphic nomenclature of the Netherlands – Verh. Kon. Ned. Geol. Mijnb. Gen. 32: 77 pp

Nickel, E. 1978 The present status of cathodoluminescence as a tool in sedimentology – Min. Sci. Engineering 10 : 73–100

Nielsen, H. 1978 Sulphur isotopes in nature. In: Wedepohl, K.H. (ed) Handbook of geochemistry 16B – Springer-Verlag, Berlin: 1–40

Roos, B.M. & B.J. Smits 1982 Rotliegend and Main Buntsandstein gas fields in Block K/13 – a case history – Geol. Mijnbouw 62: 75–83

Starkey, H.C., P.D. Blackmon & P.L. Hauff 1984 The routine mineralogical analysis of clay bearing samples – Bull. U.S. Geol. Surv.: 1563

Sullivan, M.D., R.S. Haszeldine, A.J. Boyce, G. Rogers & A.E. Fallick 1994 Late anhydrite cements mark basin inversion: isotopic and formation water evidence, Rotliegend Sandstone, North Sea – Mar. Petrol. Geol. 11: 46–54

Taylor, S.R. 1983 A stable isotope study of the Mercia Mudstones (Keuper Marl) and associated sulphate horizons in the English Midlands – Sedimentology 30: 11–31

Vreeken, A. & V. Kong 1993 The Bunter Gas Play and its seismic expression in the West Netherlands Basin. In: 1993 AAPG International Conference and Exhibition, The Hague, Abstract Volume: 84

Watson, A. 1983 Gypsum crusts. In: Goudie, A.S. & K. Pye (eds) Chemical sediments and Geomorphology; precipitates and residua in the near-surface environment – Academic Press, London: 133–161

Ziegler, P.A. 1978 North-western Europe: tectonics and basin development – Geol. Mijnbouw 57: 589–626

Ziegler, P.A. 1982 Triassic rifts and facies patterns in Western and Central Europe – Geol. Rundschau 71: 747–772

Rondeel et al. (eds), Geology of gas and oil under the Netherlands, 191–209, 1996.

Hydrocarbon habitat of the West Netherlands Basin

J. de Jager[1], M.A. Doyle[1,2], P.J. Grantham[†] & J.E Mabillard[1,3]
[1] *Nederlandse Aardolie Maatschappij BV, Postbus 28000, 9400 HH Assen, the Netherlands;* [2] *Present address: Petroleum Development Oman, P.O. Box 81, Mina Al Fahal, Muscat, Sultanate of Oman;* [3] *Present address: Shell Petroleum Development Company of Nigeria Ltd., PMB 2418, Lagos, Nigeria*

Key words: biodegradation, charge, maturity, modelling, retention, source rock

Abstract

The traditional hydrocarbon charge model for the West Netherlands Basin was that the Carboniferous coal measures have generated only gas, and the Jurassic Posidonia Shale only oil. However, it is now concluded that both have generated oil and gas, and that an additional possible source rock for oil occurs in the lower part of the Lower Jurassic Aalburg Formation. Geochemical 'fingerprints' have been established to distinguish the different groups of gas and oil, the occurrence of which is consistent with their geological setting and possible migration routes. The main phase of hydrocarbon generation and expulsion occurred prior to the mid-Cretaceous and the occurrence of biodegraded oils indicates that significant amounts of hydrocarbons were already in place at the onset of the Tertiary. Previously, it had been considered that charge ended during the Late Cretaceous inversion. However, recent modelling studies indicate that along the south-west margin of the basin and in the lows between inversion highs, charge continued during the Tertiary at rates that should have been sufficient to at least compensate for any loss of hydrocarbons through imperfect seals. The distribution of biodegraded oils in Jurassic and Cretaceous reservoirs suggests that biodegradation must have occurred during the earliest Tertiary, when meteoric fresh waters had access to the reservoirs. Significantly, reservoirs that were at that time so deeply buried that temperatures exceeded 70 to 80 °C, and that were thus at levels where bacteria cannot survive, do not contain biodegraded oils. For the fault-bounded traps in Triassic reservoirs, juxtaposition with Triassic and Jurassic sandstones and shales is the principal control of trap integrity and determines the spill point. Traps bounded by faults that have been reactivated during the Tertiary have clearly been breached. Traps at Upper Jurassic and Cretaceous levels are usually faulted dip-closures and have always been found to be full to their spill points within the limits of seismic resolution.

Introduction

Objectives

An integrated study of the hydrocarbon habitat of the West Netherlands Basin has been carried out in order to better understand the factors controlling the distribution of gas and oil in the sub-surface. The study comprised the following elements:

1. a review of the distribution of hydrocarbons and source rocks, together with a geochemical study to establish hydrocarbon – source rock correlations;
2. source rock maturation and generation modelling, using a Shell proprietary basin modelling software package, in order to establish the quantity and timing of oil and gas generation with respect to timing of trap formation; and
3. a review of retention and of characteristics of cap rocks to establish their capacity to retain hydrocarbons.

Regional geology

The West Netherlands Basin is a small, rather complex, inverted basin in the southern North Sea province (Fig.

[†] Deceased 23 March 1993; ex KSEPL/Shell Research, Rijswijk, the Netherlands

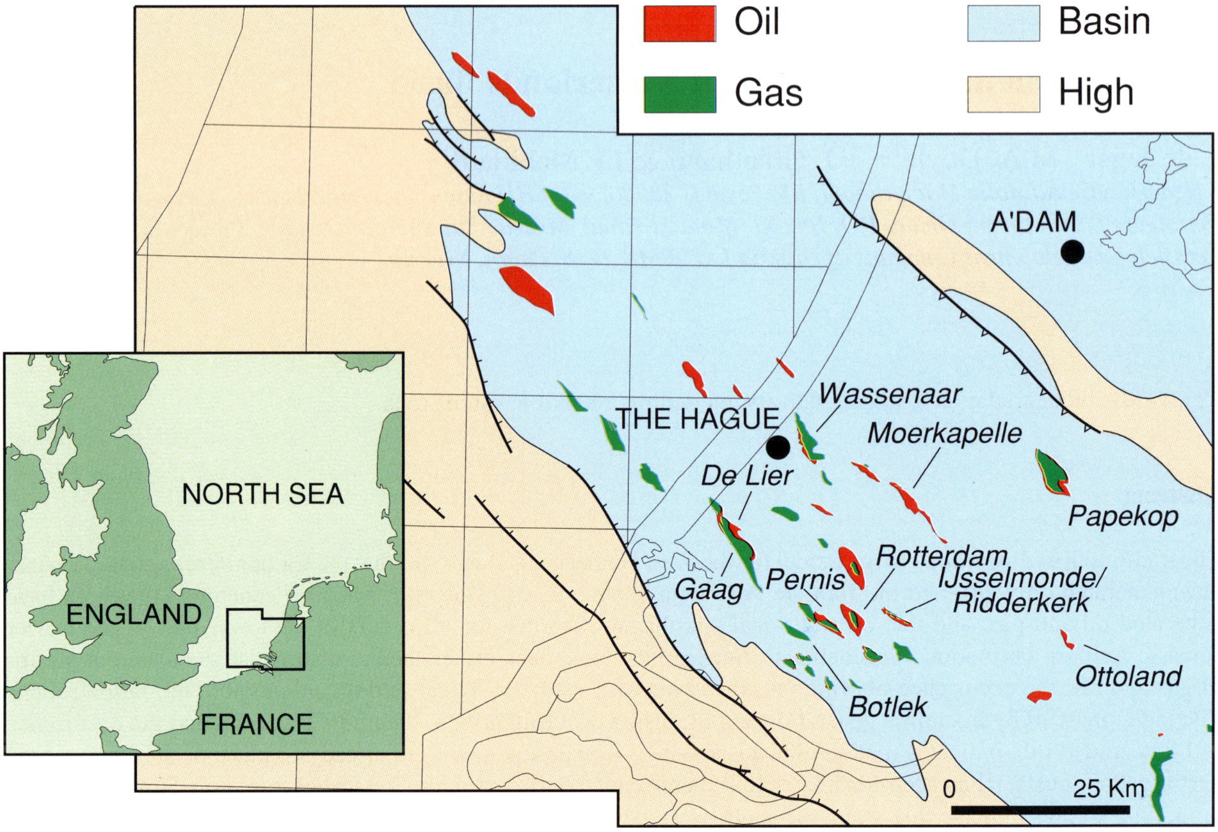

Fig. 1. West Netherlands Basin, location map of oil and gas fields.

1; Bodenhausen & Ott 1981). Sediments range in age from Carboniferous to Recent (Figs 2, 3).

A sequence of mainly continental clastics, several kilometres thick, and including extensive coal measures, was deposited during the Carboniferous. The 'Saalian tectonic event' at the end of the Carboniferous resulted in major faulting, uplift and erosion.

Deposition resumed during the Late Permian, and continued until the Middle Jurassic without major interruption. Mainly clastics, with only minor thickness variations, were deposited during this period. The marine, organic-rich source rocks of the Posidonia Shale were deposited basin-wide during the Toarcian.

A major rifting event, which created tilted fault blocks, was initiated during the Middle Jurassic. Synrift continental sands and shales of the Delfland Subgroup were deposited in subsiding half-grabens, while adjacent highs were subjected to erosion (Den Hartog Jager 1995). Major faulting came to an end during the Early Cretaceous, when the Rijnland Group was deposited regionally in marine to coastal environments onlapping the basin margins.

An inversion event, which occurred during deposition of the Chalk Group in the Late Cretaceous, modified old structures and formed new ones. Uplift and erosion were most pronounced in the centre of the basin. Thick clastics of the Tertiary North Sea Group were deposited during a period of relative tectonic quiescence, although inversion continued locally until the mid-Tertiary. Subsidence during deposition of the Chalk and North Sea Groups also imposed a regional tilt to the south upon the basin, which may have further modified pre-existing traps.

Hydrocarbon plays

The two major exploration objectives in the West Netherlands Basin are:

1. The Triassic Bunter Sandstone in tilted fault blocks, which were formed during Late Jurassic rifting and modified during Late Cretaceous to Early Tertiary

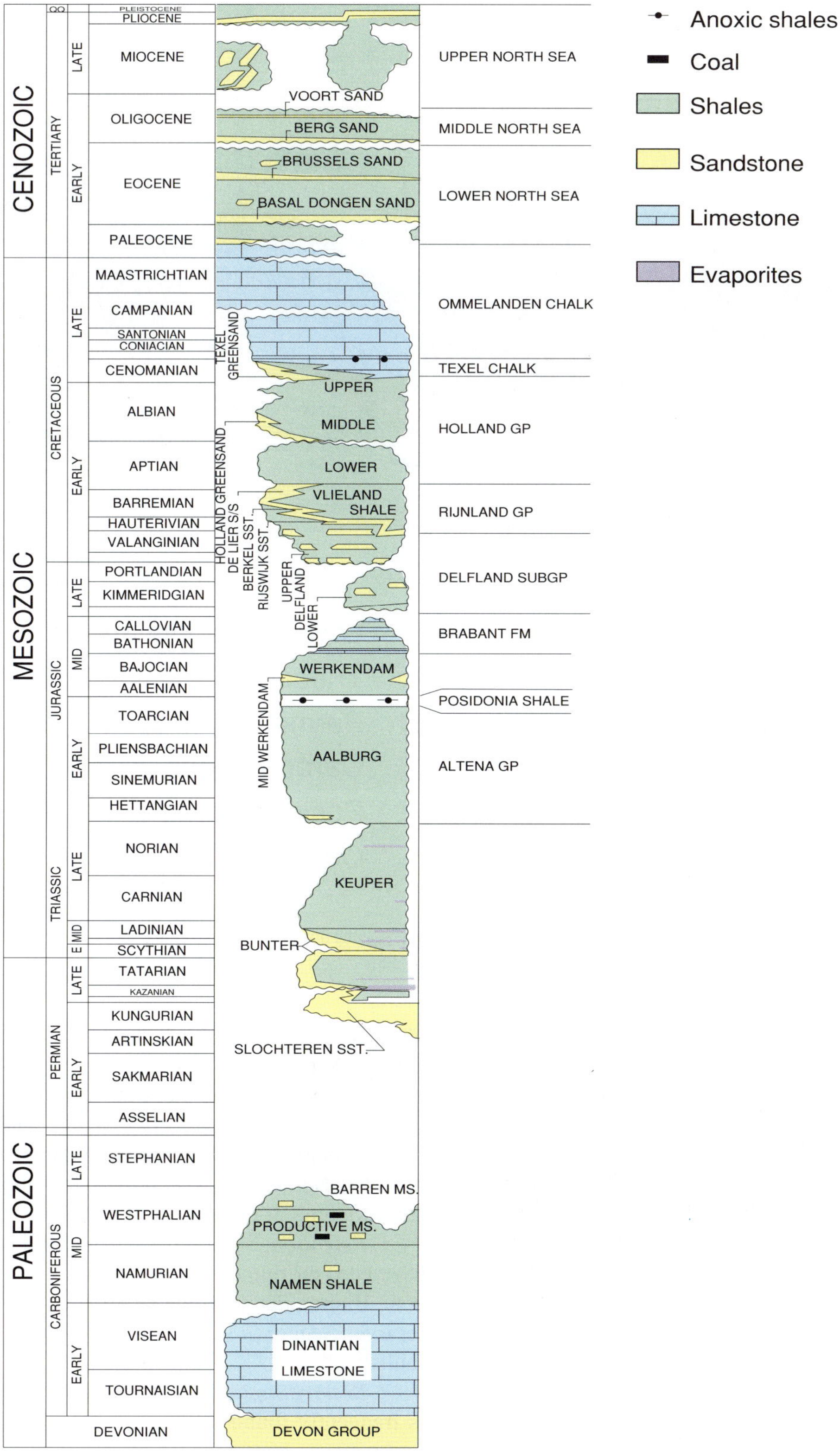

Fig. 2. Stratigraphy of the West Netherlands Basin area.

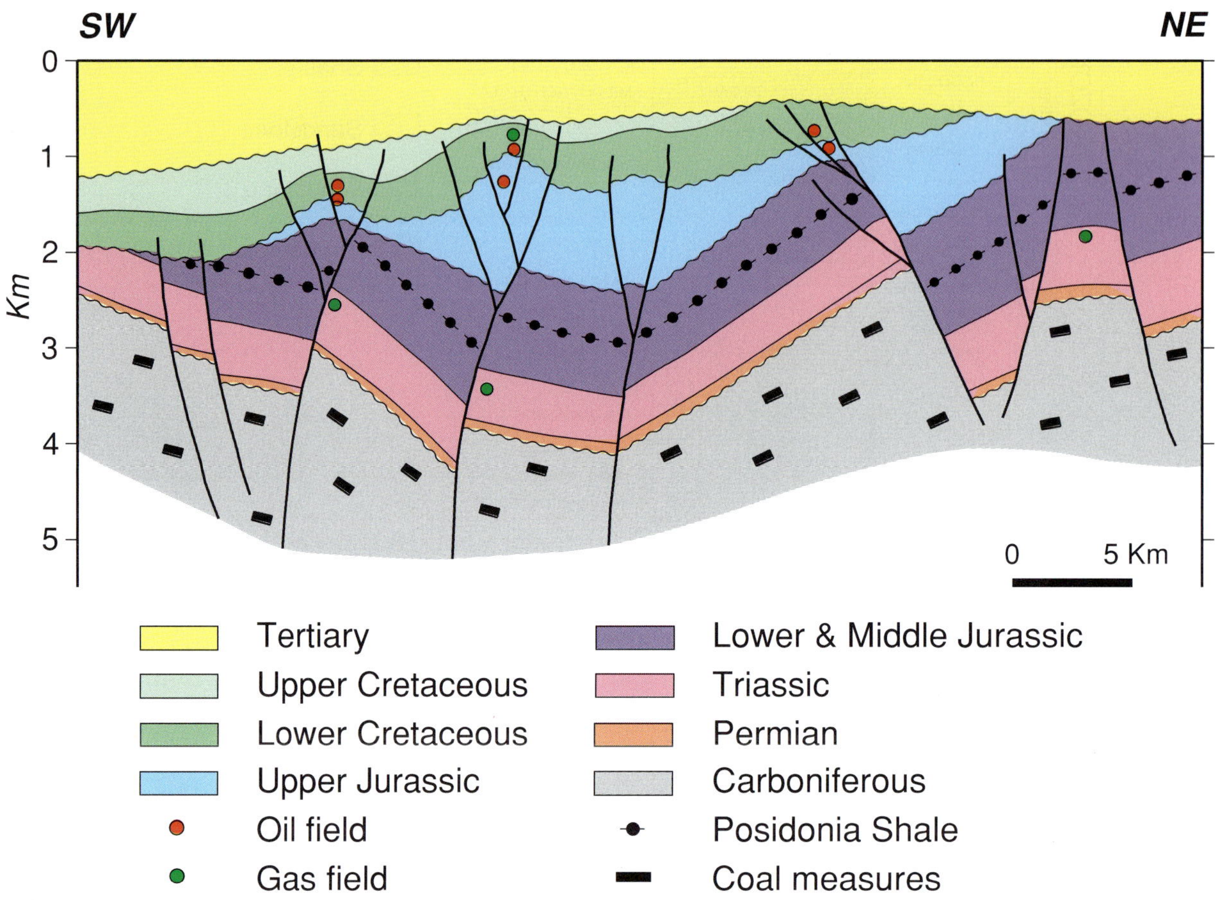

Fig. 3. Schematic SW – NE cross-section through the West Netherlands Basin, passing approximately over the Pernis, IJsselmonde/Ridderkerk and Moerkapelle hydrocarbon fields.

inversion. These traps, which rely on a top seal and on favourable juxtaposition of sealing lithologies across fault planes, contain mainly gas (initial reserves 65×10^9 m^3) with some oil (Fig. 4).

2. Upper Jurassic to Lower Cretaceous sandstone reservoirs of the Delfland Subgroup and Rijnland Group in faulted dip-closures, which were formed during Late Cretaceous to mid-Tertiary inversion (Racero-Baena & Drake 1995). These traps (Fig. 5) contain primarily oil (initial reserves including scope for recovery ca. 68×10^6 m^3) but also significant quantities of gas (initial reserves 5×10^9 m^3).

Secondary objectives, where minor quantities of gas and/or oil have been found, are the clastics of the Permian Zechstein Fringe Formation, the Middle Jurassic Brabant and Werkendam Formations, the Upper Cretaceous Chalk Group and the Lower Tertiary Basal Dongen Sand.

With increasing exploration maturity, remaining prospects are small and often subtle, and many can only be matured with 3D-seismic data. Structures in the West Netherlands Basin may be water-bearing or sub-economic for a number of reasons:

1. *Retention*. Several dry and underfilled structures have been encountered, notably at the level of the Triassic Bunter, due to unfavourable juxtaposition of the reservoir sequence with non-sealing lithologies across faults.

2. *Reservoir quality*. The Triassic Bunter Sandstone is frequently found to be tight in central parts of the basin.

3. *Hydrocarbon charge*. The Late Cretaceous inversion has resulted in significant variations in timing of charge across the basin.

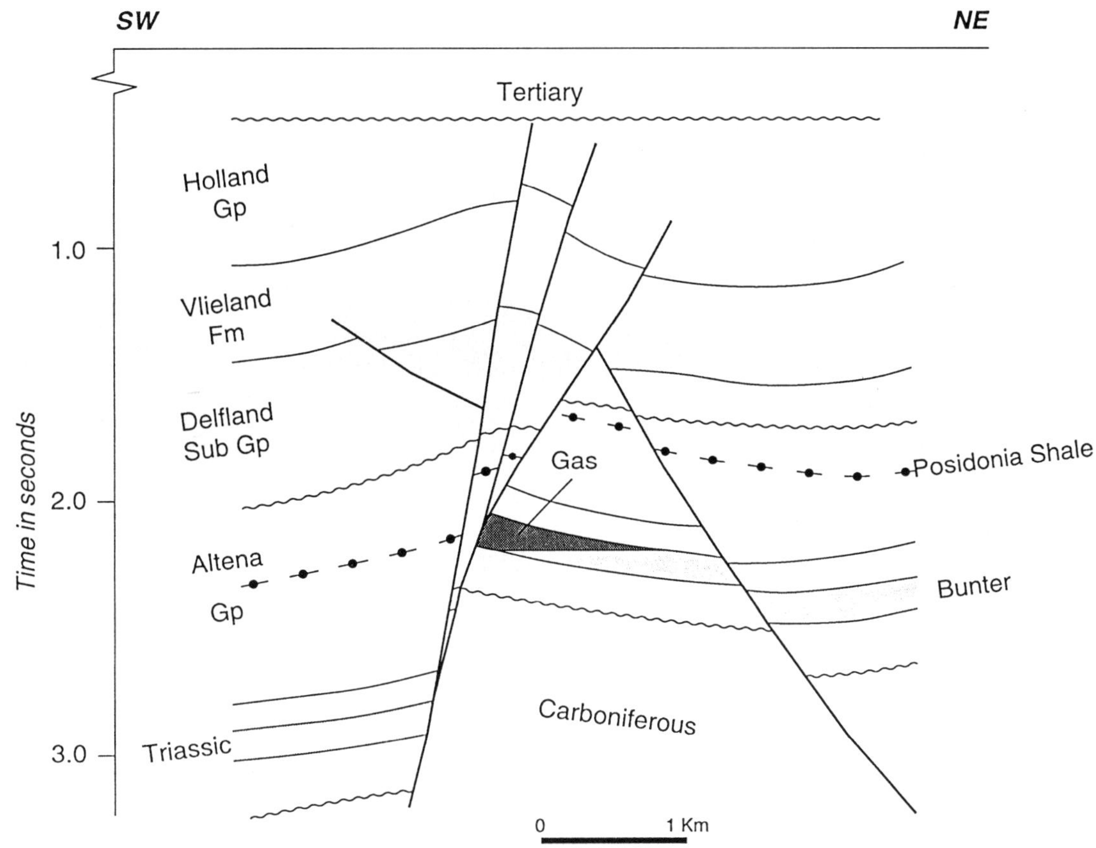

Fig. 4. The Wassenaar Deep trap as example of the Triassic Bunter play (see Fig. 1 for field location).

4. *Biodegradation.* The poor producibility of strongly biodegraded oils in small traps may render these accumulations uneconomic.

5. *Mixed hydrocarbon columns.* Small prospects may only be economic if one hydrocarbon phase is present. With mixed hydrocarbon columns, the volumes of gas and oil may be too small individually to be profitably developed.

This hydrocarbon habitat study addresses all these aspects with the exception of reservoir quality.

Hydrocarbon habitat

Source rocks

There are two main source rock intervals in the West Netherlands Basin:

1. The Carboniferous sequence, several kilometres thick, containing humic (Type III) source rocks in coals and carbonaceous shales, particularly in the Westphalian A and B.
2. The Lower Jurassic (Toarcian) Posidonia Shale, which is an approximately 30 m-thick marine, kerogenous, Type II source rock. Parts of the Lower Jurassic Aalburg Shale are geochemically similar.

The traditional concept that all oil in the basin was generated from the Posidonia Shale, and that all gas was generated from Westphalian source rocks has been revised as a result of new geochemical analyses.

Oils

Based on various geochemical criteria, three groups of oils can now be distinguished (Table 1). Comparison with analyses of extracts from the Posidonia and Westphalian source rocks shows that in particular the amount of biphenyls (a component of the aromatic frac-

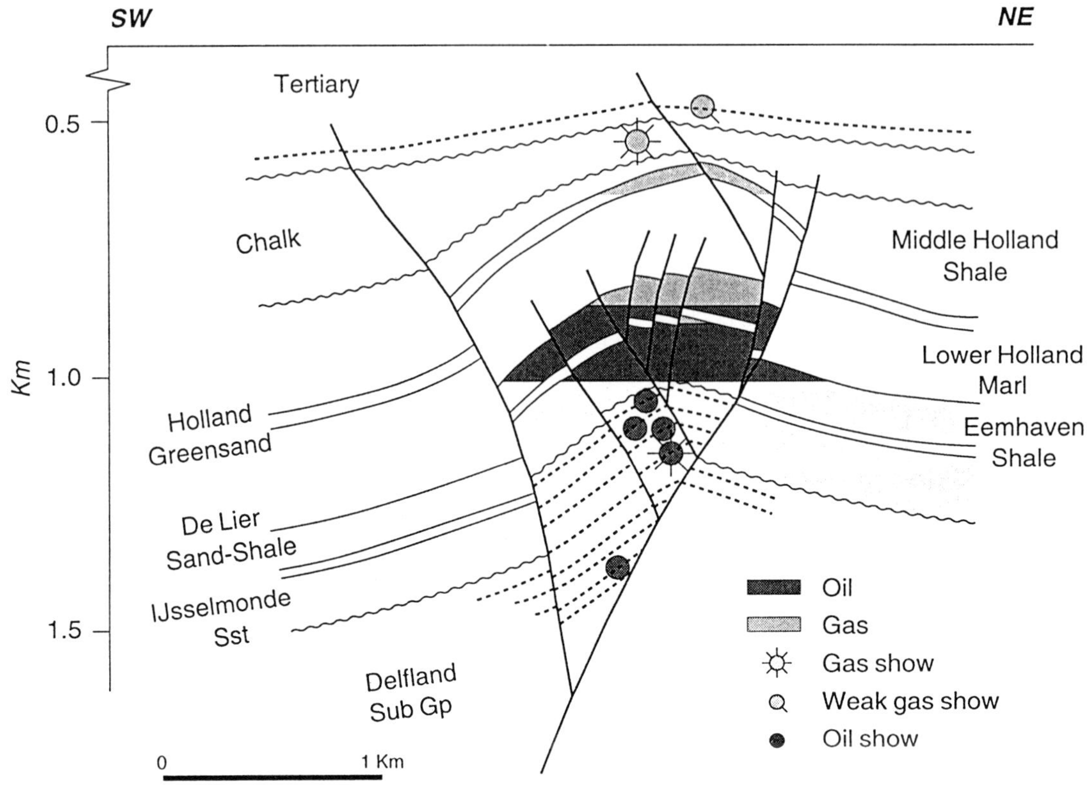

Fig. 5. The IJsselmonde/Ridderkerk trap as example of the Upper Jurassic – Lower Cretaceous play (see Fig. 1 for field location).

Table 1. Geochemical characteristics of three groups of oils in the West Netherlands Basin.

reservoirs	Jurassic and Cretaceous (mostly Posidonia oil)	Triassic reservoirs accessible for Posidonia oil	Triassic reservoirs not accessible for Posidonia oil
C_7 *Fraction*:			
benzene & toluene	low	intermediate	elevated
Aromatics:			
biphenyl	low	intermediate	elevated
Steranes C_{29}/C_{27}	1.1–1.6	1.2	0.6–1.2
^{13}*C-Isotopes*	−29.6‰	−29.5‰	−29.7‰
Whole oil	non waxy	waxy	waxy

tion of the oil), expressed in the biphenyl/naphthalene ratio and plotted against the C_{29}/C_{27} sterane ratio, is a generally reliable criterion to distinguish the various oils and to correlate them with source rocks (Fig. 6a,b). Most oils contained in Upper Jurassic and Lower Cretaceous reservoirs contain only small amounts of biphenyls, and can be correlated directly with extracts from the Posidonia Shale. Oils found in the Triassic Bunter Formation at Papekop and Ottoland contain much higher amounts of biphenyls, similar to many extracts from the Westphalian coal measures. An alternative possibility is that marine source rocks near the base of the Lower Jurassic Aalburg Shale are the source for these non-Posidonia oils. However, no conclusive

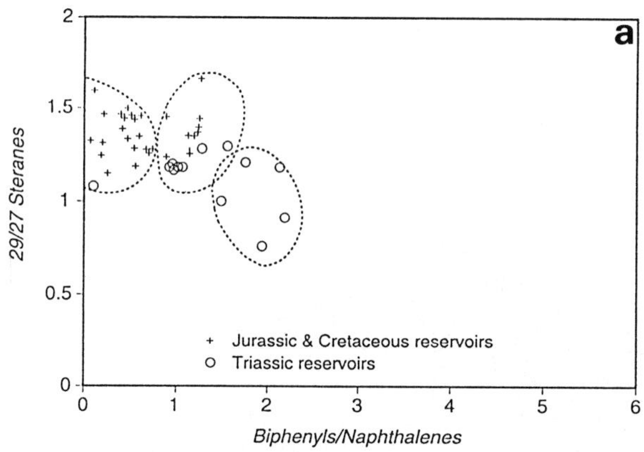

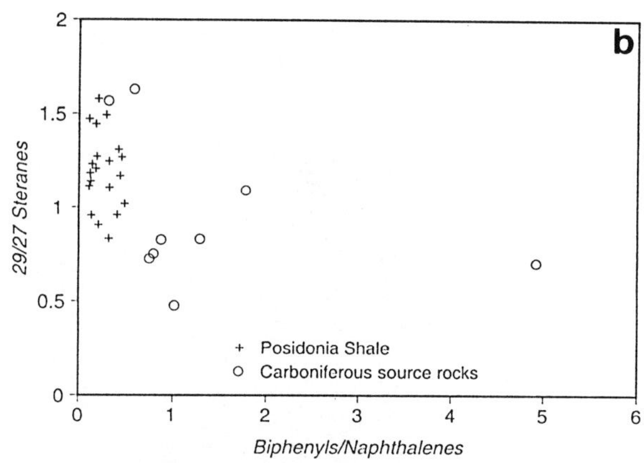

Fig. 6. West Netherlands Basin, oil characterisation. (a) Oil typing diagram, C_{29}/C_{27} steranes versus Biphenyls/Naphthalenes. Three groups of oils can be recognised: a group with low amounts of biphenyls, mainly consisting of oils from Jurassic and Cretaceous reservoirs, a group with high amounts of biphenyls of oils from Triassic reservoirs, and an intermediate group with oils from Jurassic and Cretaceous as well as from Triassic reservoirs. (b) Oil extracts typing diagram, C_{29}/C_{27} steranes versus Biphenyls/Naphthalenes. Oil extracts from the Jurassic Posidonia Shale invariably contain low amounts of biphenyls, whereas most oil extracts from Carboniferous source rocks contain higher amounts of biphenyls.

geochemical correlations with these potential source rocks can be established with the data currently available. A third group of oils, characterised by intermediate levels of biphenyls, is found in some Bunter traps (e.g. Pernis West, Gaag) and in some Upper Jurassic and Lower Cretaceous accumulations. This group appears to represent a mixture of Posidonia and Westphalian oils.

These correlations of oils with source rocks are compatible with the geological setting of the oil accumulations. The Bunter reservoirs at Papekop and Ottoland occur well below the level of the Posidonia Shale, which appears not to have been able to charge the Bunter in these traps (Fig. 7). The Westphalian as well as the lower Aalburg Shale do occur below the level of the Bunter in the drainage areas of these traps, which could have received charge from both these source rocks. Jurassic and Cretaceous reservoirs with mixed oils are accessible for hydrocarbons generated from the Posidonia, Westphalian as well as lower Aalburg source sequences. Mixed oils also occur in the Bunter of Pernis West and Gaag. In these structures, faulting has either juxtaposed the Posidonia with the Bunter Sandstones, or has brought the Posidonia to even deeper levels from where it could easily have contributed to the charge of the older Triassic reservoirs (Fig. 8).

Gases

Establishing reliable correlations between gases and their source rocks is generally more difficult than establishing oil – source rock correlations. Plots of the C_1/C_2 ratio versus the ^{13}C isotope value of gases are routinely used to distinguish gases derived from kerogenous source rocks, gases derived from humic source rocks, and bacterial gases (Fig. 9a). In the West Netherlands Basin, this standard plot can be used to distinguish gases derived from the kerogenous Posidonia Shale or the humic Westphalian coal measures.

Gases with ^{13}C isotope values less negative than ca. -33 to $-34‰$ were, in the past, generally considered to be derived from the Westphalian, with the trend of increasing C_1/C_2 ratios and less negative ^{13}C isotope values indicating increasing maturity. This trend is now interpreted to represent mainly gas mixing. The isotopically heaviest Bunter gases occur at Papekop and Ottoland (Fig. 9b), which have been charged by the Westphalian (Fig. 7). The Bunter gases with intermediate isotope values, such as at Pernis West and Gaag, are associated with oil rims of mixed origin, and are likely to have been charged with gas from both the Westphalian and Posidonia (Fig. 8). The isotopically light gas in the Vlieland Subgroup at Rotterdam is associated with pure Posidonia-derived oil, and is considered to represent pure Posidonia-derived gas.

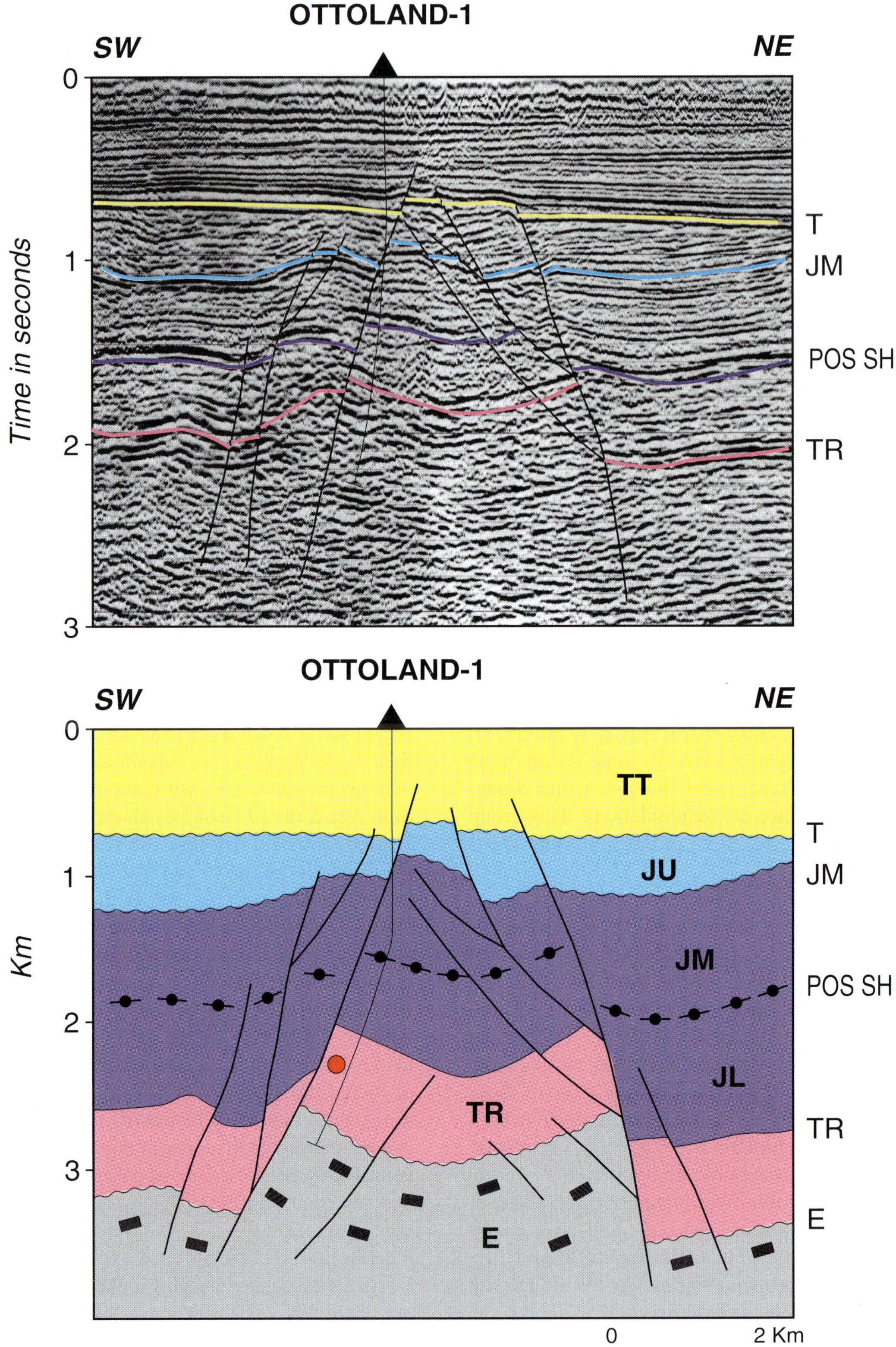

Fig. 7. Seismic and geological cross-section through the Ottoland oil field (see Fig. 1 for field location). TT = Tertiary, JU = Upper Jurassic, JM = Middle Jurassic, JL = Lower Jurassic, TR = Triassic, E = Permian and Carboniferous, POS SH = Posidonia Shale.

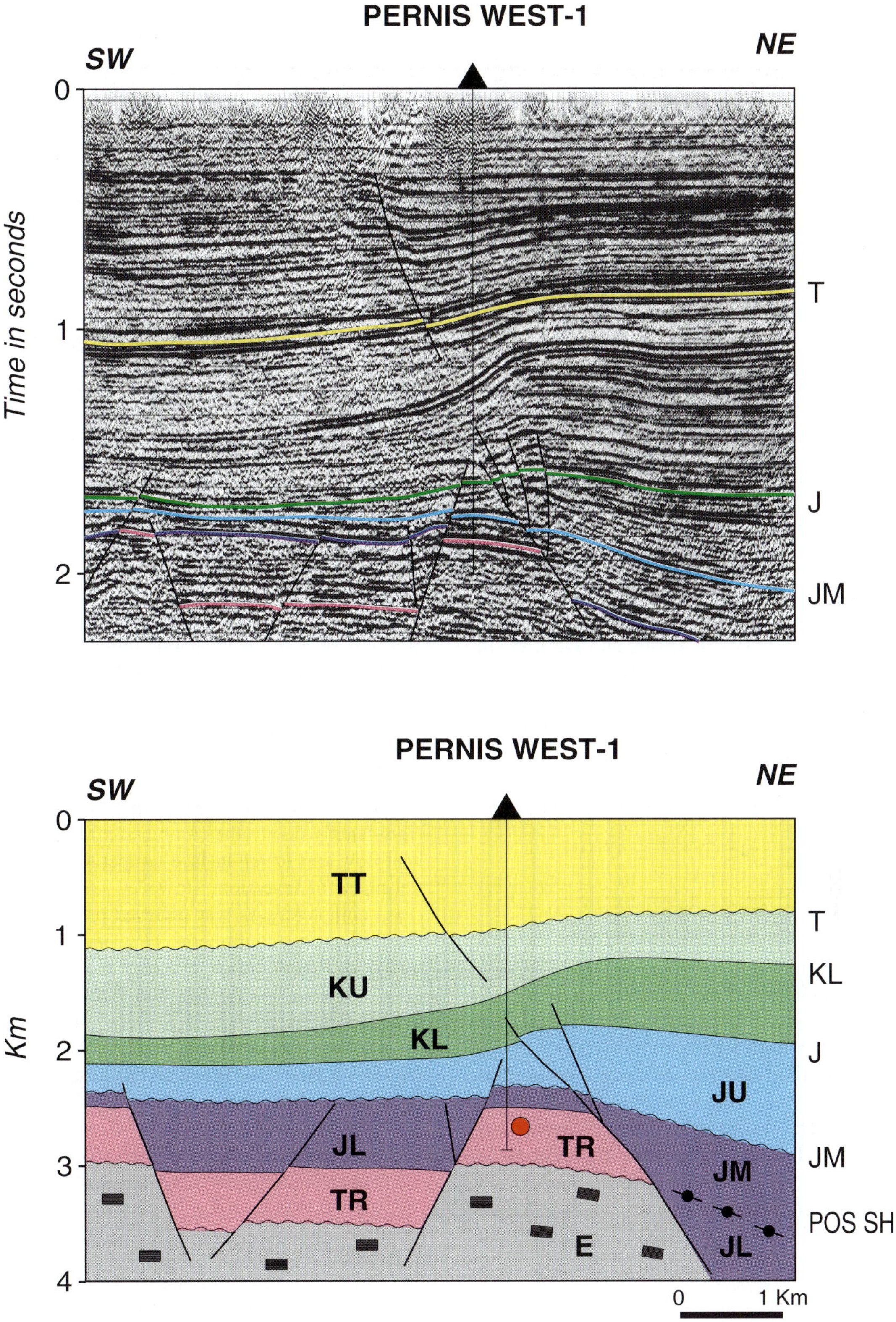

Fig. 8. Seismic and geological cross-section through Pernis West (see Fig. 1 for field location). KL = Lower Cretaceous, KU = Upper Cretaceous; further abbreviations as in Fig. 7.

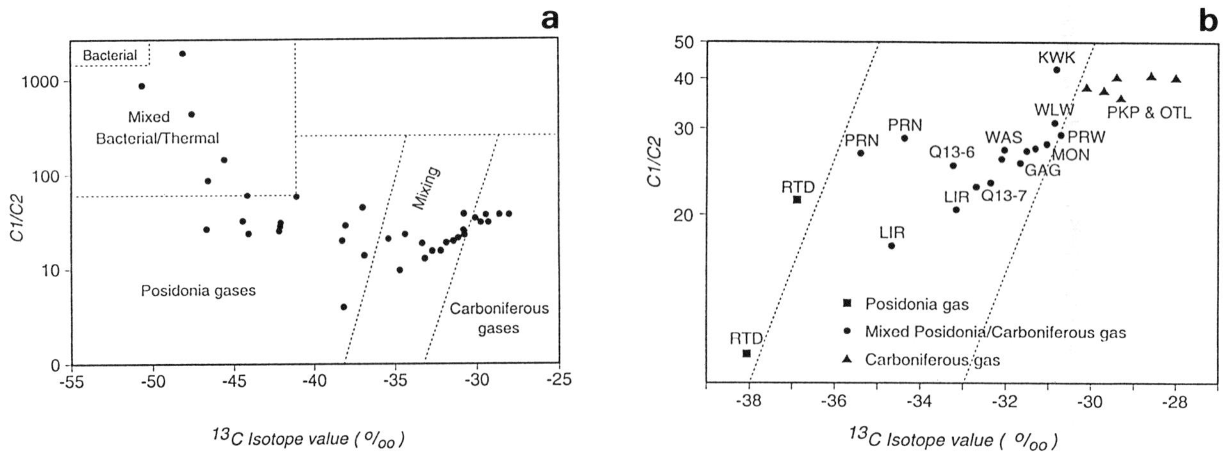

Fig. 9. West Netherlands Basin, gas typing. (a) Gas typing diagram – C_1/C_2 versus ^{13}C isotope value. (b) Gas mixing diagram – C_1/C_2 versus ^{13}C isotope value. Abbreviated well names: RTD = Rotterdam, PRN = Pernis, LIR = De Lier, WAS = Wassenaar, GAG = Gaag, MON = Monster, PRW = Pernis West, WLW = Waalwijk, KWK = Kerkwijk, PKP = Papekop, OTL = Ottoland.

Gases in the Jurassic and Cretaceous of the West Netherlands Basin can confidently be correlated to the Posidonia Shale. Only at Pernis and De Lier, in the southwest of the basin, can some contribution of Westphalian-derived gas be noted at these levels.

Hence, it is now possible to distinguish different groups of hydrocarbons from different source rocks in a manner consistent with the geological setting of the hydrocarbon accumulations and possible migration paths.

Hydrocarbon charge

The Late Cretaceous inversion of the West Netherlands Basin had important implications for the maturation and generation history of the main source rock horizons. Previously, it was believed that the temperature drop associated with this inversion effectively terminated all hydrocarbon charge in the basin. This implied a risk of no charge for young structures. Detailed modelling of the generation history of the Westphalian coal measures and the Posidonia Shale over the whole basin has considerably modified this concept. Generation was modelled using a Shell proprietary basin modelling package, developed at KSEPL, Shell's exploration and production research laboratory in Rijswijk.

Westphalian coal measures

Present-day maturities at the top of the Westphalian range from early mature (vitrinite reflectance (VR) ca. 0.6) on the flanks of the West Netherlands Basin to over mature (VR > 2.4) locally in the structural lows (Fig. 10). The base of the sequence is super mature with estimated VR (VR/E) values locally in excess of 5. The base of the Westphalian remains within the gas window only in a narrow zone on the southwest margin of the basin.

Much gas was generated from the Westphalian prior to the Late Jurassic phase of rifting and trap formation (Fig. 11). Gas generation continued at a high rate until the mid-Cretaceous, when volumes decreased significantly due to the combined effects of declining heat flow and lower surface temperature, and the initial phase of inversion. However, generation did not cease completely, as was believed previously. During the Tertiary, gas continued to be generated at a reduced rate along the southwest margin of the basin, where the effects of inversion are less and where Tertiary burial is more significant (Fig. 3). Generation also continued locally in the more central parts of the basin, and in the lows between inversion highs where relatively less uplift and erosion occurred.

The potential ultimate gas yield of the Westphalian varies with the thickness of the preserved section below the Saalian unconformity. It is calculated to range from 3000 to 10 000 $\times$ 10^9 m^3 per 100 km^2 (ca. 100 to 300 10^{12} cubic feet (tcf) per 100 km^2). In total, some 150 000 to 200 000 $\times$ 10^9 m^3 (ca. 5000 to 7000 tcf) of gas has probably been generated to date from the Westphalian in the West Netherlands Basin (50 to 70% of ultimate yield), while only some 100 $\times$ 10^9 m^3 of gas-in-place (ca. 3 tcf or 0.05% of the total yield to date) has been discovered up till now. A further 5000 $\times$ 10^9 m^3 of gas (ca. 150 tcf) may be dissolved

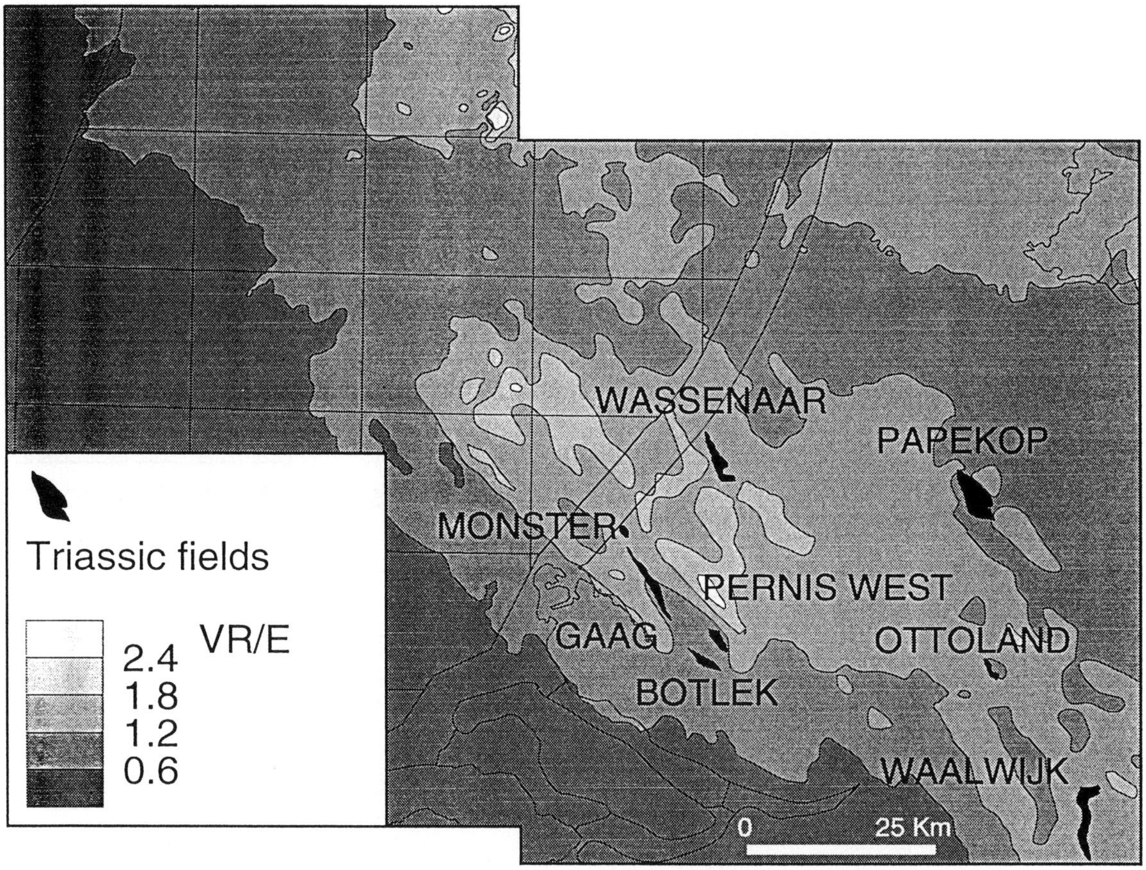

Fig. 10. Maturity at top Carboniferous (VR/E: modelled vitrinite reflectance value).

in formation water (some 3% of total yield). Given that the West Netherlands Basin is relatively mature in exploration terms, the implication is that the bulk of the generated gas has leaked to the surface over geological time.

Although the Westphalian is primarily a source rock for gas, it may also have generated significant quantities of oil, as discussed above. It is estimated that the average total yield for oil generation from the Westphalian is 125 to 200 $\times$ 10^6 m^3 per 100 km^2 (ca. 1 billion bbls/100 km^2).

Posidonia Shale

The Posidonia Shale occurs within the oil window over most of the basin, and only locally reaches maturities exceeding 1.2 VR/E (Fig. 12).

Hydrocarbons from the Posidonia Shale are trapped mainly in structures formed during Late Cretaceous inversion. The Posidonia Shale has not only generated oil but also significant quantities of gas (Figs 13, 14). Most hydrocarbons were generated before trap formation during the Late Cretaceous inversion. However, during the Tertiary, generation continued at a reduced rate in a substantial part of the basin, most notably in the south-west and in the lows between inversion axes. It is believed that expulsion from the Posidonia Shale is very efficient and that charge is rapidly available.

The Rotterdam field near the south-west margin of the basin confirms the effectiveness of Tertiary charge. This structure (Fig. 15) was created almost entirely during the Tertiary, as can be seen when the top of one of the reservoir horizons (De Lier 'Sand-Shale' Member) is flattened on the Base Tertiary reflection. Figure 16a shows the present-day size of the structure in seismic reflection time. On Fig. 16b the same horizon is flattened on the Base Tertiary reflector, showing that at the beginning of the Tertiary the Rotterdam structure was not yet in place.

The total ultimate oil yield of the Posidonia Shale is estimated to be some 350 $\times$ 10^6 m^3 per 100 km^2

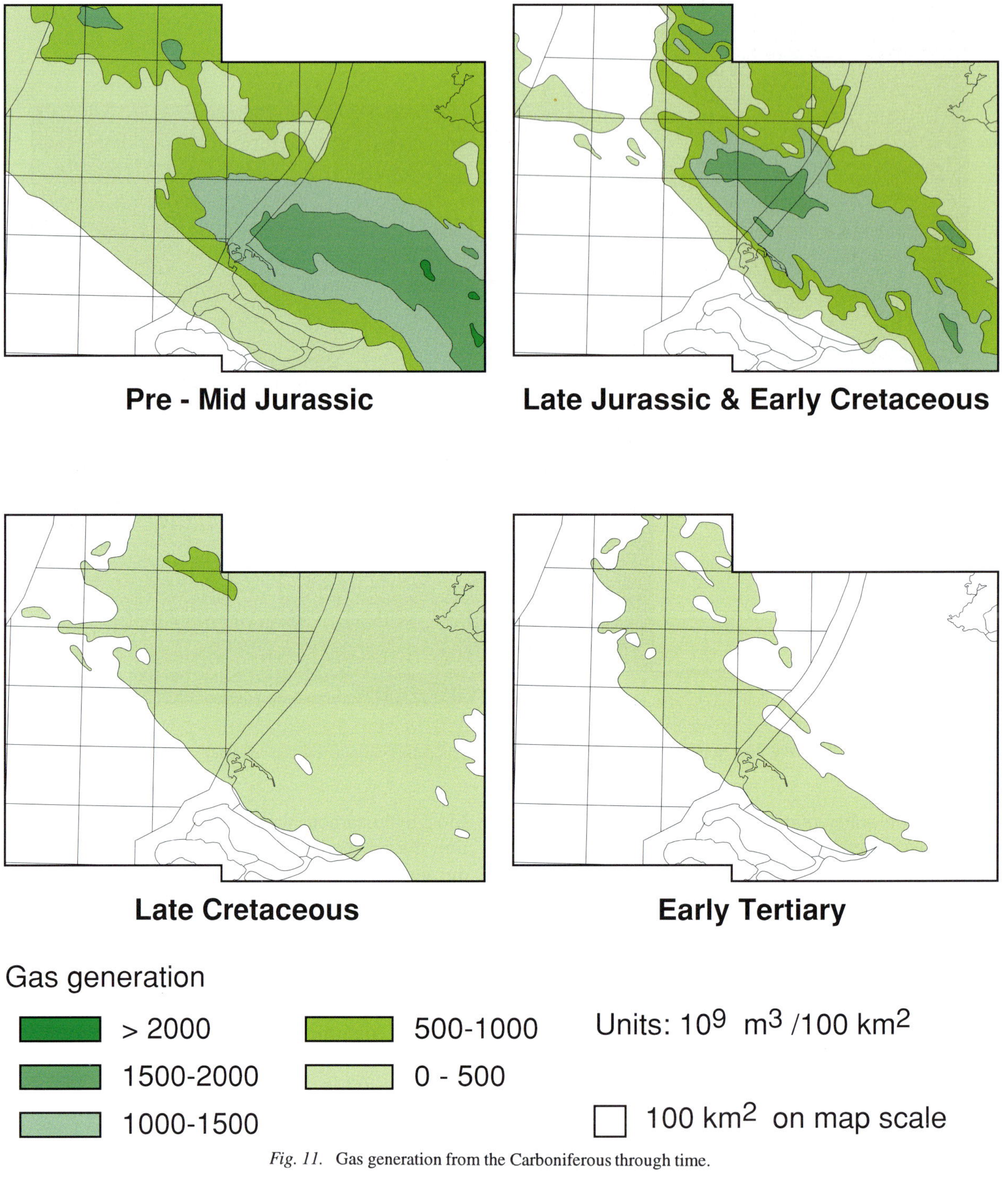

Fig. 11. Gas generation from the Carboniferous through time.

(> 2 billion barrels/100 km^2). About 50% of this has been generated to date. This implies that some 190 billion barrels of oil have been generated, of which 1.3 billion barrels have been found trapped (in-place volumes).

Retention

Seal quality in the West Netherlands Basin has been assessed by comparing the hydrocarbon column lengths with mapped spill or leak points. In this basin,

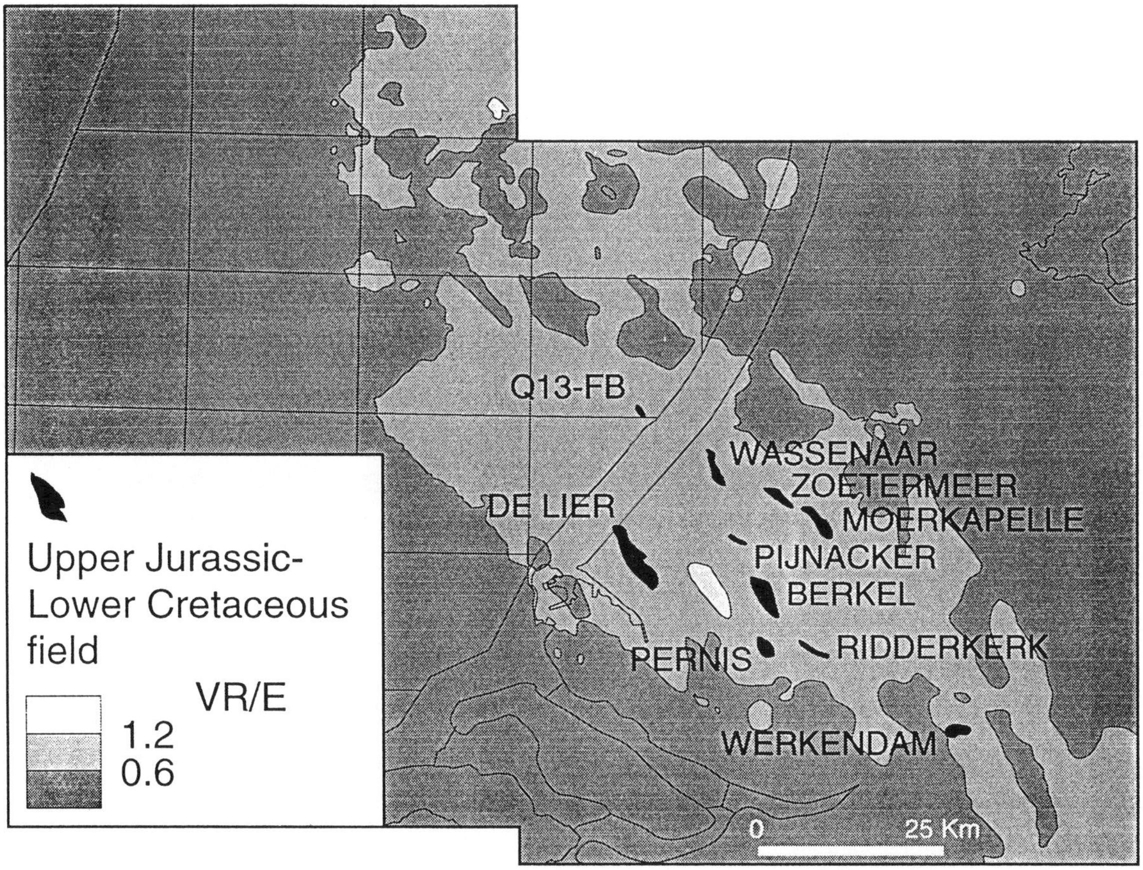

Fig. 12. Maturity of the Posidonia Shale (VR/E: modelled vitrinite reflectance value).

with an enormous excess of hydrocarbon charge, any underfill which cannot be explained by timing problems must indicate a sealing deficiency.

At the level of the Triassic Bunter Sandstones, all hydrocarbon-bearing traps are fault-bounded structures which rely on top and lateral seals provided by marls, carbonates and claystones of the Muschelkalk and Keuper Formations, and by shales of the Lower and Middle Jurassic Altena Group (Fig. 4). Traps at the level of the Bunter are water-bearing when:

1. the throw of the bounding fault(s) is so large that the reservoir is juxtaposed with sandstones of the Upper Jurassic Delfland Subgroup (e.g. Moerkapelle); and
2. the top of the Bunter reservoir is within 100 to 150 m of the base of the Delfland Subgroup or the Brabant Formation in the adjacent down-faulted block, and the bounding fault has a long history of reactivation with crestal faulting in the upthrown block. In this instance crestal faults and fractures link the reservoir in the footwall with a leak point in the hanging wall, and allow gas and oil to escape (e.g. IJsselmonde Deep). Interestingly, no leakage occurs in traps bounded by faults that have not been reactivated, and which display less crestal faulting and fracturing (e.g. Botlek, where base Delfland is within 100m of top Bunter).

Traps at the Upper Jurassic to Lower Cretaceous level are generally faulted dip-closures with multiple pay zones, which are full to spill point within the limits of seismic resolution. Breaching of seals by faulting only occurs when the top seal is very thin, such as the Eemhaven Member in the Rotterdam, Pernis, IJsselmonde/Ridderkerk area (Fig. 5).

Structures in the Upper Jurassic Delfland Subgroup are generally rather strongly faulted, and intra-Delfland shale seals are typically thin. While these thin shales may be effective as top seals, most traps at this level also rely on favourable cross-fault juxtaposition.

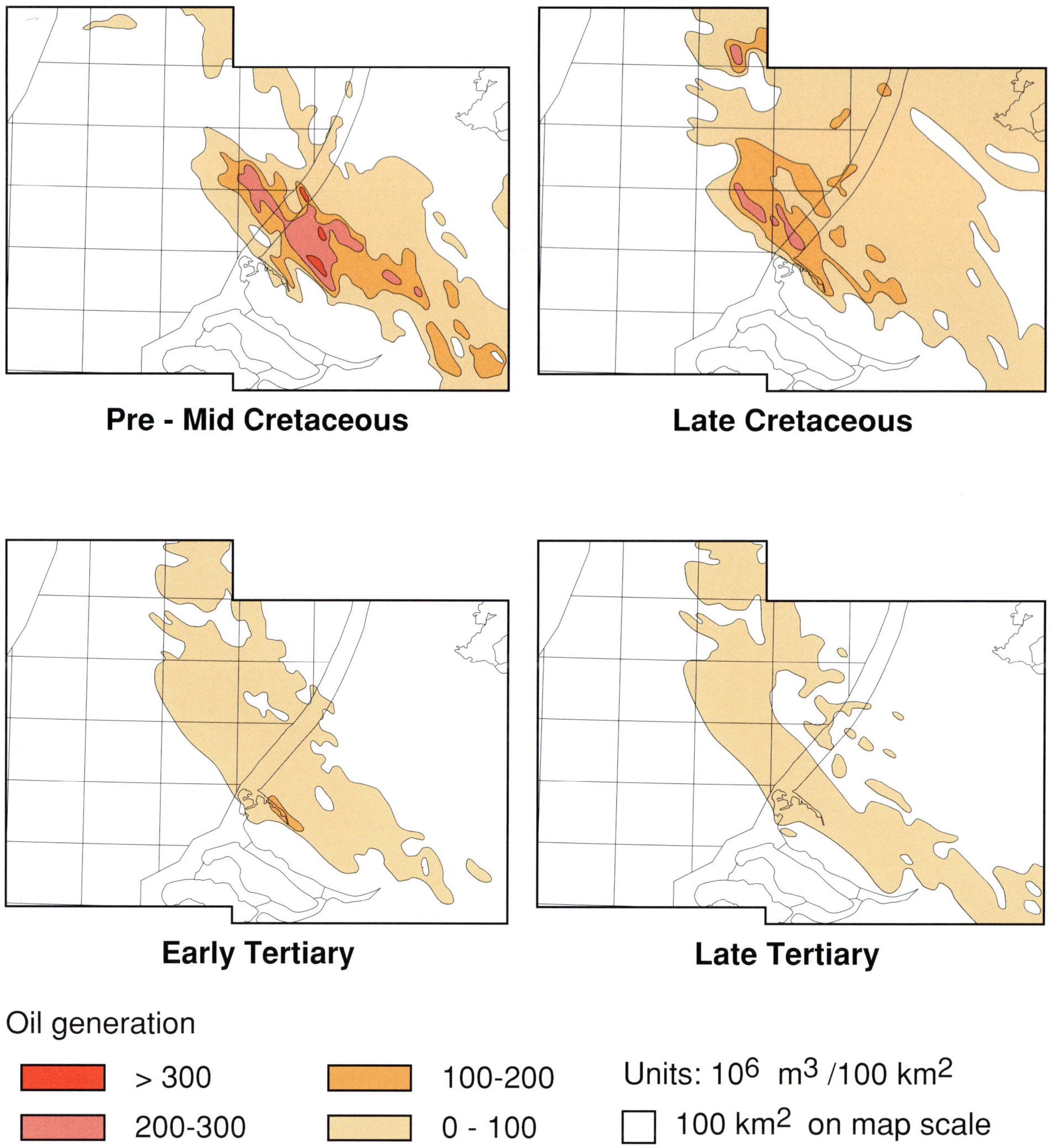

Fig. 13. Oil generation from the Posidonia Shale through time.

In faulted settings, irregular hydrocarbon distributions result (e.g. in IJsselmonde/Ridderkerk, Fig. 5).

Biodegradation

The API gravity of oils in Upper Jurassic and Lower Cretaceous reservoirs varies from 36 to less than 15 degrees. The low-gravity oils are biodegraded and are difficult to produce, rendering some accumulations uneconomic.

Biodegradation of oils occurs when aerobic bacteria feed on oils in the reservoir. These bacteria require a fresh supply of oxygen and nutrients, and cannot survive at temperatures higher than 65 to 80 °C. Con-

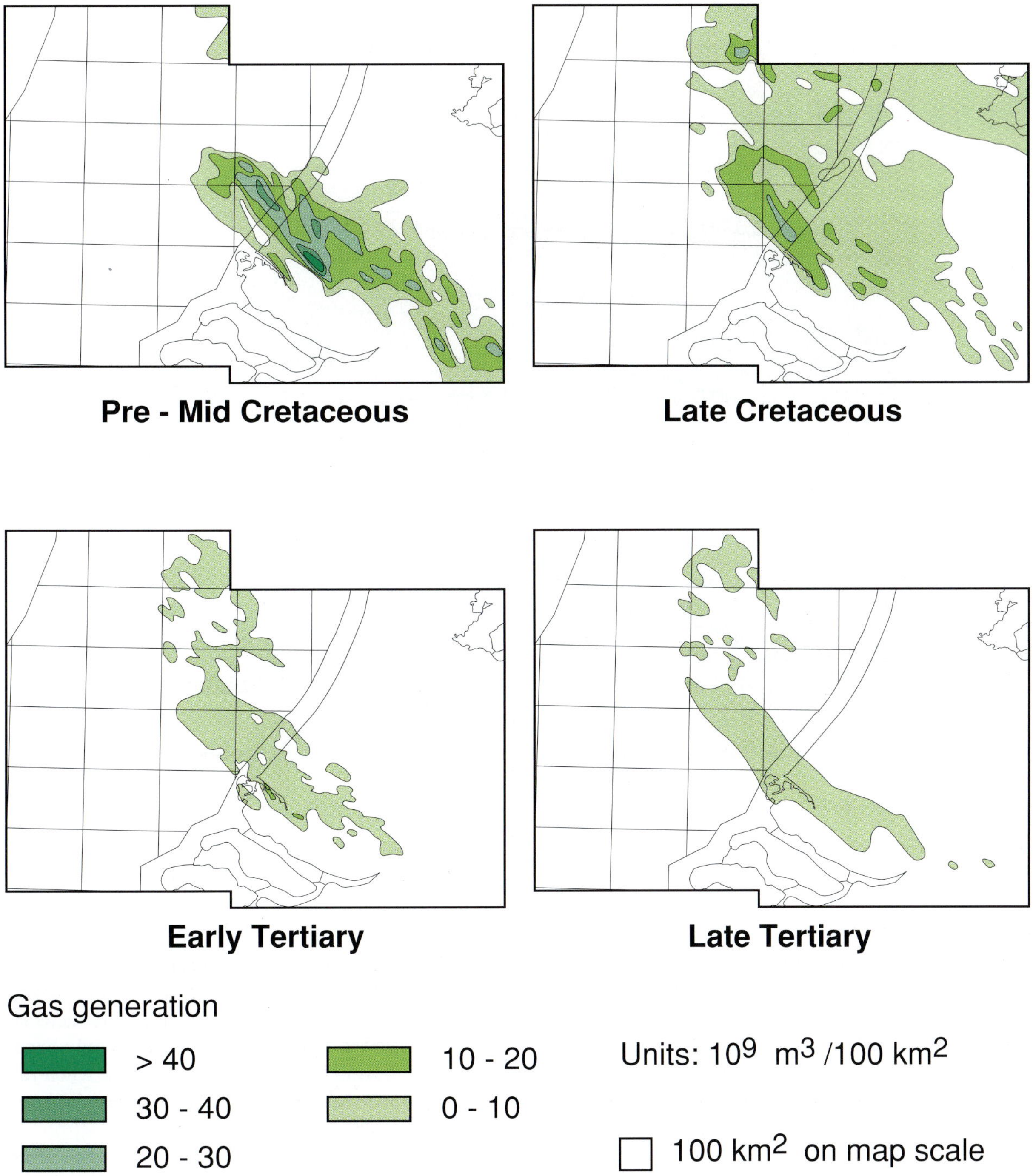

Fig. 14. Gas generation from the Posidonia Shale through time.

sequently, biodegradation is restricted to shallow reservoirs (less than 800 to 1000 m). Shallow reservoirs that are isolated from the main aquifer system by faulting or shale-out will not undergo biodegradation.

In the West Netherlands Basin, biodegraded oils occur in the Upper Jurassic Delfland Subgroup, the Rijswijk, IJsselmonde and Berkel Sandstones and in the De Lier 'Sand-Shale'. At present these reservoirs are at depths where temperatures are higher than 80

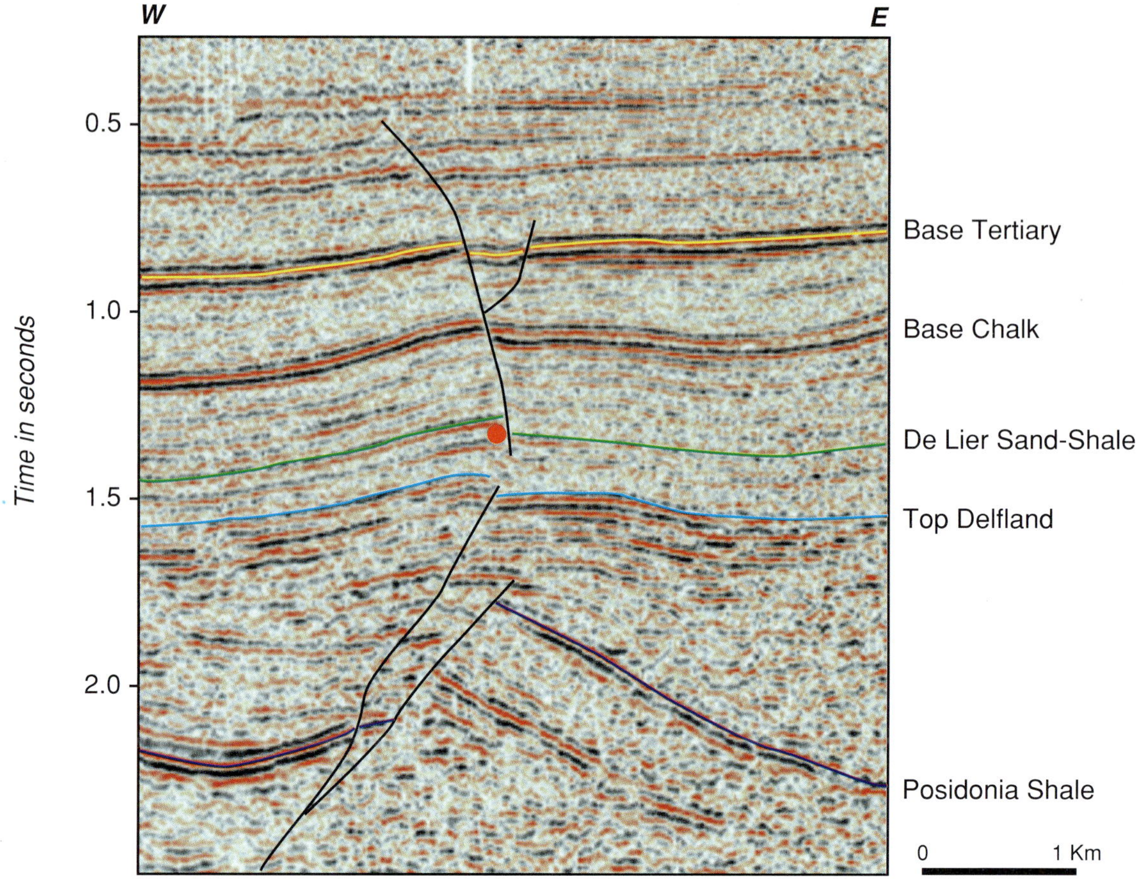

Fig. 15. Seismic section through the Rotterdam oil field (Fig. 16).

→*Fig. 16.* Rotterdam oil field. Location of seismic line of Fig. 15 is indicated. (a) Time map at the top of the De Lier 'Sand-Shale' reservoir (Lower Cretaceous), showing the size of the Rotterdam structure. (b) Time map at the top of the De Lier 'Sand-Shale', flattened on the Base Tertiary reflector, showing that at the onset of the Tertiary the Rotterdam structure had not yet developed.

°C, and have no connection to the surface. Hence, biodegradation must have occurred at some time in the past.

At the onset of the Tertiary, the Delfland Subgroup was exposed at the surface in the east of the basin (Figs 3, 17), such that meteoric waters could enter the reservoir. At the same time, reservoir temperatures were relatively low throughout much of the basin. In fact, no biodegraded oils have been encountered where reservoir temperatures during the Early Tertiary exceeded 65–70 °C (Fig. 17). In addition, there is a good relationship between the depth below the Base Tertiary unconformity and biodegradation as expressed in API gravity (Fig. 18).

The Rotterdam field contains non-biodegraded Posidonia oils in the De Lier Sand-Shale and Holland Greensand reservoirs. This is unusual, as at the beginning of the Tertiary, temperatures in both reservoirs were less than 70 °C and both had access to meteoric waters. However, the Rotterdam structure only began to develop during the Tertiary (Fig. 16a,b), when further burial increased reservoir temperatures and interrupted the access of meteoric water.

In summary, biodegraded oils are found in reservoirs that: (1) had a connection to the surface during the Tertiary via Delfland aquifers, (2) occur less than 800 to 1000 m below the Base Tertiary unconformity, and (3) occur in early (pre-Tertiary) traps.

These conclusions imply that structures containing biodegraded oils (i.e. Moerkapelle, Wassenaar, IJsselmonde/Ridderkerk) were charged prior to the Tertiary. In these structures there is no geochemical evidence for

a) Present day time map

b) Flattened on Base Tertiary

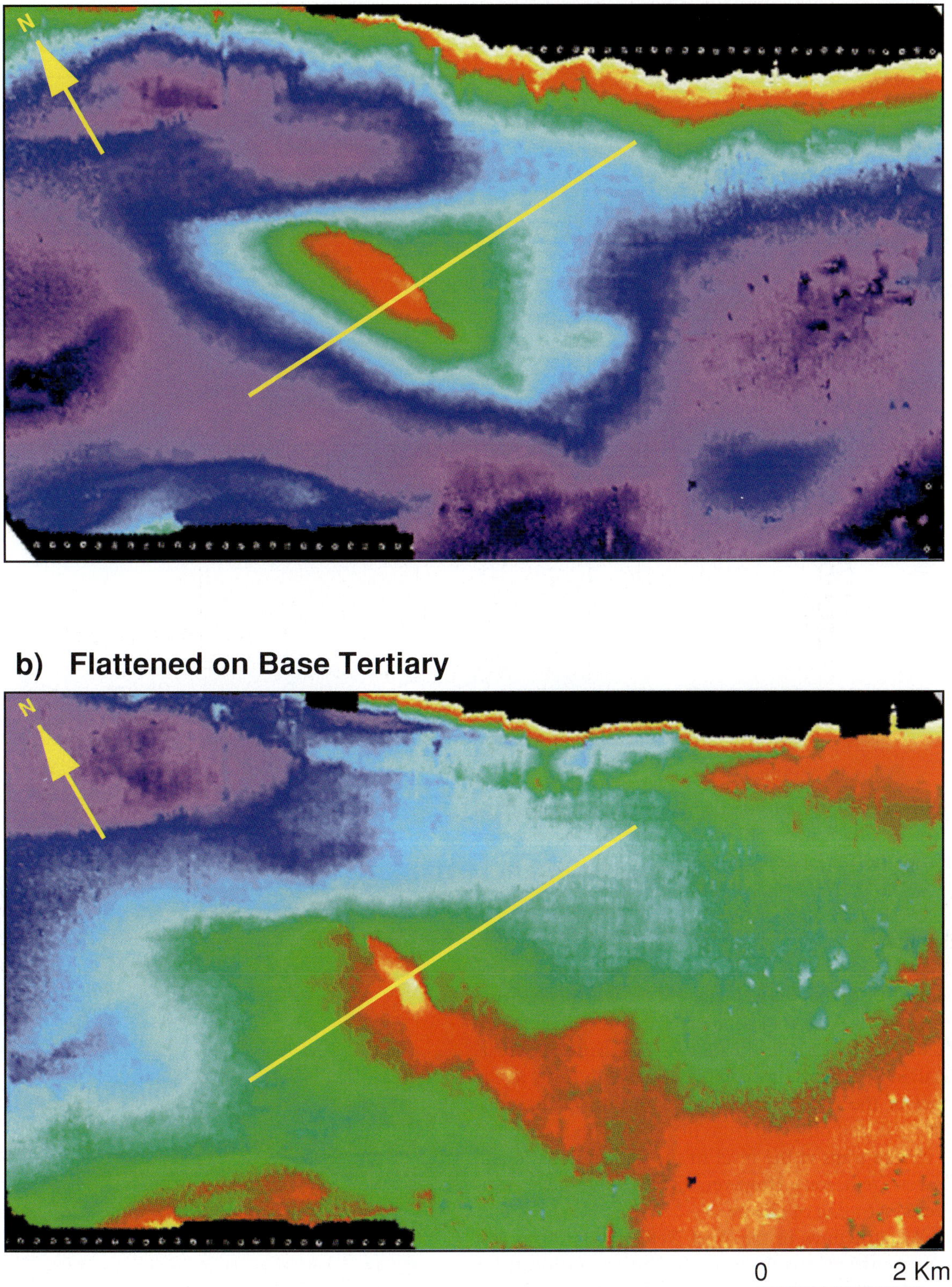

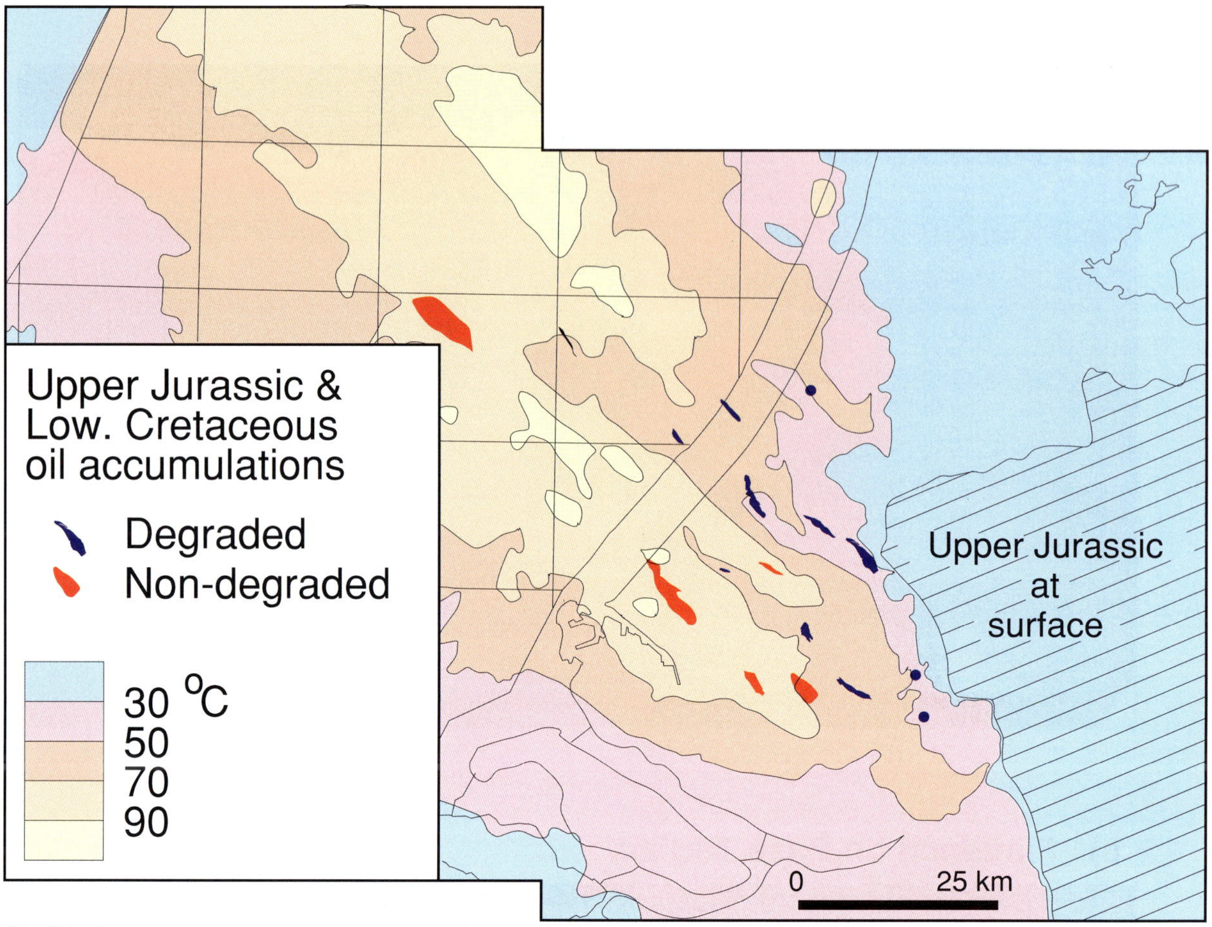

Fig. 17. Temperatures at Base Cretaceous at the beginning of the Tertiary. Biodegraded oils occur where these temperatures were < 70 °C.

any recent charge of non-degraded oil. As these traps are presently full to spill point, it appears that leakage during the Tertiary was minimal.

Discussion

Modelling of the charge history of the Westphalian and Posidonia source rocks indicates that less than 0.1 to 1% of the potential ultimate gas yield from the Westphalian is required to completely charge traps at the level of the Bunter. Less than 1 to 10% of the ultimate gas and oil yield from the Posidonia is required to charge traps at the level of the Upper Jurassic and Lower Cretaceous. However, there are significant variations in timing of charge.

All structures in the West Netherlands Basin that had significant growth during the Tertiary are located in areas of Tertiary charge in the south-west of the basin. Traps in this area are all full to maximum capacity. This

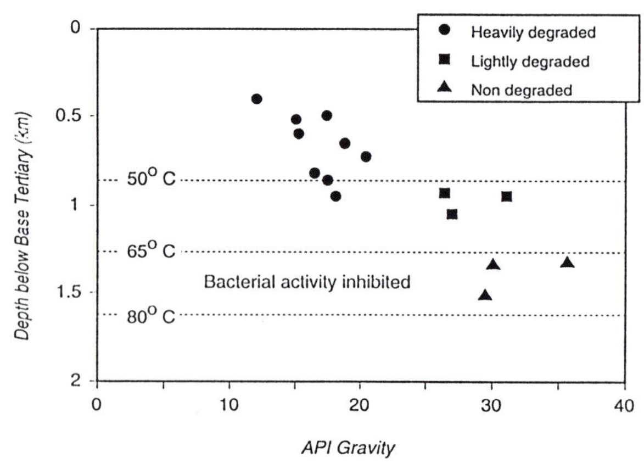

Fig. 18. API gravity of oil (as measure of biodegradation) of Upper Jurassic and Lower Cretaceous reservoirs versus depth (and paleotemperature) below the Base Tertiary unconformity.

is in agreement with the results of the modelling of the charge history, which indicates continuing gas charge in the south-west of the basin to the present day.

Bunter traps that did not have access to Tertiary charge, and that were affected by the Late Cretaceous inversion, appear locally to be underfilled. This is probably the result of loss of hydrocarbons during the inversion, after which effectively no further gas charge occurred.

Nearly all Jurassic and Cretaceous structures are full to spill point and appear to have been accessible for Tertiary charge. The occurrence of biodegraded oils suggests that many of the traps were already fully charged at the onset of the Tertiary, whereas others, such as the Rotterdam field, where only charged during the Tertiary.

Conclusions

Reliable fingerprints have been established to distinguish hydrocarbons generated from the Westphalian coal measures and the Jurassic Posidonia Shale. The total amounts of gas and oil generated from these source rocks were several orders of magnitude more than the hydrocarbon volumes found to date. The main phases of hydrocarbon charge were early. In the centre of the basin, charge ceased as a result of the Late Cretaceous inversion, while charge continued to the present day in the less inverted south-west of the basin.

The presence of both gas and oil charge from the Posidonia Shale and the Carboniferous, has resulted in the presence of mixed hydrocarbon columns in the Triassic Bunter and in Upper Jurassic and Lower Cretaceous reservoirs. Hydrocarbon charge to the Triassic Bunter Formation was dominated by gas from the Westphalian. Traps where the Posidonia Shale is downfaulted to levels below the Bunter within their drainage area, were also accessible to charge from the Posidonia Shale. Other traps were accessible for charge from the Westphalian, or possibly from the lower Aalburg Formation. Jurassic and Cretaceous reservoirs were charged primarily with oil, but also with substantial amounts of gas from the Posidonia Shale. Contributions of gas charge from the Westphalian coal measures appear to be less important.

Biodegradation of crude oil in Jurassic and Cretaceous reservoirs occurred at the onset of the Tertiary. Biodegraded oils occur in reservoirs at less than 800 to 1000 m below the Base Tertiary unconformity which were accessible for meteoric waters via Delfland aquifers at the beginning of the Tertiary. Reservoirs more than 1000 m below the Base Tertiary unconformity do not contain biodegraded oil. Neither do traps formed during the Tertiary, such as the Rotterdam field, which were charged when biodegradation was no longer possible.

Structures that do not contain hydrocarbons, usually display unfavourable juxtaposition across faults. Traps at the level of the Lower Cretaceous are almost all full to spill point, within the limits of seismic resolution. Breaching of seals by faulting only occurs when the top seal is very thin, such as the shales of the Eemhaven Member in the Rotterdam – Pernis – IJsselmonde/Ridderkerk area and intra-Delfland shales.

Cross-fault leakage in Triassic Bunter Formation traps occurs when the reservoir sandstones are juxtaposed with Delfland sandstones. Even when there is no direct contact between these units, crestal faulting and fracturing, enhanced by late reactivation of the main bounding faults, may provide communication and leakage may occur. Not reactivated structures without crestal faulting and fracturing, and with a well-developed lateral seal, are typically full to spill or leak point.

Acknowledgements

This paper is published by permission of the Nederlandse Aardolie Maatschappij BV (NAM), Shell Internationale Petroleum Maatschappij BV (SIPM) and Exxon Company International (ECI). Special thanks are due to A. Kantsler, M. Jagger, V. Noual, G. Louwaars and A. Speksnijder, who have contributed significantly to this study.

References

Bodenhausen, J.W.A. & W.F. Ott 1981 Habitat of the Rijswijk oil province, onshore, The Netherlands – In: Illing, L.V. & G.D. Hobson (eds) Petroleum geology of the continental shelf of NW Europe – Institute of Petroleum, London: 301–309

Den Hartog Jager, D. 1995 Fluviomarine sequences in the Lower Cretaceous of the West Netherlands Basin: Correlation and seismic expression. In: Rondeel, H.E., D.A.J. Batjes & W.H. Nieuwenhuijs (eds) Geology of gas and oil under the Netherlands – this volume

Racero-Baena, A. & S. Drake 1995 Structural style and reservoir development in the West Netherlands oil province. In: Rondeel, H.E., D.A.J. Batjes & W.H. Nieuwenhuijs (eds) Geology of gas and oil under the Netherlands – this volume

Structural style and reservoir development in the West Netherlands oil province

Alvaro Racero-Baena[1,2] & Stephen J. Drake[1]
[1] *Nederlandse Aardolie Maatschappij BV (Business Unit Oil, Dept. ODP/1), Postbus 33, 3100 AA Schiedam, the Netherlands;* [2] *Present address: Shell Venezuela SA, Urdaneta West, Edificio Maelga, Avenida 3F (80–81), Maracaibo, Estado Zulia, Venezuela*

Key words: basin inversion, Lower Cretaceous reservoirs, oilfields, rifting, West Netherlands Basin

Abstract

The area of the Rijswijk Concession largely coincides with the onshore part of the West Netherlands Basin. To date, oil has been produced from ten fields, with a total Stock Tank Oil Initially In Place (STOIIP) of ca. 210 × 10^6 m^3 (1.3 × 10^9 bbls). Reservoirs comprise continental and shallow marine clastics of Late Jurassic to Early Cretaceous age, deposited in syn-rift and post-rift settings. The West Netherlands Basin was created during Late Jurassic and Early Cretaceous rifting, characterised by divergent oblique-slip faulting. This created a NW – SE trending block-faulted depression between the London-Brabant Massif and the Zandvoort Ridge. Shallow marine clastics and marls were deposited in the subsequent post-rift stage. During the latest Cretaceous and Early Tertiary, regional uplift and convergent oblique-slip faulting resulted in basin inversion and reverse reactivation of pre-existing normal faults. Due to the transpressional nature of the basin inversion, narrow asymmetrical anticlines were formed, often bounded by upwardly divergent reverse faults, in the hanging-wall blocks of former normal faults. These structures constitute the general trap geometry of all Cretaceous oil fields in the basin. The acquisition of 3D seismic data has resulted in significantly enhanced structural definition and has led to the development of new structural and depositional models which improve the prediction of reservoir development and risk assessment.

Introduction

The geological history of the mature Rijswijk Concession in the West Netherlands oil province is described with reference to three particular oil fields each of which illustrates important aspects of the structural and sedimentological evolution of the area.

The Rijswijk Concession corresponds approximately to the onshore part of the West Netherlands Basin (PGK 1993: Fig. 10, Van Wijhe 1987). The first significant oil find was at Rijswijk in 1953. The Rijswijk Concession was awarded in 1956 and extended in 1957 (Knaap & Coenen 1987). To date, oil has been produced from ten fields with a total production (as at 1.1.94) of 32 × 10^6 m^3 (200 × 10^6 bbls). Total estimated Stock Tank Oil Initially In Place (STOIIP) in the concession is 210 × 10^6 m^3 (1.3 × 10^9 bbls). Production from the concession reached a peak of around 4600 m^3/day (29 000 bbl/day) in the mid-1960s. Present production is 2400 m^3/day (15 000 bbl/day) from the Rotterdam, Berkel and IJsselmonde/Ridderkerk Fields (Fig.1). The remaining oil fields have been (or are in the process of being) abandoned. Oil gravities vary systematically from light (30–36° API) in the south (e.g. Rotterdam, Pijnacker) to heavy in the shallower, northern fields (15–18° API; e.g. Zoetermeer, Wassenaar, Moerkapelle) as a result of Tertiary biodegradation (De Jager et al. 1995). Some 60% of the STOIIP in the Rijswijk Concession is $< 20°$ API.

This highly populated area, which includes the cities of The Hague and Rotterdam, is now largely covered by 3D seismic. Three energy sources were used: airgun around harbours and wider canals, vibroseis in towns, and dynamite on the 'polders' and inside greenhouses.

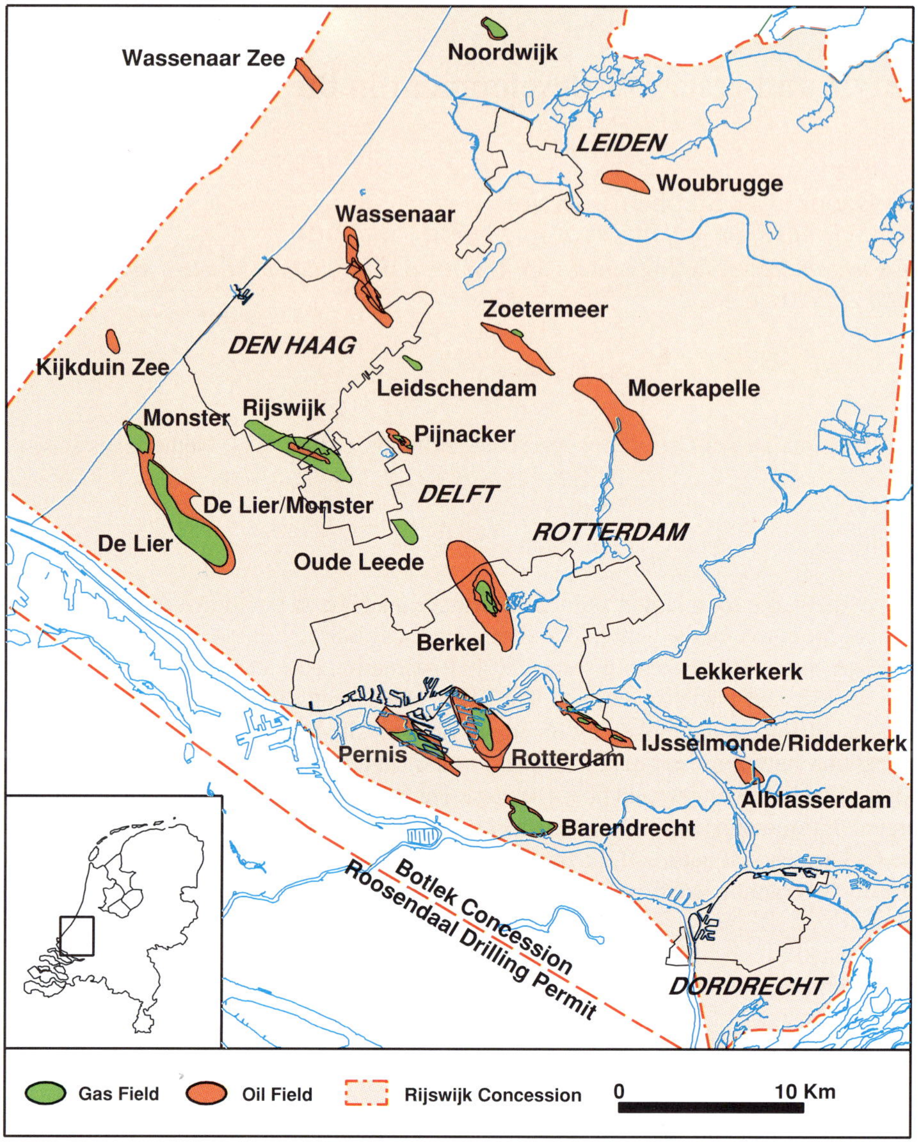

Fig. 1. Gas and oil fields in the Rijswijk Concession. Oil at present is being produced from the Rotterdam and Berkel Fields. The map includes fields currently being abandoned as well as undeveloped and non-commercial accumulations.

Geological history

The West Netherlands Basin is located between the London-Brabant Massif and the Broad Fourteens and Central Netherlands Basins (PGK 1993: fig. 10). The Zandvoort Ridge separates the basin from the Central Netherlands Basin.

The West Netherlands Basin is an inverted rift basin created during Late Jurassic–Early Cretaceous rifting when uplift, erosion and divergent oblique-slip faulting took place. In the Late Cretaceous to Early Tertiary, compressional movements resulted in convergent oblique-slip faulting, general uplift, erosion and basin inversion which reactivated pre-existing normal (rift) faults (Van Wijhe 1987, Zijp 1987, Burgers & Mulder 1991, Dronkers & Mrozek 1991). By taking these two main tectonic events as references, the post-Variscan stratigraphy (Fig. 2) may be subdivided into four sequences: i) a pre-rift clastic sequence from the Permian to the Early – Mid Jurassic, ii) a syn-rift sequence of Late Jurassic to Early Cretaceous age, iii) a post-rift sequence of Early to Late Cretaceous age,

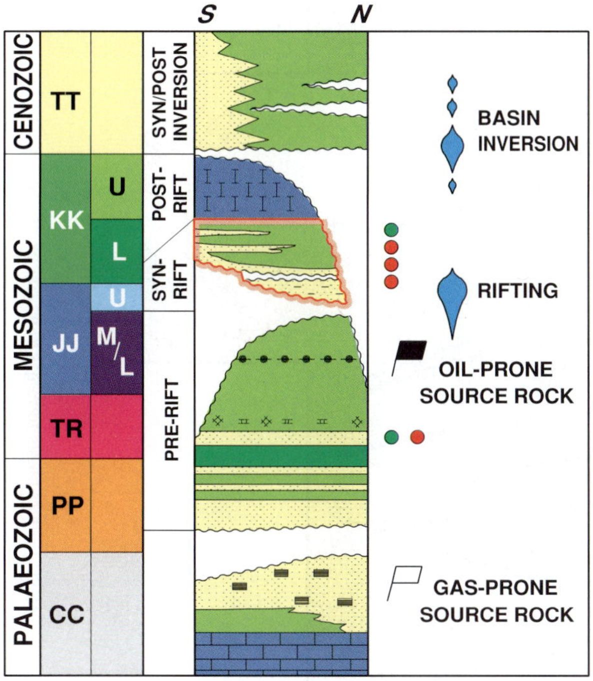

Fig. 2. Generalised stratigraphy of the West Netherlands Basin. The post-Variscan stratigraphy has been subdivided into four sequences (pre-, syn- and post-rift followed by inversion) using the two main tectonic episodes (rifting and basin inversion) as references. The diagram also shows the distribution of oil and gas reservoirs (red and green circles respectively) and source rocks. CC = Carboniferous; PP = Permian; TR = Triassic; JJ = Jurassic; KK = Cretaceous; TT = Tertiary.

and iv) a syn- and post-inversion sequence of Late Cretaceous – Tertiary to Quaternary age.

The pre-rift section starts with Permo-Triassic continental clastics (Zijp 1987). These contain gas reservoirs in the West Netherlands Basin, sourced by gas-prone source rocks within the Westphalian coal measures (De Jager et al. 1995). The Lower and Middle Jurassic consist of mainly marine shales including the Lower Jurassic kerogen-rich Posidonia Shale, the main source of the oil in the West Netherlands Basin as well as in the Broad Fourteens and Vlieland basins (Bodenhausen & Ott 1981, Roelofsen & De Boer 1991, Herngreen et al. 1991).

The syn-rift sequence is divided into two groups. The lower part comprises the continental clastics of the Upper Jurassic – Lower Cretaceous Delfland Subgroup. The upper part consists of the remaining Lower Cretaceous which comprises two marine sequences: the Holland and Vlieland Sandstone Formations. The shallow marine clastics of this interval form the main oil reservoirs in the West Netherlands Basin.

The post-rift Lower – Upper Cretaceous sequence consists mainly of thick chalk deposits accumulated over the basin and adjacent highs. This was followed by basin inversion and deposition of the clastics of the Lower North Sea Group.

The structural style of the West Netherlands Basin is a result of these two tectonic events; rifting and basin inversion. This is illustrated in a regional seismic line in Fig. 3. The pre-rift sequence has been block faulted; the horsts forming traps for the Triassic gas play. The Lower and Upper Cretaceous syn- and post-rift sequences exhibit a folded overburden geometry over the rift basin. Asymmetrical fault-bounded anticlines form traps for oil. The syn- and post-orogenic clastics of the North Sea Group unconformably overlie the folded Cretaceous strata. Evidence for inversion is clearly visible with the Tertiary clastics thinning over the former Cretaceous basin and thickening towards the margins of this basin.

Due to the reactivation of some of the pre-existing rift faults, some faults (e.g. the western boundary of the De Lier oil field, Fig. 3) have normal offset in the pre-rift sequence and reverse offset in the syn- and post-rift sequence. This reactivation also gives a corresponding structuration between the rift structures and the inversion anticlines. For example, the Triassic Monster gas field, a rift horst, underlies an inversion anticline, the De Lier oil field (Fig. 3).

A regional time map at Top Vlieland Subgroup level (Base Aptian) illustrates the continuity of the central NW – SE oriented inverted anticlinal trend which reflects the continuity of the rift faults (Fig. 4). This central trend includes the Rijswijk, Oude-Leede, Berkel and IJsselmonde/Ridderkerk Fields.

Reservoir rocks

Oil-shows have been reported in the West Netherlands Basin from all levels between the Tertiary and the Triassic Bunter Group. Oil-bearing reservoirs, which are at present economically attractive, are restricted to the Lower Cretaceous (Holland and Vlieland Sandstone Formations and Delfland Subgroup; Figs 5, 6) and, where juxtaposed against the Posidonia source rocks, the Triassic (Bunter Group).

Three excellent Lower Cretaceous reservoirs are distinguished, containing some 65% of the total ultimate recovery. These are the Berkel and IJsselmonde

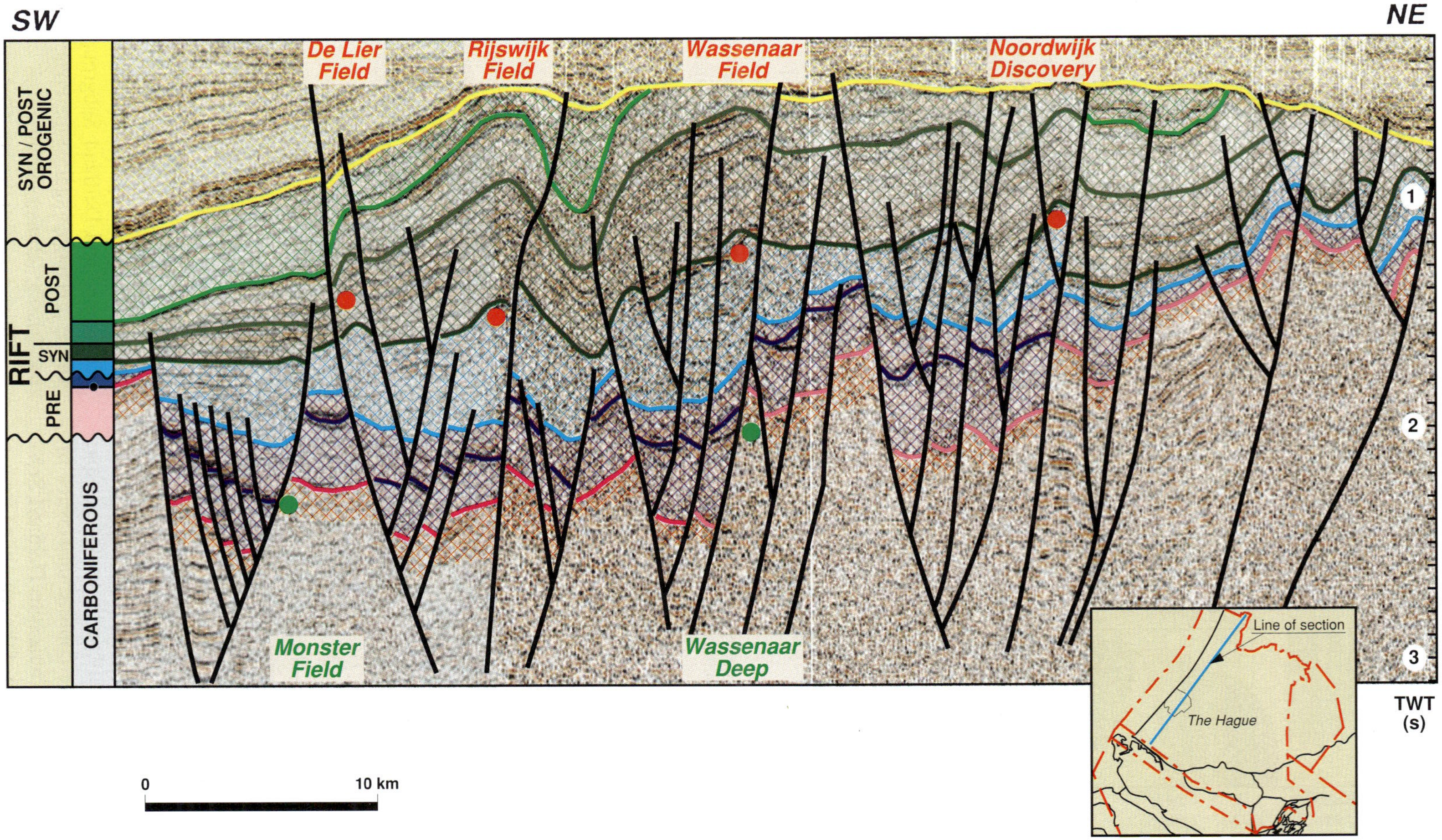

Fig. 3. Regional seismic line through the West Netherlands Basin. The line illustrates both rifting and basin inversion. Bunter horsts form gas traps and oil has accumulated in assymetrical fault-bounded anticlines in the syn- and post-rift sequences. See Fig. 2 for time stratigraphy.

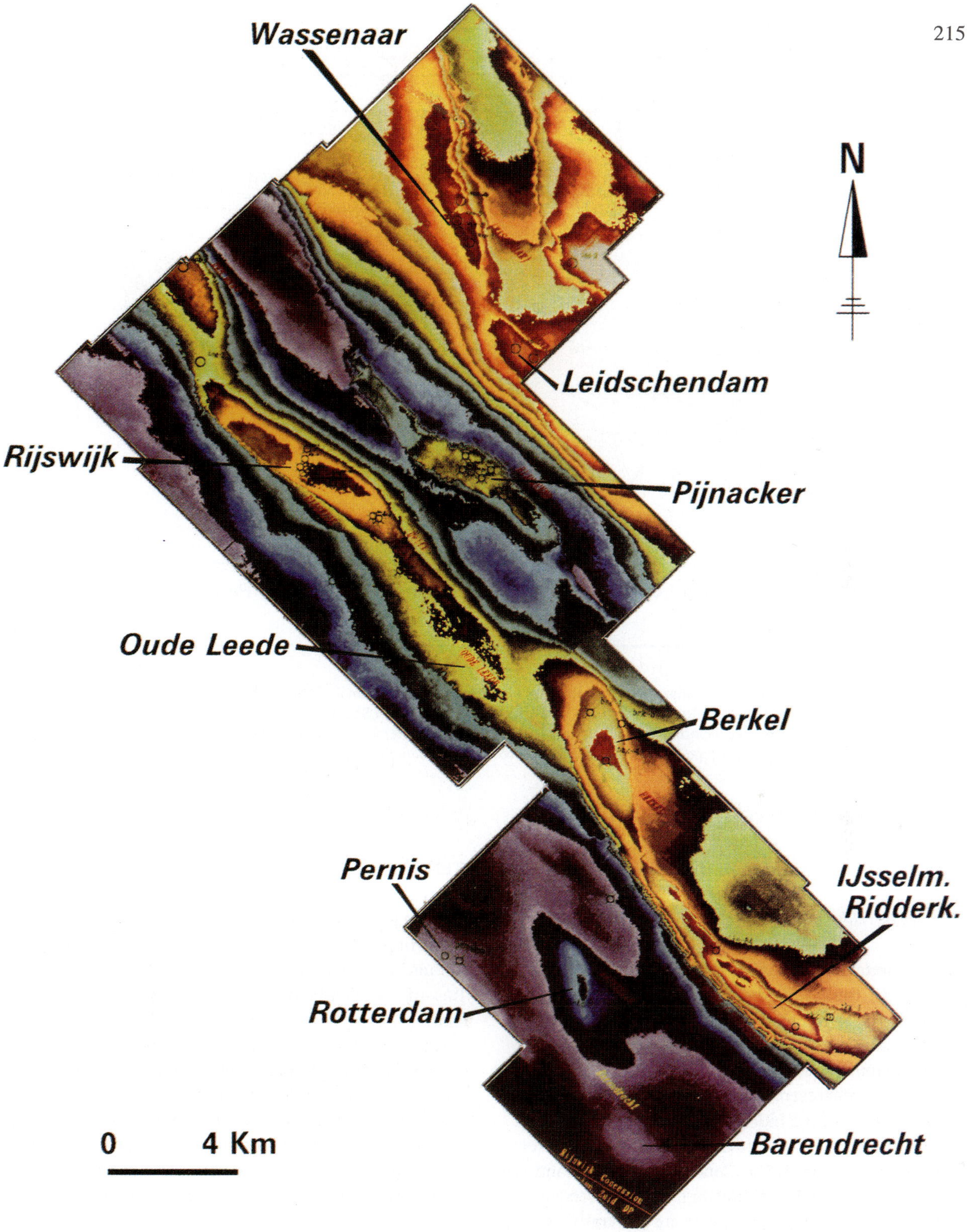

Fig. 4. Top Vlieland Subgroup time map (Base Aptian). The map illustrates the continuity of the central NW – SE oriented inverted anticlinal trend which reflects the continuity of the rift faults. Colour code ranges from 700 (red) to 1700 ms TWT (dark purple).

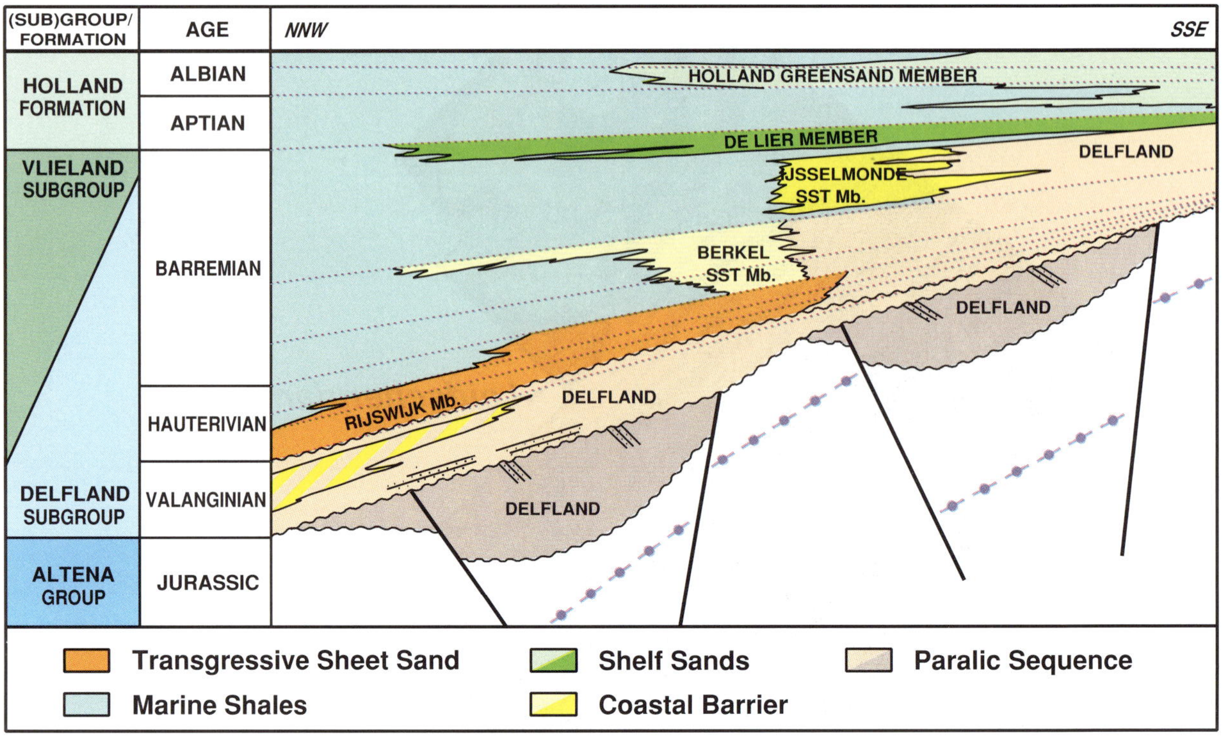

Fig. 5. Schematic Lower Cretaceous stratigraphy of the Rijswijk Concession. The diagram illustrates the stratigraphic position and geometry of the oil-bearing reservoirs in the West Netherlands Basin. The IJsselmonde and Berkel Sandstone Members are barrier complexes, whilst the Rijswijk Member is a transgressive deposit. The De Lier and Holland Greensand Members represent shelf deposits of poor reservoir quality. See Fig. 6 for the location of this conceptual section. Blue symbols below Delfland indicate Posidonia Shale.

Sandstone Members, both coastal barrier complexes, and the transgressive Rijswijk Member (Den Hartog Jager 1995). They are described below.

Coastal barrier sand complexes

Two separate complexes are distinguished; the IJsselmonde and Berkel Sandstone Members, comprising generally E – W oriented (locally NNW – SSE) coastal barrier sand complexes deposited along the fringe of the West Netherlands Basin (Figs 5, 6). They represent the most prolific reservoirs in the basin, where the net sand thickness of the individual complexes may reach 120 m. The permeability of the sands is good to excellent (400 to > 3000 mD). Cementation is low and is generally restricted to nodules in thin, but often correlatable layers. Average porosities are generally in the range of 20 to 30%. Lateral sand continuity is good. Most of the oil presently being produced comes from these sediments.

Transgressive sheet sands

This comprises the Rijswijk sandstone (Rijswijk Member), a diachronous, transgressive marine rock unit representing the onset of the Vlieland marine ingression (Figs 5, 6). It was deposited unconformably over an uneven substratum of (mainly) continental clastics of the Delfland Subgroup. The effects of this marine transgression were a reworking of the underlying Delfland Subgroup and a progressive infill of the existing palaeotopography. This topography was strongly related to the existing active syn-sedimentary faults. A clear relationship between Rijswijk sandstone thickness and palaeotopography has been observed in the Rijswijk, Pijnacker, Zoetermeer and Moerkapelle Fields, where the reservoir thickness varies between 0 and 70 m.

This transgressive sand sheet consists of fine to coarse-grained, cross-bedded, often bioturbated sandstone with some thin lagoonal shale intercalations, marine fossils and glauconite. Occasionally, calcare-

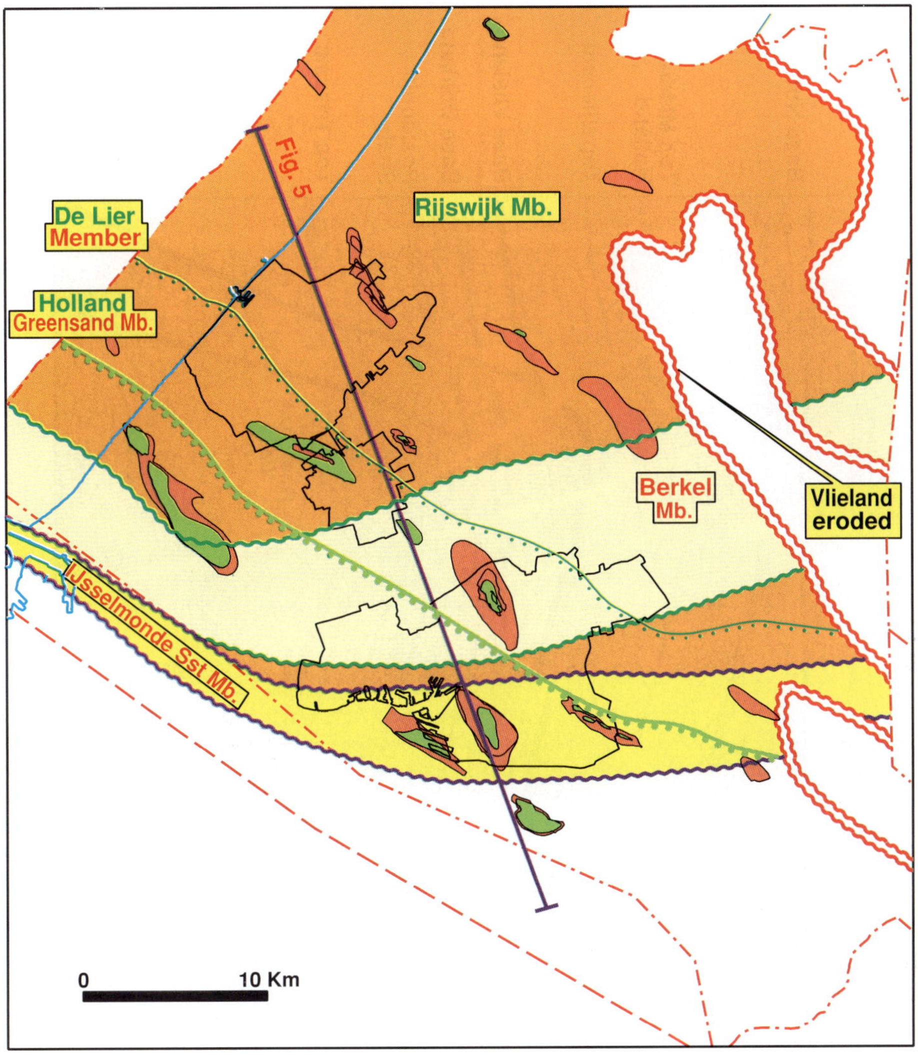

Fig. 6. Lower Cretaceous reservoir distribution in the Rijswijk Concession. The Berkel and IJsselmonde Sandstone Member barrier complexes which are excellent reservoirs in the West Netherlands Basin, are orientated approximately E – W (Fig. 15 shows the distribution of the IJsselmonde Sandstone in the Rotterdam area). The reservoir quality of these members deteriorates towards the north. The northern limits of reservoir quality in the De Lier and Holland Greensand Members are illustrated. The reservoir quality of the Rijswijk Member also deteriorates northward over the basin.

ous cementation is observed at the top of the sequence. The lateral continuity of this sand is excellent due to its depositional origin, with typical average porosities ranging between 15 and 28%, net/gross ratios of 60 to 85% and permeabilities up to 4000 mD. This reservoir, developed over the north and northwest of the concession, has been a prolific producer in the past. All oil producing wells from this reservoir have been or are being abandoned.

In addition, two reservoir sections of rather poor reservoir quality are distinguished:

Shelf sand deposits

These comprise the De Lier and Holland Greensand Members, consisting of thin, argillaceous sands and silts with good to excellent lateral continuity (Figs 5, 6). Their permeability is low to moderate (from < 10 up to 400 mD); cementation is restricted to thin layers. The average porosities generally range between 16 and 28%. Strong bioturbation is often present and gross reservoir thickness can be up to 100 m. Net/gross ratios are generally low, decreasing rapidly northwards

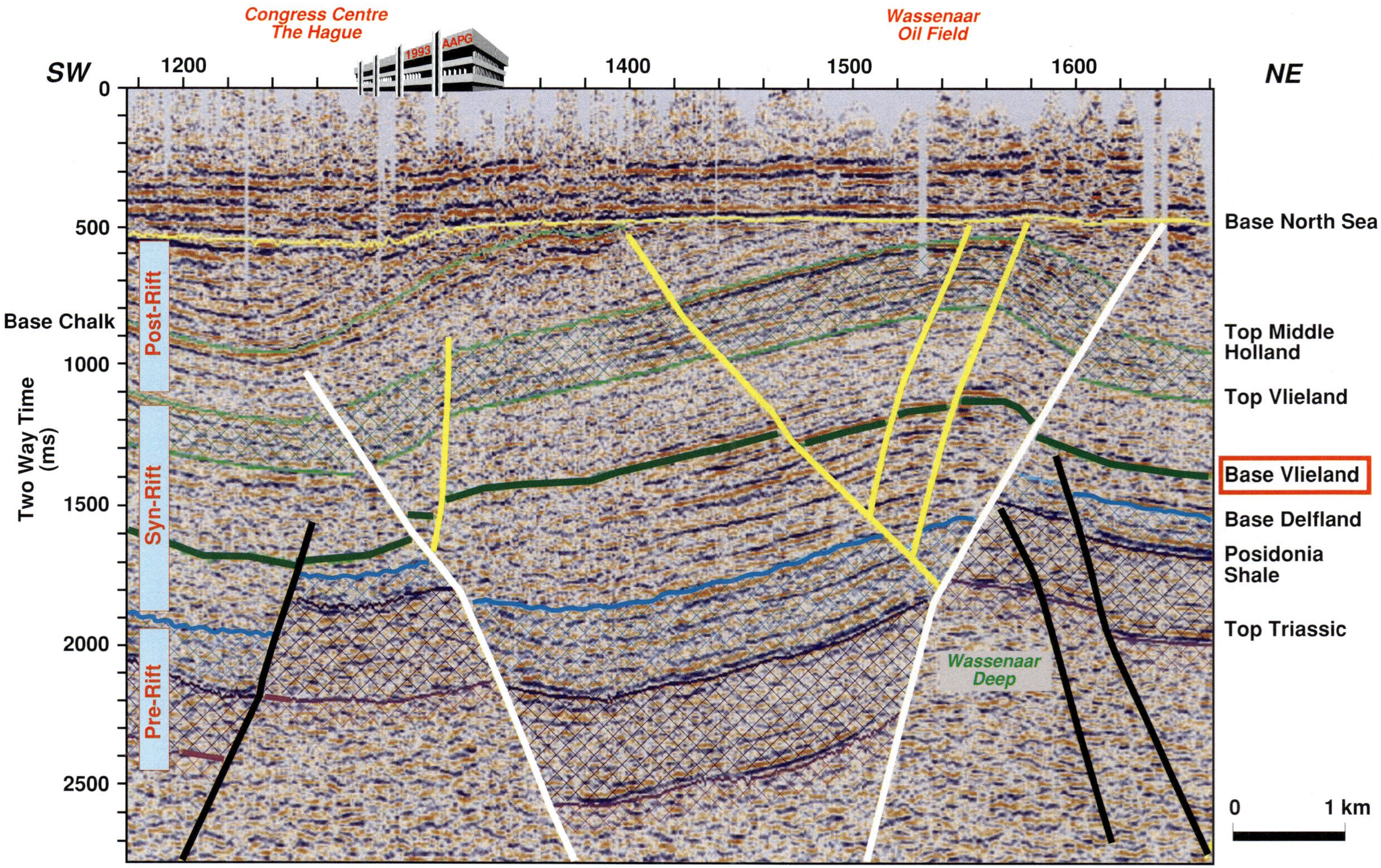

Fig. 7. Seismic line across the inverted Wassenaar Graben. (For location Figs 8, 9). The Wassenaar Field is essentially an inverted graben. The yellow faults have throws which increase with depth in the syn-rift sequence and are therefore syn-sedimentary in nature. This implies they are contemporaneous with the white graben-bounding faults. The syn-rift sequence thickens dramatically between the two white faults, both of which were re-activated during basin inversion. The yellow faults have not been reactivated. This illustrates the structural correspondence between rift structures and inversion anticlines. The white fault to the northeast bounds the Triassic Wassenaar Deep gas accumulation in the foot-wall and the Lower Cretaceous oil accumulation in the Rijswijk sandstone at Base Vlieland in the hanging wall.

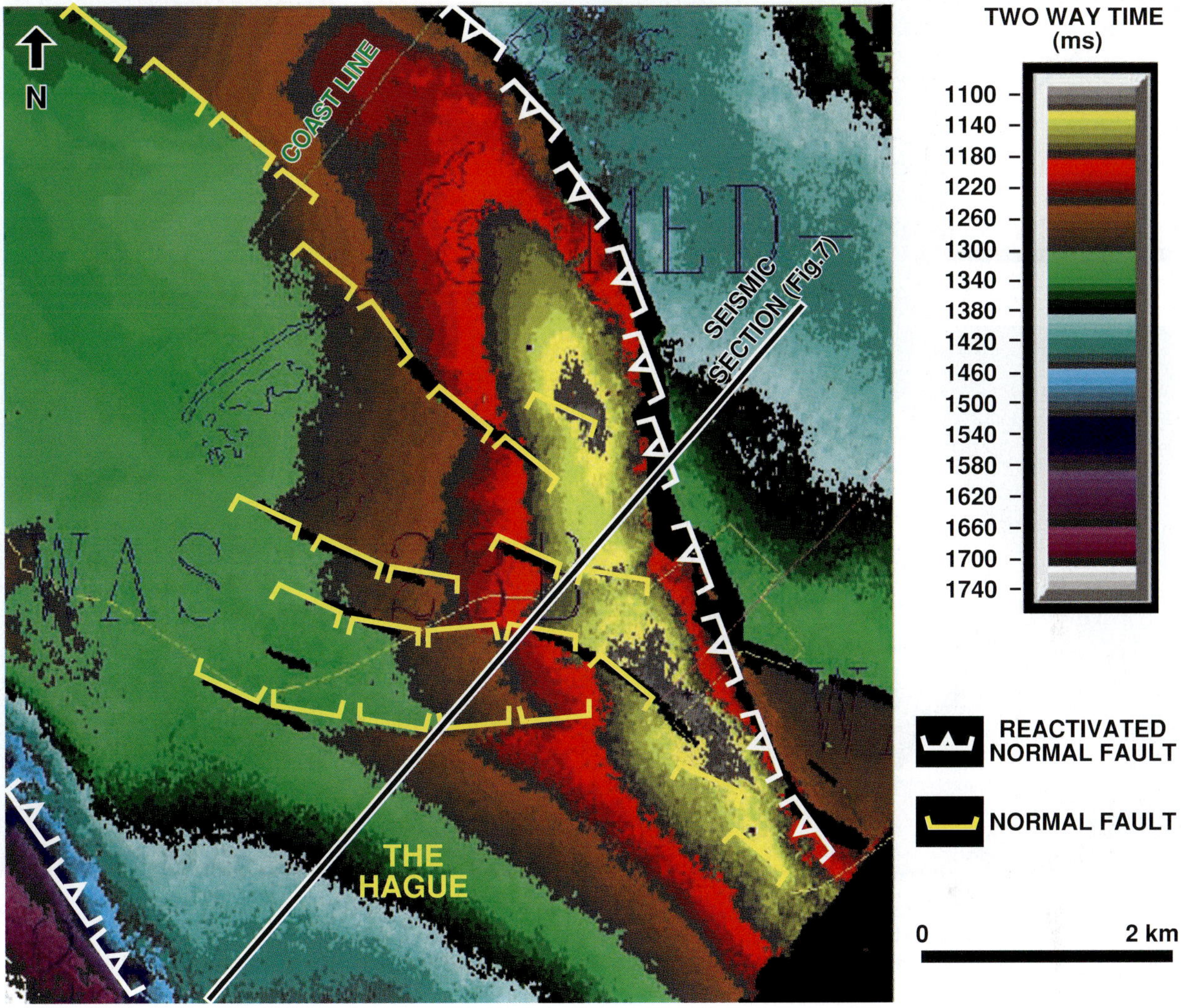

Fig. 8. Wassenaar Field; Base Vlieland time map. The white and yellow coloured faults display different orientations; the white reactivated faults have a NNW – SSE trend, whilst the yellow normal faults have a WNW – ESE trend. The indicated line is shown as Fig. 7.

from the basin margin (e.g. Rotterdam Field) towards the basin centre where these sections are of very poor to non-reservoir quality (e.g. Pijnacker Field). Minor production comes from these sediments at Rotterdam and Berkel.

Coastal plain deposits

The paralic deposits of the Delfland Subgroup comprise stacked, laterally discontinuous, sand bodies alternating with shales and coalbeds (Fig. 5). Large variations in permeability (< 100 to 3000 mD) and cementation are typical for this subgroup. The porosity of the sands varies between < 10 to 27%. The total net sand thickness of the subgroup can amount to more than 300m. Oil production from these complex sand bodies presently comes from the IJsselmonde/Ridderkerk Field and formerly from the Moerkapelle, Zoetermeer, Wassenaar and Pijnacker Fields.

Field examples

Three case examples (Wassenaar, Pijnacker and Rotterdam Fields, Fig. 1) are discussed below. These illustrate different facets of the tectonic evolution and reservoir development of the West Netherlands Basin.

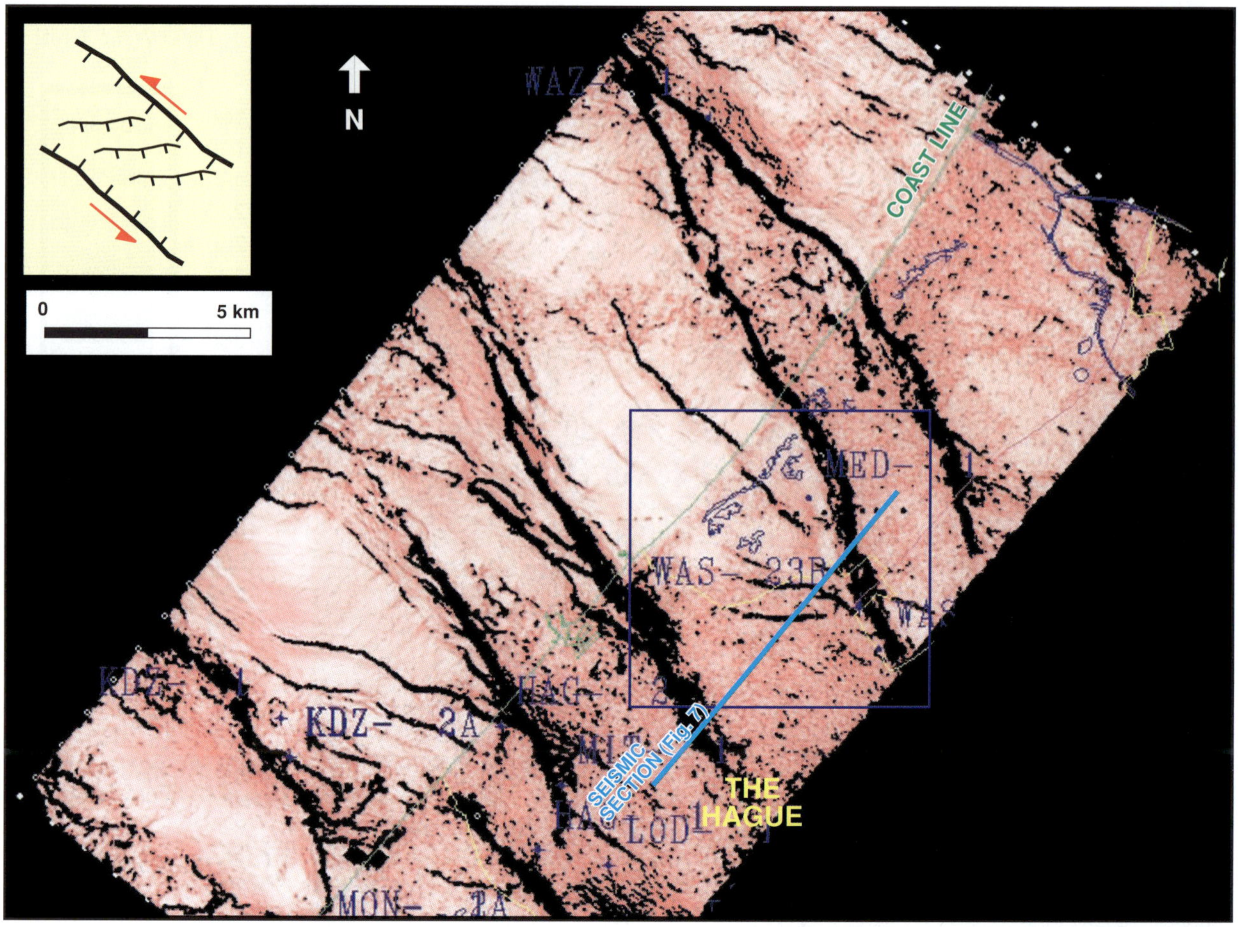

Fig. 9. Base Vlieland dip map, The Hague area. This horizon display map shows a transtensional (left-lateral) fault arrangement. Faults are shown as steeply dipping events (in black). Lower structural dips are shown as red, grading through pink to white for horizontal strata (zero dip). The indicated seismic line is shown in Fig. 7.

The Wassenaar Field shows evidence of Early Cretaceous transtensional rifting; Pijnacker is an example of a small inverted pull apart basin, a classic 'pop-up' structure (as a result of compression and inversion) with implications for reservoir development, whilst the Rotterdam Field, located on the fringe of the West Netherlands Basin, displays multiple reservoirs.

Wassenaar Field

The now abandoned Wassenaar Field consists of an elongated, asymmetrical NNW – SSE trending faulted anticline. A seismic line illustrates the structure which is essentially an inverted graben (Fig. 7). The syn-rift sequence shows dramatic thickness increase between the two faults, indicated in white in Fig. 7. These two faults were re-activated during basin inversion. The shoulders of this graben are two Triassic horst blocks, one of which comprises the Wassenaar-Deep gas field (Figs 3, 7). The Triassic gas accumulation is located in the foot-wall and the Lower Cretaceous oil accumulation in the hanging-wall.

A time map at base Vlieland shale level (top Rijswijk sandstone reservoir) level is shown as Fig. 8.

All these faults (white and yellow) acted as normal faults during the Early Cretaceous rifting phase. Transtensional (left-lateral) fault arrangement is also illustrated in a dip extraction map of the base Vlieland Subgroup covering a larger area around The Hague (Fig. 9).

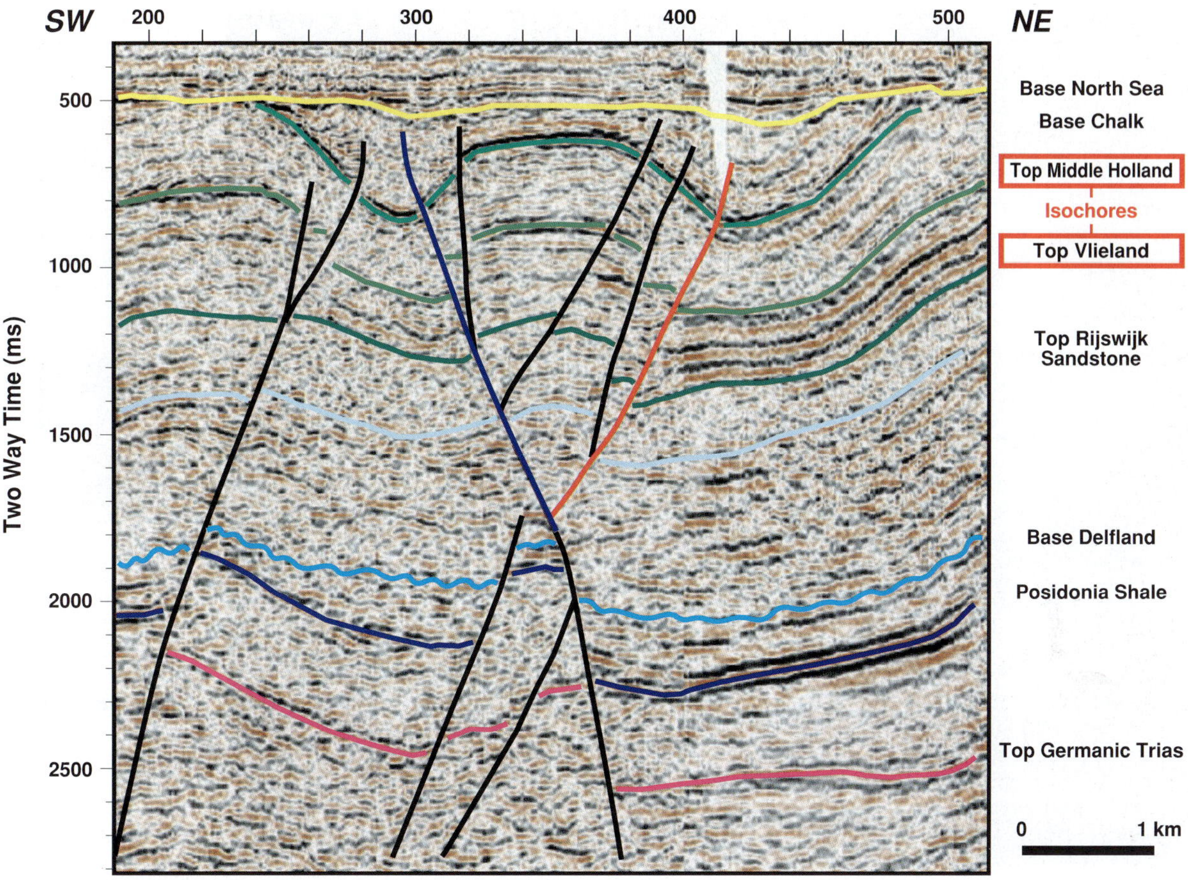

Fig. 10. Seismic line across the Pijnacker Field. The section illustrates the inverted 'pop-up' anticline of the Pijnacker structure. Note the thickening of the Top Middle Holland to Top Vlieland interval over the crest (see also Fig. 11). The bounding faults are shown in red and blue (see also Fig. 12), a set of reverse faults in between in black.

Pijnacker Field

The Pijnacker Field is located approximately 10 km SE of the Wassenaar Field. It comprises a NW – SE trending, inverted, S-shaped 'pop-up' anticline. A seismic profile is shown in Fig. 10. The correspondence between the rift and inversion structuration is again noted. The structure is bounded by two faults with, in addition, a set of reverse faults intersecting the structure. Thickness variations (time isochores) of the Top Vlieland to Top Middle Holland interval show clear evidence of thickening between the two bounding faults in the central crestal part of the Pijnacker structure (Figs 10, 11). Two main phases in the structural evolution of Pijnacker are distinguished (Fig. 12):

Phase I (Early Cretaceous) comprises the development of the sinistral 'pull apart' basin and the development of two half grabens within this 'pull apart' basin. The transgressive Rijswijk sandstone reservoir displays syn-sedimentary thickness variations with thicker reservoir development on the downthrown sides of the main boundary faults (i.e. in the 'pull apart' basin, Fig. 13). The 'pull apart' basin was active during deposition of the Rijswijk sandstone, resulting in a thickening of the reservoir at the depocentres.

Phase II (Early Tertiary). Development of reverse faulting, which did not influence the thickness of the Rijswijk Member reservoir. From the orientation of these faults, it is concluded that the Tertiary compression had a dextral strike-slip component. At Pijnacker, there is evidence of dual inversion in the vertical and lateral senses. In the vertical sense, the pull apart basin, a former 'low', is now a 'pop up' anticline. In the lateral sense, the master fault, active in the Early Cretaceous as a left-lateral

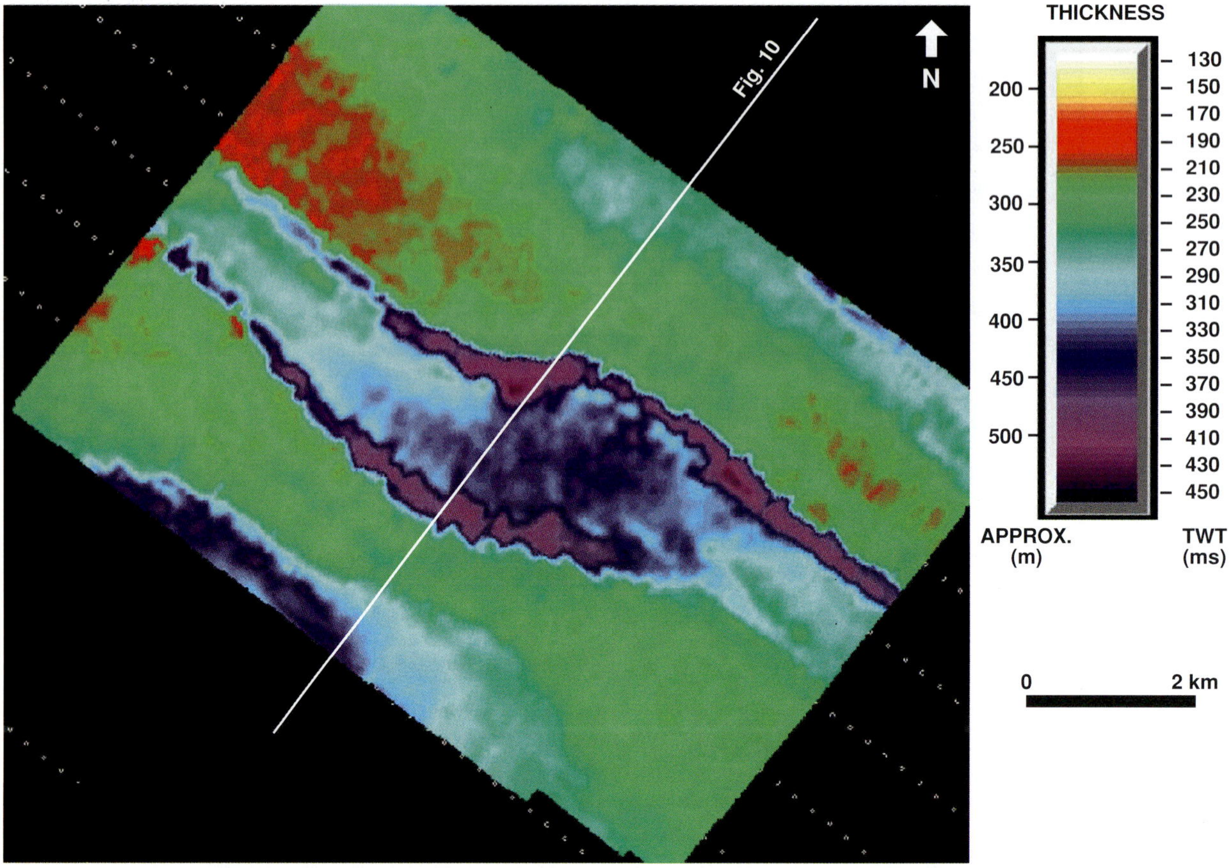

Fig. 11. Time isochores Top Middle Holland to Top Vlieland in the Pijnacker Field. Note the clear thickening of the time isochores over the crest between the bounding faults (see also Fig. 10). The thickness along (close to) the reverse boundary faults is exaggerated since the top and the base of the isochore interval are strongly dipping and in different blocks.

transtensional feature, acted during the later compressional phase as a right-lateral transpressional fault.

Rotterdam Field

The Rotterdam Field, located on the fringe of the Lower Cretaceous basin, comprises a mildly faulted, N – S trending inversion anticline with a maximum relief of some 190 m. A seismic profile is shown in Fig. 14. The field produces around 1450 m^3/day (9000 bbl/day), mainly from the IJsselmonde Sandstone Member although most of the oil is contained in the poor-quality shelf sand deposits (Fig. 1). The seismic section shows three kinds of faults; a Cretaceous non-reactivated normal fault (labelled ‘A’ in Fig. 14), a Cretaceous reactivated normal fault (‘B’) and an inversion-related normal fault (‘C’). These suggest a transpressional regime for this structure.

The Rotterdam Field contains three separate Cretaceous oil reservoirs; the excellent IJsselmonde Sandstone Member and the poorer De Lier and Holland Greensand Members. Minor hydrocarbon volumes have also been encountered in the Delfland Subgroup.

A seismic amplitude extraction map at the top of the 115 m thick IJsselmonde Sandstone is shown in Fig. 15. This clearly shows the E – W trend of the coastal barrier complex. Well data confirm a regional shale-out northwards to marine shales (Berkel Field) and southwards to lagoonal deposits (e.g. Barendrecht Field). The Rotterdam Field is situated in the zone of maximum sand thickness. This amplitude extraction map can be used as a palaeogeographical or as a facies distribution map. This is important in order to

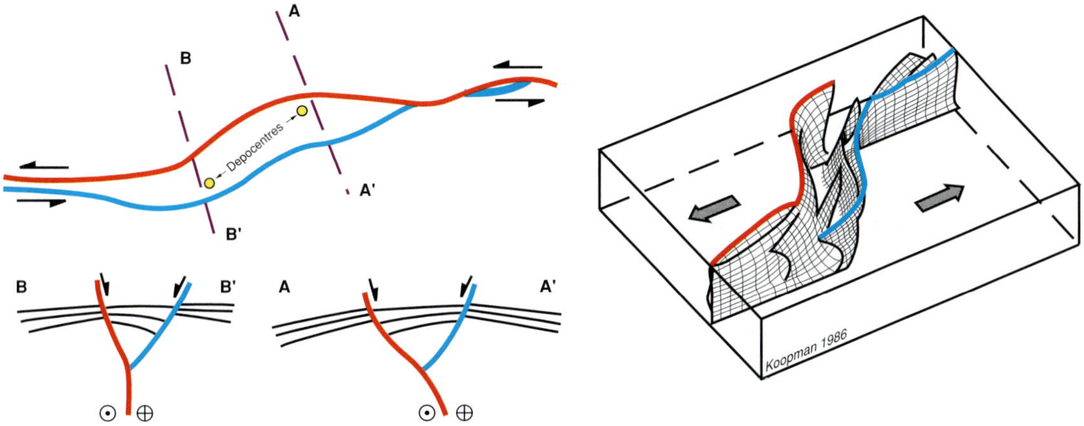

PHASE I:

- Lower Cretaceous.
- Transtensional left-lateral regime (pull-apart).
- Development of half-grabens within the pull-apart basin.
- Synsedimentary thickness variations and development of negative flower structures (tulip type).

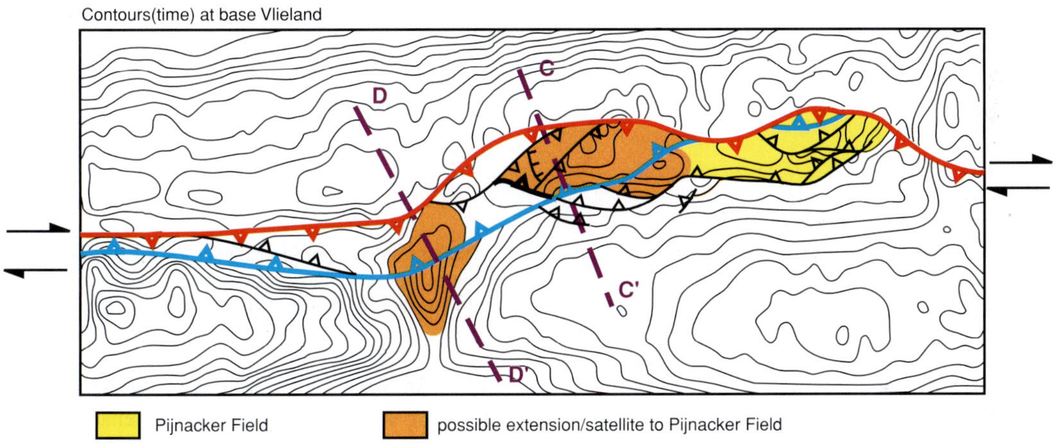

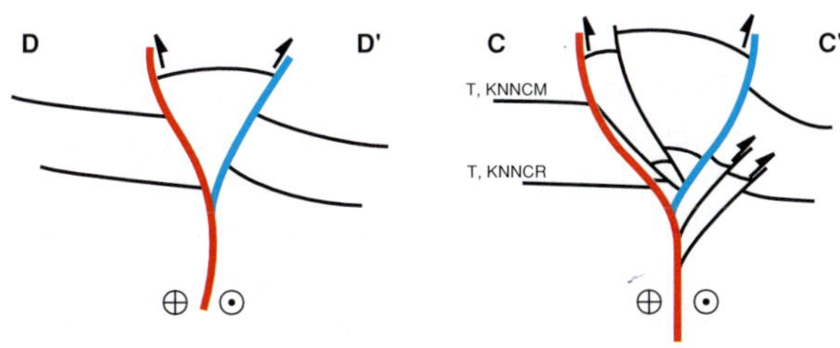

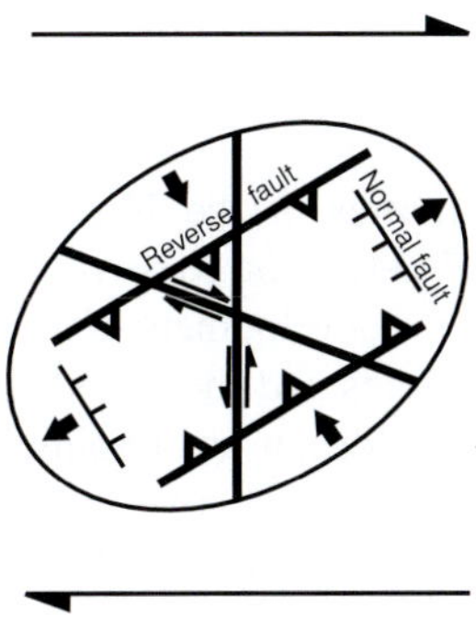

PHASE II:

- Lower-Middle Tertiary.
- Transpressional right-lateral regime (pop-up).
- Development of positive flower structures (palm-type) within the former pull-apart basin.
- Reactivation of (some) Lower Cretaceous faults.

Fig. 12. Structural evolution of the Pijnacker Field. The red- and blue-coloured faults represent the Pijnacker 'pull apart' basin boundary faults (see text). The yellow and orange colours on the time map represent structural highs. The yellow one is the Pijnacker Field. The orange structures are a possible field extension and/or satellite. Top KNNCM and top KNNCR in the sketch sections C – C′ and D – D′ refer to Top Vlieland Subgroup (Vlieland Claystone Fm) and Top Rijswijk sandstone, respectively.

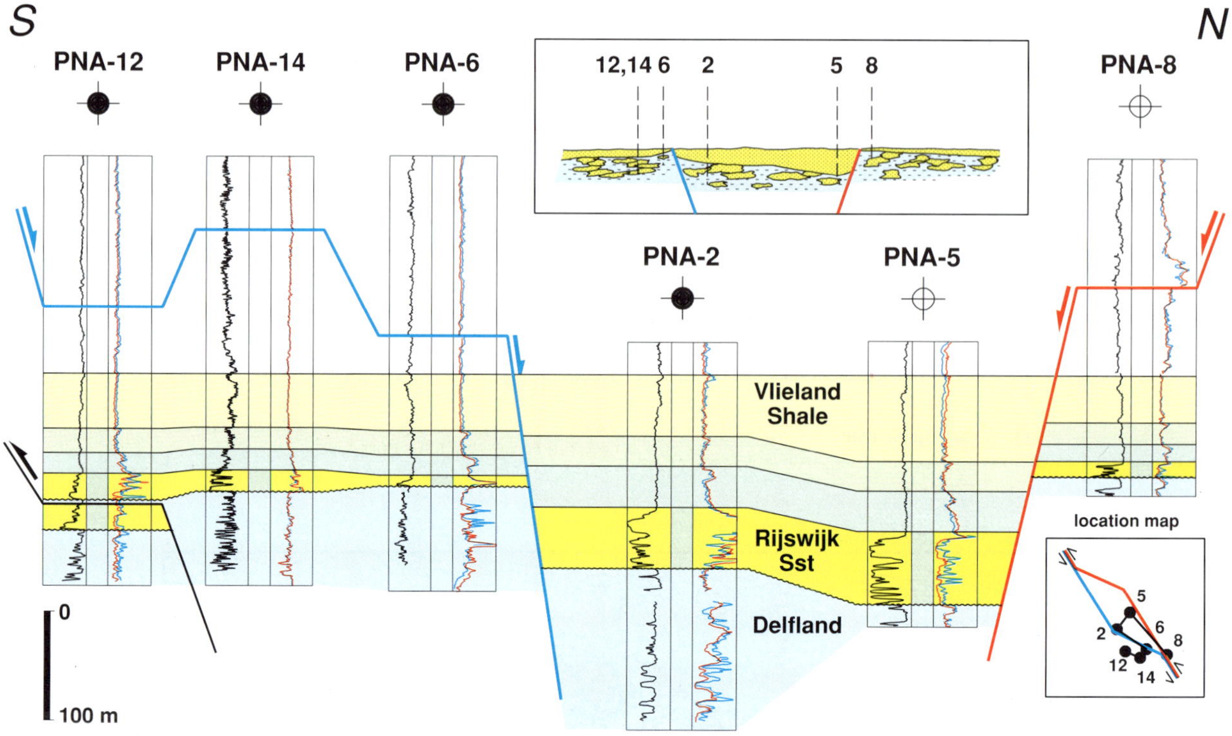

Fig. 13. Log correlation of the Rijswijk sandstone in the Pijnacker Field. The logs shown are Spontaneous Potential (SP)/Resistivity logs, except in wells PNA-12 and PNA-14 where they are Gamma Ray (GR)/Resistivity. The datum horizon is an intra Vlieland Subgroup marker. Note the thickening of the Rijswijk sandstone on the downthrown sides of the main boundary faults (wells PNA-2 and PNA-5).

fully optimise reservoir geological models as part of maximising oil production and assessment of reservoir development for prospect evaluation.

The future

The present activities of the Nederlandse Aardolie Maatschappij (NAM) in the Rijswijk Concession concern completing the development of the 'newer' Berkel and Rotterdam Fields and the abandonment of older, depleted fields. In addition, an ongoing exploration and petroleum engineering initiative is evaluating further exploration and appraisal prospects in the region following the acquisition of 3D seismic over most of the concession. Amplitude studies will be used to further delineate the E – W trending barrier sand complexes. This is important to refine reservoir geological models as part of optimising production from existing fields and prospect evaluation.

One of the main future challenges is to unlock the considerable STOIIP (ca. 87×10^6 m^3 or 550×10^6 bbls) contained in the poor-quality shelf sand deposits (De Lier sand-shale; Holland Greensand) which contain 41% of the STOIIP in the Rijswijk Concession but which to date have hardly been produced. The use of horizontal well technology is expected to play an ever increasing role (Murphy 1990), as well as slim hole drilling techniques and sharing production facilities for future discoveries ('clustering'). NAM has recently successfully drilled horizontal wells at Berkel. 3D-Reservoir modelling studies are ongoing to help further develop the large volumes of STOIIP (heavy oil) contained in the 'channel-type' Delfland Subgroup (Nieuwerkerk Formation) reservoir of complex 'channel' sand geometry.

Conclusions

Regional 3D seismic coverage has resulted in much improved understanding of the tectonically and stratigraphically complex West Netherlands Basin.

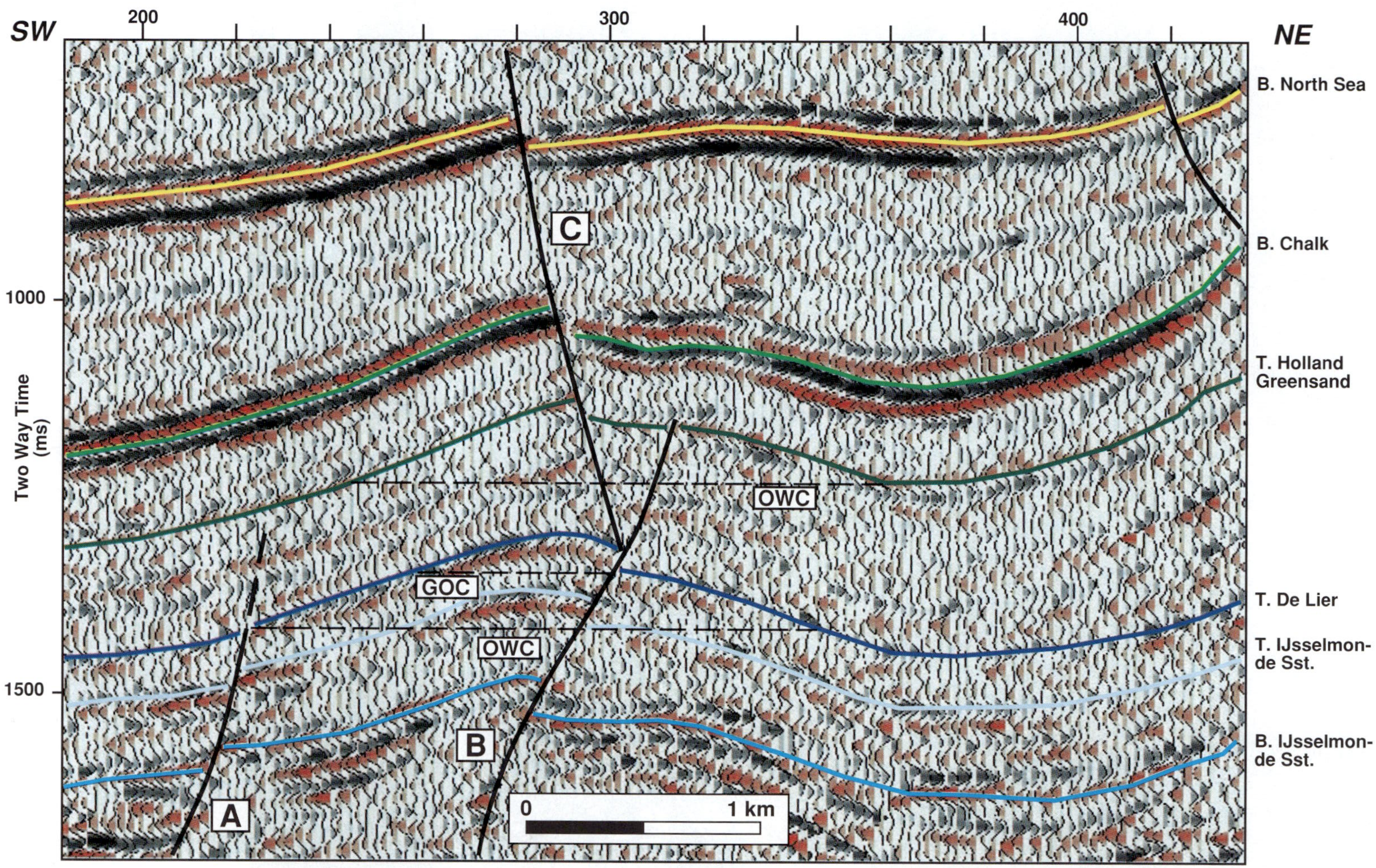

Fig. 14. Seismic line across the Rotterdam Field (For location see Fig. 15). A = non-reactivated normal fault; B = reactivated normal fault; C = inversion-related normal fault. The fault arrangement suggests a transpressional nature for the Rotterdam structure. OWC = Oil – Water Contact; GOC = Gas – Oil Contact.

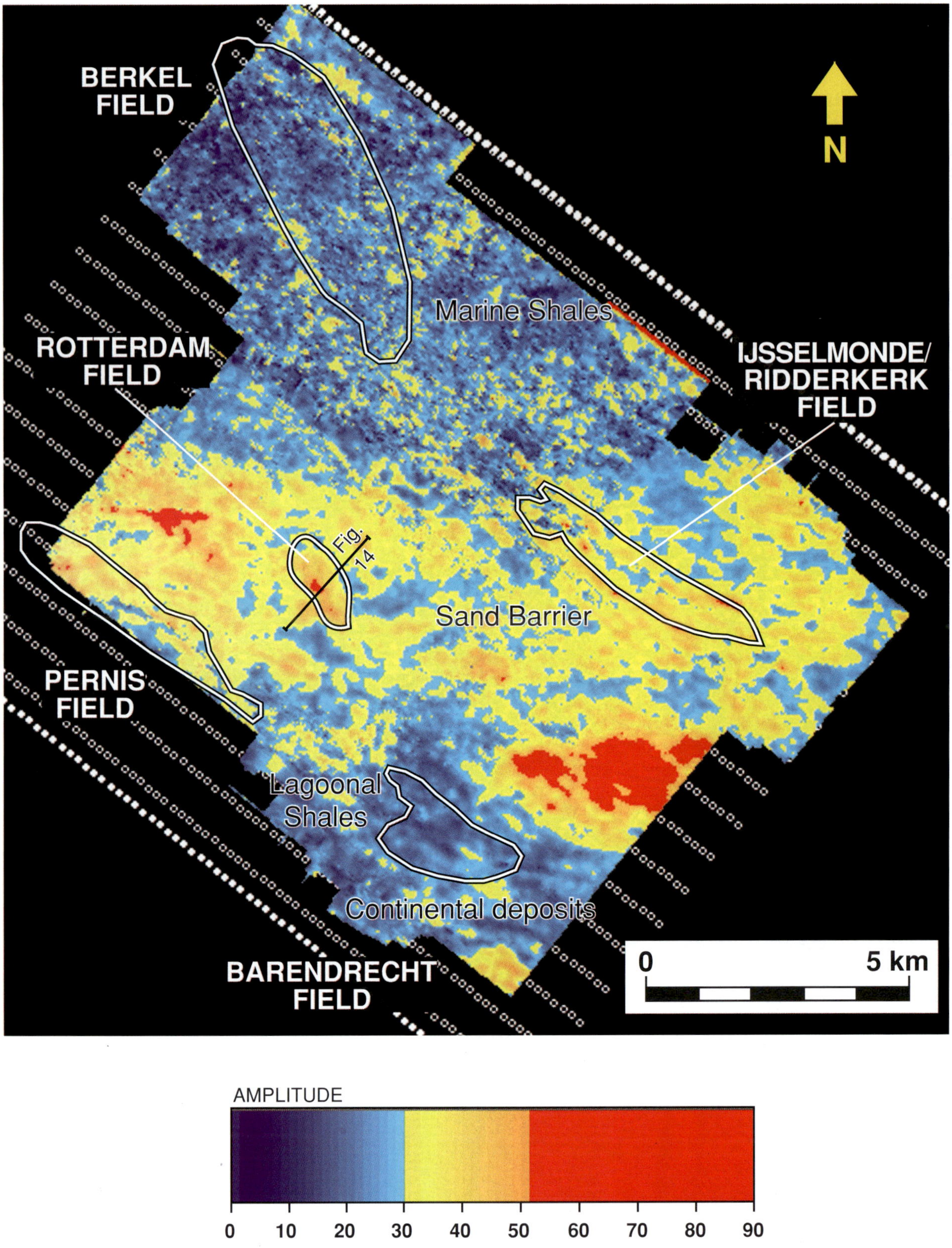

Fig. 15. Waalhaven 3D survey; Top IJsselmonde Sandstone amplitude map showing reservoir development. The E – W trending 'yellow'-coloured area shows the excellent quality of the IJsselmonde Sandstone Member at Rotterdam, shaling-out northwards towards the Berkel Field and southwards towards the Barendrecht Field. See Fig. 6 for the regional distribution of the IJsselmonde Sandstone Member.

Two main tectonic events, Late Jurassic to Early Cretaceous rifting and Late Cretaceous to Early Tertiary inversion dominate the post-Variscan geological history of the basin. A left-lateral component developed during the rifting phase, at least during the Hauterivian – Barremian (probably also during the Aptian and part of the Albian; e.g. Wassenaar Field). Right-lateral transpression is noted during the subsequent inversion phase in some structures (e.g. Pijnacker Field).

Palaeotopography controls the variable thickness of the transgressive Rijswijk Member as noted at Pijnacker where palaeotopography is strongly influenced by left-lateral rifting faults. Palaeogeography (location of the fringe of the basin) controls the distribution of the barrier sand complexes (e.g. the IJsselmonde Sandstone Member) and shelf sand deposits.

Acknowledgements

The authors would like to thank all NAM Exploration and Production staff, too numerous to mention individually, who have contributed to a better understanding of the West Netherlands Basin. We would also like to thank NAM, Shell Internationale Petroleum Maatschappij (SIPM) and Exxon Company, International (ECI) for granting permission to publish this paper, originally presented at the 1993 AAPG International Conference and Exhibition in The Hague.

References

Bodenhausen, J.W.A. & W.F. Ott 1981 Habitat of the Rijswijk Oil Province, Onshore, The Netherlands. In: Illing, L.V. & G.D. Hobson (eds) Petroleum geology of the continental shelf of North-West Europe – Institute of Petroleum, London: 301–309

Burgers, W.F.J. & G.G. Mulder 1991 Aspects of the Late Jurassic and Cretaceous history of The Netherlands – Geol. Mijnbouw 70: 347–354

Den Hartog Jager, D.G. 1995 Fluviomarine sequences in the Lower Cretaceous of the West Netherlands Basin: correlation and seismic expression. In: Rondeel, H.E., D.A.J. Batjes & W.H. Nieuwenhuijs (eds) Geology of gas and oil under the Netherlands – this volume

Dronkers, A.J. & F.J. Mrozek 1991 Inverted basins of The Netherlands – First Break 9: 409–425

De Jager, J., M.A. Doyle, P.J. Grantham & J.E. Mabillard 1995 Hydrocarbon habitat of the West Netherlands Basin. In: Rondeel, H.E., D.A.J. Batjes & W.H. Nieuwenhuijs (eds) Geology of gas and oil under the Netherlands – this volume

Herngreen, G.F.W., R. Smit & Th.E. Wong 1991 Stratigraphy and tectonics of the Vlieland basin, The Netherlands. In: Spencer, A.M. (ed) Generation, accumulation and production of Europe's hydrocarbons – Eur. Ass. Petroleum Geosc. Eng. Spec. Publ. 1 – Oxford University Press, Oxford: 175–192

Knaap, W.A. & M.J. Coenen 1987 Exploration for oil and natural gas. In: Visser, W.A., J.I.S. Zonneveld & A.J. van Loon (eds) Seventy-five years of geology and mining in the Netherlands – (1912–1987). Roy. Geol. Mining Soc. Netherlands (KNGMG), The Hague: 207–231

Murphy, P.J. 1990 Performance of horizontal wells in the Helder Field – J. Petroleum Technology 42: 792–800

PGK (Petroleum Geological Circle) 1993 Synopsis: Petroleum geology of the Netherlands – 1993. In: Rondeel, H.E., D.A.J. Batjes & W.H. Nieuwenhuijs (eds) Geology of gas and oil under the Netherlands – this volume

Roelofsen, J.W. & W.D. de Boer 1991 Geology of the Lower Cretaceous Q/1 oil-fields, Broad Fourteens basin, The Netherlands. In: Spencer, A.M. (ed) Generation, accumulation and production of Europe's hydrocarbons – Eur. Ass. Petroleum Geosc. Eng. Spec. Publ. 1 – Oxford University Press, Oxford: 203–216

Van Wijhe, D.H. 1987 Structural evolution of inverted basins in the Dutch offshore – Tectonophysics 137: 171–219

Zijp, F.R. 1987 Structural evolution, stratigraphic sequences and subsurface reservoir horizons. In: Visser, W.A., J.I.S. Zonneveld & A.J. van Loon (eds) Seventy-five years of geology and mining in the Netherlands – (1912–1987). Roy. Geol. Mining Soc. Netherlands (KNGMG), The Hague: 269–285

Rondeel et al. (eds), Geology of gas and oil under the Netherlands, 229–241, 1996.

Fluviomarine sequences in the Lower Cretaceous of the West Netherlands Basin: correlation and seismic expression

D.G. den Hartog Jager
Nederlandse Aardolie Maatschappij BV (Business Unit Exploration, Dept. XEX/2), Postbus 28000, 9400 HH Assen, the Netherlands

Key words: Delfland Subgroup, Vlieland Subgroup, clastic deposition, sedimentology, sequence stratigraphy, seismic facies

Abstract

The Lower Cretaceous of the West Netherlands Basin is characterised by fluvial deposits of the Delfland Subgroup, overlain by shallow marine sediments of the Vlieland Subgroup. Sequence stratigraphy was applied to study the distribution of reservoirs and seals within both successions. A total of six depositional sequences has been identified, using extensive newly acquired biostratigraphy and sedimentology. Three of the sequences contain fluvial sediments only, two are mixed fluvial and marine, and one is fully marine. The typical thickness per sequence is 200 to 400 m. Characteristic for the fluvial sequences is an overall fining-upward pattern. The marine sequence starts with a fining-upward interval (Transgressive Systems Tract), followed by a prolonged coarsening-upward interval (Highstand Systems Tract). The sequence boundaries have been correlated on logs and on regional seismic. Each of the main depositional settings displays a characteristic seismic facies, which has been used to reconstruct the sedimentological facies distribution. The results demonstrate that the main transport direction for sediments of the Delfland Subgroup was SE – NW, controlled by the tectonic grain. Strong thickness variations within the lowermost sequence indicate syn-depositional rifting, which confined the main channel systems. The major intervening floodplain shales correspond to base-level highstands, and it appears that these can be correlated on a regional scale. The marine transgression entered the basin from the north during the Late Valanginian. By the Early Barremian, it covered the entire basin. The overall transgressive regime was punctuated by periods of coastline progradation, which determine the sequence boundaries. Good reservoirs for oil are found in marine transgressive sands (Rijswijk Member), and in prograding shoreface sands (Berkel Sandstone, IJsselmonde Sandstone and De Lier Members), as well as in the fluvial sands of the Delfland Subgroup.

Introduction

The West Netherlands Basin, which underlies the cities of Rotterdam, The Hague, Leiden and their surroundings, has been an area of oil production since 1954 (Racero-Baena & Drake 1995, De Jager et al. 1995; Fig. 1). The basin *sensu stricto* existed from the Late Jurassic to the Late Cretaceous. It opened in a series of NW – SE trending rift basins with a strike-slip component, reflecting stress relaxation after Late Kimmerian compression (Ziegler 1990). While being formed, these rifts were filled with fluvial sediments. Around Hauterivian times, active faulting decreased and the basin entered a post-rift sag phase. This was characterised by onlap onto the basin margins, and development of a classical 'steer's head' basin-fill geometry. Continued subsidence led to an ingression of the sea, and most of the post-Hauterivian sediments were deposited in a marine setting. The basin gradually became deeper until, during the Late Cretaceous, the Laramide compressional phase started a new period of inversion and uplift. The structural evolution of the West Netherlands Basin, and the style of its faulting, are further discussed by Racero-Baena & Drake (1995).

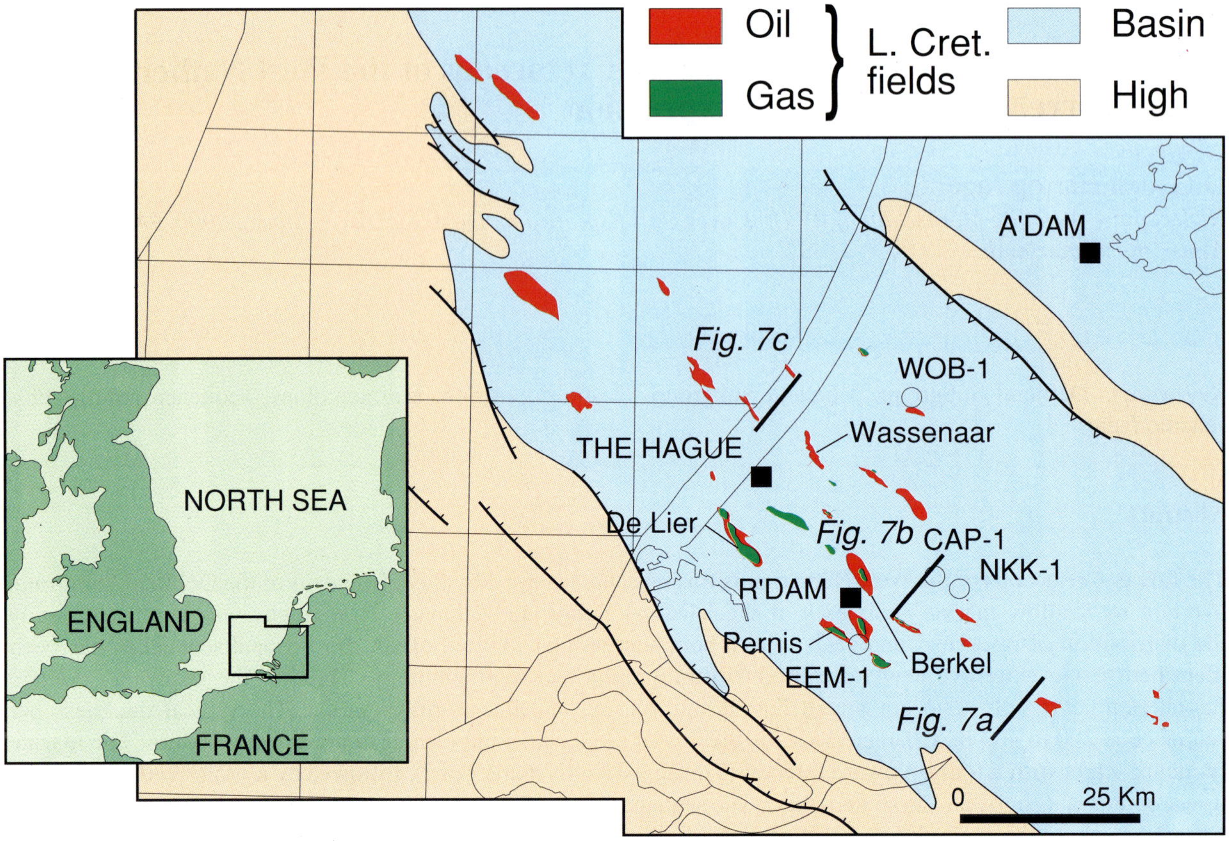

Fig. 1. Map of the south-western Netherlands with locations of relevant fields, wells and seismic lines. CAP-1 = Capelle-1, EEM-1 = Eemhaven-1, NKK-1 = Nieuwerkerk-1, WOB-1 = Woubrugge-1.

The tectonic history as outlined above resulted in an Upper Jurassic and Cretaceous basin fill of up to ca. 3000 m thick, much of which was eroded during the Laramide inversion. The main stratigraphic subdivision of this package is threefold, with the main facies changes as boundaries: fluvial clastics (Delfland Subgroup, max. 1500 m thick, ?Portlandian – Hauterivian), marine clastics (Rijnland Group, max. 1200 m, Hauterivian – Albian) and pelagic carbonates (Chalk Group, max. 900 m, Cenomanian–Paleocene). The Rijnland Group has been subdivided into a partially sand-rich lower part (Vlieland Subgroup, Hauterivian – Barremian) and a clay-dominated upper part (Holland Formation, Aptian–Albian). An example of a typical West Netherlands Basin well, Capelle-1, with wireline logs and a synthetic seismogram, is shown in Fig. 2.

The Delfland Subgroup and the Vlieland Subgroup are of particular interest, because they contain prolific oil reservoir sandstones. The geometry and correlation of particular sandstone bodies depends on their depositional setting and on the nature of synsedimentary tectonism. The Lower Cretaceous of the West Netherlands Basin has been reviewed previously (Haanstra 1963, NAM & RGD 1980, Van Adrichem Boogaert & Kouwe 1993), but with the introduction of 3D-seismic and the concept of sequence stratigraphy, more detailed work is now possible. Using these two new techniques, this paper presents an updated view on the stratigraphy and sedimentology of the Delfland and Vlieland Subgroups.

Sedimentology

The depositional settings for the Delfland and Vlieland Subgroups have been reconstructed using ca. 1500 m of core from 37 wells, from depths varying between 650 m and 2350 m below the NAP ordnance level. In addition, sand-shale ratios for complete formations were determined from wireline logs.

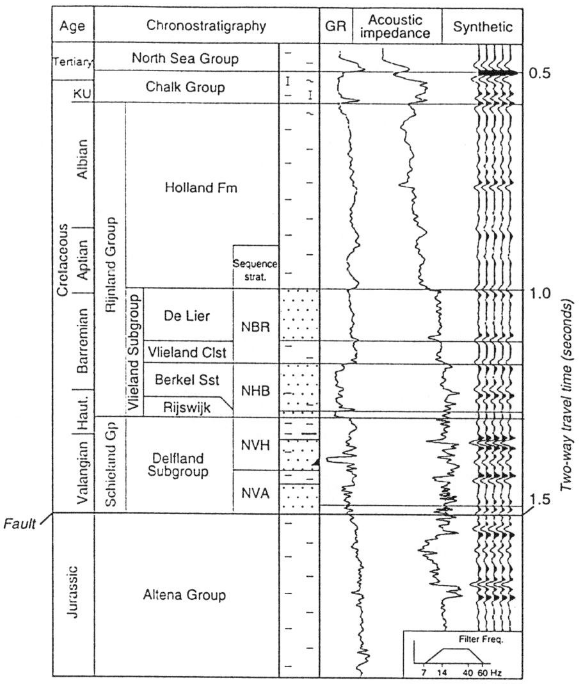

Fig. 2. Well Capelle-1: stratigraphy and synthetic seismogram. Group, formation and member names after Van Adrichem Boogaert & Kouwe 1993. See Fig. 5 for explanation of sequence names and Fig. 7b for a seismic profile through the well location.

The Delfland Subgroup is characterised by fluvial deposits, as evidenced by a lack of marine flora and fauna, the presence of channel sands, *in situ* coal beds and paleosols. Deposition of sandstones in channels is indicated by the presence of cross-bedding and erosive bases, and a poor lateral correlation of individual sandbodies. Within the Delfland, an upward evolution is recognised: from primary red beds, deposited in a semi-arid climate, to coal-bearing grey beds, deposited in a more humid climate. The presence of red beds in the basal part makes palynological age-dating difficult. The oldest firm datings indicate a Ryazanian age (i.e. earliest Cretaceous, using Boreal age names), but some of the Delfland deposits are likely to be latest Jurassic in age. Sand/shale ratios within the red beds are very variable with an average around 50%. Many of these sandstones have been interpreted as braided river deposits, which were probably deposited very rapidly.

In the higher part of the Delfland Subgroup, a transition can be seen to grey-coloured organic-rich beds, which may contain well-preserved plant fossils (Van Amerom et al. 1976). Sand/shale ratios became somewhat lower (down to 25% on average), and the fluvial channel type evolved from braided to meandering.

It has been demonstrated that the source of the rivers which deposited the Delfland clastics was consistently from the southeast, and that they flowed through the newly formed grabens towards the northwest. A schematic depositional model is shown in Fig. 3. Three aspects may be observed from this figure:

1. A decrease of the sand/shale ratio occurs from proximal to distal, i.e. from southeast to northwest.

2. Sedimentation and tectonic subsidence took place simultaneously. The main fluvial channels followed the axes of the graben systems. Observed sand/shale ratios therefore decrease towards the horst blocks.

3. The grabens in the southeast were relatively narrow and deep. The oldest Delfland sediments, and the thickest red bed intervals, occur in the southeast. The strong confinement in this area contributed to the persistence of braided river systems.

The Vlieland Subgroup has been deposited in a marine setting, as proven by the presence of marine flora and fauna, the common presence of glauconite, the abundance of bioturbation, and the lack of paleosols and *in situ* coal (although floated coal, without seat earth, may be present). In most of the marine sandstones, intense bioturbation has obliterated the primary sedimentary structures.

Four distinct sandstone units have been recognised within the Vlieland Subgroup: the Rijswijk, Berkel Sandstone, IJsselmonde Sandstone and De Lier Members. The sandstone of the Rijswijk Member mainly consists of material reworked during transgression; the three others are the result of coastline progradation followed by marine transgression. Most of these sands are very fine- to fine-grained, with a slight coarsening upward towards the top; within the top part, a relatively coarse lag deposit may have formed during transgression ('ravinement surface'). Further details on the reservoir geology of these sandstones, which frequently have been found oil-bearing, are described by Racero-Baena & Drake (1995). The stratigraphic nomenclature has recently been updated by Van Adrichem Boogaert & Kouwe (1993), a publication which also contains detailed lithological descriptions.

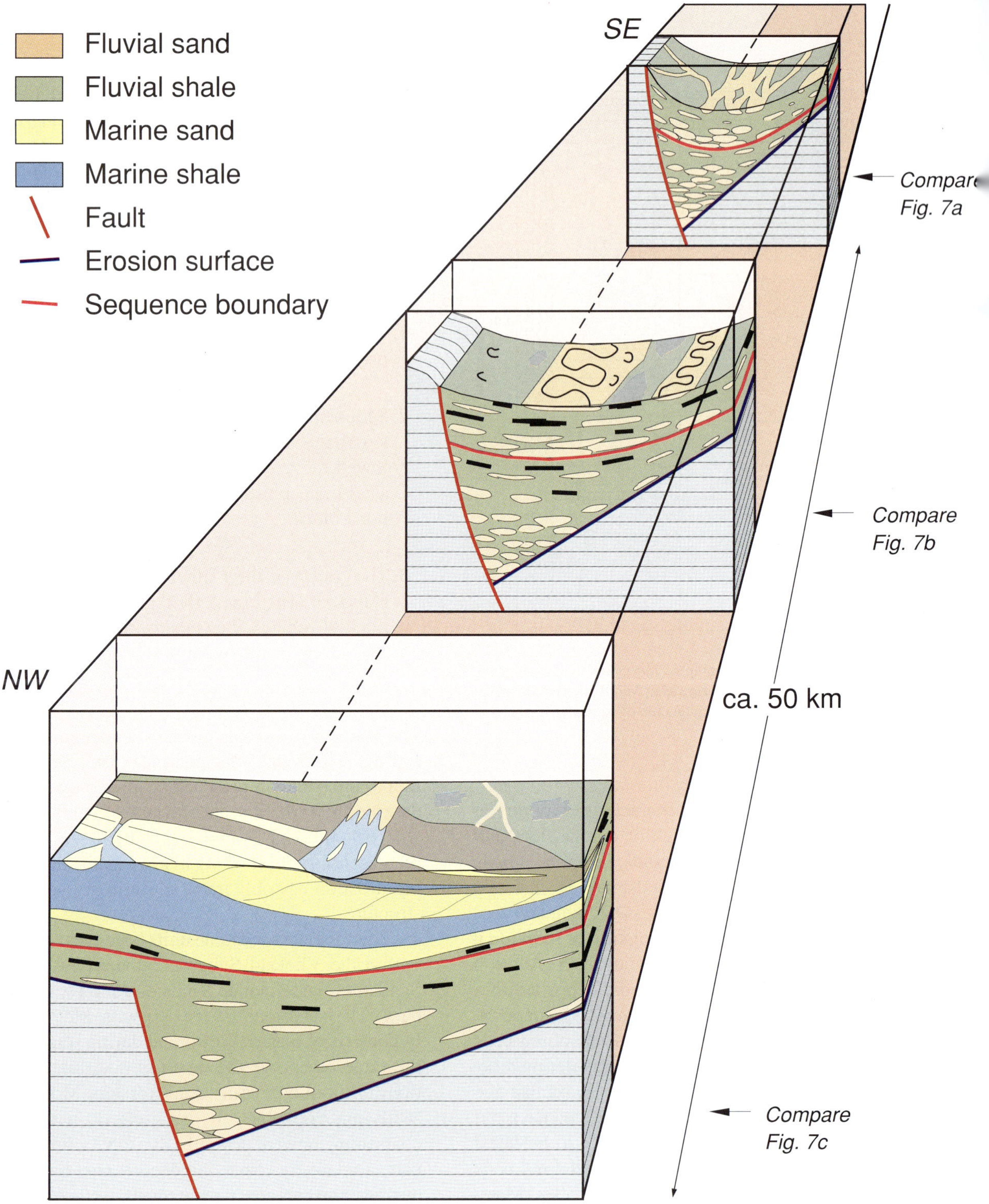

Fig. 3. Model for the deposition of the Delfland Subgroup in the West Netherlands Basin. The source of the fluvial sands is further southeast.

Sequence stratigraphy

From the available data, it has become clear that the boundary between the continental Delfland Subgroup and the marine Vlieland Subgroup is diachronous, marking a stepwise marine transgression. For the mapping of facies variations, it was decided to use a

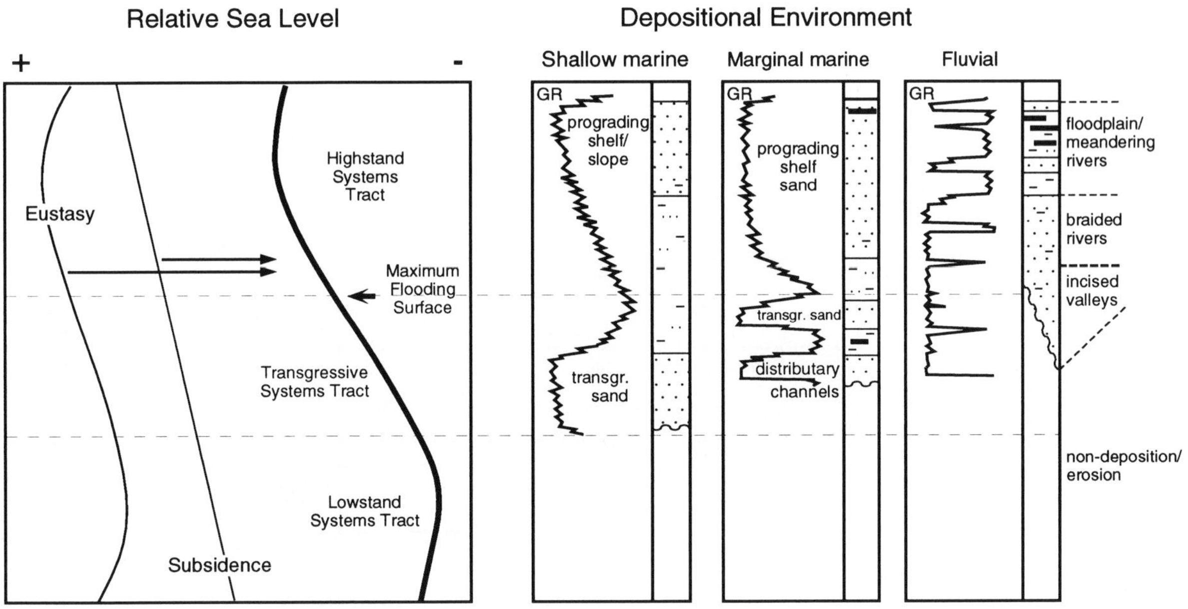

Fig. 4. Concept for the definition of sequence boundaries and systems tracts using the natural gamma-ray log (GR).

sequence stratigraphic approach, which is outlined in Fig. 4. This provides a systematic method of drawing time lines through the various fluvial and shallow marine facies belts.

It has been assumed that, both in the marine and the fluvial settings, the depositional pattern is determined by rise and fall in sea level. The cause of this may either be eustasy or tectonism; however, eustatic changes probably occur at a higher rate than tectonic subsidence (cf. Vail 1987). In the fluvial setting, a fall in sea level will most likely be represented by erosion or non-deposition, with formation of paleosols. In the meantime, the river profile will have steepened, and with resuming sea level rise, relatively coarse material will be deposited in incised valleys. With continuing rise, the profile flattens and an overall fining-upward sequence is expected to develop (Fig. 4; cf. Miall 1991).

In the shallow marine setting, each depositional sequence is expected to start with a transgressive sand. An underlying 'forced regression' sand (cf. Posamentier et al. 1992) could be present, although it appears that such sands in this area have mostly been reworked by the subsequent transgression. With ongoing sea level rise, a fining-upward sequence is expected to form. The top of this is the maximum flooding surface, usually represented by an organic-rich marine claystone, which can be recognised on wireline logs by a high gamma ray-peak. Thereafter, the rate of sedimentation overtakes the rate of subsidence, the coastline starts to build out, and a coarsening-up pattern develops, typically first with storm sands (lower shoreface) and then with barrier sands (upper shoreface). The top of the coarsening-up interval, which may be marked by lagoonal shales, forms the next sequence boundary (Fig. 4).

The transgressive sand may be coarser grained, with better reservoir quality, than a prograding sand (Abbott 1985). The Berkel Sandstone Member, for example, can be subdivided into 1) a relatively fine-grained lower part, which displays a coarsening-up trend, and 2) a relatively coarse-grained upper part, which displays a fining-up trend. The boundary between these is a sequence boundary.

Correlation

A sequence stratigraphic scheme for the Lower Cretaceous of the West Netherlands Basin (Fig. 5) has been constructed using the available biostratigraphic (mainly palynological) data, plus the sequence concept for picking the exact boundaries. The combined Delfland and Vlieland Subgroups, the boundary of which is

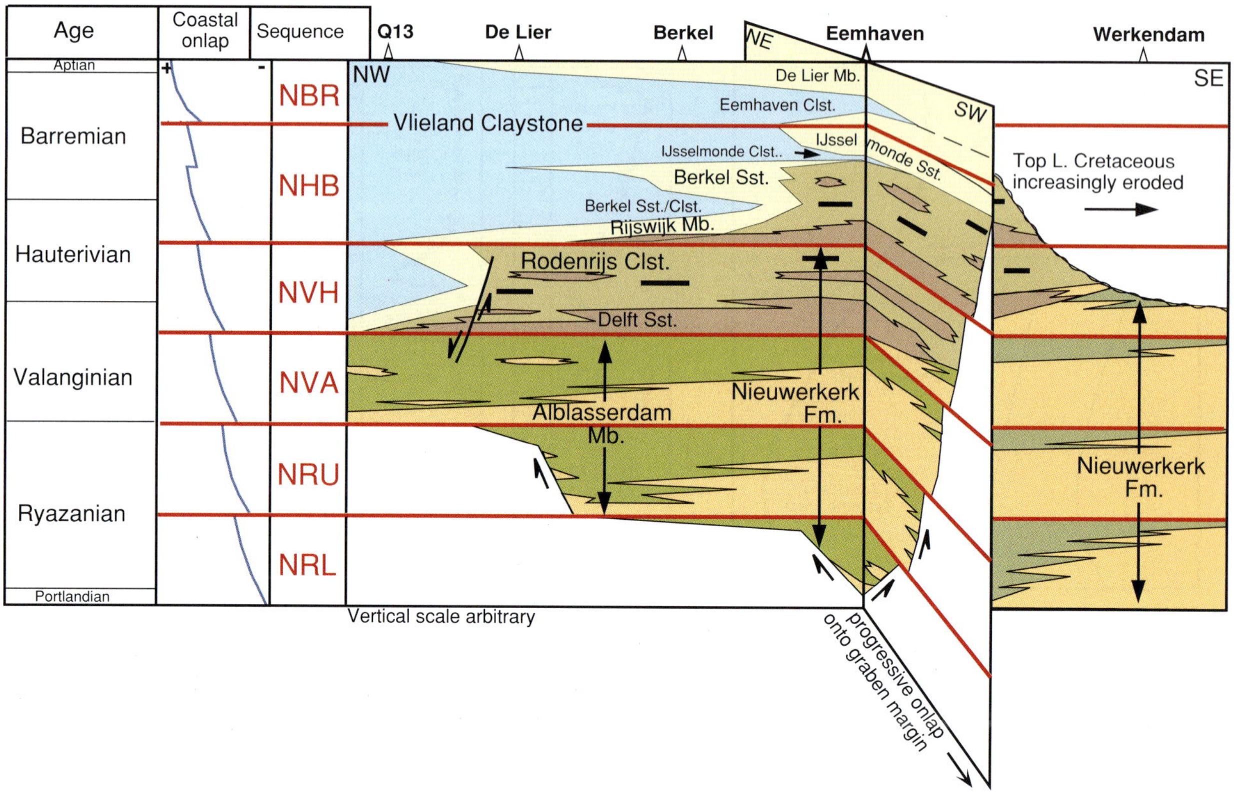

Fig. 5. Sequence stratigraphic scheme for the Lower Cretaceous of the West Netherlands Basin. For each of the six depositional sequences used in this article, relative time span and expected lithology are indicated. Sequence names: NRL = Neocomian/ Ryazanian Lower, NRU = Neocomian/ Ryazanian Upper, NVA = Neocomian/ Valanginian, NVH = Neocomian/ Valanginian – Hauterivian, NHB = Neocomian/ Hauterivian – Barremian, NBR = Neocomian/ Barremian.

diachronous, have been subdivided into six depositional sequences. These were given three-letter names, beginning with 'N' (for 'Neocomian'), and followed by two letters indicating the dominant age (e.g. RL = Lower Ryazanian; HB = Hauterivian – Barremian). During the Valanginian, the northern part of the basin was entered by a marine transgression while, on the southern basin margin, fluvial sedimentation continued until the Early Barremian (Fig. 5).

The two wireline log correlation panels (Fig. 6a,b) demonstrate how the sequence concept has been applied to log data. A correlation of three wells through the Delfland (Fig. 6a) shows a predominance of thinning-upward sequences, which are variable in thickness from ca. 100 to 800 m. Thickness variations within a single sequence are explained by differences in subsidence rate, both within a single (half-)graben and among different grabens.

Most of the wells drilled through the Delfland Subgroup penetrated at least one major fault. In wells Eemhaven-1 and Woubrugge-1, the base of the Delfland is marked by a fault. In well Nieuwerkerk-1, the type well for the Delfland Subgroup (NAM & RGD 1980; 'Nieuwerkerk Formation', Van Adrichem Boogaert & Kouwe 1993), a major normal fault is penetrated at 1595 m along hole (Fig. 6a). This fault separates 'downthrown block facies', with a relatively high sand/shale ratio, from 'upthrown block facies', which contain much more claystone and thinner bedded sandstones.

A representative correlation through the Vlieland Subgroup is shown in Fig. 6b. Within this marine section, each depositional sequence starts with a fairly thin fining-upward interval, and usually ends with a thick coarsening-upward interval. The two are separated by a distinct high gamma ray peak, which represents the maximum flooding surface. The Wassenaar well shows

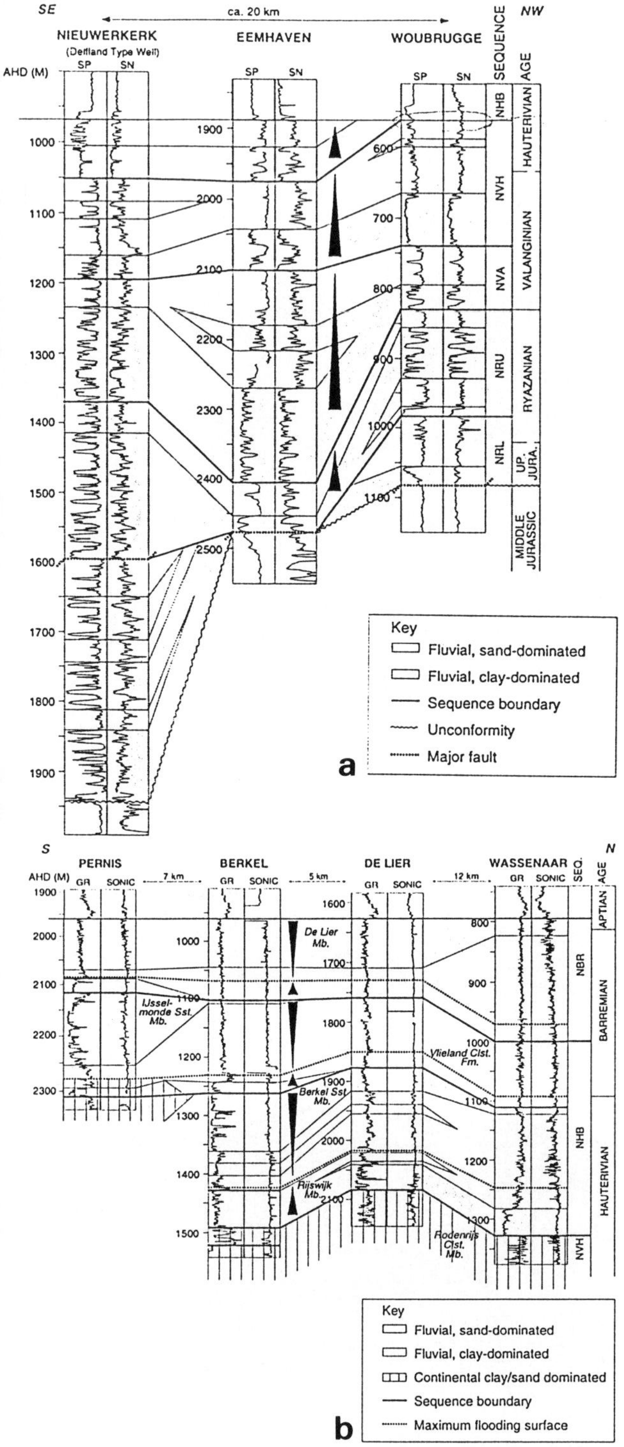

Fig. 6. (a) Log correlation panel for the Delfland Subgroup (fluvial deposits), with sequence subdivision. Depth along hole (AHD) in metres. SN = Short Normal, SP = Spontaneous Potential log. Note the fining-upward signature in many sequences. (b) Log correlation panel for the Vlieland Subgroup (marine deposits). Depth along hole (AHD) in metres. GR = Gamma-Ray log. Many sequences are characterised by a relatively thin fining-up, followed by a much thicker coarsening-up pattern.

that, even in the absence of sand, the typical fining- and coarsening-up cycles can still be recognised in the time-equivalent basinal shales.

Seismic facies

In many places it has been possible to recognise sequence boundaries on seismic, and the seismic character can be correlated to the depositional facies. A dip-line through well Capelle-1, taken from the Gouda 3D-seismic survey, is shown in Fig. 7b. This well does not penetrate the NRL- and NRU-sequences, and the position of their boundaries was correlated from nearby wells such as Nieuwerkerk-1. Examination of the seismic expression of the basin fill shows very little reflectivity in the basal part of the Delfland. The amplitudes increase towards the top of the Delfland. In the Vlieland, both high and low amplitudes may occur. The continuity of reflections in the Vlieland is clearly higher than in the Delfland. Within the Delfland, the continuity increases upward.

These patterns can be related to the sedimentology (Figs 5, 8). The basal part of the Delfland is dominated by braided river channels, with rapid lateral variations in sand body thickness causing a low seismic continuity. The amplitudes observed here are also often low, probably for two different reasons. Firstly, in a depositional setting with such strong lateral facies changes, the stacking of seismic traces will tend to cancel peak amplitudes. Secondly, the lower part of the Delfland is a red bed section, without large impedance contrasts between sand and shale.

The upper part of the Delfland was deposited when active rifting started to be replaced by sag subsidence. In addition, the climate became more humid. Fluvial deposition proceeded in wider, less inclined valleys, with an increasing amount of vegetation and swamps on the overbank area. The channel type became more sinuous, with more lateral migration, resulting in better continuity of seismic reflections. The coals and organic-rich claystones form a larger impedance contrast with the sandstones, which explains the high amplitudes seen in the upper part of the Delfland.

In general, the marine sandstones have a more sheet-like geometry, and marine clastics can be correlated over a larger distance than fluvial ones (compare Fig. 6a with 6b). On seismic, marine deposits are easily distinguished from fluvial deposits by their much better reflection continuity. A basal transgressive sand like the Rijswijk Member produces a reflection which

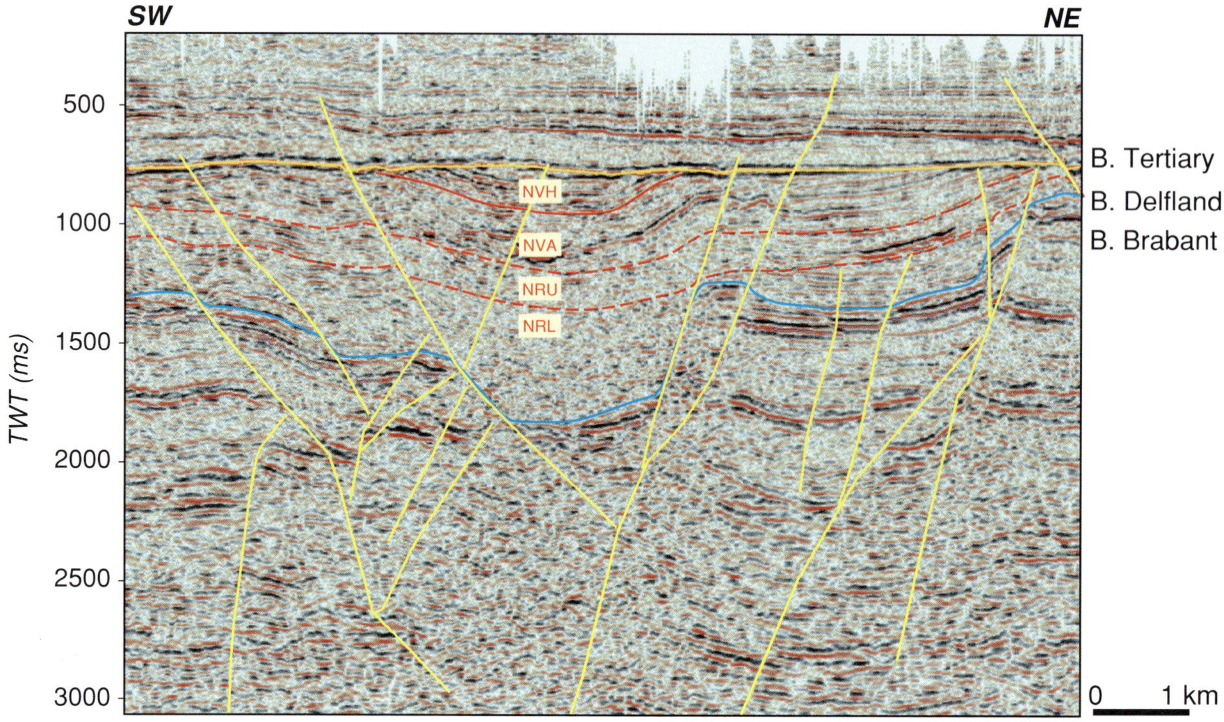
SW
NE
NVH
NVA
NRU
NRL
B. Tertiary
B. Delfland
B. Brabant
TWT (ms)
500
1000
1500
2000
2500
3000
0 1 km

a

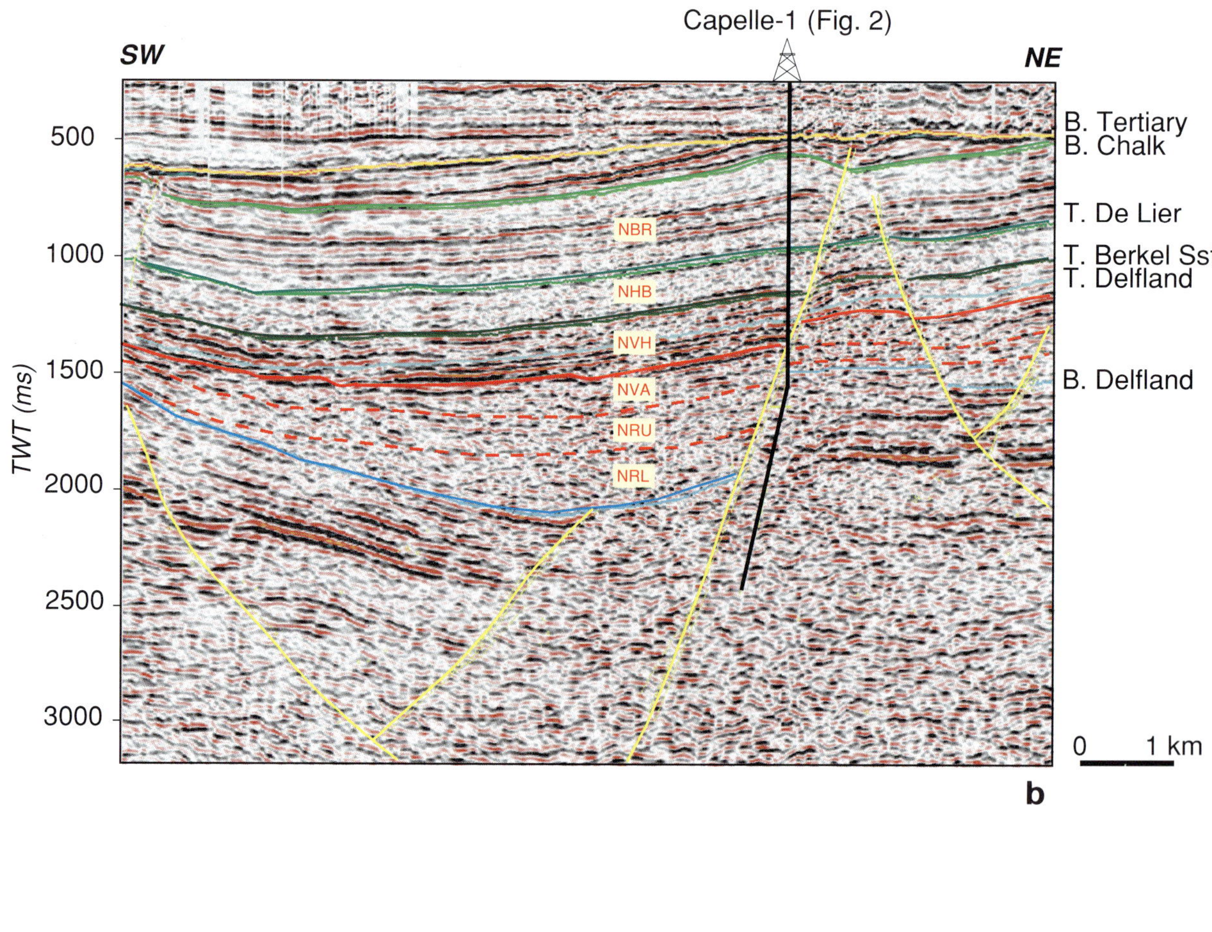
Capelle-1 (Fig. 2)
SW
NE
NBR
NHB
NVH
NVA
NRU
NRL
B. Tertiary
B. Chalk
T. De Lier
T. Berkel Sst.
T. Delfland
B. Delfland
TWT (ms)
500
1000
1500
2000
2500
3000
0 1 km

b

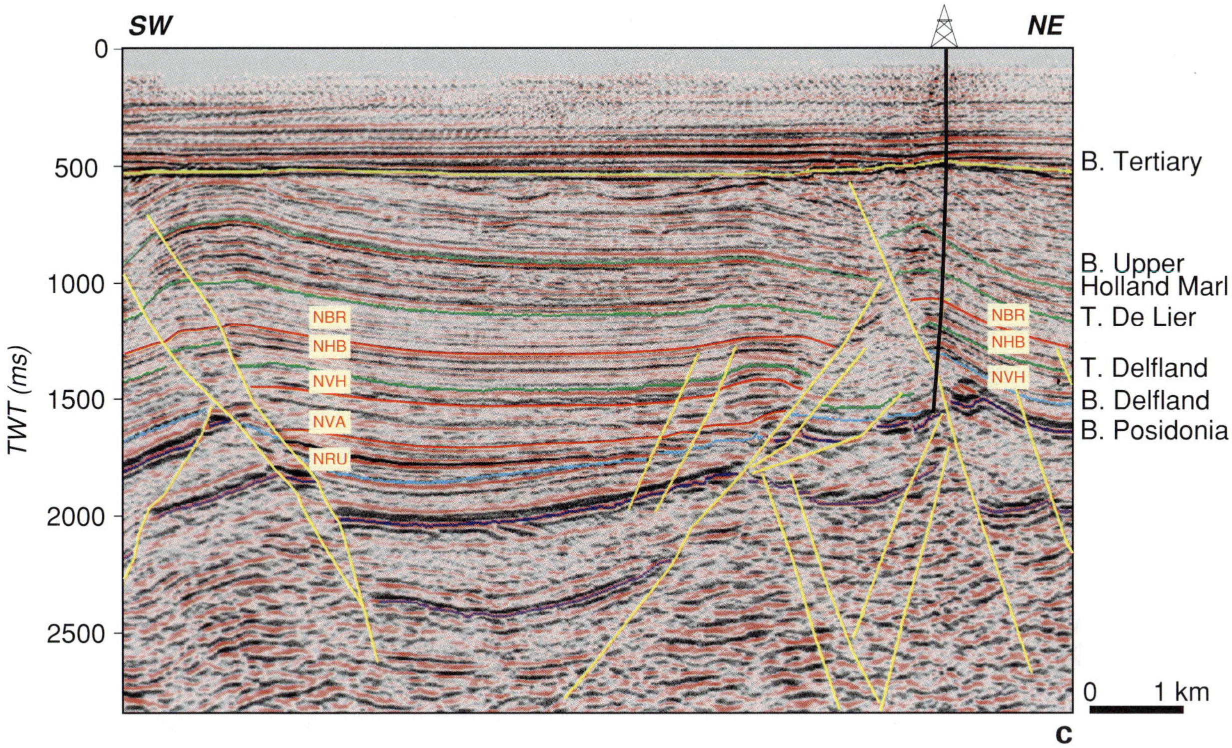

Fig. 7. Seismic examples West Netherlands Basin. See Fig. 1 for location and Fig. 5 for sequence names. (a) 'proximal' rift basin fill. Note low reflection continuity and a thick NRL-sequence. The post-Valanginian Cretaceous has been eroded here. (b) 'intermediate' rift basin fill. Note the general increase in reflection continuity both upward and laterally from the graben centre. See Fig. 2 for details on well Capelle-1. (c) 'distal' rift basin fill. Note that sequences are thinner and reflections are more continuous than in Fig. 7b.

is much more continuous than underlying reflections (see Abbott 1985, for further examples). The strength of the reflection depends on thickness and lithological contrast; a maximum flooding surface is usually represented by a low-velocity, organic-rich claystone, which will appear as a soft loop on seismic.

The seismic expression of the basin fill in respectively a proximal, an intermediate and a distal position is shown in Figs 7a-c (compare the cartoons in Fig. 3). In Fig. 7a, a 'proximal' Delfland rift basin fill, from the southeast of the basin, is displayed. The Delfland in the rift centre may be some 1500 m thick. Reflection continuity is generally low, but increases upwards. The continuity also increases laterally, towards the basin edge, where the rate of deposition was slower. The low-continuity seismic facies in the deepest part of the rift ('wormy facies') is likely to contain rapidly deposited braided-river sediments, with a high sand/shale ratio. In this part of the basin, the top of the Delfland plus the entire Vlieland have been eroded. The strong amplitudes seen within the NVA-sequence are probably due to igneous rocks.

An 'intermediate' position within the basin is shown in Fig. 7b. Again, the continuity within the Delfland Subgroup increases both upwards and sideways. The low-continuous part, which largely corresponds to the NRL-sequence, is thinner in this 'intermediate' position than in the proximal position. This is partially explained by onlap towards the northwest of the NRL-sequence, and for another part by a change in depositional facies. Within the NRL-, NRU- and NVA-sequences, the fluvial system tends to change from braided to more meandering in a SE – NW direction. The (marine) Vlieland Subgroup has been preserved in the 'intermediate' location, which can be recognised on Fig. 7b by the increased continuity in the NHB- and NBR-sequences.

The 'distal' basin fill is illustrated on a seismic line near The Hague (Fig. 7c). In this location, seismic reflections within the Delfland are both stronger and

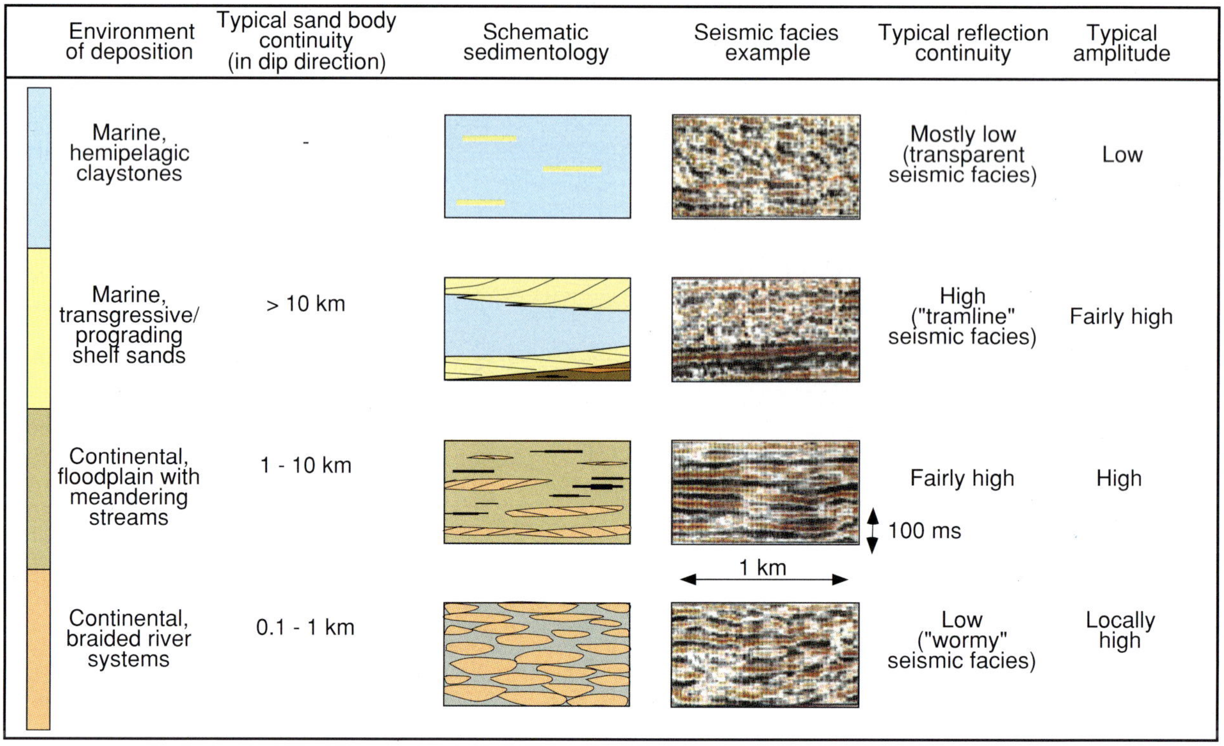

Fig. 8. Comparison between *depositional* facies and *seismic* facies found in the Lower Cretaceous of the West Netherlands Basin.

more continuous than those shown on the other two, more proximal sections. Again, this is due both to onlap of the NRL-sequence (which has almost disappeared here) and to a lateral facies change within the NRU-, NVA- and NVH-sequences.

Differences in acquisition and processing are expected to have an influence on the appearance of the seismic facies. Nevertheless, the seismic facies characteristics described above have been recognised on numerous vintages of both 2D- and 3D-seismic. The three lines in Fig. 7 are from 3D-surveys which were processed with similar parameters.

Facies distribution

The combination of well results and seismic facies interpretation has resulted in a series of six facies distribution maps, one per sequence, depicting the nature of the infill of the West Netherlands Basin during the Early Cretaceous (Fig. 9).

Deposition of the Delfland Subgroup probably started during the latest Jurassic, with the syn-rift, fluvial NRL-sequence. The main sediment source may have been the Ardennes and Rhenish Massif, some 150 km to the southeast. The main transport direction was SE – NW, parallel to the newly formed graben boundary faults. Typical for this sequence are very rapid lateral variations in thickness (0–800 m) and sand/shale ratio, with very sand-rich deposits probably concentrated in the graben centres.

The NRU-sequence was developed in a similar syn-rift fluvial setting, although with less pronounced thickness changes. The proximal part of the underlying NRL-sequence was partly overstepped, and deposition of fluvial sandstones extended further to the northwest. The NVA-sequence shows a similar pattern, but the position of the major fluvial channels shifted towards the southwest. The cause of this shift in depositional axis was probably a variation in subsidence rate for the various grabens at different times. The ongoing tectonic activity is also reflected by the presence of extrusive rocks in the Lower Valanginian.

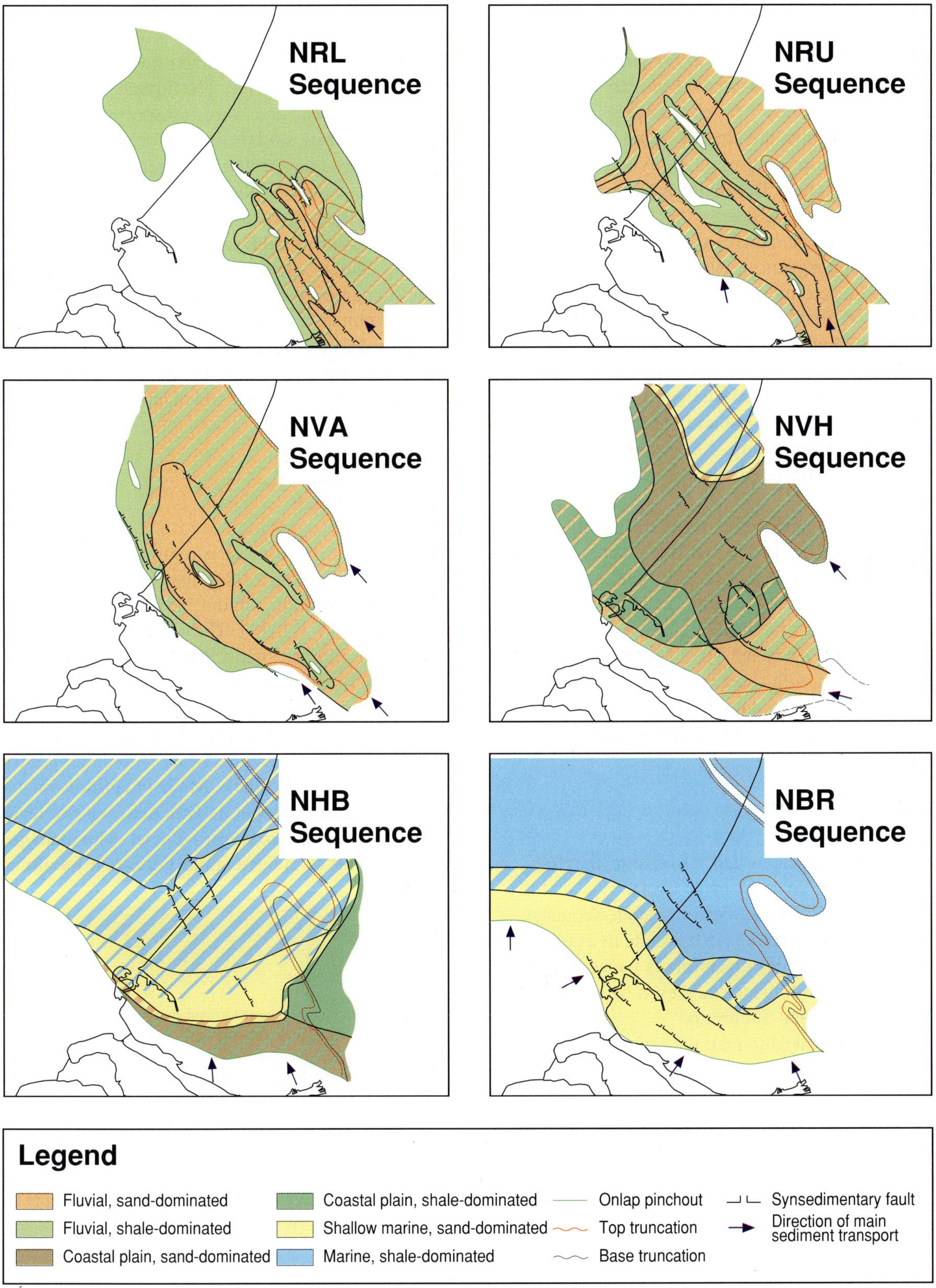

Fig. 9. Series of facies distribution maps showing the nature of the infill of the West Netherlands Basin from the Early Ryazanian (NRL-sequence) to the Barremian (NBR-sequence). See Fig. 5 for sequence names.

During deposition of the NVH-sequence, the degree of active faulting decreased and an extensive coastal plain was formed. Marine transgressive sands of the Rijswijk Member are already found in the north of the basin, while more proximal fluvial deposition continued in the south. The major marine transgression, however, occurred during the Hauterivian. In most of the basin, the base of the NHB-sequence is formed by the transgressive Rijswijk Member *sensu stricto*. Tectonic activity at this time was subdued but still ongoing, as demonstrated by thickness changes of the Rijswijk Member across some faults. An example from the Pijnacker Field has been described by Racero-Baena & Drake (1995). Ongoing transgression was punctuated by the progradation of the Berkel and IJsselmonde Sandstones (Fig. 5). Meanwhile, in the south of the basin, coastal plain sedimentation continued (Fig. 9).

The marine transgression covered the entire basin during the Barremian. In the NBR-sequence, which ends with the progradation of the De Lier sand-shale Member, no continental deposits have been found. Thereafter, a pattern of continuing transgression and onlap of the London – Brabant Massif persists throughout the Cretaceous, with short intercalated periods of outbuilding, e.g. the Early Albian (Holland Greensand) and the Early Cenomanian (Texel Greensand).

Conclusions

Using sequence stratigraphy and 3D-seismic, the stratigraphic framework for the Lower Cretaceous of the West Netherlands Basin has been updated. Sequence stratigraphy provides the tools to correlate in a consistent way time lines across diachronous facies boundaries. A sequence subdivision could even be applied to the (terrestrial) Delfland Subgroup.

3D-Seismic was used to map the sequence boundaries. In addition, the seismic facies could be linked to the depositional facies, providing a means to predict the environment of deposition in the deeper, undrilled parts of the rift basin. For fluvial deposits, a relatively high gradient during deposition is characterised by low seismic continuity. Lower coastal plain deposits are more continuous on seismic, and show significantly higher amplitudes. Marine deposits show a strong continuity, with variable amplitudes.

The interval between the latest Jurassic and the Aptian has been subdivided into six depositional sequences. Within the study area, three of these are completely fluvial, two are mixed fluvial-marine, and the youngest is fully marine. The fluvial deposits came from the southeast, and were transported in channel systems which followed the newly formed graben system. Synsedimentary tectonism determined the facies in the oldest sequence (NRL), and continued to influence the facies pattern until the widespread marine transgression during the Hauterivian (NHB). The transgression, which came from the north during the Valanginian, moved stepwise towards the south until, in the Barremian, the entire basin had drowned.

The models and maps discussed here can be used to predict more accurately the occurrence of high-quality reservoir sandstones and of laterally continuous intraformational seals.

Acknowledgements

This paper has its roots in a regional study carried out within the Geological Services department, Business Unit Exploration, of the Nederlandse Aardolie Maatschappij BV (NAM). The author is indebted to NAM, Shell Internationale Petroleum Maatschappij (SIPM) and Exxon Company International (ECI) for granting permission to publish these results.

The study was inspired by Arie Speksnijder and coordinated by Kees van der Zwan, with major contributions from Chris de Klerk, Michel Gaillard, Pieter-Jan Pestman and Greg van de Bilt. Last but not least, thanks are due to Alvaro Racero-Baena, who managed to persuade me to include some important modifications.

References

Abbott, W.O. 1985 The recognition and mapping of a basal transgressive sand from outcrop, subsurface, and seismic data. In: Berg, O.R. & D.G. Woolverton (eds) Am. Ass. Petroleum Geol. Mem. 39 'Seismic Stratigraphy II: An integrated approach to hydrocarbon exploration': 156–169

De Jager, J., M.A. Doyle, P.J. Grantham & J.E. Mabillard 1995 Hydrocarbon habitat of the West Netherlands Basin. In: Rondeel, H.E., D.A.J. Batjes & W.H. Nieuwenhuijs (eds) Geology of gas and oil under the Netherlands – this volume

Haanstra, U. 1963 A review of Mesozoic geological history in the Netherlands – Verh. Kon. Ned. Geol. Mijnb. Gen. 21-I: 35–55

Miall, A.D. 1991 Stratigraphic sequences and their chronostratigraphic correlation – J. Sed. Petr. 61: 497–505

NAM & RGD (Nederlandse Aardolie Maatschappij & Rijks Geologische Dienst) 1980 Stratigraphic nomenclature of the Netherlands – Verh. Kon. Ned. Geol. Mijnb. Gen. 32

Posamentier, H.W., G.P. Allen, D.P. James & M. Tesson, 1992 Forced regressions in a sequence stratigraphic framework: concepts, examples, and exploration significance – Am. Ass. Petroleum Geol. Bull. 76: 1687–1709

Racero-Baena, A. & S. Drake 1995 Structural style and reservoir development in the West Netherlands oil province. In: Rondeel, H.E., D.A.J. Batjes & W.H. Nieuwenhuijs (eds) Geology of gas and oil under the Netherlands – this volume

Vail, P.R. 1987 Seismic stratigraphy interpretation using sequence stratigraphy, part I: seismic stratigraphy interpretation procedure. In: Bally, A.W. (ed) Am. Ass. Petroleum Geol. Studies in Geology 27 'Atlas of seismic stratigraphy': 1–10

Van Adrichem Boogaert, H.A. & W.F.P. Kouwe (compilers) 1993 Stratigraphic nomenclature of the Netherlands, revision and update by RGD and NOGEPA – Med. Rijks Geol. Dienst 50, section G: Upper Jurassic and Lower Cretaceous

Van Amerom, H.W.J., G.F.W. Herngreen & B.J. Romein 1976 Palaeobotanical and palynological investigation with notes on the microfauna of some core samples from the Lower Cretaceous of the West Netherlands Basin – Med. Rijks Geol. Dienst NS 27 (2): 41–79

Ziegler, P.A. 1990 Geological atlas of Western and Central Europe, 2nd ed. – Shell Internat. Petroleum Mij., The Hague/Geol. Soc. Publ. House, Bath, 239 pp, 56 encls

Rondeel et al. (eds), Geology of gas and oil under the Netherlands, 243–253, 1996.

De Wijk gas field (Netherlands): reservoir mapping with amplitude anomalies

Abraham N. Bruijn
Nederlandse Aardolie Maatschappij (Business Unit Exploration), Postbus 28000, 9400 HH Assen, the Netherlands

Key words: 3D seismic, seismic attribute displays

Abstract

The De Wijk gas field was discovered in 1949 and commercial gas production started in 1955. The GIIP (Gas Initially In Place) is currently estimated at 19.8×10^9 m^3. The field is unique in the Netherlands as it contains gas in Tertiary, Cretaceous, Triassic, Permian and Carboniferous reservoirs. The main gas accumulation is contained in Triassic claystones, where post-depositional leaching of anhydrite has significantly enhanced the reservoir properties. 3D seismic attribute measurement techniques show the effects of gas fill, sand distribution and leaching in the various reservoirs.

Introduction

The De Wijk Field was discovered by exploration well De Wijk-1 in 1949, ten years before the giant Groningen gas field in 1959. The field is located entirely within the Schoonebeek Concession (Fig. 1) and is fully operated by the Nederlandse Aardolie Maatschappij BV(NAM).

The objective of this paper is to demonstrate the uniqueness of the field by the interpretation of 3D seismic data.

Stratigraphic and reservoir development

A brief description of the stratigraphy as encountered in the De Wijk area is given below. The stratigraphic development is illustrated by a generalised stratigraphic column (Fig. 2).

Carboniferous and Rotliegend

The oldest penetrated rocks in the De Wijk area are the fluvial sands, shales and coals of Westphalian-B age. Although the sands have been evaluated as gas-bearing, reservoir deliverability is considered to be too poor for commercial production. The Carboniferous strata are separated from the overlying Permian Rotliegend deposits by the Saalian Unconformity. The Rotliegend is developed as a maximum 10 m thick conglomeratic sandstone of no commercial interest.

Zechstein

The Permian Zechstein 2 Carbonate Member consists of laminated carbonate mudstones deposited in a basinal setting. The reservoir was evaluated as gas-bearing but production proved to be of no commercial interest.

Triassic

The Lower Triassic Rogenstein Member and underlying Main Claystone Member form part of the Lower Buntsandstein Formation. These members subcrop against the Base Rijnland Unconformity in the western part of the field (Fig. 3). The best reservoir properties are present in the Rogenstein Member, the full sequence being some 160 m thick of which only 40 m contain oolite grainstone facies. The oolite beds have a maximum thickness of 5 m and are interbedded with sand, silt and silty claystones. Post-depositional leaching has removed the anhydrite present as nodules and

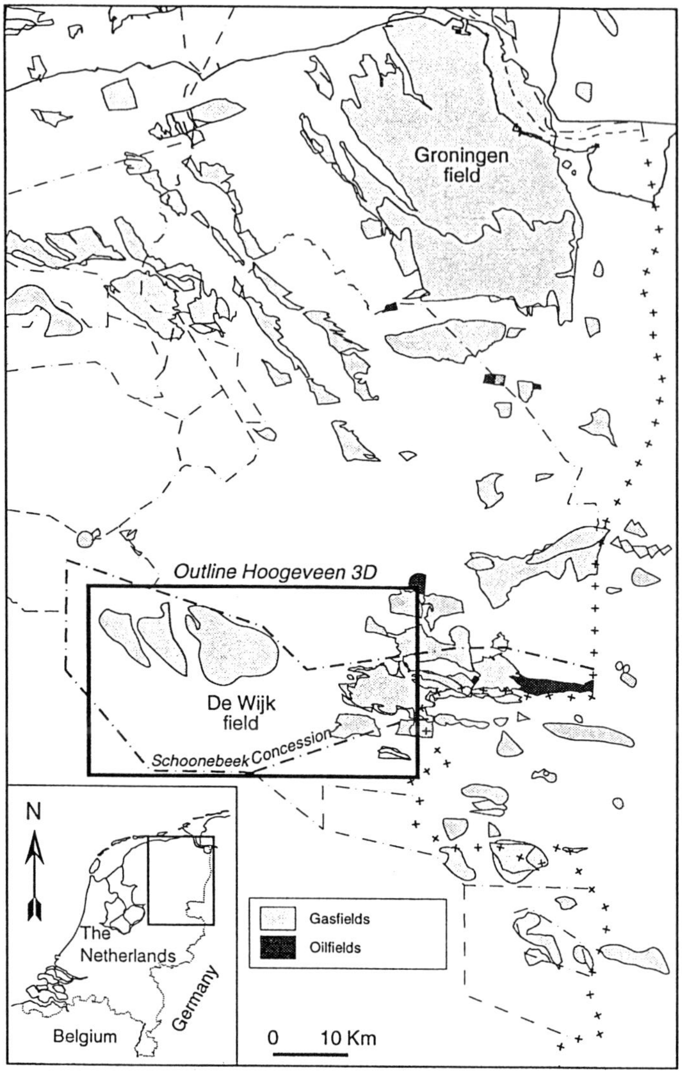

Fig. 1. De Wijk Field, location map.

cement in all lithologies, which resulted in both vuggy and intergranular porosities averaging 20%.

The Middle Triassic Solling Sandstone Member of the Röt Formation directly overlies the Lower Triassic Volpriehausen Sandstone Member of the Main Buntsandstein Formation in all wells except De Wijk-17, where erosion associated with the Hardegsen Unconformity has removed the remainder of the Main Buntsandstein Formation. The sandstones are deposited in a fluvial/aeolian environment and are the only Triassic reservoirs in the De Wijk Field with primary reservoir properties. Average porosities vary from 15% in the Solling Sandstone to 30% in the Volpriehausen Sandstone.

The Upper Triassic Lower Muschelkalk Member of the Muschelkalk Formation subcrops against the Base Rijnland Unconformity in the eastern part of the field. It reaches a maximum thickness of over 100 m on the eastern flank of the field before being progressively onlapped by Jurassic sediments. The interval consists of finely crystalline dolomite layers interbedded with marls. Porosities can be as high as 20% and are created by post-depositional leaching of the evaporite minerals such as nodular anhydrite and gypsum.

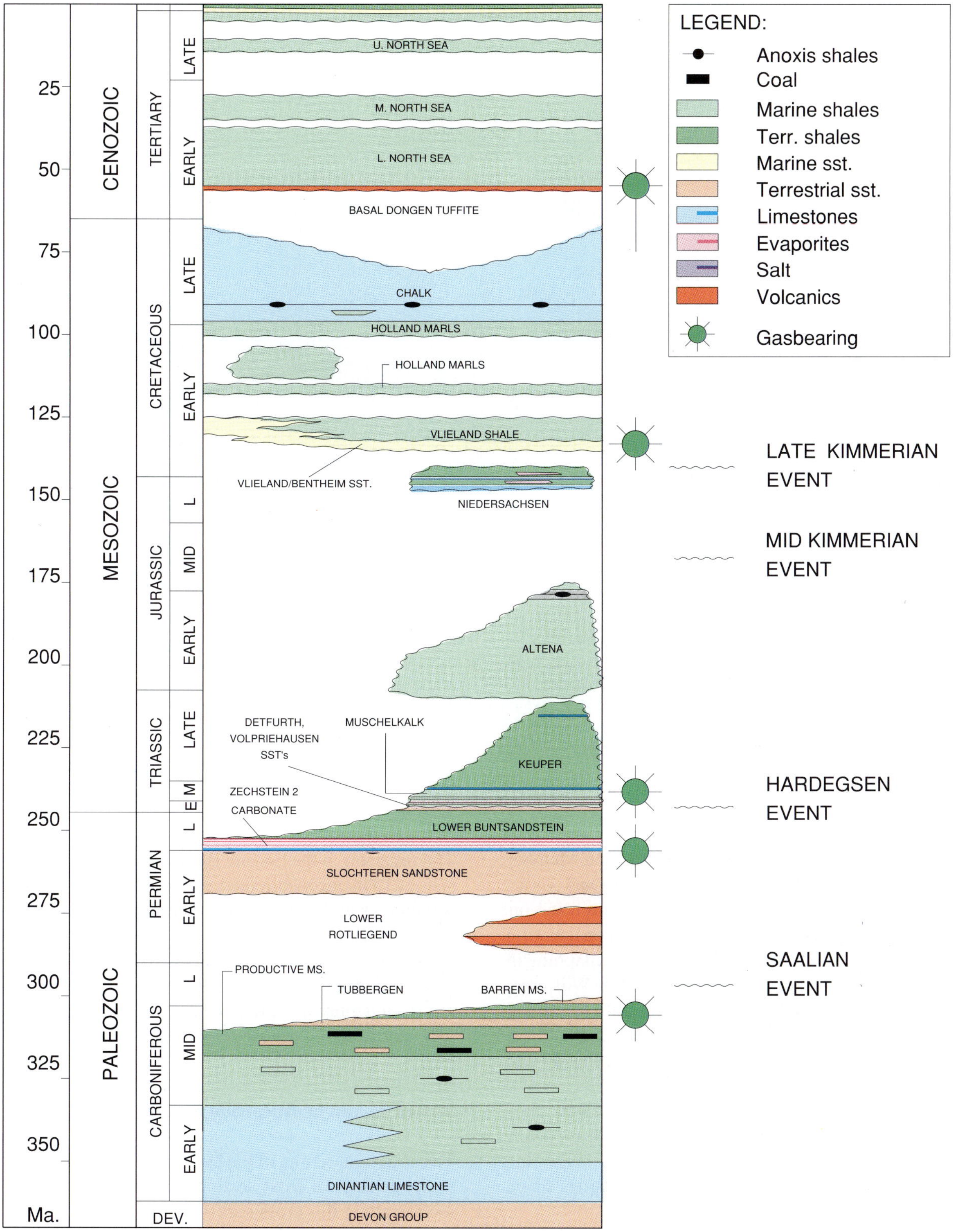

Fig. 2. Generalised stratigraphy of the NE Netherlands.

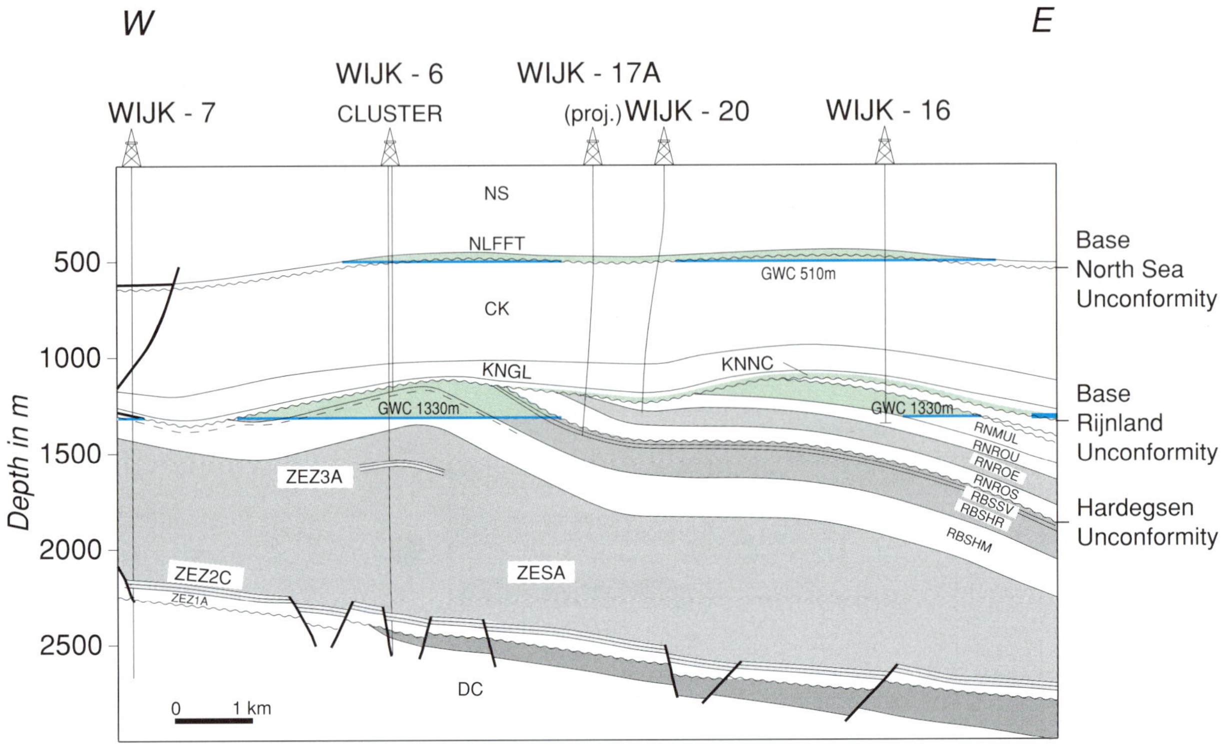

Fig. 3. Geological cross section over the De Wijk Field. The location of the line coincides with the seismic section of Fig. 4 and is indicated on Fig. 9. NS = North Sea Group; NLFFT = Basal Dongen Tuffite; CK = Chalk Group; KNGL = Holland Formation; KNNCZ = Vlieland Sandstone Member; RNMUL = Lower Muschelkalk Member; RNROU = Upper Röt Claystone Member; RNROE = Röt Evaporite Member; RNROS = Solling Claystone/Sandstone Member; RBSSV = Volpriehausen Sandstone Member; RBSHR = Rogenstein Member; RBSHM – Main Claystone/Basal Buntsandstein Member; ZEZ3A = Zechstein 3 Anhydrite floater; ZESA = Zechstein Salt; ZEZ2C = Zechstein 2 Carbonate Member; ZEZ1A = Zechstein 1 Anhydrite Member; DC = Carboniferous (Westphalian-B); GWC = Gas-water contact.

Cretaceous

The Lower Cretaceous Vlieland Sandstone Member overlies the Base Rijnland Unconformity and varies considerably in both thickness (0–11m) and reservoir quality. The lower part of the sandstone consists of a clean sand (porosity about 25%) which is onlapping in a westerly direction onto the salt-uplifted De Wijk area. It is only present in the east and is possibly the lateral equivalent of the more easterly deposited Bentheim Sandstone. The shallower unit is very rich in clay and has an average porosity of around 20%.

The Upper Cretaceous Ommelanden Chalk Formation consists of chalky limestones, with reservoir properties strongly enhanced by leaching below the Base North Sea Unconformity. The porosities range between 35–40%.

Tertiary

The reservoir in the Eocene Basal Dongen Tuffite consists of 15 m unconsolidated, tuffaceous siltstones with some clays and minor sands. The porosities are typically 30–35%.

The Upper Cretaceous and Eocene reservoirs are not being produced as the expected subsidence resulting from depletion carries the risk of endangering the recovery from the deeper reservoirs.

Structure and hydrocarbon habitat

The structural setting of the De Wijk Field is illustrated in the geological cross section (Fig. 3) and a seismic section along this cross section (Fig. 4).

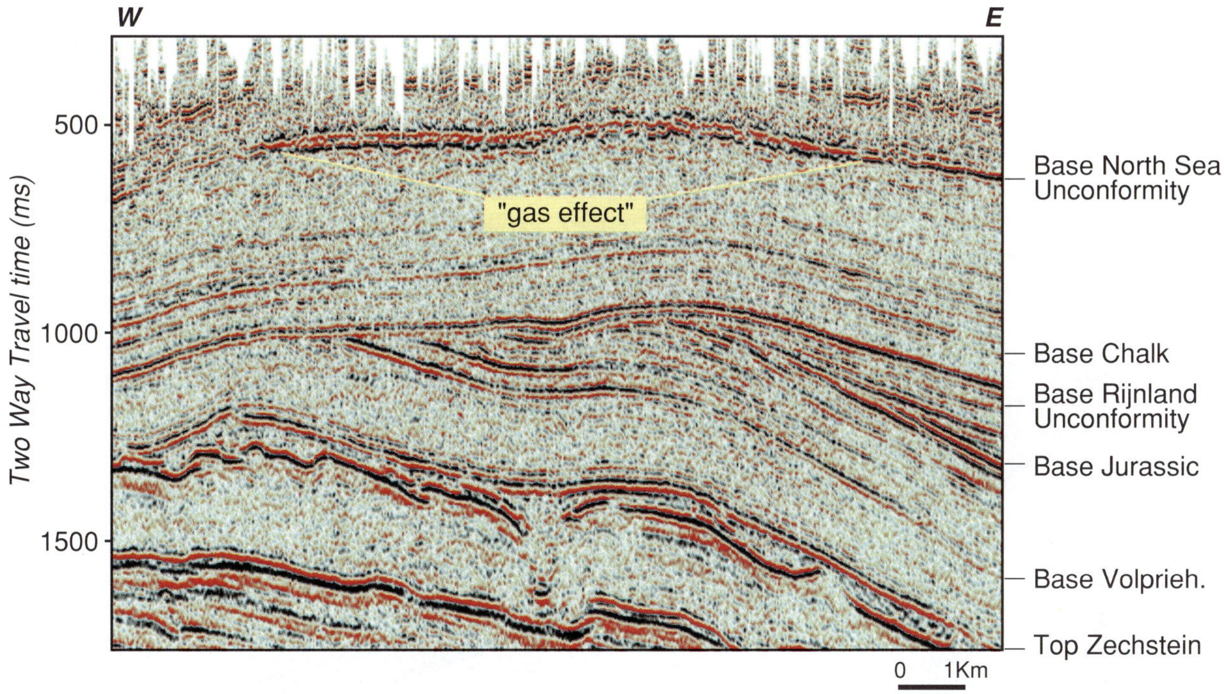

Fig. 4. Seismic section over the De Wijk Field. The location of the line coincides with the geological cross section of Fig. 3 and is indicated on Fig. 9. A black loop represents a positive amplitude ('soft kick') and a red loop represents a negative amplitude ('hard kick'). Note the amplitude brightening of the Base Basal Dongen Tuffite reflection, representing a 'gas effect'.

The Base Rijnland Unconformity is believed to be related to the Kimmerian uplift, representing a major tectonic phase during the Jurassic. This event is thought to be responsible for the structuration of the sub-salt (Carboniferous and Zechstein 2 Carbonate) fault blocks. In addition, it resulted in an eastward tilting of the Triassic and Lower Jurassic sequences, which were subsequently eroded and leached. Due to the increasing overburden pressure, Zechstein salt movements became widespread after the Early Triassic. Prolonged halokinesis during Cretaceous and Tertiary times created local thinning of Cretaceous and Tertiary strata over the De Wijk area and is interpreted to be responsible for the structuration of the Triassic–Cretaceous and Cretaceous–Tertiary gas accumulations.

The gas accumulation contained within the Lower Cretaceous and Triassic reservoirs is trapped in a dip closure of about 80 km^2 areal extent (Fig. 3). Within this structure, Triassic sediments, ranging from the Lower Triassic Main Claystone Member in the west to the Upper Triassic Lower Muschelkalk Member in the east, subcrop with marked angularity below the Base Rijnland Unconformity. The Lower Cretaceous Vlieland Sandstone Member overlies the unconformity and also forms part of the gas-bearing accumulation (gas-water contact at 1330 m below NAP ordnance datum). The vertical seal is provided by the Lower Cretaceous shales and marls which overlie the Vlieland Sandstone Member.

The trap configuration of the Upper Cretaceous and Tertiary reservoirs consists of a low-relief dip closure with the gas-water contact at 510 m below NAP. The vertical seal is provided by the Tertiary Ieper Clay Member.

The Carboniferous Coal Measures are generally accepted as being the source rock for the gas trapped in the De Wijk Field. In the area west of the De Wijk Field, the sealing Zechstein salts are not present, due to truncation at the Base Rijnland Unconformity. The current interpretation is that the gas in the Triassic, Cretaceous and Tertiary reservoirs has migrated along major faults to the Base Rijnland and Base North Sea Unconformities, where it has migrated updip along the unconformity surfaces and subsequently charged the present-day gas accumulations.

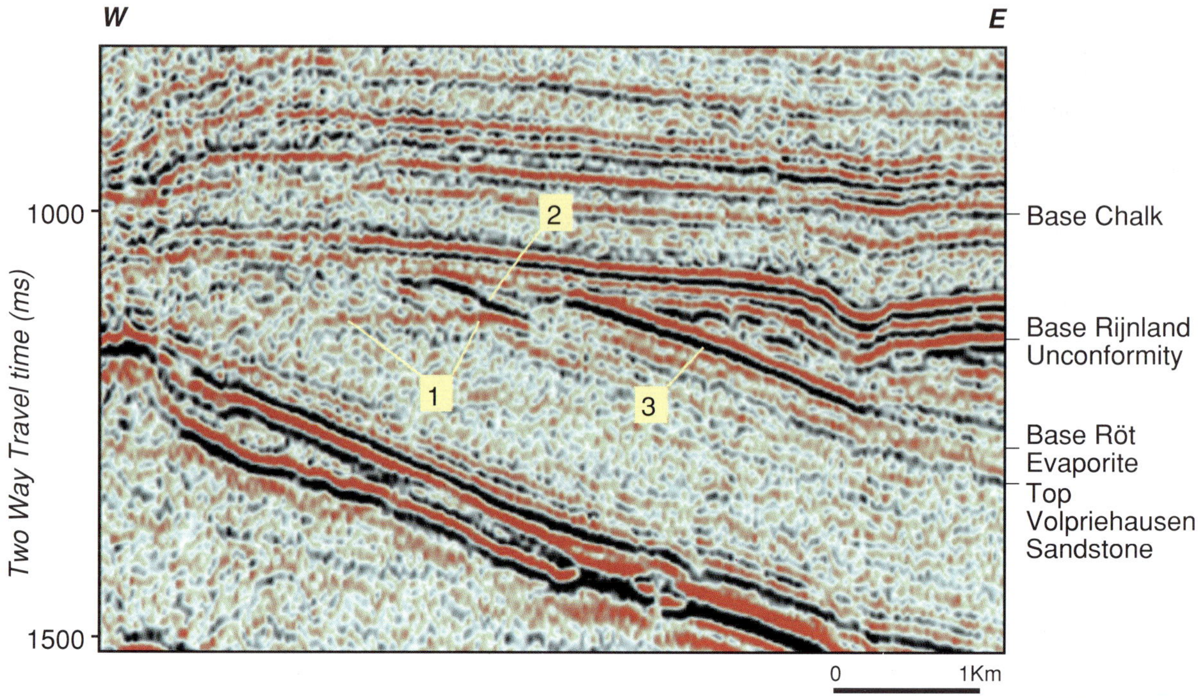

Fig. 5. Seismic section showing effects of gas fill, sand distribution and leaching of Triassic and Early Cretaceous reservoirs. The location of the line is indicated on Fig. 6. A black loop represents a positive ('soft kick') and a red loop a negative amplitude ('hard kick'). 1 = 'flat event' in Rogenstein Member; 2 = Amplitude brightening top Volpriehausen reflection; 3 = Amplitude brightening base Röt Evaporite reflection. Events 1 and 2 are related with hydrocarbon fill. Event 3 is due to leaching associated with the Base Rijnland Unconformity.

Seismic interpretation

The De Wijk Field was covered by the Hoogeveen 3D seismic survey in 1989 (Fig. 1). The data are of good quality and were interpreted on a Landmark Interpretation System.

It was only after the interpretation of the 3D data that the uniqueness of the field was fully realized. Various attribute measurement techniques resulted in a series of spectacular displays, which are discussed below. The examples shown include hydrocarbon effects on the Volpriehausen Sandstone Member and the Rogenstein Member, the Basal Dongen Tuffite Member, leaching effects on the Röt Evaporite Member and sand distribution in the Vlieland Sandstone Member.

Amplitude map Top Volpriehausen Sandstone

The Top Volpriehausen Sandstone reservoir can be recognized on seismic as a clear 'soft kick' reflection (i.e. a decrease in acoustic impedance, displayed as a black loop) with varying amplitudes (Fig. 5). A distinct brightening of the event can be observed above a level coinciding with a well-developed 'flat event'. This 'flat event' coincides with the gas-water contact of the field, but is only developed in the area of good reservoir (i.e. leaching) in the Rogenstein Member.

In addition, the amplitude map of the Top Volpriehausen Sandstone reflection (Fig. 6) shows the presence of a high-amplitude area (represented by the red colours) over the De Wijk Field which is structurally conformable with the Top Volpriehausen time map.

Amplitude map Base Röt Evaporite

Apart from the distinct amplitude brightening of the Top Volpriehausen Sandstone reflection, which clearly can be related to the presence of hydrocarbons, a marked amplitude brightening of the Base Röt Evap-

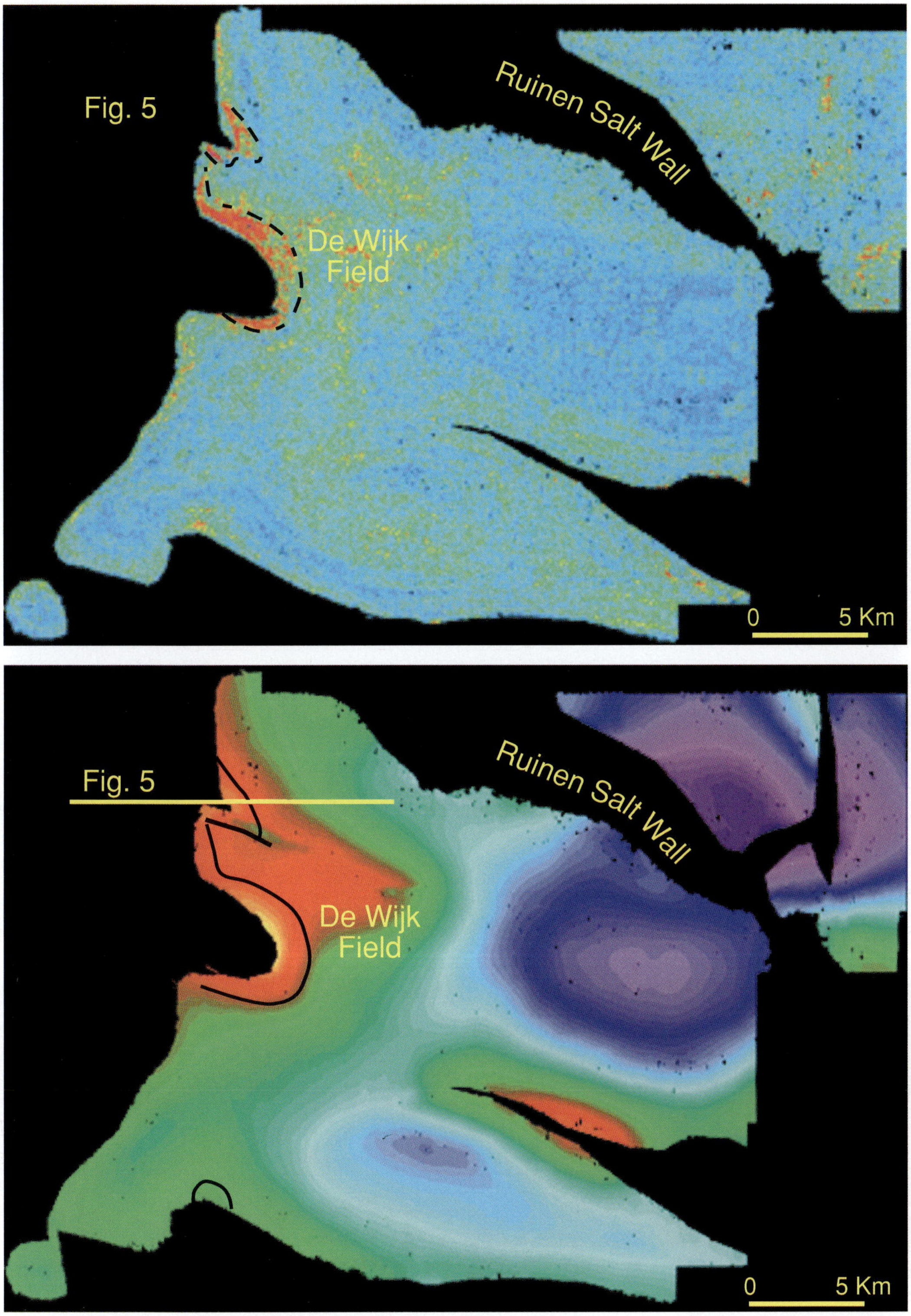

Fig. 6. Amplitude map Top Volpriehausen Sandstone (above) and time map Top Volpriehausen Sandstone (below). The red colours on the amplitude map indicate high seismic amplitude values; the blue colours represent low seismic amplitude values. The red colours on the time map indicate low seismic time values; the blue colours represent high seismic time values. The amplitude map shows the presence of high amplitude values over the De Wijk Field, structurally conformable with the time map.

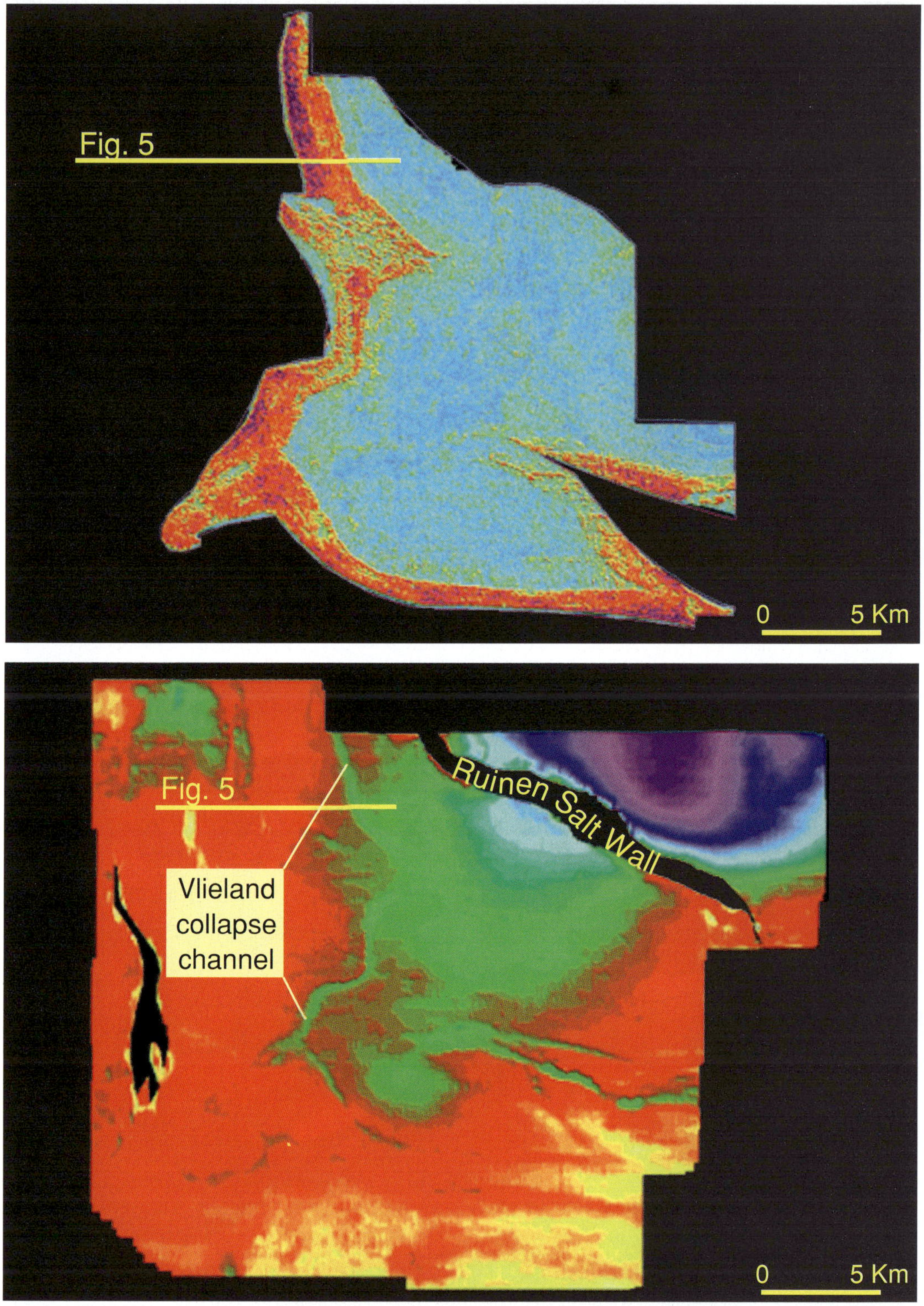

Fig. 7. Amplitude map Base Röt Evaporite (above) and isotime map Rijnland Group (below). The red colours on the amplitude map indicate high seismic amplitude values; the blue colours represent low seismic amplitude values. The increase in amplitude is related to salt dissolution of the Röt Evaporite Member below the Base Rijnland Unconformity.The red colours on the isotime map indicate low seismic isotime values, indicating a thin Rijnland Group in the west; the blue colours represent high seismic isotime values, reflecting a thickening of the package to the east. A subtle thickening of the sequence can be observed in the Vlieland collapse channels, conformable with the maximum depth of salt dissolution of the Röt Evaporite Member.

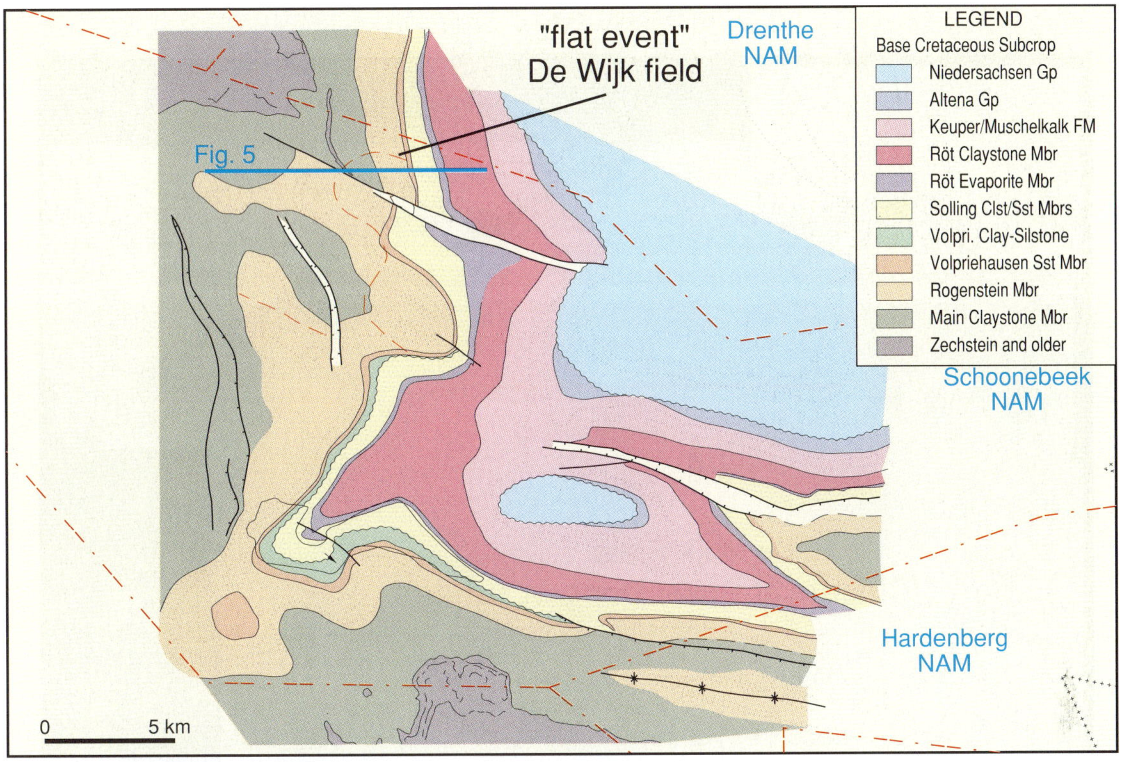

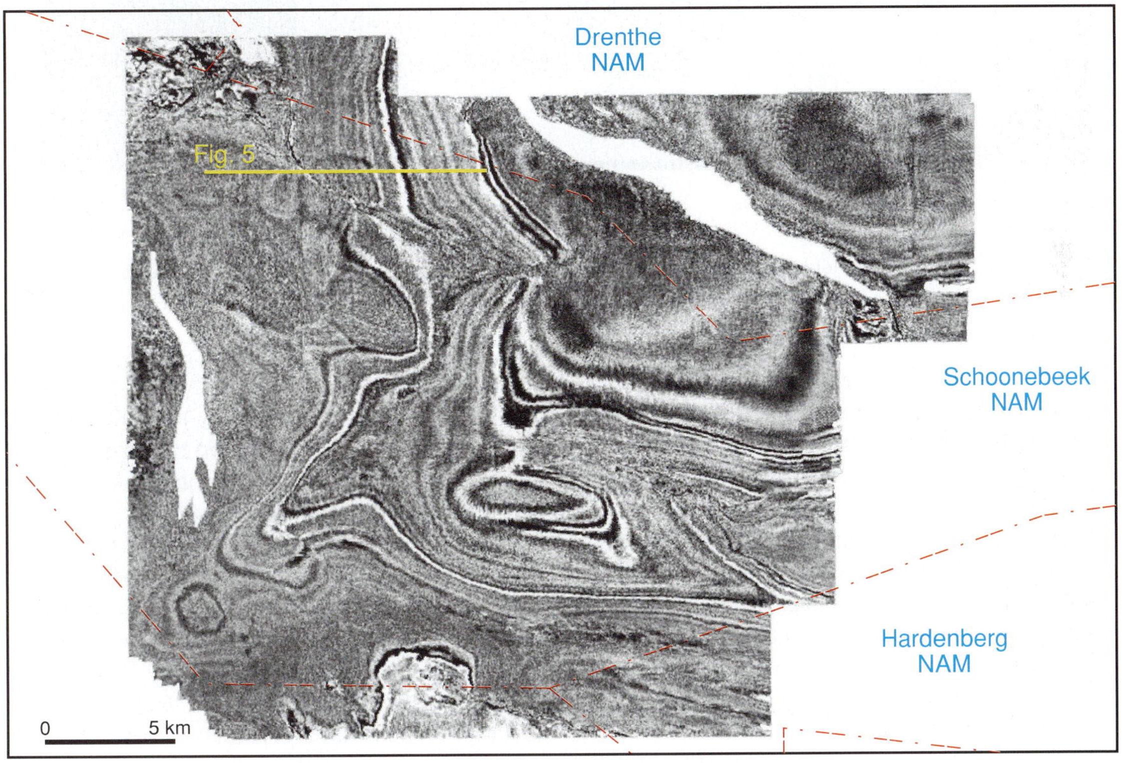

Fig. 8. Subcrop map Base Rijnland Unconformity (below) and geological interpretation of this map (above). A black loop represents a positive amplitude ('soft kick') and a white loop represents a negative amplitude ('hard kick').

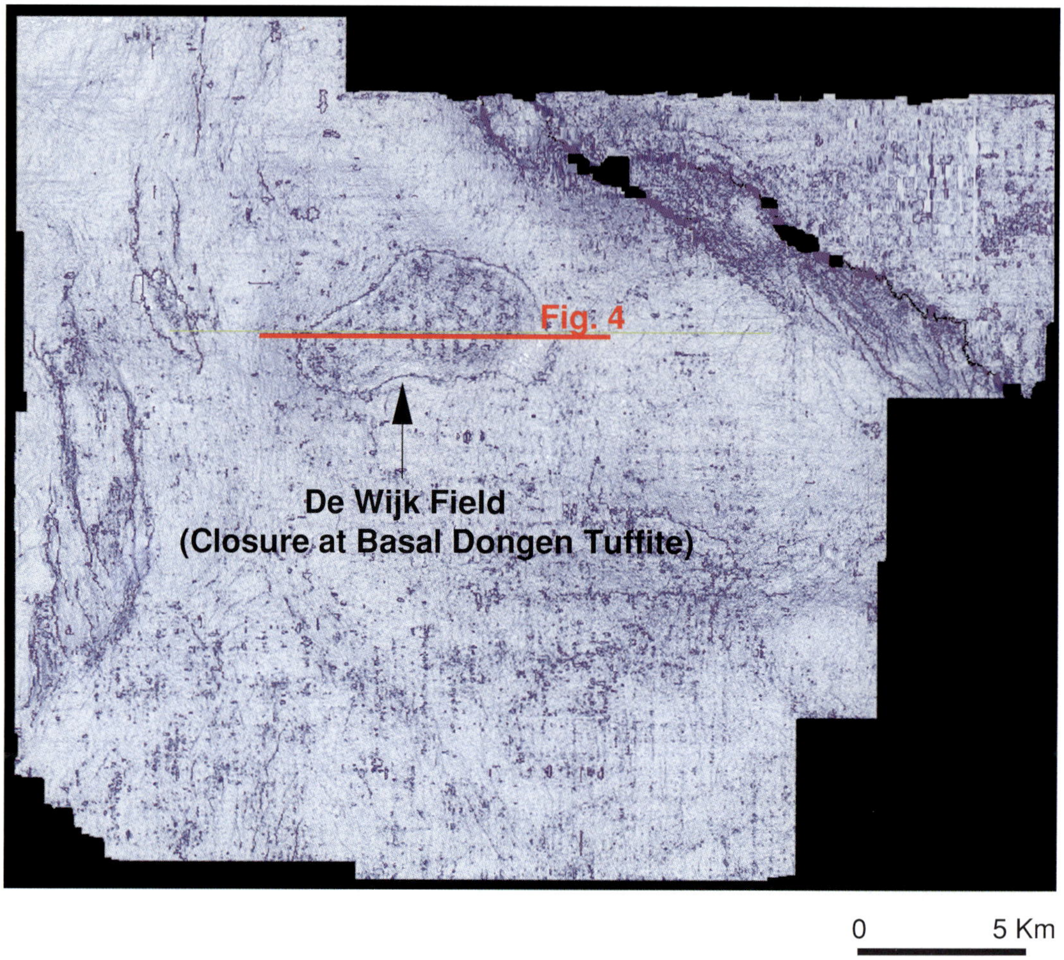

Fig. 9. Dip map Base North Sea, showing the outline of the gas bearing closure of the Basal Dongen Tuffite.

orite reflection can be observed (Fig. 5). The Base Röt Evaporite reflection is interpreted as a black loop.

Since the marker cannot be related to the presence of any reservoir, it is interpreted that this effect is due to salt dissolution below the Base Rijnland Unconformity. This mechanism would have removed the thin salt layers from the Röt Evaporite Member which normally consists of a rapid alternation of rock salt and anhydrite. The remaining anhydrite package is thick enough to be resolved by seismic and the acoustic impedance contrast with the underlying claystones causes a strong 'soft kick' (i.e. decrease of acoustic impedance) of the Base Röt Evaporite reflection.

Areal mapping of the reflection reveals the high amplitudes (as displayed by the red colours) being parallel to the truncation by the Base Rijnland Unconformity (Fig. 7).

Subcrop map Base Rijnland Unconformity

The subcrop map is derived by amplitude extraction along a pseudohorizon at a fixed two-way-time difference (24 ms) below the Base Rijnland Unconformity. It reveals the various subcropping reflections, e.g. the Volpriehausen Sandstone and the Base Röt Evaporite Members (Fig. 8). In addition, gas effects like the amplitude brightening of the Volpriehausen Sandstone and the 'flat event' in the Rogenstein Member can be observed over the De Wijk Field (Fig. 5).

The interpretation and identification of the various subcropping reflections eventually leads to a geological subcrop map, essentially reflecting the regional field geological map at Early Cretaceous times (Fig. 8).

Isotime map Rijnland Group

Due to the dissolution of the salt layers of the Röt Evaporite Member, space is created since the salt is removed by fluids. From the seismic section (Fig. 5), it becomes evident that above a zone of deepest leaching, a small collapse graben is formed in the Lower Cretaceous Rijnland sequence.

In addition, this sequence shows a marked thickening to the east. Areal mapping of the time difference (isotime) between the Base Chalk reflection and the Base Rijnland Unconformity clearly confirms the thickening of the Rijnland Group to the east (Fig. 7).

A subtle thickening of this package is observed in the collapse channels (displayed as green colours on Fig. 7), conformable with the maximum depth of salt dissolution of the Röt Evaporite. Well data confirm that the collapse grabens contain a thicker, sandy to conglomeratic channel fill, indicating active dissolution of the salt during Early Cretaceous times.

Dip map Base North Sea Group

The Base North Sea Unconformity is expressed on seismic as a strong 'hard kick' (increase in acoustic impedance) which is displayed on the seismic section as a red loop (Fig. 4). The weak red loop preceding the Base Tertiary reflection is interpreted as the Base Basal Dongen Tuffite marker.

The seismic character of the two reflections changes from a relatively weak amplitude doublet in the water zone to the high amplitude doublet in the gas-bearing zone. Due to the interference of the gas-water contact with the doublet, one broad strong hard loop is visible at the edge of the gas-bearing zone. As seismic interpretation uses techniques which can automatically pick the peak amplitude (autotrack) of a specific reflection, a small 'jump' will be interpreted at the boundary of the gas accumulation. Dip extraction techniques will result in a high value at this boundary.

The areal mapping of this reflector and the subsequent dip extraction clearly outlines the gasfield at Base North Sea level (Fig. 9).

Conclusions

The interpretation of 3D seismic data has led to a full understanding of the unique geological complexity of the De Wijk Field by a series of spectacular seismic attribute displays, highlighting the effects of gas fill, leaching and sand distribution in the various reservoirs.

Acknowledgements

The author is indebted to the Nederlandse Aardolie Maatschappij BV (NAM), Shell Internationale Petroleum Maatschappij (SIPM) and Exxon Company International (ECI) for granting permission to publish this paper. Particularly thanked is Bob van Wees (currently with Petroleum Development Oman) who contributed to the paper.

Rondeel et al. (eds), Geology of gas and oil under the Netherlands, 255–263, 1996.

The Logger oil Field (Netherlands offshore): reservoir architecture and heterogeneity

L.S. Goh
Continental Netherlands Oil Company, Weigelia 25, 2260 BC Leidschendam, the Netherlands

Key words: reservoir characterization, carbonate concretions, shoreface sequences, Vlieland Sandstone, Lower Cretaceous

Abstract

The Logger Field is a small oil field (51 million barrels oil in place) located in Block L16a in the Dutch offshore. The oil is trapped in a complexly faulted, overthrust anticline. The producing reservoir is the Lower Cretaceous Vlieland Sandstone. In the Logger Field, this is a thin (10 to 30 m thick) barrier sand with horizontal permeabilities in the order of hundreds of millidarcies to a few darcies. A reservoir geological study, aimed at providing the input for a reservoir simulation model, showed that, contrary to earlier interpretations, the reservoir is strongly layered. Layering reflects stacked, shoaling-upwards and deepening-upwards sequences. The presence of carbonate concretions and cemented layers causes additional vertical heterogeneity within the individual layers. The more detailed description of the reservoir in the reservoir simulation model gave rise to a better history match and hence to more confidence in production predictions.

Introduction

The Logger Field is located in Block L16a in the offshore Netherlands, some 35 km west of Den Helder (Fig. 1). Currently, Conoco, 42.125% license holder, operates the field on behalf of three partners: L.L. & E. Netherlands Petroleum Co. 37.5%, Nederlandse Aardolie Maatschappij 12.5%, and Elf Petroland BV 7.875%.

The Logger Field is small by North Sea standards and was developed as a satellite to the larger Kotter Field located nearby in Block K18b. Discovered in the spring of 1982, Logger contains an estimated 51 million barrels of oil in place. The ultimate recovery is expected to be 29 million barrels. The oil accumulation at Logger is trapped in a complexly faulted, overthrust anticline. The producing reservoir is the Vlieland Sandstone, a thin Lower Cretaceous coastal barrier sand which is mostly less than 20 m thick.

No previous reservoir geological characterization study had been undertaken on the Logger sandstone. In the past, it had been simply regarded as a thin, clean and homogeneous reservoir sand, with no internal shale layers present. In the only previous reservoir simulation study in 1988, the Logger reservoir was arbitrarily subdivided into three layers.

In 1989–1990, a detailed geological study of the Logger sandstone was carried out, to provide the reservoir engineers with an accurate geological model for reservoir simulation and field optimization (Goh 1992). The study used logs, cores and production data, and was integrated with a seismic re-interpretation of the field. It included petrography, SEM (scanning electron microscopy), XRD (X-ray diffraction), 'Mineralog' (infrared spectroscopy) and minipermeameter measurements on cores. The study has provided a layered geological model, and has resulted in an understanding of the geological controls on fluid flow. Aspects on Logger reservoir anatomy and heterogeneity from the study have been adapted for this paper.

Field history

The Logger Field was discovered in March 1982 by well L16–6, drilled near the crest of the structure. The

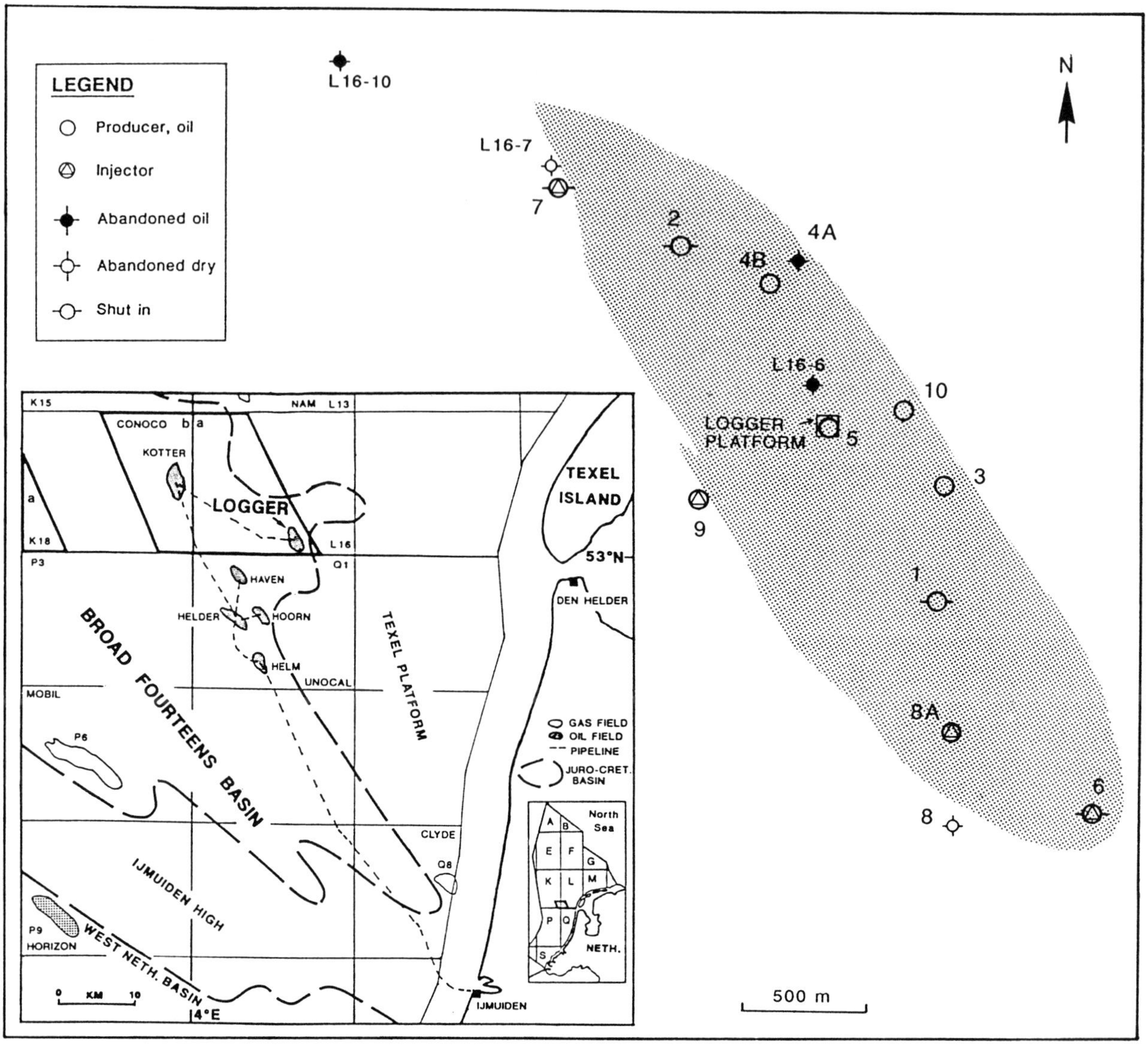

Fig. 1. Logger Field location map.

well found 16.4 m of Lower Cretaceous Vlieland Sandstone (at 1836 m sub-sea) completely oil-bearing. The sandstone tested 3031 barrels per day of 34.1 degree API oil on a 1 inch choke. An outstep delineation well, L16–7 (drilled in April 1982 1.4 km to the NW), encountered the Vlieland Sandstone water-bearing at 2004 m sub-sea.

Following the success of the second delineation well (Logger-1), the Logger development was approved in July 1984. A 3D seismic survey was shot over the Logger structure. The original development plan called for five producers and three injectors. This was completed by September 1986. Two more wells were added subsequently, an injector in 1989 and a producer in 1991. The Logger Field came on stream in August 1985. Water production commenced about one year later. Overall, water production has been below forecast. Logger has moved into gradual decline since late 1987.

The field has no gas cap, and has a very low gas oil ratio (GOR). Field development is based on down-dip water injection and up-dip oil production, and the drive mechanism is by edge water drive. Due to the low GOR, electric submersible pumps have been used to

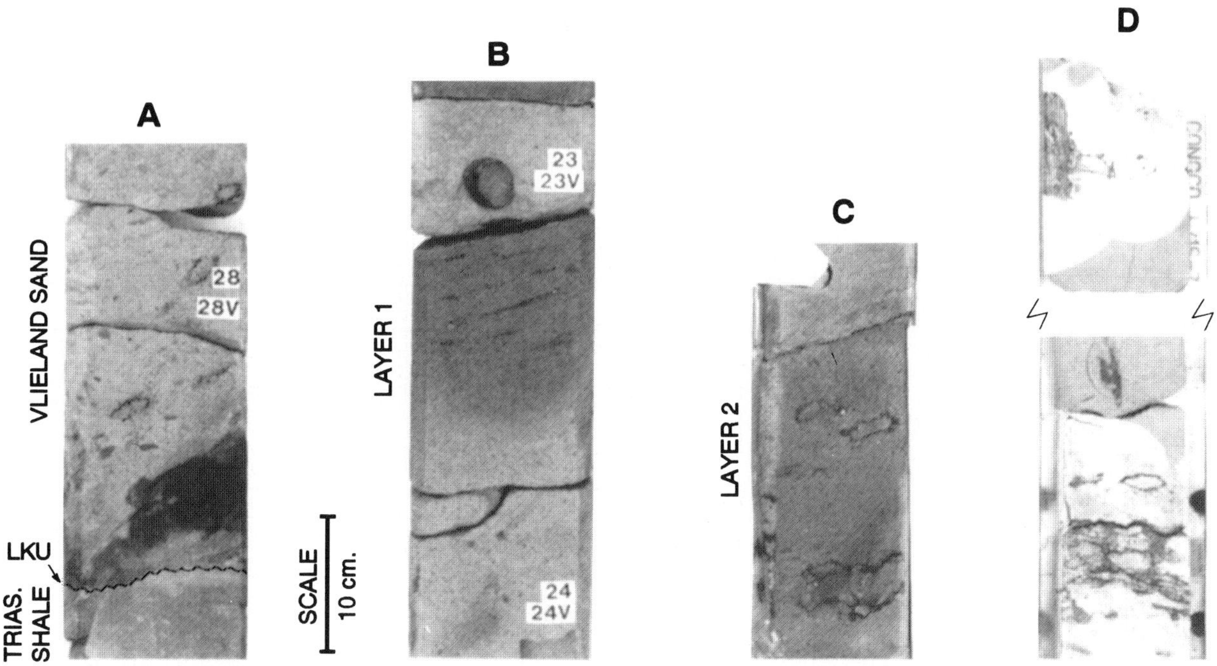

Fig. 2. Examples of core facies in Logger sandstone. (A) Close-up view of Late Kimmerian Unconformity (LKU) showing basal lithoclasts, (B) Upper shoreface-foreshore sand, laminated, (C) Middle shoreface sand, strongly bioturbated, (D) View of carbonate-cemented Concretionary Bed 1, showing colloform outline.

sustain high production rates. At the time of submission of this paper, the Logger Field was producing a total of about 13 500 barrels of fluids per day from four producers, of which about 3600 barrels were oil.

Reservoir geology

Geological setting

The Logger Field is situated on the northeastern margin of the Broad Fourteens Basin (Fig. 1), one of a series of NW – SE oriented, Mesozoic extensional rifts in the southern North Sea. The Vlieland Sandstone, the focus of this study, was deposited as a transgressive body on the Late Kimmerian Unconformity during post-rift subsidence in the Valanginian (Van Wijhe 1987). It progressively onlapped the unconformity, as the basin widened. The Vlieland Sandstone was eventually buried under a thick sequence of marine shales (Vlieland Shale), which form the regional top seal for the oil accumulations in the basin.

The oil in Logger has been sourced by the Lower Jurassic Posidonia Shale (Goh 1992). As the Posidonia Shale is absent in the Logger Field area (due to erosion), oil is assumed to have undergone long-distance lateral migration from a kitchen area to the west. Recently, the geochemistry of the Logger crude oil has been analysed and correlated with that of a source rock extract taken from the Posidonia Shale beneath the Kotter Field. The correlation, based on carbon isotope values, biomarkers, pristane/phytane ratios, and n-alkane distributions, was conducted by the Koninklijke/Shell Exploratie en Produktie Laboratorium (KSEPL).

Sedimentological model

Detailed sedimentological descriptions were carried out on some 160m of sandstone cores from ten Logger wells. The core study reveals the Vlieland Sandstone to be a coastal barrier sand consisting of (preserved) lower shoreface to foreshore facies (Fig. 2; cf. McCubbin 1982). The diagnostic features observed on the cores that characterize the shoreface environment are the fine to medium grain size, the good sorting, the occurrence of horizontal to low-angle laminations, and bioturba-

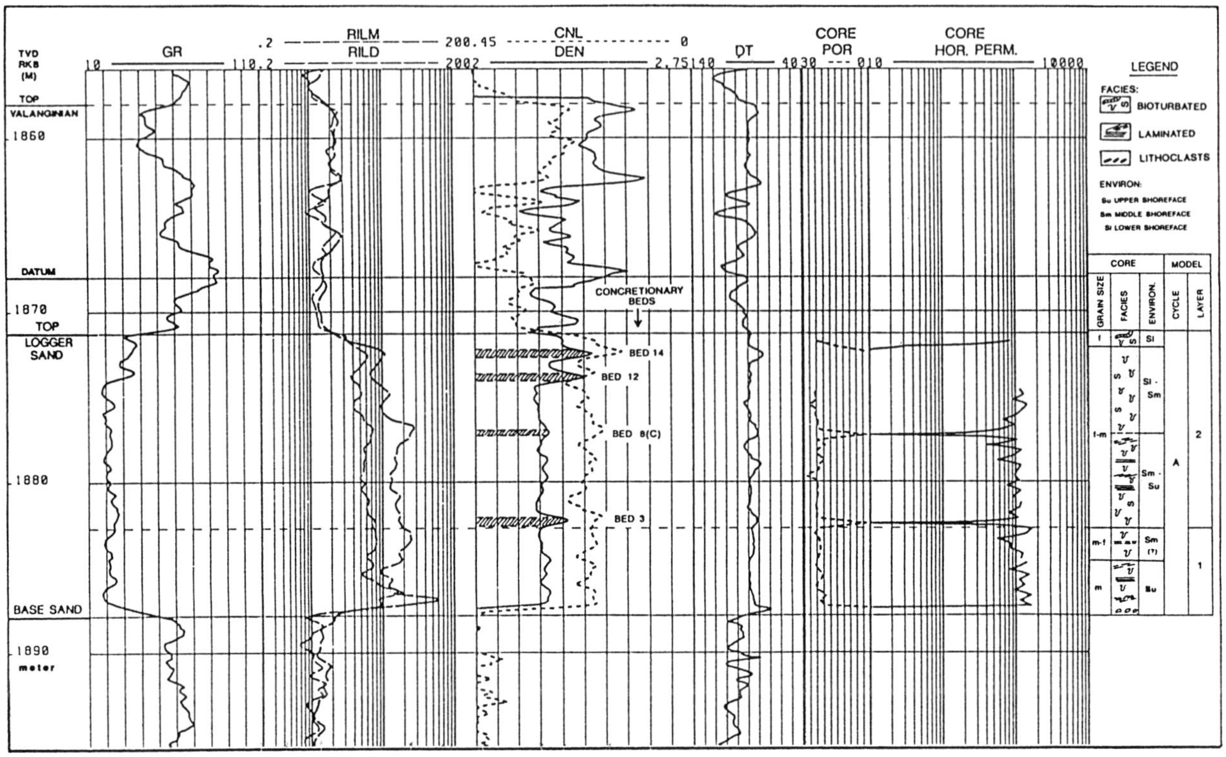

Fig. 3. Example of core to log correlation and Logger sandstone subdivision.

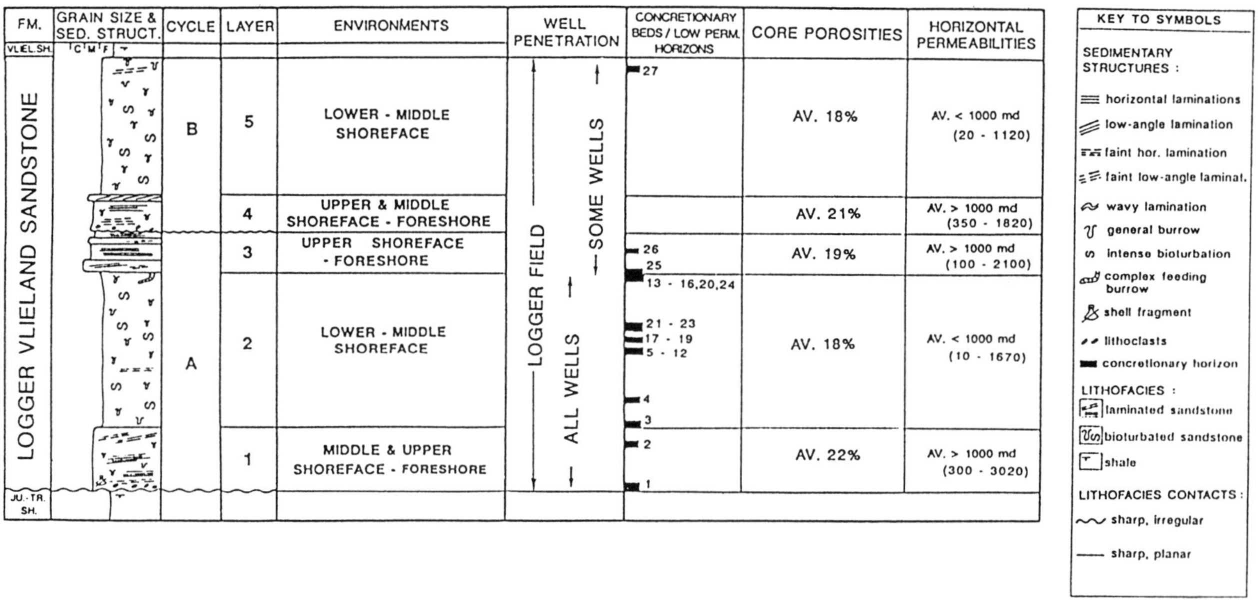

Fig. 4. Lithofacies, layers and cycles of the Logger Vlieland Sandstone. The reconstruction is not to scale, and is intended to illustrate the general vertical stacking or facies succession identified in the cores.

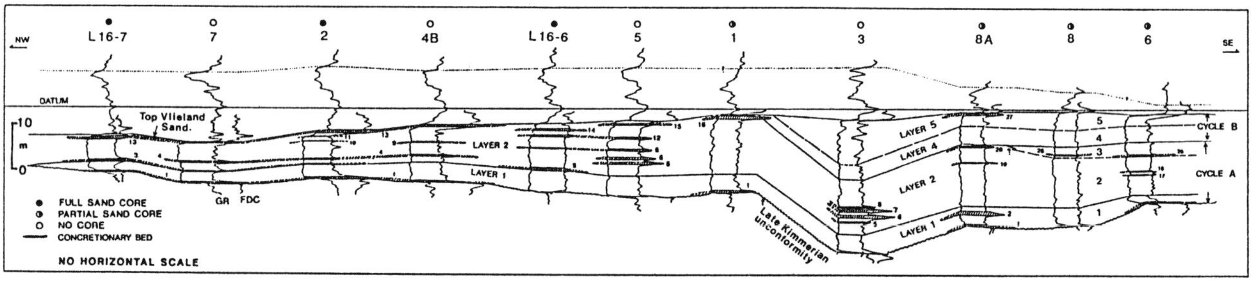

Fig. 5. Well correlation section showing the lens-like (external) geometry and the internal layer-cake architecture of the Logger sandstone along strike. The presence of concretionary beds has given the reservoir anatomy a slightly more complex character. For well locations see Fig. 1.

tions. No fluvial or tidal channels, nor lagoonal facies were recognized.

Two principal sand types or depositional facies were recognized from the core study. They form the basic 'building blocks' in the Logger sand and facilitate a subdivision into correlatable layers. They are:

a) medium – coarse grained sands, low-angle to horizontally laminated, clean, deposited in upper shoreface to foreshore environments; and
b) fine-grained sands, bioturbated, 'dirty' (i.e. rich in detrital clays, fine organic matter and pyrite), deposited in lower – middle shoreface environments.

Reservoir architecture

The integration of core facies and wireline logs (Fig. 3) has allowed the subdivision of the Logger sand into five sheet-like layers (numbered from bottom to top). The layering, based on depositional facies differences (Figs 4, 5), gives rise to a layer-cake architecture expected of a composite barrier sand (Weber & Van Geuns 1989). Layers 1, 3 and 4 are characterized by the presence of coarser-grained, laminated sands, with rare or no bioturbation and high horizontal permeabilities (exceeding 1 Darcy). By contrast, Layers 2 and 5 are clay-rich and organic-rich, finer-grained units, characterized by intense to moderate burrowing and moderate to low horizontal permeabilities (generally below 1 Darcy). In general, the boundaries between the layers coincide with significant changes in reservoir properties. A number of concretionary horizons were documented at or near the boundaries.

The entire Vlieland Sandstone is genetically divisible into two depositional cycles, separated by an erosional surface. Cycle A consists of Layers 1, 2 and 3, and Cycle B of Layers 4 and 5. Each cycle can be regarded as an overall transgressive unit, starting with a basal lag. The Cycle A sand directly onlaps the Late Kimmerian Unconformity and has accumulated more or less parallel to the paleo-shoreline with good strike continuity. The Cycle B sand is believed to result from a lateral (east to southeasterly) migration or shift of the shoreline (Fig. 6).

Mineralogy and diagenesis

Petrographic, 'Mineralog', XRD and SEM analyses have been carried out on selected core samples to investigate mineralogy and diagenetic history, and their importance as controls on reservoir quality. The Logger Vlieland Sandstone is shown to be a predominantly fine to medium-grained quartz arenite (Fig. 7). Its good to excellent reservoir quality is attributable to initially high depositional porosities (resulting from good sorting), moderate compaction and minor quartz cementation. Internal variations in reservoir properties directly reflect the occurrence of the bioturbated facies versus the laminated facies.

Carbonate concretions

The Logger sandstone contains numerous carbonate-cemented concretions. These are easily recognizable in cores and on logs, and they range from 10 cm to 2 m thick. Twenty-seven such concretionary beds and low-permeability horizons have been identified (Fig. 4).

In slabbed cores, the concretionary beds have colloform outlines, light colours (due to lack of oil staining) or a patchy appearance. Petrographical studies show floating sand grains and shell fragments surrounded by

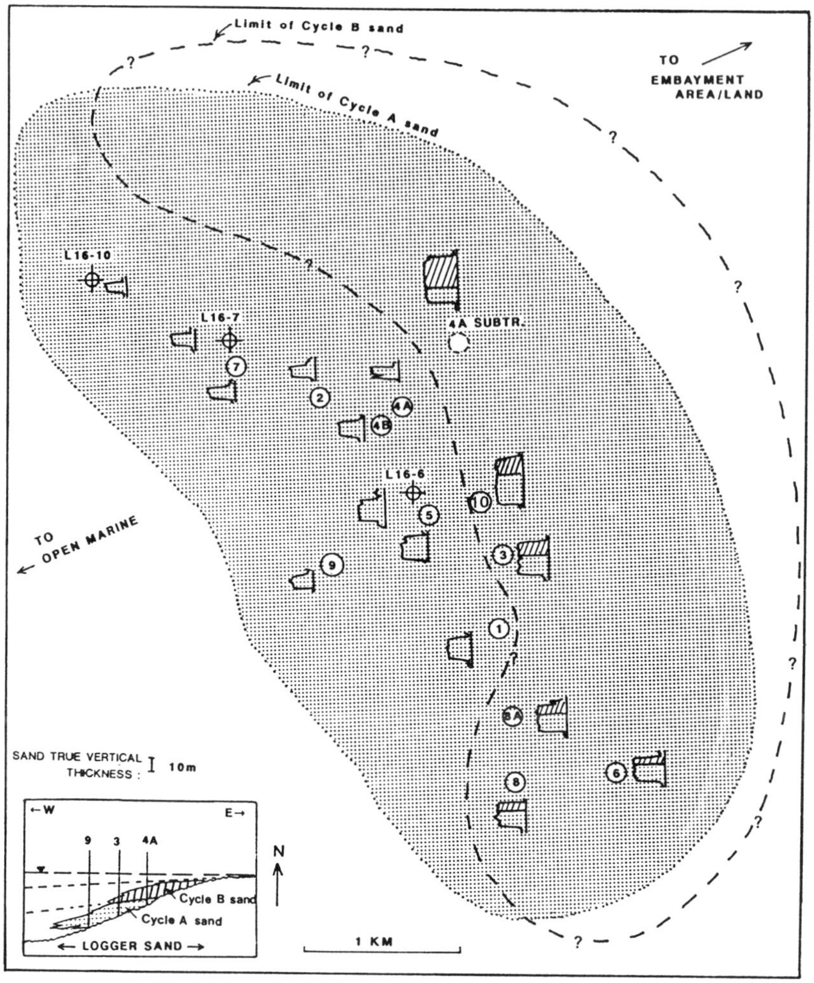

Fig. 6. Depositional setting of the Logger sandstone. Although the thicker sand development reflects the stacking of units from the two depositional cycles, the fundamental cause is syn-sedimentary tectonic subsidence in the underlying Triassic graben.

Fig. 7. Thin-section and SEM photomicrographs. (A) to (D) show micro-scale reservoir permeability heterogeneity inherent in the original sedimentary fabric; (E) and (F) show macro-scale reservoir permeability barriers of secondary origin. (A) Fine-grained sand, showing good sorting, moderate compaction and organic debris and/or pyrite in disseminated and laminar distribution. (B) Closer view of the same sand, highlighting pores filled with organic remains, pyrite and/or detrital clays (dark brown areas). (C) Clear, open pores lined with minor quartz overgrowths, clays and pyrite grains. (D) Close-up view of an impervious streak in a pore, consisting of an organic debris/pyrite/detrital clay mixture. (E) Sample from a concretionary bed, completely cemented by early diagenetic calcite. The cement probably initially nucleated upon the clay lithoclasts and bioclasts. (F) Another example of a calcite-cemented concretionary bed, showing loose packing of detrital grains and replacement by calcite. →

early carbonate cements. Mineralogically, the carbonate composition is calcite, but with some Mg and/or Fe substitution.

As the concretionary beds are tight zones which form permeability barriers (or semi-barriers) to vertical fluid flow within the reservoir, an attempt has been made to correlate some of the concretionary beds between wells in order to determine their lateral extent (Fig. 5). This interpretation does not imply that correlatable beds occur as lenticular sheets. The beds could vary from sheet-like to mesh-like in their morphology. No probabilistic modellling was carried out. The presence of the concretions has given the reservoir anatomy a more complex character.

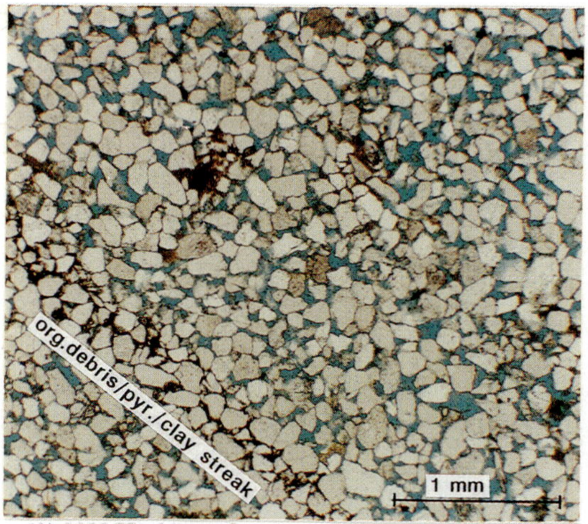

(A) 2798.75m Logger-8 PPL

(B) 2798.75m Logger-8 PPL

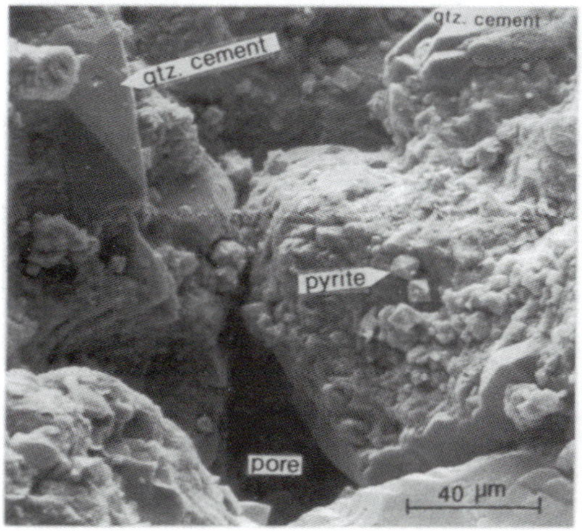

(C) 2919.15m Logger-6 SEM

(D) 2901.58m Logger-6 SEM

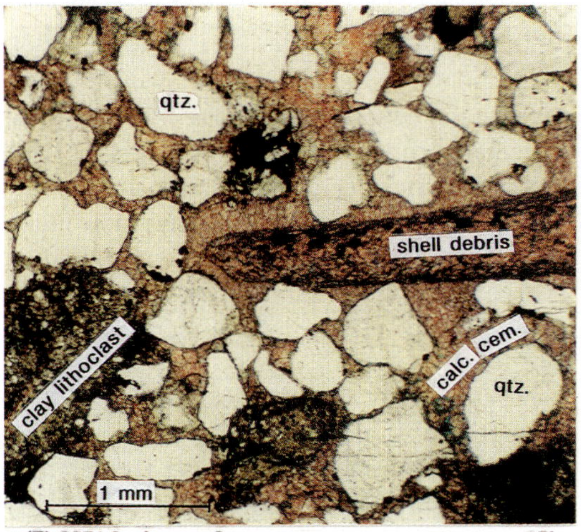

(E) 2271.0m Logger-2 PPL

(F) 1876.6m L16-6 XPL

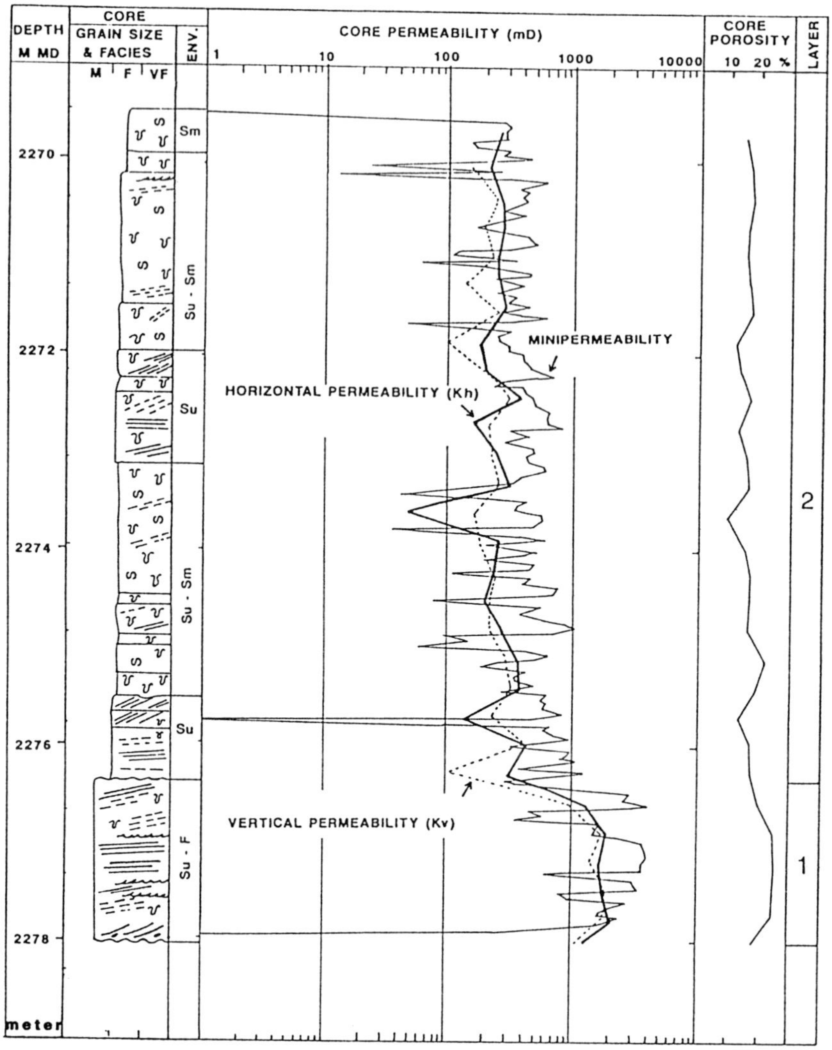

Fig. 8. Reservoir heterogeneity in the Logger sandstone. Minipermeability measurements can reveal permeability variations 'overlooked' by conventional core plugs. MD = Measured depth below derrick floor.

Reservoir heterogeneity

Reservoir heterogeneity in the Logger sandstone is closely associated with primary depositional features or facies. Heterogeneity can be viewed in at least two scales. The 'macro-scale' heterogeneity is expressed by the layering and the concretionary beds. The 'micro-scale' heterogeneity is much more difficult to quantify or define, but is clearly inherent in the original sedimentary fabric in each layer.

Mini-permeameter studies show the local permeability variations to be associated with facies changes within the reservoir layers, i.e. low mini-perm values generally correspond to bioturbated, organic and/or clay-rich zones (Fig. 8). Bioturbation is believed to have introduced some detrital clays (now lining the burrows) into sandstone and also brought about a destruction of the original sedimentary fabric. Both processes are believed to have locally increased the flow path tortuosity.

Petrography and SEM studies (Fig. 7D) reveal the occurrence of fine, organic, clay and/or pyrite-rich streaks within the pore throats. These streaks locally hinder flow or increase flow tortuosity.

Conclusions

The Vlieland Sandstone at the Logger field is thin but is not homogenous. It has a layer-cake architecture. Primary controls on reservoir quality and heterogeneity are related to depositional facies (i.e. grain size, bioturbation and matrix composition in pores).

High-permeability layers and carbonate concretionary horizons are two key geological controls on fluid flow and field performance in Logger.

Epilogue

The reservoir geological model has been upgraded in 1993 concurrent with the seismic re-interpretation of the Logger Field. The most important upgrade is the subdivision of Layer 2 into two sub-layers. This allows more accurate modelling of the effect of well completions and fluid offtake. Another upgrade to the model is the introduction of lateral horizontal permeability variation or anisotropy within individual layers. Both changes have resulted in some improvement to the history match.

Acknowledgements

I thank Conoco Inc., Conoco Netherlands, and the Logger Field partners for permission to publish my original study. I thank Josie Hendrikx for typing the manuscript.

References

Goh, L.S. 1992 The Logger Field: Geology and reservoir characterization. North Sea oil & gas reservoirs – III, NTH, Trondheim, Norway, Nov. – Dec. 1992, Conference Proceedings – Kluwer Academic Publishers, Dordrecht 1994: 75–93

McCubbin, D.G. 1982 Barrier-Island and Strand-Plain Facies. Am. Ass. Petroleum Geol. Mem. 31

Weber, K.J. & L.C. Van Geuns 1989 Framework for Constructing Clastic Reservoir Simulation Models – Preprint, 64th Soc. Petroleum Engineers (SPE) Annual Technical conference and Exhibition, San Antonio, Texas, October 1989

Van Wijhe, D.H. 1987 Structural evaluation of inverted basins in the Dutch offshore. Tectonophysics 137: 171–219

Rondeel et al. (eds), Geology of gas and oil under the Netherlands, 265–284, 1996.

Sedimentary and structural history of the Texel-IJsselmeer High, the Netherlands

R.H.B. Rijkers & M.C. Geluk
Geological Survey of the Netherlands, Postbus 157, 2000 AD Haarlem, the Netherlands

Key words: Friesland Platform, Central Netherlands Basin, Permian, Late Jurassic, fault reactivation, structural model, natural gas

Abstract

The Texel-IJsselmeer High, in the northern part of the Netherlands, is a NW – SE trending fault block, slightly tilted to the northeast. It affected sedimentation patterns and the structural development of the area from the Late Carboniferous – Early Permian to the Tertiary. The high influenced both the Permian (Upper Rotliegend and Zechstein) facies distribution and the sedimentary processes during the Late Jurassic and the Early Cretaceous. Uplift of the high during the Late Jurassic and contemporaneous subsidence of adjacent basins can be linked with crustal extension. Late Jurassic faulting at the southern edge of the high was accompanied by hanging-wall subsidence (Central Netherlands Basin) and footwall uplift (Texel-IJsselmeer High and Friesland Platform). This resulted in significant erosion on the high, which can be modelled with an isostatic model of the crust. During the Late Cretaceous – Early Tertiary inversion phase, reverse faulting occurred along pre-existing Jurassic faults in a zone directly to the south of the high. On both sides of the high, gas-producing Permian sandstones and carbonates have been found with good reservoir characteristics. Locally, Lower Cretaceous sandstones are gas-producing on the Friesland Platform.

Introduction

The Texel-IJsselmeer High is a pre-Cretaceous fault-bounded structural unit, situated in the northwestern part of the Netherlands (Fig. 1). The high is defined by Late Jurassic – Early Cretaceous uplift and erosion, as expressed by a hiatus at the crest of the structure where Cretaceous sediments unconformably overlie the Carboniferous. The stratigraphy and tectonic history of the area studied are summarized in Fig. 2.

The high is approximately 150 km long and 30 km wide and trends NW – SE. In a broader sense it may be considered as the expression of a large basement fault block that also includes the Friesland Platform. It has been assumed that the Texel-IJsselmeer High already originated during the Variscan orogeny as a result of compressional foreland stresses (Bless et al. 1977, Ziegler 1987). Faults at the southern edge of the high trend WNW – ESE and have been reactivated during several tectonic phases. Not only normal and reverse faulting, but also wrench-faulting participated in the history of the high.

Most of the findings presented here are the result of the 1:250 000 regional mapping programme carried out by the Rijks Geologische Dienst (RGD), the Geological Survey of the Netherlands (RGD 1993a,b). The limits of the two sheets involved are given in Fig. 1.

The Texel-IJsselmeer High and adjacent structural elements

The overall structural setting of the Texel-IJsselmeer High shown in Fig. 1, covers the area from the London-Brabant Massif in the south to the basin areas in the north. The structural elements in the northern Netherlands have been described by RGD (1991a,b, 1993a,b) and Van Adrichem Boogaert & Kouwe (1993–1995). The Broad Fourteens Basin, the Central Netherlands Basin, the Lower Saxony Basin and the Vlieland Basin

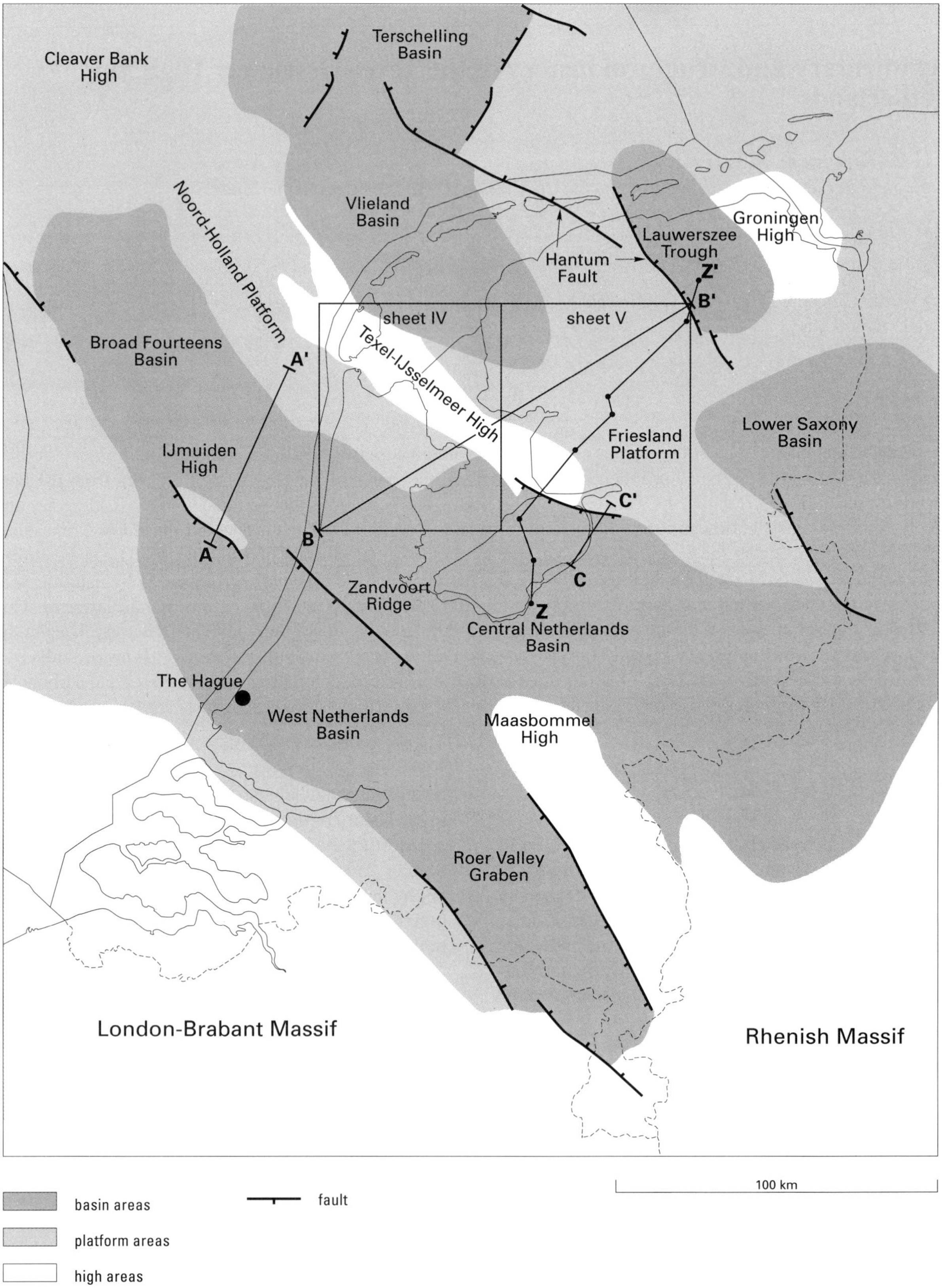

Cleaver Bank High
Terschelling Basin
Vlieland Basin
Noord-Holland Platform
Hantum Fault
Lauwerszee Trough
Groningen High
Broad Fourteens Basin
sheet IV
sheet V
Texel-IJsselmeer High
IJmuiden High
Friesland Platform
Lower Saxony Basin
A
A'
B
B'
C
C'
Z
Z'
Zandvoort Ridge
Central Netherlands Basin
The Hague
West Netherlands Basin
Maasbommel High
Roer Valley Graben
London-Brabant Massif
Rhenish Massif
100 km
basin areas
platform areas
high areas
fault

←

Fig. 1. Location of study area (map sheets IV and V, RGD 1993a,b) and Late Jurassic – Early Cretaceous structural elements. Most elements trend NW – SE and are strongly accentuated by erosion during the Jurassic. Z-Z′: Zechstein correlation profile (Fig. 5); A-A', B-B', C-C′: interpreted seismic lines and depth sections (Fig. 13).

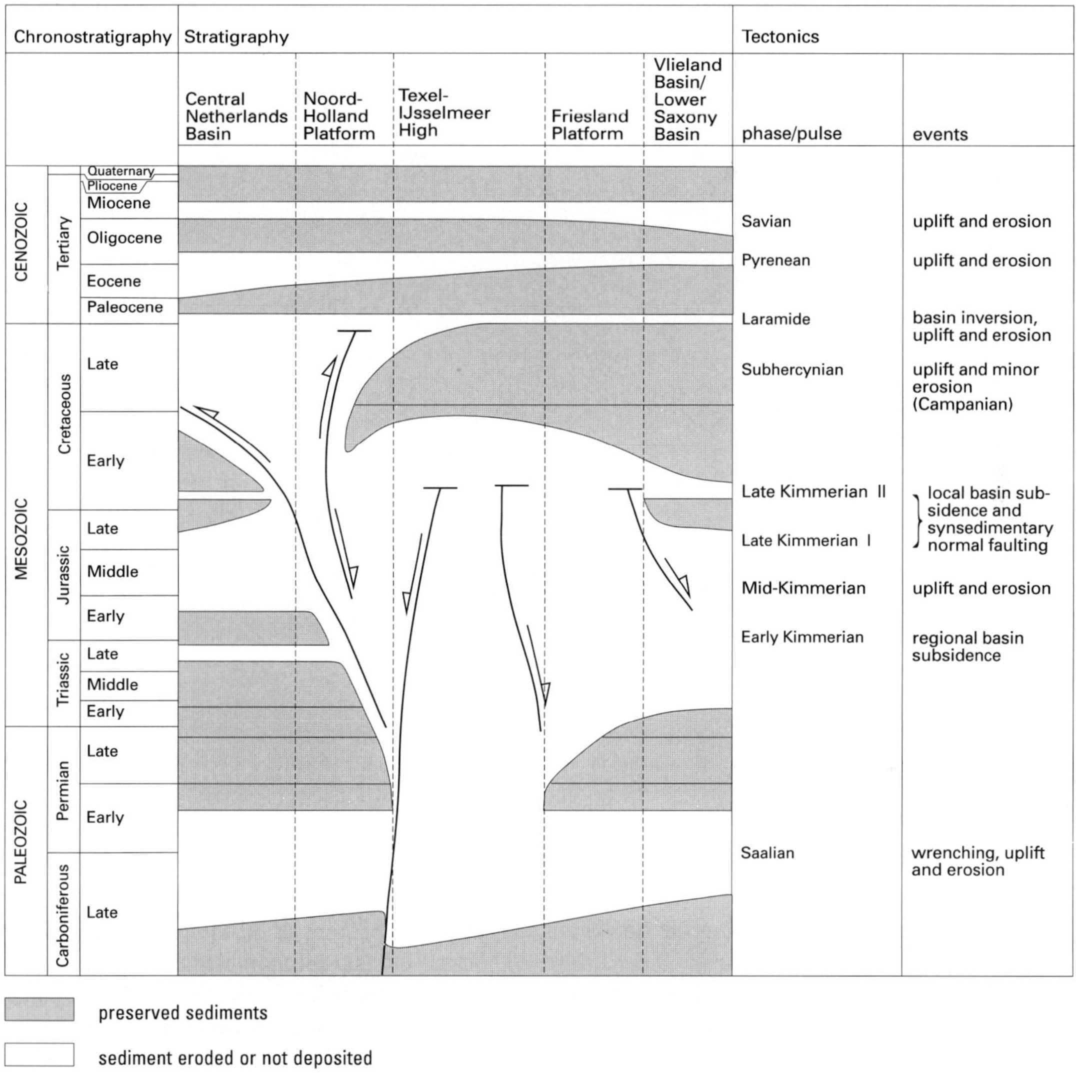

Fig. 2. Summarized stratigraphy and tectonic events. Faults on the southern edge of the Texel-IJsselmeer High were reactivated during several tectonic phases. The Middle – Upper Triassic and Lower Jurassic deposits are absent in the Vlieland Basin, but present in the Lower Saxony Basin.

surround most of the Texel-IJsselmeer High. These basins contain Jurassic and Lower Cretaceous sediments. The Noord-Holland Platform is an intermediate area between the Central Netherlands Basin and the Texel-IJsselmeer High. In this area Lower and Upper Cretaceous rest upon deeply truncated Triassic and Permian. In the southeast, the Texel-IJsselmeer High is directly bordered by the Central Netherlands Basin. To the northeast the high passes into the Friesland Platform without fault contact. Their boundary is defined by the subcrop of the Upper Rotliegend. The Texel-IJsselmeer High and the Friesland Platform can together be regarded as one fault block.

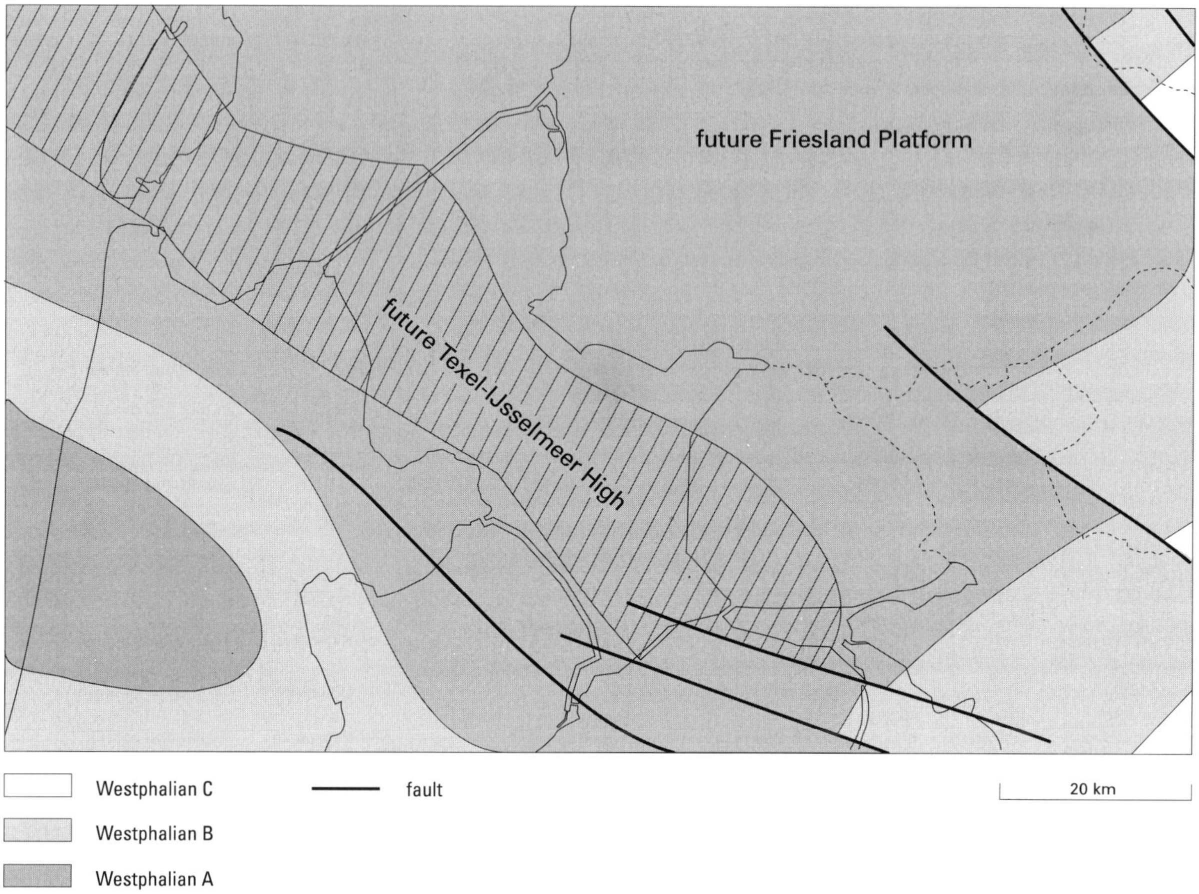

Fig. 3. Subcrop map of the Carboniferous below the Saalian unconformity. The contours of the future Texel-IJsselmeer High and Friesland Platform are also shown.

Tectonic events

The main fault systems originated during the Saalian phase of the Variscan orogeny and during the Kimmerian and the Subhercynian-Laramide phases of the Alpine orogeny (Fig. 2). Variscan tectonics during the Late Carboniferous and Early Permian created NW – SE trending fault zones which were reactivated during later tectonic phases (Arthaud & Matte 1977, Van Wijhe 1987a). The southern edge of the Texel-IJsselmeer High represents such a NW – SE trending Variscan fault zone.

Kimmerian tectonic events are characterized by extension and are expressed by normal faulting. These events lasted from the Late Triassic to Early Cretaceous. The Kimmerian deformation phase had an episodic character and led to several unconformities, the Early Kimmerian, Mid-Kimmerian, and Late Kimmerian I and II unconformities. These unconformities can only be distinguished in the basins and merge into one unconformity near the high (Fig. 2). The Kimmerian events accompanied the break-up of the Permian – Triassic basin into smaller depocentres such as the Broad Fourteens Basin, the Central Netherlands Basin, the Lower Saxony Basin and the Vlieland Basin.

Reversal of the stress regime during the Late Cretaceous from extension to compression resulted in tectonic inversion by reverse faulting on Kimmerian normal faults. In the study area, Alpine compression lasted from the Late Cretaceous to the Miocene and was expressed as several pulses, known as the Subhercynian, Laramide, Pyrenean and Savian phases (Fig. 2). In the Central Netherlands and Broad Fourteens Basins, Cretaceous and Upper Jurassic sediments were locally eroded. Transpressional strike-slip fault systems can be recognized in seismic sections of the northern margins of the Central Netherlands Basin (see under 'Structural development').

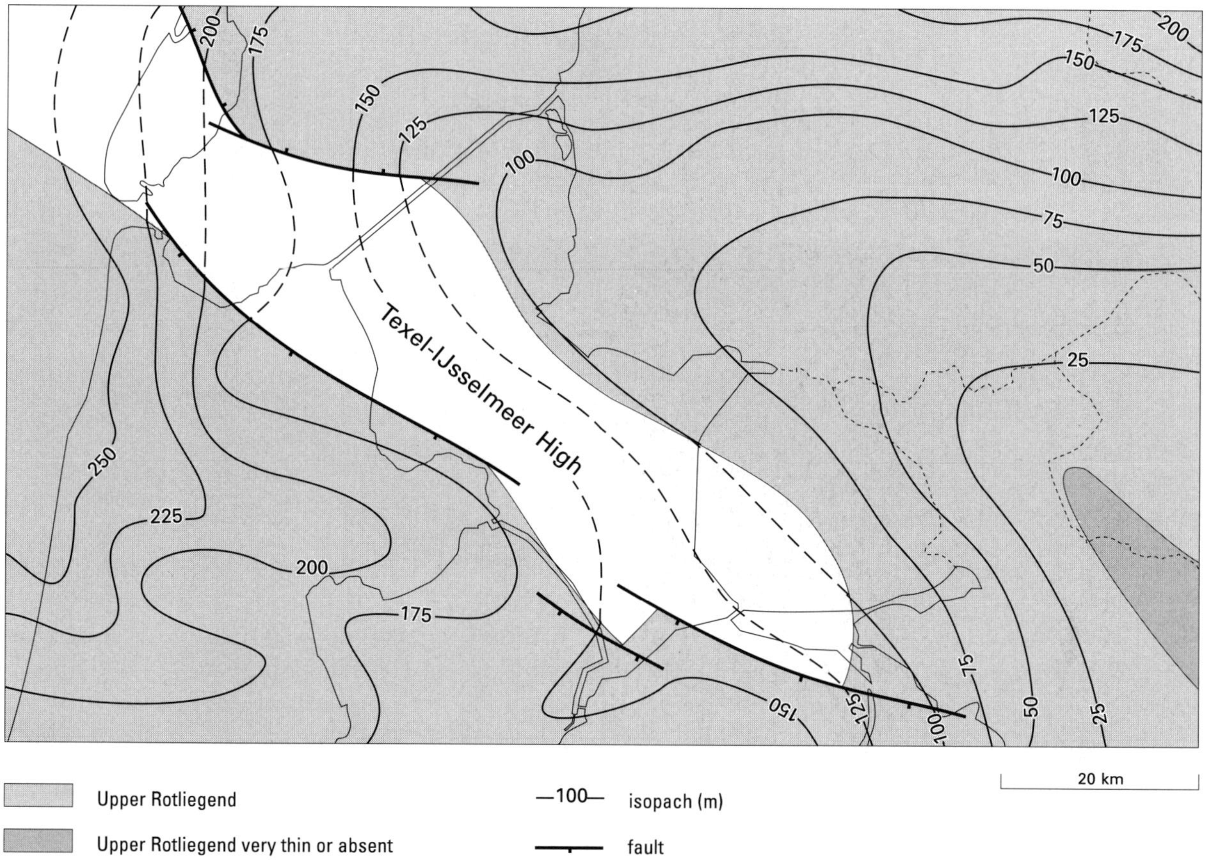

Fig. 4. Thickness map of the Upper Rotliegend showing the Texel-IJsselmeer High on which the Upper Rotliegend has been eroded owing to Jurassic uplift.

Geological evidence for reactivation of the Texel-IJsselmeer High

As will be discussed below, the Texel-IJsselmeer High repeatedly affected sedimentary processes in the area studied.

Variscan origin

The earliest indication for the existence of the Texel-IJsselmeer High is the deep truncation of Westphalian sediments below the Saalian unconformity in the area of the high (Fig. 3). The contours of the incipient Texel-IJsselmeer High and Friesland Platform are visible owing to the Saalian erosion. Although the Carboniferous was also eroded at the crest of the high during the Kimmerian phase, severe erosion must have occurred during the Saalian phase of the Variscan orogeny.

A Permian barrier

During the early Late Permian the Texel-IJsselmeer High influenced the distribution of Upper Rotliegend sediments. The high marks a boundary zone where playa and lacustrine sediments of the Silverpit Formation (Ameland, Hollum and Ten Boer Claystone Members) interfinger southward with the fluvial and aeolian Slochteren Sandstone Formation. The southernmost extent of the clayey deposits of the Ameland and Ten Boer Claystone Members appears to have been controlled by the northern edge of the high. Furthermore, the Upper Rotliegend Group onlaps onto the high; the oldest part of the Slochteren Sandstone was not deposited on the northern flank of the high. The high functioned as a barrier within the Southern Permian Basin. This barrier separated the more central parts of this basin in the north from the incipient Central Netherlands Basin in the south. At the crest of the

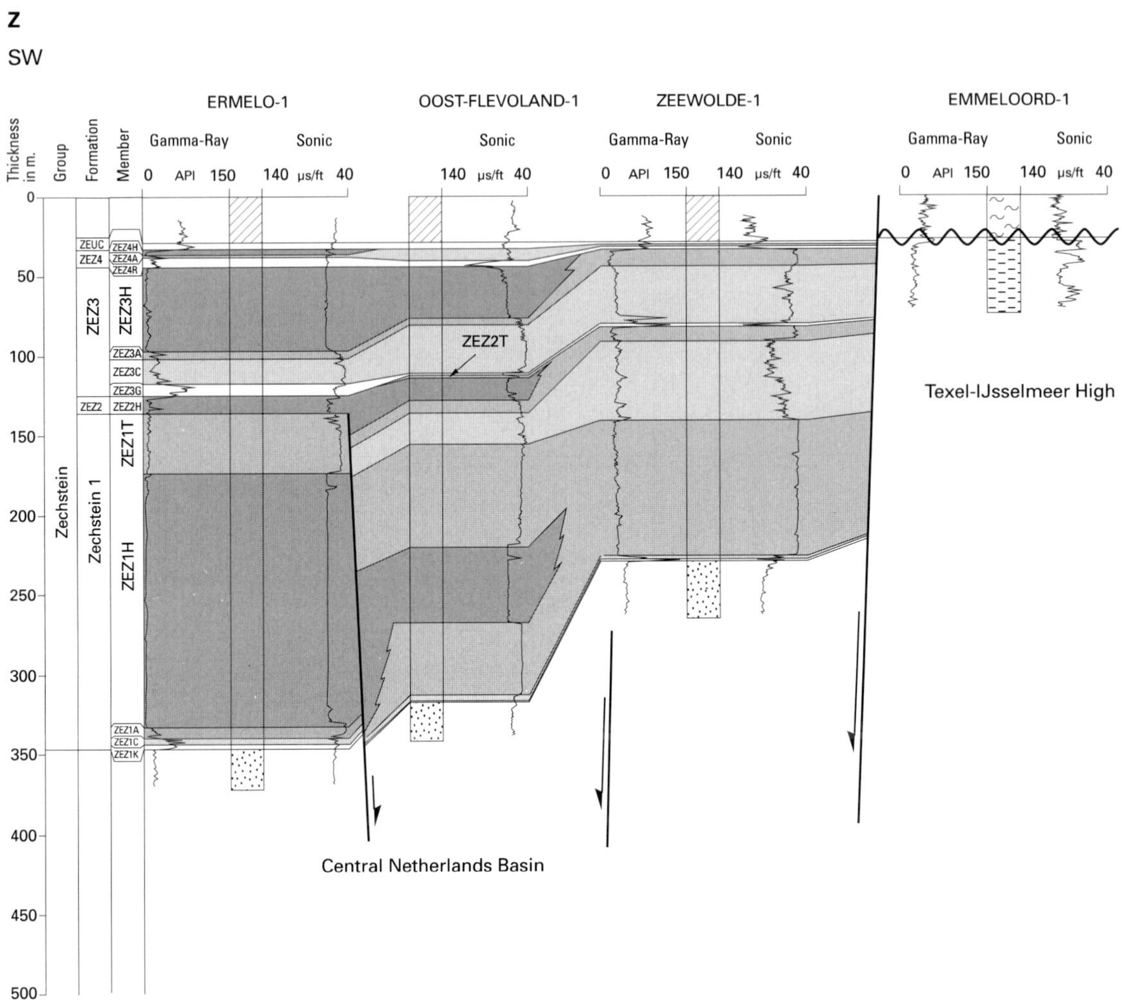

- salt
- anhydrite
- carbonate
- claystone
- Rijnland Group (Lower Cretaceous)
- Lower Germanic Trias Group (Triassic)
- Upper Rotliegend Group (Permian)
- Limburg Group (Carboniferous)
- hiatus
- fault

ZEZ1K Coppershale
ZEZ1C Zechstein 1 Carbonate
ZEZ1W Zechstein 1 Anhydrite
ZEZ1A Zechstein 1 Lower Anhydrite
ZEZ1H Zechstein 1 Salt
ZEZ1T Zechstein 1 Upper Anhydrite
ZEZ2C Zechstein 2 Carbonate
ZEZ2A Zechstein 2 Basal Anhydrite
ZEZ2H Zechstein 2 Salt
ZEZ2T Zechstein 2 Roof Anhydrite
ZEZ3G Gray Salt Clay
ZEZ3C Zechstein 3 Carbonate
ZEZ3A Zechstein 3 Main Anhydrite
ZEZ3H Zechstein 3 Salt
ZEZ4R Red Salt Clay
ZEZ4A Zechstein 4 Pegmatite-Anhydrite
ZEZ4H Zechstein 4 Salt
ZEUC Upper Zechstein
P potassium-magnesium salt

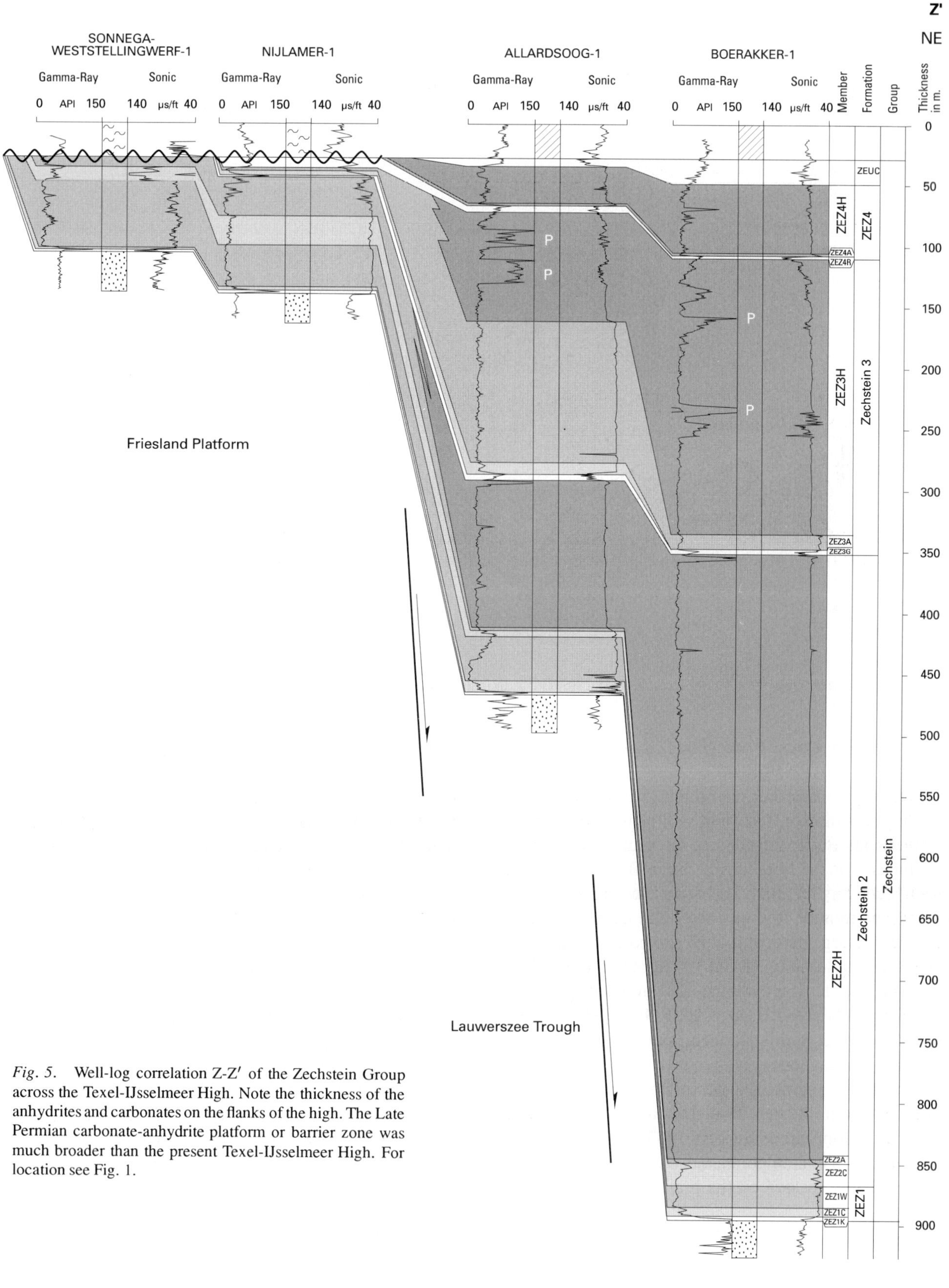

Fig. 5. Well-log correlation Z-Z′ of the Zechstein Group across the Texel-IJsselmeer High. Note the thickness of the anhydrites and carbonates on the flanks of the high. The Late Permian carbonate-anhydrite platform or barrier zone was much broader than the present Texel-IJsselmeer High. For location see Fig. 1.

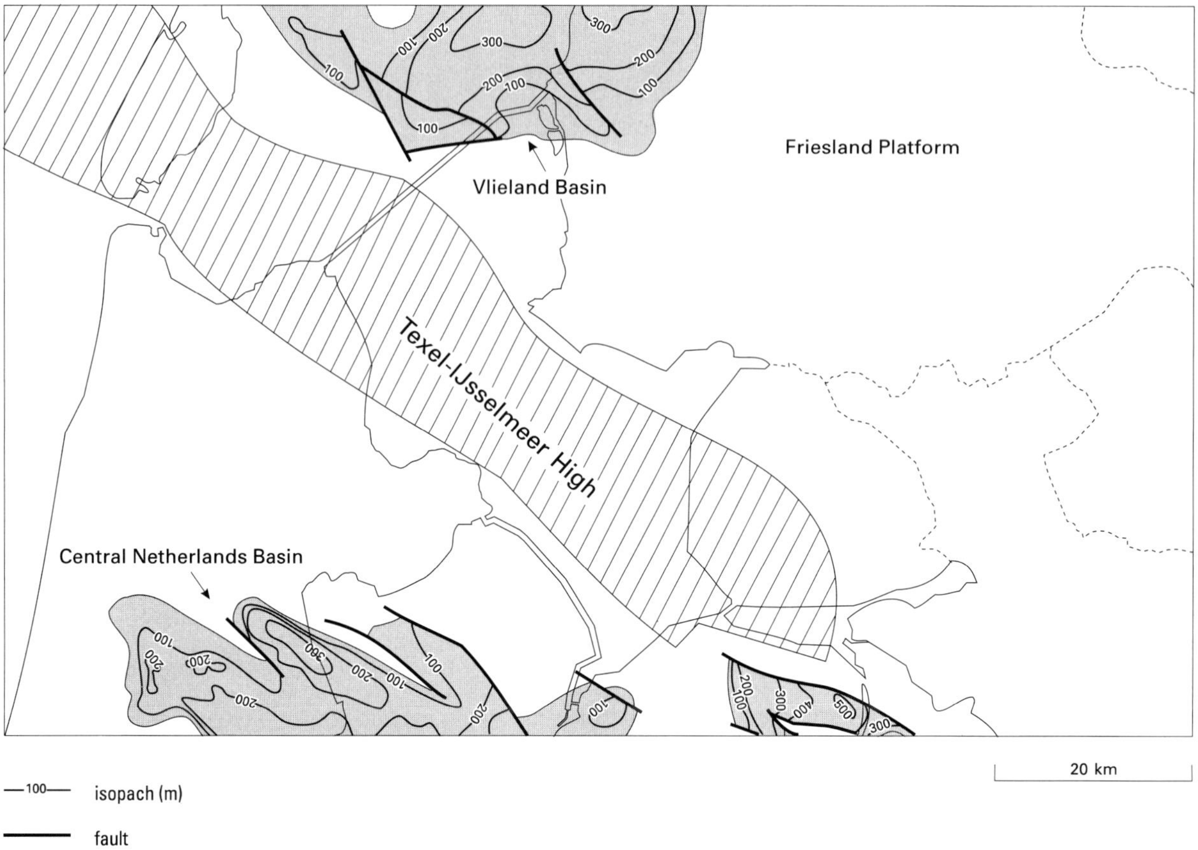

Fig. 6. Thickness map of the Upper Jurassic. Upper Jurassic sediments occur only in the Central Netherlands Basin and Vlieland Basin.

high the entire Upper Rotliegend is absent due to later Kimmerian erosion (Fig. 4).

During the Late Permian the high strongly affected the deposition of Zechstein sediments. The Central Netherlands Basin subsided in the south and became a subbasin of the Southern Permian Basin. The high was flooded by the main Zechstein transgressions, but during periods of low sea level and evaporite sedimentation, large parts of it were subjected to subaerial exposure and leaching (RGD 1991a). On the flanks of the high, an extensive carbonate and anhydrite platform was formed (Van der Baan 1990, RGD 1993b). To the north and south of this platform, salt was deposited in highly saline lagoons (Fig. 5).

Argillaceous limestone successions occur in a zone directly south of the Texel-IJsselmeer High (Van Adrichem Boogaert & Burgers 1983) and suggest the presence of a contemporaneous source area for clastics, probably located on the high itself. The argillaceous sediments do not belong to the clay-dominated Zechstein basin-fringe sediments present in the south of the Netherlands, because they are separated from this fringe by the salt-lake facies of the Central Netherlands Basin. At the end of the Permian, differential subsidence ceased and tectonic processes played only a subsidiary role. Mudflats eventually spread over the high and its surroundings during the Triassic.

Triassic: the Netherlands Swell

During the Early Triassic, the Texel-IJsselmeer High was initially involved in continuous subsidence in an entirely different tectonic setting. Two depocentres, the Ems Low and the Off Holland Low border the high area of the Netherlands Swell (Van Adrichem Boogaert & Kouwe, 1993–1995). These structural units trend NNE – SSW, perpendicular to the trend of the past and future Texel-IJsselmeer High, and show different sedimenta-

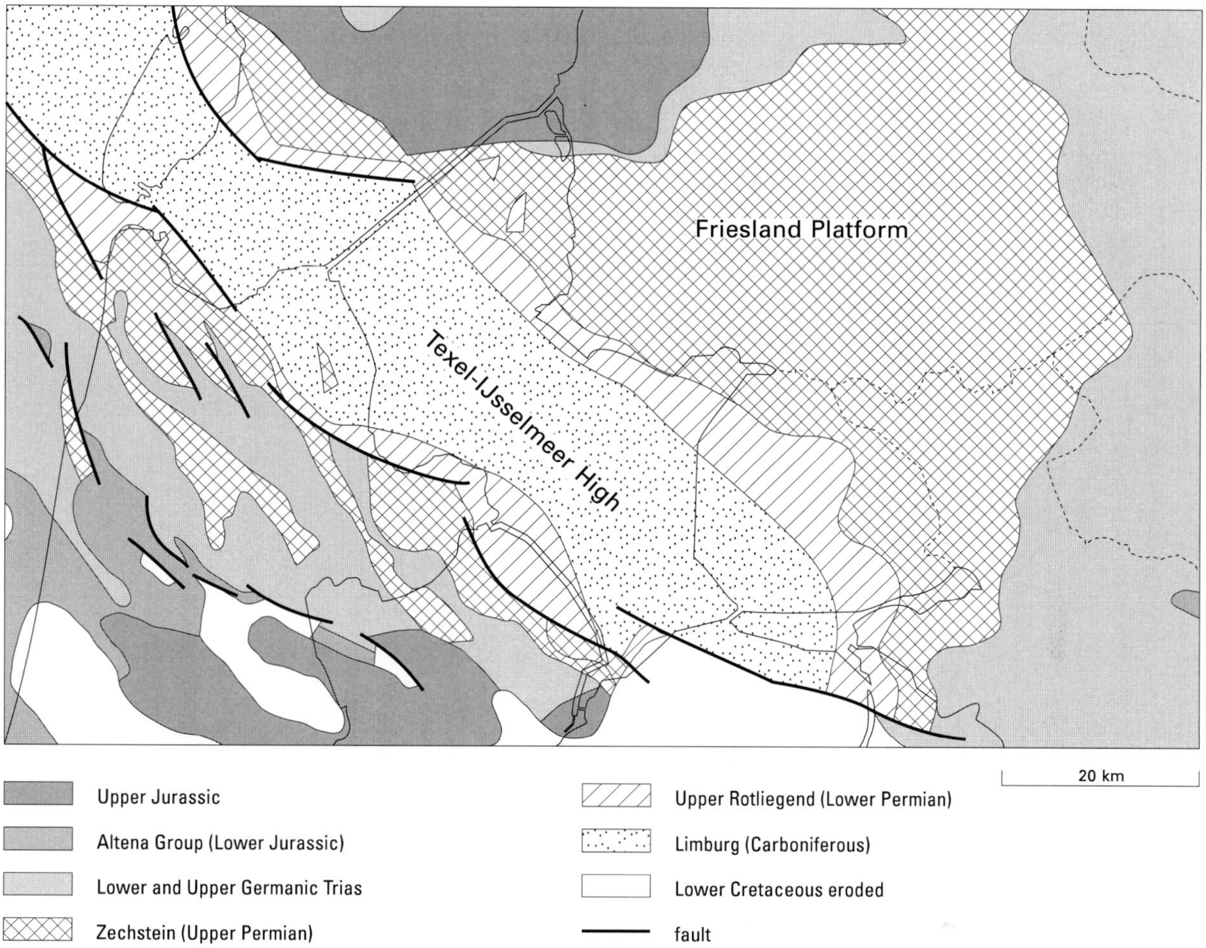

Fig. 7. Subcrop map at base Lower Cretaceous. The Late Kimmerian erosion pattern of the Upper Rotliegend, Zechstein and Triassic indicates that the Texel-IJsselmeer High together with the Friesland Platform was a single fault-block tilted to the northeast.

tion rates. In the basin centres, the thickness of the Triassic can reach over 1000 m. On the Netherlands Swell a reduced Triassic sequence was deposited, with an estimated thickness of some 500 m. The differences in development between the depocentres and the swell increased during the Late Triassic, mainly as a result of Early Kimmerian uplift and erosion.

Jurassic uplift and erosion

During the Early Jurassic the tectonic constellation remained essentially the same as in the Late Triassic. Reconstruction of the palaeogeography is difficult because the Jurassic has been eroded in a large part of the Netherlands. Lower Jurassic sequences in the basins around the Netherlands Swell are dominated by marine shales. This overall shalyness and correlations between the basins indicate that the swell must have been covered by the sea.

During the Late Jurassic and Early Cretaceous, sedimentation concentrated in fault-bounded basins. Seismo-stratigraphic patterns indicate syntectonic sedimentation in the northern part of the Central Netherlands Basin during the Late Jurassic. The Late Kimmerian I tectonic pulse (Oxfordian) caused the break-up of the northern part of the Netherlands into different units, such as the Texel-IJsselmeer High, the Friesland Platform and the Vlieland Basin. Claystones of the Upper Jurassic suggest that the high areas still had a low relief. The original distribution of these claystones probably was much greater than shown on Fig. 6. Subsidence was differential and most pronounced in the basin centres.

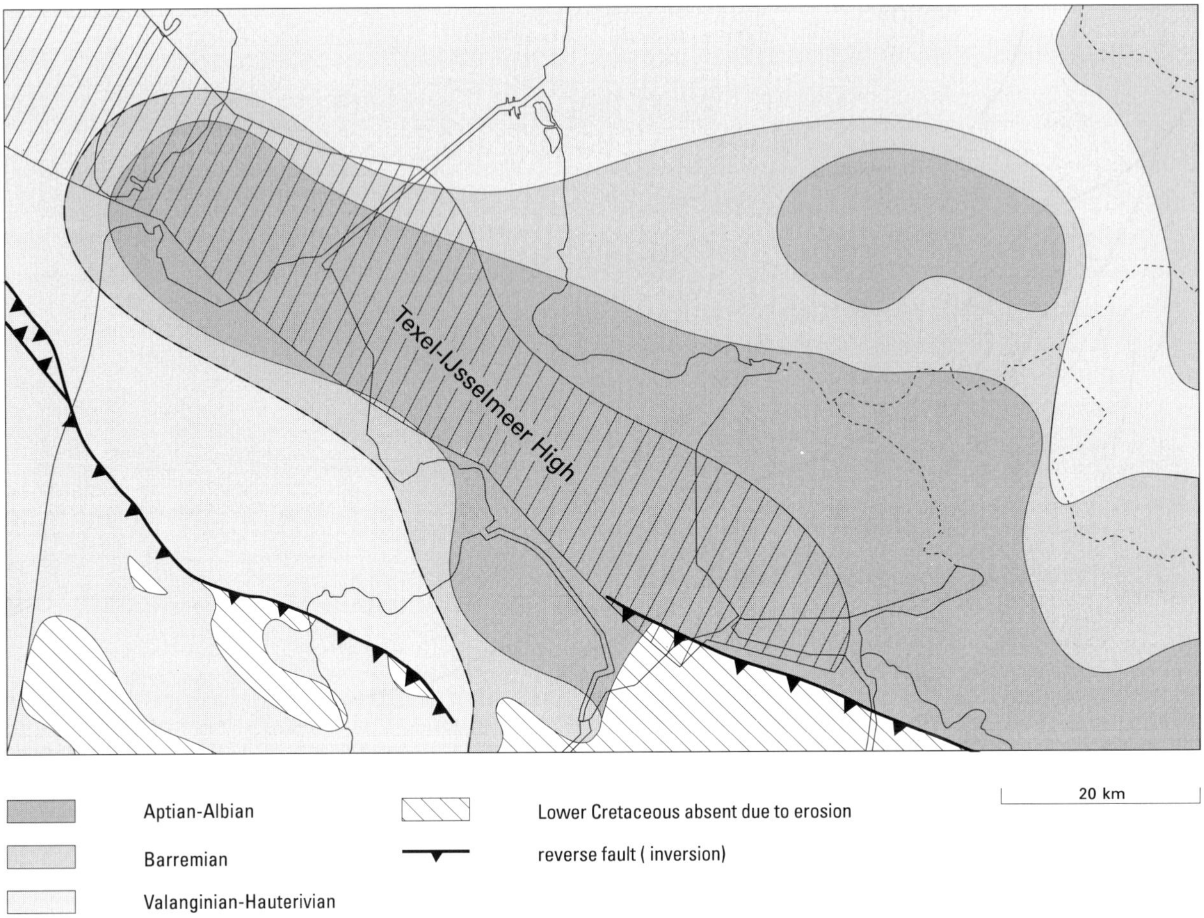

Fig. 8. Map showing the ages of the basal Lower Cretaceous sediments. South of the Texel-IJsselmeer High the extent of the Lower Cretaceous is obscured by later faulting and erosion.

Cretaceous transgression and tectonic inversion

During the Late Kimmerian II tectonic pulse (Ryazanian) the Texel-IJsselmeer High and the Friesland Platform became accentuated and deep erosion occurred. Fault activity ceased completely during the later Early Cretaceous. During this period the Upper Rotliegend Slochteren Sandstone was eroded at the crest of the Texel-IJsselmeer High and its erosion products were mainly redeposited in the Vlieland Basin and the Lower Saxony Basin. The subcrop map below base Lower Cretaceous clearly shows the Texel-IJsselmeer High and Friesland Platform (Fig. 7). Marine sands and clays were diachronously deposited during the Hauterivian to Albian on the Late Kimmerian palaeorelief. The age of basal sedimentary units of the Lower Cretaceous reveals the Early Cretaceous transgression in the area (Fig. 8). Valanginian sands were deposited only in the basin centres while younger Albian marls cover the entire area including the crest of the Texel-IJsselmeer High. From this, together with onlap features on seismic data, it has been concluded that a relative sea-level rise occurred which resulted in flooding of the Friesland Platform and the Texel-IJsselmeer High (Fig. 9; Herngreen et al. 1991, RGD 1993a,b).

Tectonic inversion of the Vlieland, Central Netherlands and Lower Saxony Basins started during the Campanian: the so-called Subhercynian phase (Betz et al. 1987, Van Wijhe 1987a,b, RGD 1991a, 1993a,b). In the shallow Vlieland Basin inversion was limited to the Campanian, while the Central Netherlands and Broad Fourteens Basins show major tectonic movements and uplift during both the Campanian (Subhercynian phase) and the Early Tertiary (Laramide phase). Inversion of these basins occurred under a transpressional regime and caused reverse faulting along pre-

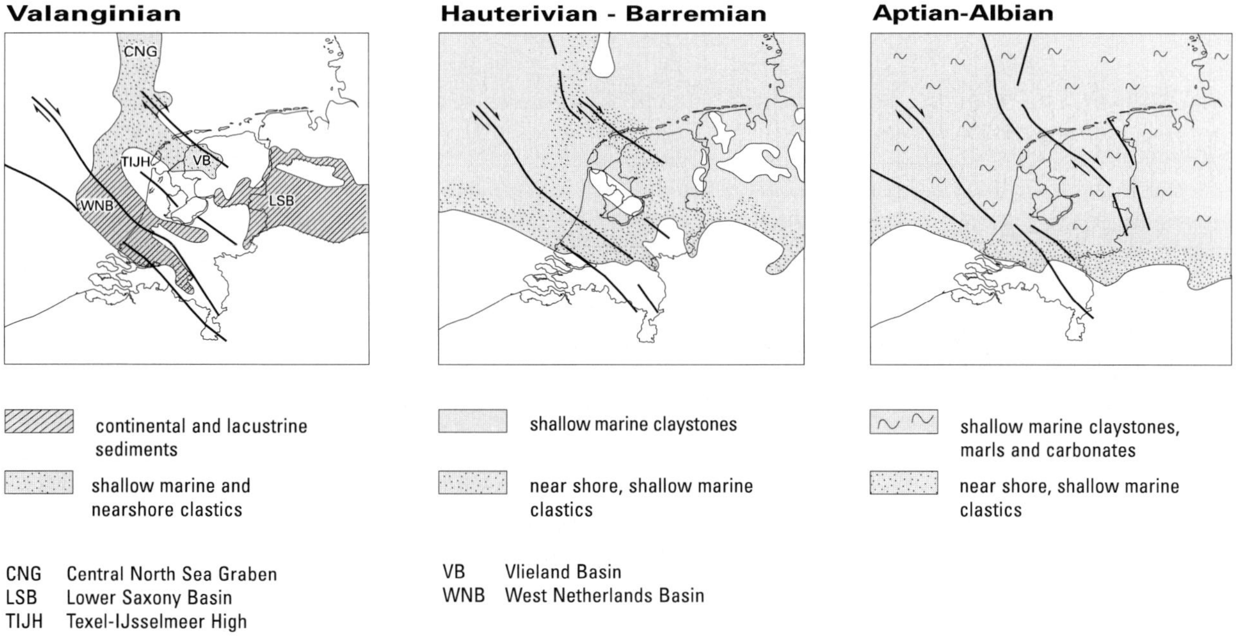

Fig. 9. Palaeogeography during the Early Cretaceous. A relative sea-level rise inundated the crest of the Texel-IJsselmeer High during the Albian.

existing Kimmerian faults. Strong erosion resulted in the total removal of Upper Cretaceous sediments in the Central Netherlands Basin (Fig. 10). The Texel-IJsselmeer High and the Friesland Platform were also affected by inversion tectonics; on these highs the Maastrichtian was truncated.

Cenozoic stability

Shallow marine sedimentation prevailed during the Paleocene, the Eocene and the Oligocene in the entire area. The area south of the high was split into several structural elements as the result of moderately weak differential subsidence. During the Oligocene this area was uplifted once more (Savian phase). The Texel-IJsselmeer High and the Friesland Platform were tectonically stable. Strong differential subsidence in the area studied did not occur until the Miocene when a low originated at the southern and central parts of the Texel-IJsselmeer High (Zuiderzee Low).

Drilling activity, gas fields and potential reservoirs

Since 1959 more than 100 exploration wells have been drilled in the surroundings of the Texel-IJsselmeer High (Fig. 11). Eighty-one wells reached the Carboniferous. Only three were drilled on the high itself. Potential reservoir rocks, such as Upper Rotliegend sandstone, Zechstein carbonate, and Triassic and Lower Cretaceous sandstones are present directly around the high, but do not occur on top of the high (Figs 11, 12). Here, they are eroded or were never deposited.

Initially, after discovery of the Groningen field in 1959, oil companies considered the Rotliegend sandstone reservoirs as the main objective. In the sixties and seventies gas-discoveries were also made in Lower Cretaceous sandstones and Zechstein carbonates. Some of the gas fields produce from both the Rotliegend and the Zechstein. In the Zechstein 2 Carbonate minor amounts of oil were found, probably

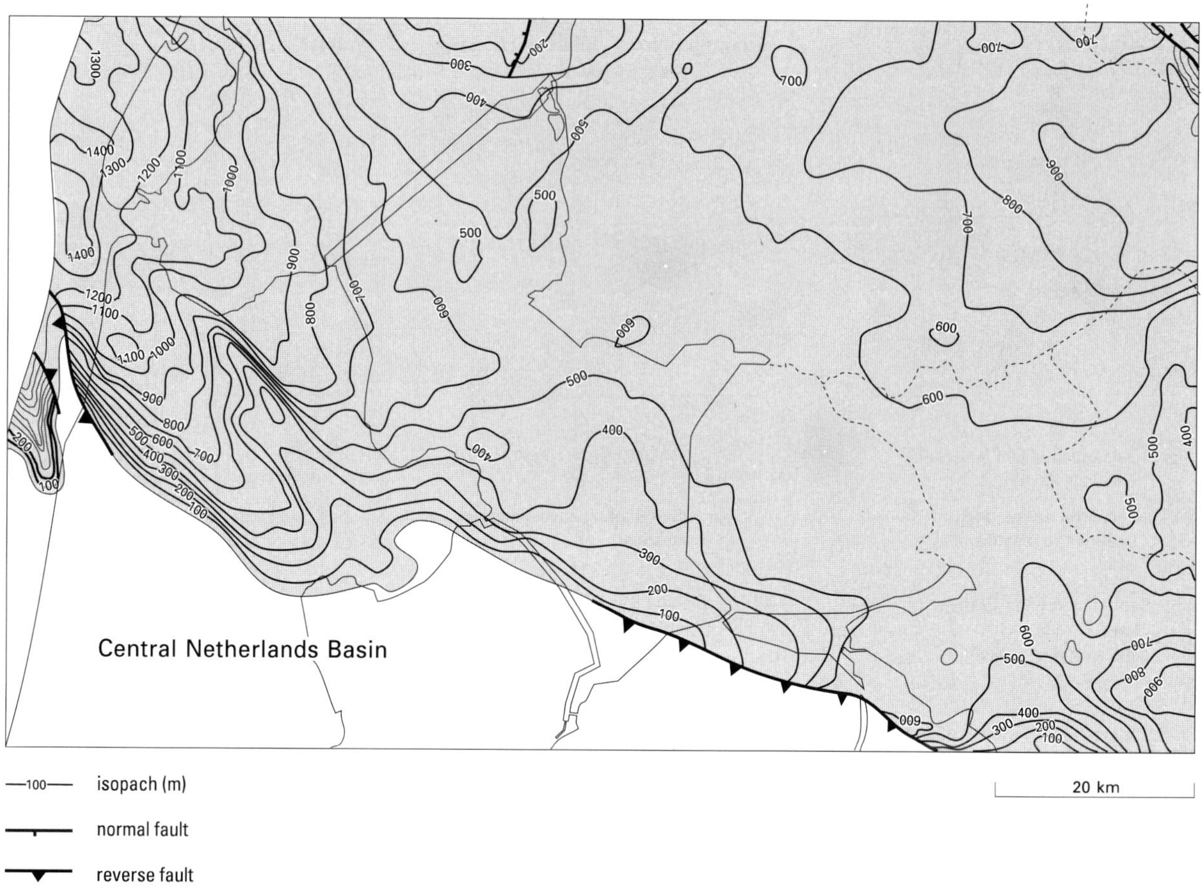

Fig. 10. Thickness map of the preserved Upper Cretaceous. In the Central Netherlands Basin the Upper Cretaceous is absent owing to erosion after Late Cretaceous-Early Tertiary uplift.

generated from algae-rich Zechstein carbonates (*in-situ* oil). The source rock for the gas is the Carboniferous.

Figure 12 gives a schematical overview of different types of reservoir-seal configurations in the area studied. Westphalian C sandstones have produced gas in the De Wijk field at the southeastern margin of the Friesland Platform (Fig. 12: type A). At the crest of the Texel-IJsselmeer High the Westphalian C does not occur (Fig. 3). Westphalian A/B sandstone bodies are hard to locate and moreover have poor reservoir qualities.

Gas-bearing Upper Rotliegend reservoirs are common in the northern part of the Friesland Platform. Here, and also in the Central Netherlands Basin, the Upper Rotliegend is ultimately top-sealed by Zechstein salt (Fig. 12: type B). Upper Rotliegend sandstones in the southern part of the Friesland Platform have high average porosities of 25%. In this high-porosity zone, the reservoirs are unconformably overlain by Cretaceous claystones. The high-porosity zone on the Friesland Platform is probably due to leaching during Kimmerian erosion (RGD 1993b).

Zechstein carbonates thicken towards the Texel-IJsselmeer High and are partly dolomitized. The degree of dolomitization increases towards the high. Leaching of these carbonates occurred during Kimmerian erosion. Average porosities vary from 4 to 12% (RGD 1993b). Several Zechstein gas fields are found in the immediate vicinity of the high on the Noord-Holland and Friesland Platforms (Fig. 12: types C and D). Gas-bearing Zechstein carbonates on the Friesland Platform are mostly top-sealed by Lower Cretaceous claystones.

Triassic Volpriehausen and Solling sandstones are only found gas-bearing in the western part of the Central Netherlands Basin (Fig. 12: type E).

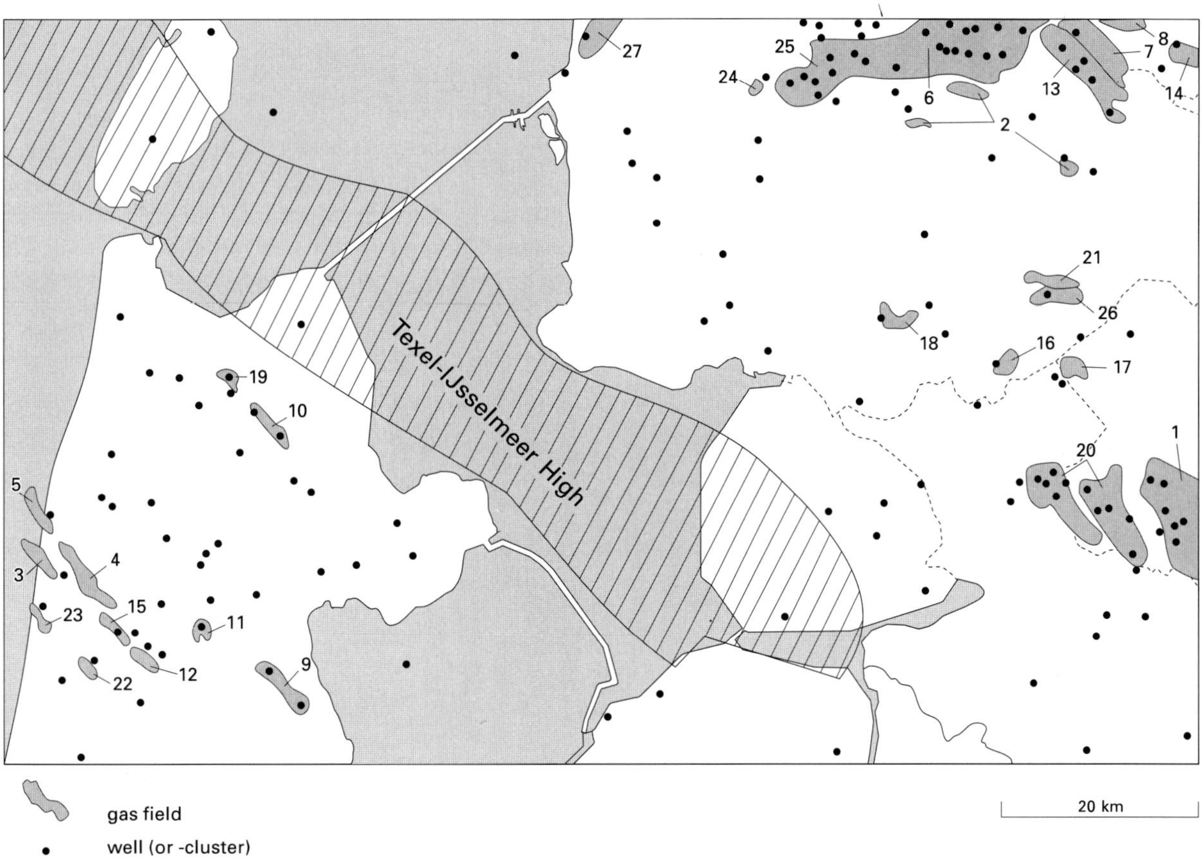

Gas-bearing reservoir	No.	Gas field	Type *
Carboniferous sandstone	1	De Wijk	A
Rotliegend sandstone	2	Akkrum	B
	3	Bergen	B
	4	Bergermeer	B
	5	Groet	B
	6	Leeuwarden cluster	B
	7	Marum	B
	8	Marumerlage	B
	9	Middelie	B
	10	Middenmeer	B
	11	Rustenburg	B
	12	Schermer	B
	13	Ureterp	B
	14	Zevenhuizen	B

Gas-bearing reservoir	No.	Gas field	Type *
Zechstein carbonate	15	Alkmaar	C
	2	Akkrum	C
	16	De Blesse	D
	6	Leeuwarden cluster	D
	9	Middelie	C
	10	Middenmeer	D
	17	Nijensleek	D
	18	Oldelamer	D
	12	Schermer	C/D
	19	Slootdorp	D
	20	Wanneperveen	D
	21	Weststellingwerf	D
Triassic sandstone	22	Heiloo	E
	23	Wimmenum	E
Lower Cretaceous sandstone	24	Bozum	F
	25	Friesland	F
	17	Nijensleek	F
	26	Noordwolde	F
	20	Wanneperveen	F
	1	De Wijk	F
Upper Cretaceous chalk	27	Harlingen	G

* Types refer to reservoir-seal configurations in Fig. 12.

Fig. 11. Drilling activity and gas fields discovered in the area studied.

Lower Cretaceous sandstone is present in the Vlieland Basin and on the Friesland Platform in palaeodepressions (Fig. 12: type F). These reservoirs unconformably cover Carboniferous, Permian, Triassic and Jurassic rocks, and are top-sealed by Lower Cretaceous claystones. Gas-bearing reservoirs with

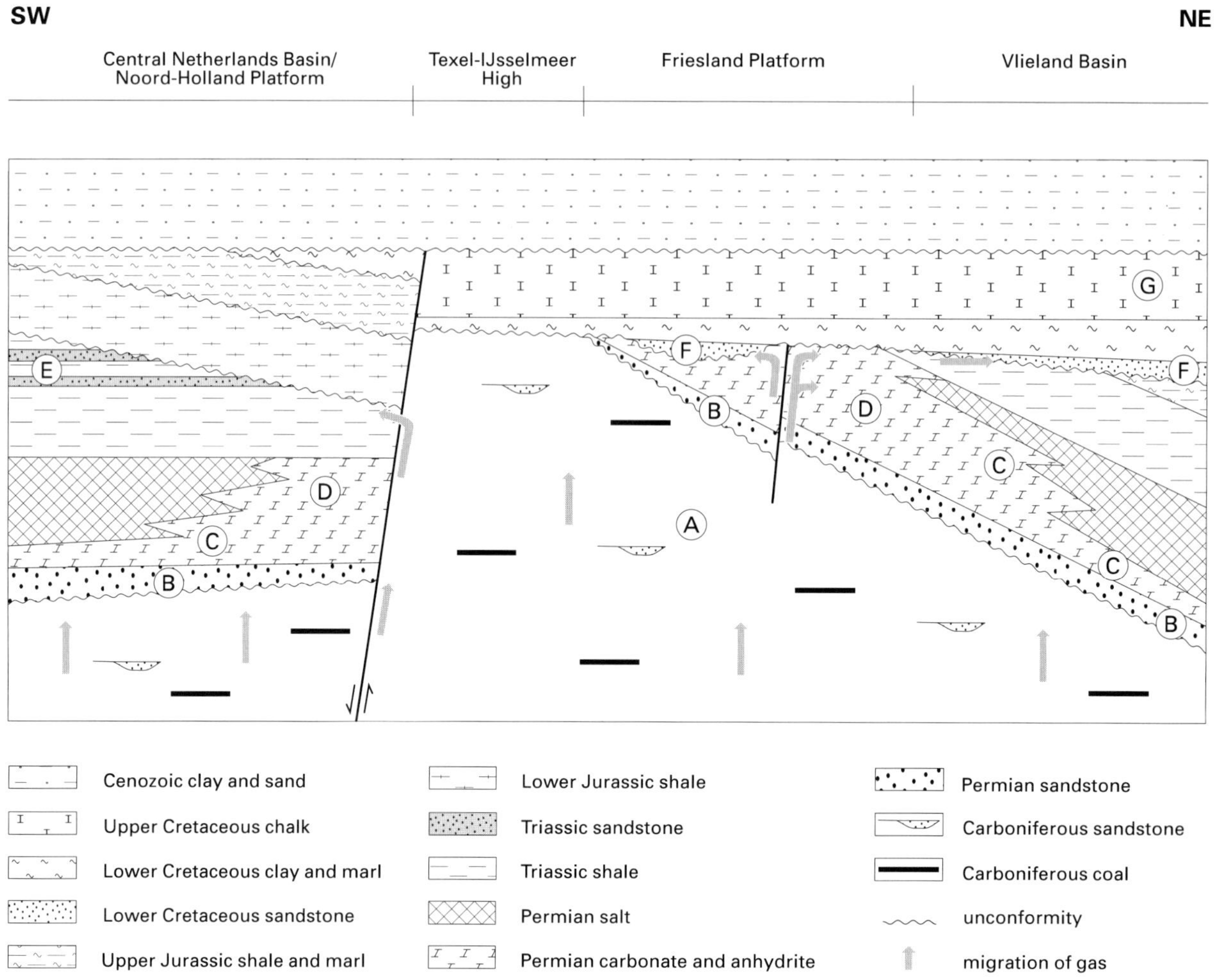

Type	Reservoir	Seal
(A)	Carboniferous sandstone	Carboniferous shales
(B)	Rotliegend sandstone	Zechstein salt and claystones and/or Lower Cretaceous claystones
(C)	Zechstein carbonate	Zechstein salt
(D)	Zechstein carbonate	Lower Cretaceous claystones and/or Triassic claystones
(E)	Triassic sandstone	Upper Triassic salt and claystones
(F)	Lower Cretaceous sandstone	Lower Cretaceous claystones
(G)	Upper Cretaceous chalk	Tertiary claystones

Fig. 12. Schematic presentation of reservoir – seal configuration types around the Texel-IJsselmeer High. Types also refer to Fig. 11.

→*Fig. 13.* Cross-sections of the Texel-IJsselmeer High and the Noord- Holland Platform. For locations of sections see Fig. 1. Note the different scales. A–A′) Northern part of interpreted seismic section MPNI-9101 across the inverted Broad Fourteens Basin and the Noord-Holland Platform. B–B′) Interpreted depth section across the study area. Note the highly deformed area south of the Texel-IJsselmeer High and the relatively little deformed strata on top of the high itself. C–C′) Interpreted seismic section NAM-813401 in the southeast of the study area; the Texel-IJsselmeer High is bounded to the south by major faults. The offset at the top Carboniferous is approximately 1500 m.

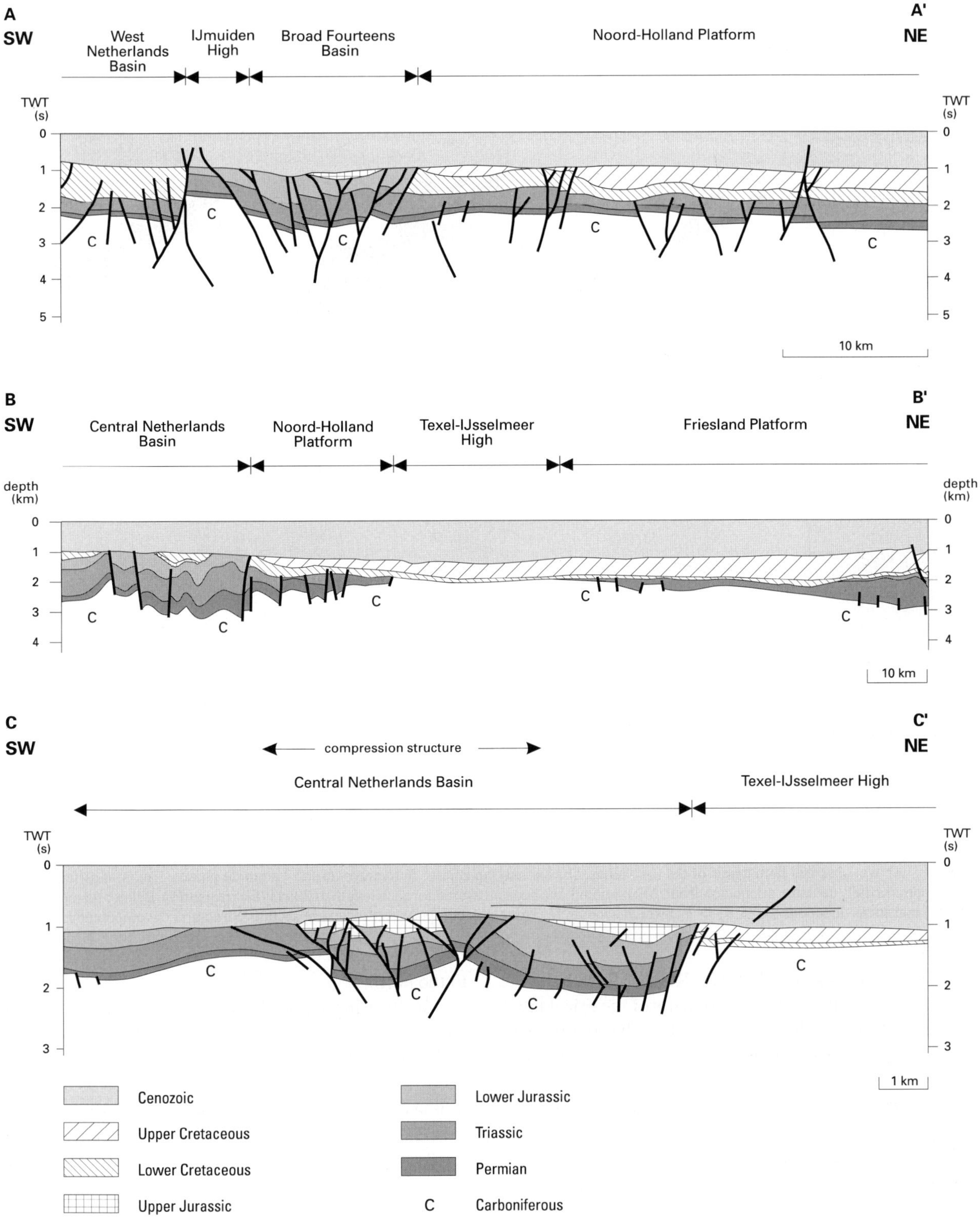
A
SW
West Netherlands Basin
IJmuiden High
Broad Fourteens Basin
Noord-Holland Platform
A'
NE
TWT (s)
10 km
B
SW
Central Netherlands Basin
Noord-Holland Platform
Texel-IJsselmeer High
Friesland Platform
B'
NE
depth (km)
10 km
C
SW
compression structure
Central Netherlands Basin
Texel-IJsselmeer High
C'
NE
TWT (s)
1 km
Cenozoic
Upper Cretaceous
Lower Cretaceous
Upper Jurassic
Lower Jurassic
Triassic
Permian
C Carboniferous

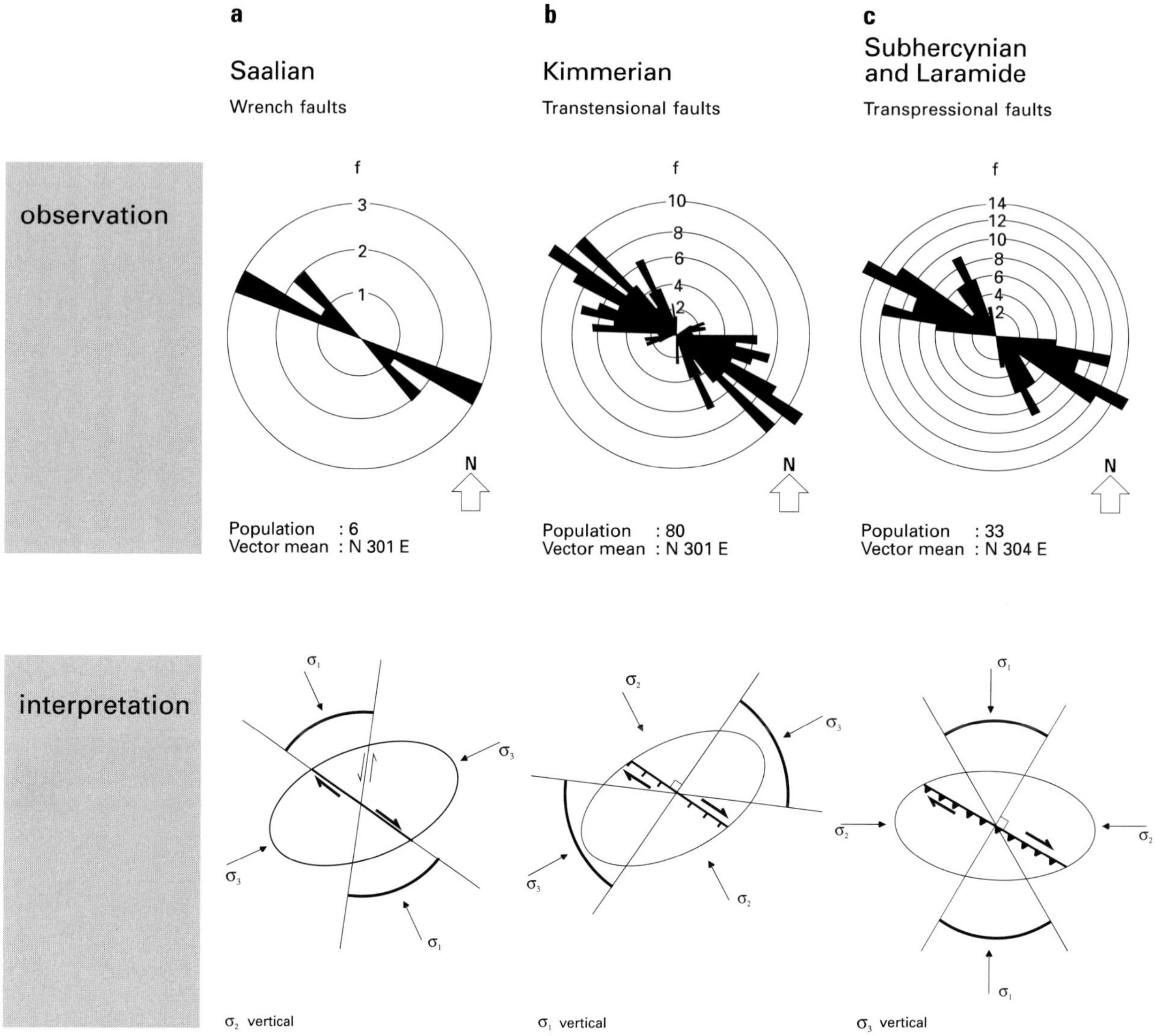

Fig. 14. Above: observed fault trends of the (a) Saalian, (b) Kimmerian and (c) Subhercynian and Laramide phases. Faults of different ages have approximately the same orientation. Fault orientations have been measured for each length of 20 km (f = frequency). Below: reconstruction of principal stress directions ($\sigma_1 > \sigma_2 > \sigma_3$) which account for strike-slip components (see text in section 'Structural development').

average porosities of 11–18% (RGD, 1993a,b) are found on the northern and eastern parts of the Friesland Platform (Fig. 11). In view of the fact that the reservoirs in the north are underlain by Zechstein salt, gas must have migrated laterally into them. In areas without salt (the southern parts of the Friesland Platform) the gas migrated directly from the Carboniferous or from the Upper Rotliegend into the Lower Cretaceous sandstone. From here the gas continued its migration to the north.

Gas-bearing Upper Cretaceous chalk is present in the north of the area studied (Figs 11, 12: type G). Several wells in the Harlingen gas field have high average porosities of 28–30%. The porosity is due to dissolution and fracturing of chalk during the Early Tertiary. Permeabilities range from 0.7–8 mD (Van den Bosch 1983).

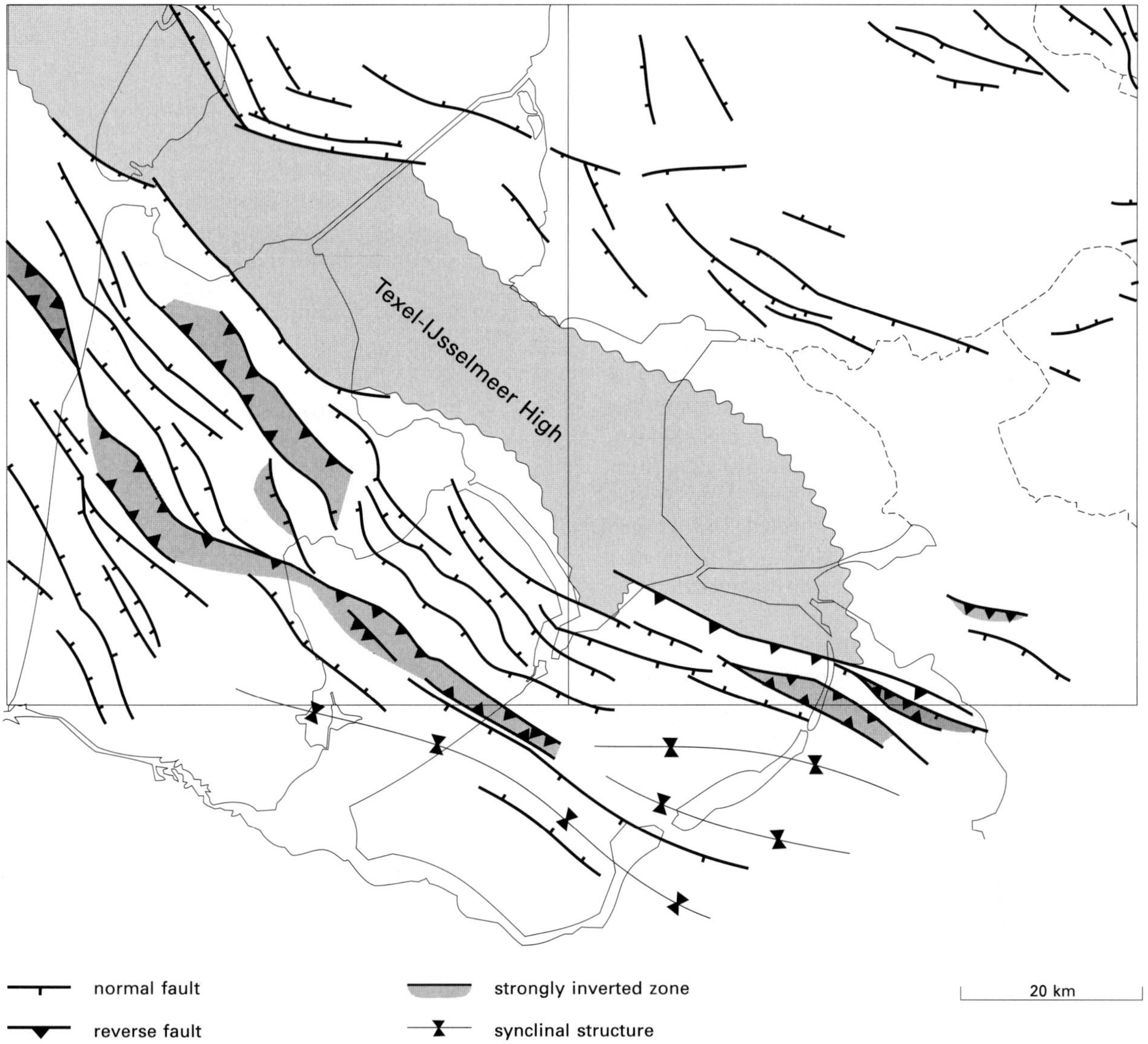

Fig. 15. Structural map showing inverted zones affected by reverse faulting. Kimmerian faults were reactivated during Late Cretaceous – Early Tertiary compression.

Structural development

Faults and stress analysis

NW – SE trending faults are present in the Carboniferous substratum and the sedimentary cover. Beneath the Zechstein, faults in the Rotliegend and underlying Carboniferous appear to be dominantly planar on seismic. Inversion faults at the southern border of the Noord-Holland Platform cutting through the Triassic, Jurassic and Cretaceous show a listric geometry. Directly around the Texel-IJsselmeer High the Zechstein salt is not sufficiently thick (< 300 m) to disconnect basement faults from the sedimentary cover or to cause halokinetic faulting.

NE – SW oriented cross-sections show the strongly inverted Broad Fourteens and Central Netherlands Basins south of the Texel-IJsselmeer High. Figure 13B indicates that this high together with the Friesland Platform formed one single fault block during the Early Cretaceous. The block was tilted to the northeast (Fig. 7). The northern part of the Central Netherlands Basin, just south of the high is defined by south-dipping Late

Jurassic normal faults (Fig. 13C). Some of these faults became reactivated later as reverse faults.

The NW – SE trending faults recognized in the Carboniferous substratum are Variscan wrench faults, formed in a compressional stress regime (Ziegler 1990). The orientation of the maximum principal stress (σ_1) belonging to this fault system is NNW – SSE (Fig. 14a).

Late Kimmerian fault tectonics generated a network of normal faults in Permian, Triassic and Jurassic rocks. These faults vary in trend from E – W to NNW – SSE (Fig. 14b). Subsidence of the basins surrounding the Texel-IJsselmeer High was contemporaneous with major rifting and dextral strike-slip deformation (Van Hoorn 1987, Van Wijhe 1987a). During this period of transtensional tectonics the maximum principal stress (σ_1) was vertical. The minimum principal stress (σ_3) in the horizontal was ENE – WSW oriented. The intermediate principal stress (σ_2) was NNW – SSE oriented. This configuration corresponds with dextral strike-slip (Fig. 14b).

During the Late Cretaceous and Early Tertiary the basins became inverted and reverse faulting took place along pre-existing faults. Thus, Kimmerian extensional faults were reactivated in a compressional stress field. Reverse faulting has been recognized in a zone of approximately 20 km width, south of the Texel-IJsselmeer High (Fig. 15). The high itself has only been slightly affected by inversion tectonics. The main trend of the reverse faults is NW – SE (Fig. 14c). A strike-slip component has been concluded from the flower structures, observed on seismic lines from the northern margin of the Central Netherlands Basin. Inversion faults are also present at the Cleaver Bank High in the North Sea and show characteristics of dextral strike-slip (Oudmayer & De Jager 1993). The maximum principal stress σ_1 in the northwestern part of the Netherlands during the inversion phase is concluded to have been approximately N – S and, given the fault pattern, corresponds with dextral strike-slip (Fig. 14c).

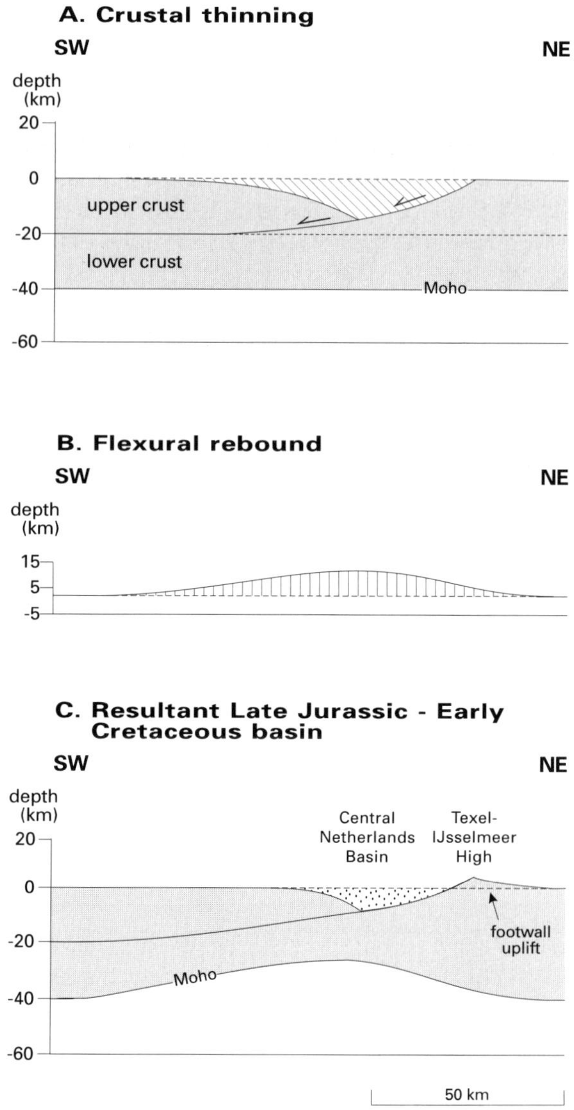

Fig. 16. Modelled isostatic response of the lithosphere after Late Jurassic – Early Cretaceous extension along a major low-angle fault (after Egan 1992). Thermal effects and basin fill have been ignored. (A) Extension along a deep listric fault which dips 30 degrees. (B) Amount of flexural rebound of the lithosphere. (C) Formation of a basin several kilometres deep (Central Netherlands Basin) and uplift of the footwall shoulder (Texel-IJsselmeer High) as the result of flexural rebound of the lithosphere.

A structural model for the tectogenesis of the Texel-IJsselmeer High

The southern edge of the Texel-IJsselmeer High is strongly faulted. Main faults dip to the south. During the Late Jurassic, faulting at this edge accompanies subsidence of the hanging wall (Central Netherlands Basin) and uplift of the footwall (Texel-IJsselmeer High). Half-graben geometries have been described by different models such as basement-involved listric faulting (Gibbs 1983) or domino block-faulting (Barr 1987). Other extensional models are based on isostatic response to stretching of the lithosphere and normal faulting of the sedimentary cover. These models generate both hanging-wall subsidence and footwall uplift (Marsden et al. 1990, Roberts & Yielding

1991, Egan 1992). Roberts & Yielding (1991) applied a flexural-cantilever model to single and multiple faulting with varying parameters (i.e. amount of extension (β), length of eroded footwall and effective elastic thickness of the lithosphere). When stretching the crust by 5 km under brittle behaviour of the upper crust and ductile behaviour of the lower crust, the flexural isostatic model of Egan (1992) predicts a footwall uplift of approximately 1200 m.

Stretching resulted in isostatic uplift and subsequent erosion of the Texel-IJsselmeer High and the Friesland Platform (Fig. 16). Uplift of these highs and subsidence of adjacent basins during the Late Jurassic were contemporaneous. The estimated Late Jurassic uplift and erosion of the Texel-IJsselmeer High, based on regional geological evidence, amounts to 1000 m.

Indications of a reduced crustal thickness beneath the Central Netherlands Basin come from regional deep seismic surveys. A Moho depth map shows a regional trend which seems to be dictated by the NW – SE orientation of the supracrustal structural grain (Rijkers & Duin, 1994). Beneath the Central Netherlands Basin, a minimum thickness of 28 km supports the model given in Fig. 16.

Conclusions

1. The Texel-IJsselmeer High dates from at least Early Permian times. The high affected the lithofacies and distribution of Permian, Upper Jurassic and Lower Cretaceous sediments. Strong uplift and erosion accentuated the high during the Late Jurassic.
2. During Late Kimmerian extension, the southern edge of the Texel-IJsselmeer High became defined by normal faulting. During the extension the high behaved like a footwall block that was strongly uplifted in a flexural isostatic rebound of the stretched lithosphere. The hanging wall subsided and represents the northern edge of the Central Netherlands Basin. Extensional models based on isostatic principles predict a footwall uplift of approximately 1200 m. The estimated Late Jurassic erosion on the Texel-IJsselmeer High amounts to 1000 m.
3. During the Late Cretaceous-Early Tertiary transpression the Texel-IJsselmeer High acted as a stable block. South of the high, Late Kimmerian normal faults have been reactivated during inversion in a compressional stress field. Several structures were generated in a NW – SE trending zone south of the high.
4. The maximum horizontal stresses of the Saalian, Kimmerian and Subhercynian-Laramide tectonic phases have approximately the same orientation, but the resulting fault styles differ radically as a consequence of the different dimensions of the principal stress acting in the vertical.
5. Permian sandstones and carbonates around the high have developed good reservoir characteristics owing to the influence of the high on sedimentary and diagenetic processes.

Acknowledgements

The authors like to express their thanks to the Director of the Geological Survey of the Netherlands for authorizing this publication. The authors also wish to thank P. van Tongeren and G. Remmelts for their valuable and constructive suggestions. Thanks are also due to D.A.J. Batjes, M. Epting and H.E. Rondeel for their thorough reviews of the paper.

References

Arthaud, F. & P. Matte 1977 Late Paleozoic strike-slip faulting in southern Europe and North Africa; results of a right-lateral shear zone between the Appalachians and the Urals – Geol. Soc. Am. Bull. 88: 1305–1320

Barr, D. 1987 Structural/stratigraphic models for extensional basins of half-graben type – J. Structural Geol. 9: 491–500

Betz, B., F. Führer, G. Greiner & E. Plein 1987 Evolution of the Lower Saxony Basin – Tectonophysics 137: 127–170

Bless, M.J.M., J. Bouckaert, M.A. Claver, J.M. Graulich & E. Paproth 1977 Palaeogeography of Upper Westphalian deposits in NW Europe with reference to the Westphalian C North of the mobile Variscan Belt – Med. Rijks Geol. Dienst 28–5: 101–147

Egan, S.S. 1992 The flexural isostatic response of the lithosphere to extensional tectonics – Tectonophysics 202: 291–308

Gibbs, A.D. 1983 Balanced cross section construction from seismic sections in areas of extensional tectonics – J. Structural Geol. 5: 344–361.

Herngreen, G.F.W., R. Smit & Th.E. Wong 1991 The stratigraphy and tectonics of the Vlieland Basin, the Netherlands. In: Spencer, A.M. (ed.) Generation, accumulation and production of Europe's hydrocarbons – Eur. Ass. Petroleum Geosc. Eng. Spec. Publ. 1: 175–192

Marsden, G., G. Yielding, A.M. Roberts & N.J. Kuznir 1990 Application of flexural cantilever simple-shear/pure shear model of continental lithosphere extension to the formation of the North Sea Basin. In: Blundell, D.J. & D. Gibbs (eds) Tectonic evolution of the North Sea Rifts – Oxford University Press, Oxford: 240–261

Oudmayer, B.C. & J. De Jager 1993 Fault reactivation and oblique-slip in the Southern North Sea. In: Parker, J.R. (ed.) Petroleum Geology of Northwest Europe, Proc. 4th Conference, Geol. Soc. London: 1281–1290

RGD (Rijks Geologische Dienst) 1991a Geological Atlas of the Subsurface of the Netherlands 1:250 000, Explanation to map sheet I: Vlieland-Terschelling, Geological Survey of the Netherlands, Haarlem: 71 pp

RGD 1991b Geological Atlas of the Subsurface of the Netherlands 1:250 000, Explanation to map sheet II: Ameland-Leeuwarden, Geological Survey of the Netherlands, Haarlem: 87 pp

RGD 1993a Geological Atlas of the Subsurface of the Netherlands 1:250 000, Explanation to map sheet IV: Texel – Purmerend, Geological Survey of the Netherlands, Haarlem: 128 pp

RGD 1993b Geological Atlas of the Subsurface of the Netherlands 1:250 000, Explanation to map sheet V: Sneek – Zwolle, Geological Survey of the Netherlands, Haarlem: 126 pp

Rijkers, R.H.B. & E.J.Th. Duin 1994 Crustal observations beneath the southern North Sea and their tectonic and geological implications – Tectonophysics 240: 215–224

Roberts, A.M. & G. Yielding 1991 Deformation around basin-margin faults in the North Sea/Mid-Norway rift – The Geometry of Normal Faults, Geol. Soc. Spec. Publ. 56: 61–78

Van Adrichem Boogaert, H.A. & W.F.J Burgers 1983 The development of the Zechstein in the Netherlands – Geol. Mijnbouw 62: 83–92

Van Adrichem Boogaert, H.A. & W.F.P. Kouwe 1993–1995 Stratigraphic nomenclature of the Netherlands; revision and update by RGD and NOGEPA – Med. Rijks Geol. Dienst 50

Van den Bosch, W.J. 1983 The Harlingen Field, the only Cretaceous gasfield in the Upper Cretaceous Chalk of the Netherlands – Geol. Mijnbouw 62: 145–156

Van der Baan, D. 1990 Zechstein reservoirs in the Netherlands. In: Brooks, J. (ed.) Classic Petroleum Provinces, Geol. Soc. Spec. Publ. 50: 379–398

Van Hoorn, B. 1987 Structural evolution, timing and tectonic style of the Sole Pit inversion – Tectonophysics 137: 239–284

Van Wijhe, D.H. 1987a Structural evolution of inverted basins in the Dutch Offshore – Tectonophysics 137: 171–219

Van Wijhe, D.H. 1987b The structural evolution of the Broad Fourteens Basin. In: Brooks, J. & K.W. Glennie (eds) Petroleum Geology of Northwest Europe – Graham & Trotman, London: 315–323

Ziegler, P.A. 1987 Late Cretaceous intra-plate compressional deformations in the Alpine foreland, a geodynamic model – Tectonophysics 137: 389–420

Ziegler, P.A. 1990 Geological Atlas of Western and Central Europe (second edition) – Shell Internat. Petroleum Mij, The Hague: 239 pp, 56 encls